Inhibitors of Molecular Chaperones as Therapeutic Agents

RSC Drug Discovery Series

Editor-in-Chief:
Professor David Thurston, *King's College, London, UK*

Series Editors:
Dr David Fox, *Vulpine Science and Learning, UK*
Professor Ana Martinez, *Medicinal Chemistry Institute-CSIC, Madrid, Spain*
Professor David Rotella, *Montclair State University, USA*

Advisor to the Board:
Professor Robin Ganellin, *University College London, UK*

Titles in the Series:
 1: Metabolism, Pharmacokinetics and Toxicity of Functional Groups
 2: Emerging Drugs and Targets for Alzheimer's Disease; Volume 1
 3: Emerging Drugs and Targets for Alzheimer's Disease; Volume 2
 4: Accounts in Drug Discovery
 5: New Frontiers in Chemical Biology
 6: Animal Models for Neurodegenerative Disease
 7: Neurodegeneration
 8: G Protein-Coupled Receptors
 9: Pharmaceutical Process Development
10: Extracellular and Intracellular Signaling
11: New Synthetic Technologies in Medicinal Chemistry
12: New Horizons in Predictive Toxicology
13: Drug Design Strategies: Quantitative Approaches
14: Neglected Diseases and Drug Discovery
15: Biomedical Imaging
16: Pharmaceutical Salts and Cocrystals
17: Polyamine Drug Discovery
18: Proteinases as Drug Targets
19: Kinase Drug Discovery
20: Drug Design Strategies: Computational Techniques and Applications
21: Designing Multi-Target Drugs
22: Nanostructured Biomaterials for Overcoming Biological Barriers
23: Physico-Chemical and Computational Approaches to Drug Discovery
24: Biomarkers for Traumatic Brain Injury
25: Drug Discovery from Natural Products
26: Anti-Inflammatory Drug Discovery
27: New Therapeutic Strategies for Type 2 Diabetes: Small Molecules
28: Drug Discovery for Psychiatric Disorders
29: Organic Chemistry of Drug Degradation
30: Computational Approaches to Nuclear Receptors
31: Traditional Chinese Medicine
32: Successful Strategies for the Discovery of Antiviral Drugs
33: Comprehensive Biomarker Discovery and Validation for Clinical Application

34: Emerging Drugs and Targets for Parkinson's Disease
35: Pain Therapeutics; Current and Future Treatment Paradigms
36: Biotherapeutics: Recent Developments using Chemical and Molecular Biology
37: Inhibitors of Molecular Chaperones as Therapeutic Agents

How to obtain future titles on publication:
A standing order plan is available for this series. A standing order will bring delivery of each new volume immediately on publication.

For further information please contact:
Book Sales Department, Royal Society of Chemistry, Thomas Graham House, Science Park, Milton Road, Cambridge, CB4 0WF, UK
Telephone: +44 (0)1223 420066, Fax: +44 (0)1223 420247
Email: booksales@rsc.org
Visit our website at www.rsc.org/books

Inhibitors of Molecular Chaperones as Therapeutic Agents

Edited by

Timothy Machajewski and Zhenhai Gao
Novartis, California, USA
Email: timothy.machajewski@novartis.com; zhenhai.gao@novartis.com

RSCPublishing

RSC Drug Discovery Series No. 37

ISBN: 978-1-84973-666-4
ISSN: 2041-3203

A catalogue record for this book is available from the British Library

© The Royal Society of Chemistry 2014

All rights reserved

Apart from fair dealing for the purposes of research for non-commercial purposes or for private study, criticism or review, as permitted under the Copyright, Designs and Patents Act 1988 and the Copyright and Related Rights Regulations 2003, this publication may not be reproduced, stored or transmitted, in any form or by any means, without the prior permission in writing of The Royal Society of Chemistry or the copyright owner, or in the case of reproduction in accordance with the terms of licences issued by the Copyright Licensing Agency in the UK, or in accordance with the terms of the licences issued by the appropriate Reproduction Rights Organization outside the UK. Enquiries concerning reproduction outside the terms stated here should be sent to The Royal Society of Chemistry at the address printed on this page.

The RSC is not responsible for individual opinions expressed in this work.

Published by The Royal Society of Chemistry,
Thomas Graham House, Science Park, Milton Road,
Cambridge CB4 0WF, UK

Registered Charity Number 207890

For further information see our website at www.rsc.org

Preface

The advancement of our understanding of the cellular functions of molecular chaperones in both normal and diseased states has come a long way. Since the first coinage of the term "molecular chaperone" by Laskey in 1978[1] to the present day, tremendous progress has been made in elucidating the structure and function of this important class of proteins and has resulted in the pursuit of inhibitors to address major unmet medical needs such as cancer and neurodegenerative diseases.

Molecular chaperones play a significant role in maintaining and adjusting protein homeostasis. They assist in the folding, maturation, and stabilization of a host of client proteins involved in health and disease. An ideal target from a drug discovery perspective is one that is essential for a diseased state but nonessential for the normal, healthy state. How the molecular chaperones, which play such an important role in normal cell function, can be useful as therapeutic targets is still the subject of some controversy but the evidence is mounting that clinical success is achievable.

The goal of this book is to provide a comprehensive examination of the field of molecular chaperones, their modulation, and application to pharmaceutical research. With over 10 small molecule Hsp90 inhibitors currently in oncology clinical development, we have accumulated significant knowledge in basic drug discovery as well as in clinical translational research for drugging molecular chaperones. With the ever increasing difficulty of bringing new therapies to market, it is imperative that we apply these collective experiences to future drug discovery efforts to improve our chances for success. It is our hope that researchers both in the field of molecular chaperones and in drug discovery will find this retrospective analysis and forward-looking concepts from bench to bedside valuable.

Written by top experts from both academia and industry, this volume brings together for the first time the current knowledge of basic structure and function

---

RSC Drug Discovery Series No. 37
Inhibitors of Molecular Chaperones as Therapeutic Agents
Edited by Timothy Machajewski and Zhenhai Gao
© The Royal Society of Chemistry 2014
Published by the Royal Society of Chemistry, www.rsc.org

of molecular chaperones, the design and development of several small molecule inhibitors, and the application of these inhibitors in clinical settings.

Section I serves as background and introduction to molecular chaperones and their associated biology and contributions to human diseases. Chapter 1 provides a comprehensive review of the many types of molecular chaperones and presents their connections with diseases and the phenotypes of genetic alterations of the genes encoding them. An overview of strategies that have been developed to modulate the activity of several molecular chaperones is also presented. The structure and function of two of the most studied molecular chaperones, Hsp90 and Hsp70 are detailed in chapter 2 and chapter 3. Chapter 4 demonstrates how the dependency of cancer cells on molecular chaperones, such as Hsp90, can be exploited to provide potential therapies for this important disease area.

Section II provides detailed case studies of the academic and industrial discovery efforts of several small molecule inhibitors. Chapters 5 through 9 detail the drug discovery leading to several inhibitors of the N-terminal domain of Hsp90 in clinical development. Chapter 10 describes preclinical development of inhibitors of the C-terminal domain of Hsp90. Chapter 11 presents an overview of recent discovery of several classes of Hsp70 inhibitors.

While most of the initial effort has been in the field of oncology, recent work in other disease areas is emerging. Chapters 12 and 13 describe the current state of Hsp90 inhibitor clinical development in oncology. Chapter 14 addresses the role of Hsp90 as an antimalarial therapeutic target. Finally, Chapter 15 describes the opportunities that exist for molecular chaperones as therapeutic targets in neurodegenerative diseases.

Such a comprehensive undertaking would not have been possible without the committed diligence of the individual chapter authors. For their scholarship, scientific rigor, hard work, dedication, and patience during the lengthy process of writing, reviewing, and editing their contributions, we are grateful.

Timothy D. Machajewski
Zhenhai Gao

Reference

1. R. A. Laskey, B. M. Honda, A. D. Mills and J. T. Finch, *Nature*, 1978, **275**, 416–420.

Contents

Chapter 1	**Overview of Molecular Chaperones in Health and Disease**		1
	Tai Wang, Pablo C. Echeverría and Didier Picard		
	1.1	Proteostasis and the Central Role of Molecular Chaperones	1
	1.2	The Major Classes of Molecular Chaperones	2
		1.2.1 Hsp100	2
		1.2.2 Hsp90	2
		1.2.3 Hsp70 and J Proteins	3
		1.2.4 Chaperonins	3
		1.2.5 Small Hsps	4
	1.3	Miscellaneous Other Molecular Chaperones	4
	1.4	Molecular Chaperones in Health and Disease	5
	1.5	Strategies for Modulating Chaperone Activities	10
		1.5.1 Hsp90 Inhibitors	10
		1.5.2 Hsp70 Inhibitors	18
		1.5.3 Inhibitors Targeting Small Molecular Chaperones	22
		1.5.4 Modulators of the HSF-1 Pathway	23
	References		26
Chapter 2	**Structural Basis of Hsp90 Function**		37
	Chrisostomos Prodromou and Laurence H. Pearl		
	2.1	Introduction	37
	2.2	Domain Structure of Hsp90	38
	2.3	Determining the Rate-limiting Step of the Hsp90 Cycle	39

RSC Drug Discovery Series No. 37
Inhibitors of Molecular Chaperones as Therapeutic Agents
Edited by Timothy Machajewski and Zhenhai Gao
© The Royal Society of Chemistry 2014
Published by the Royal Society of Chemistry, www.rsc.org

	2.4	Nucleotide Interactions of the ADP and ATP/AMP-PNP Bound States of Hsp90	40
	2.5	Structural Changes in Cytosolic Hsp90 upon ATP Binding	42
	2.6	Structural Changes in the Middle Domain of Hsp90	44
	2.7	Function of the Charged Linker	45
	2.8	Structural Changes of the C-terminal Domain of Hsp90	47
	2.9	Long-range Communication between Domains	47
	2.10	Co-chaperone Regulation and Modulation of Structural Elements of the Hsp90 ATPase Machinery	47
	2.11	Co-chaperones that Deliver Client Proteins to Hsp90 Inhibit its ATPase Activity	48
	2.12	Co-chaperones that Activate the ATPase Activity of Hsp90	51
	2.13	Co-chaperones that Modulate the ATPase Activity of Hsp90	54
	2.14	Co-chaperones that Target Hsp90 Client Proteins for Proteasome Degradation	55
	2.15	Client Proteins and Activation of Hsp90	56
	2.16	Phosphorylations and other Modifications	57
	2.17	Natural Resistance Mechanisms	58
	2.18	Concluding Remarks	59
		References	60

Chapter 3 Structure and Function of Hsp70 Molecular Chaperones   **65**
Eugenia M. Clerico and Lila M. Gierasch

3.1	Introduction		65
3.2	General Background on Hsp70 Chaperones		66
3.3	Classes of Human Hsp70s		68
3.4	Hsp70 Structure		71
	3.4.1	The Hsp70 NBD	71
	3.4.2	The Hsp70 Interdomain Linker	72
	3.4.3	The Hsp70 SBD	73
	3.4.4	Hsp70 Substrate Binding	74
	3.4.5	The Disordered Hsp70 C-terminal Domain	76
3.5	Hsp70 Allosteric Mechanism		76
3.6	Co-chaperones		79
	3.6.1	J proteins	79
	3.6.2	NEFs	84
	3.6.3	Other Co-chaperones	87
3.7	Cellular Functions of Hsp70s		89
	3.7.1	*De Novo* Protein Folding	89
	3.7.2	Protein Re-folding	90
	3.7.3	Protein Degradation	91

		3.7.4	Intracellular Protein Traffic	94
		3.7.5	ER Stress	102
		3.7.6	Apoptosis	104
		3.7.7	Senescence	108
	3.8	Perspectives		108
	Acknowledgements			109
	References			109

Chapter 4 Exploiting the Dependency of Cancer Cells on Molecular Chaperones 126
Swee Sharp, Jenny Howes and Paul Workman

	4.1	Introduction: Stepping into the Limelight	126
	4.2	Brief Essentials of Hsp90 Biology	128
	4.3	Hsp90 in Cancer and the Basis for Therapeutic Selectivity	132
	4.4	Exploiting the Hsp90 Dependency of Driver Oncoproteins by a Range of Hsp90 Inhibitors	134
	4.5	Exploiting the Non-oncogene Dependency of Cancer Cells on Stress Pathways	141
	4.6	Lessons Learned and Future Directions	142
	4.7	Concluding Remarks	145
	Acknowledgements		147
	References		148

Chapter 5 The Discovery of BIIB021 and BIIB028 158
Karen Lundgren and Marco A. Biamonte

5.1	Background		158
5.2	The Discovery of BIIB021		160
	5.2.1	BIIB021 Criteria	160
	5.2.2	BIIB021 Design	160
	5.2.3	BIIB021 Selectivity	162
	5.2.4	BIIB021 Pharmacokinetics	163
	5.2.5	BIIB021 Pharmacodynamics	164
	5.2.6	BIIB021 *in Vivo* Efficacy	165
	5.2.7	BIIB021 Summary	166
5.3	The Discovery of BIIB028		166
	5.3.1	BIIB028 Criteria	166
	5.3.2	BIIB028 Design	167
	5.3.3	BIIB028 Pharmacokinetics	173
	5.3.4	BIIB028 Pharmacodynamics	174
	5.3.5	BIIB028 *in Vivo* Efficacy	174
	5.3.6	BIIB028 Summary	175

	5.4	On the Tolerability of Hsp90 Inhibitors	176
	5.5	Conclusion	178
	References		178

Chapter 6 Discovery and Development of Ganetespib 180
Weiwen Ying

6.1	Ganetespib		180
	6.1.1	Chemical Description	180
	6.1.2	Screening Process	180
	6.1.3	Co-crystal Structure with Hsp90 N-terminal	181
6.2	Pre-clinical Pharmacology		182
	6.2.1	*In Vitro* and *In Vivo* Single Agent Activity	182
	6.2.2	Anti-angiogenic Properties of Ganetespib	183
	6.2.3	Ganetespib in Non-small-cell Lung Cancer	184
	6.2.4	Ganetespib in Hematological Malignancies	186
	6.2.5	Combination Studies with Ganetespib	187
	6.2.6	Broad Activity of Ganetespib in Different Tumor Types	188
6.3	Pre-clinical Pharmacokinetics and Toxicology		189
	6.3.1	Tissue Distribution in Tumor-bearing Mice	189
	6.3.2	Lack of Liver and Cardiac Toxicities	190
	6.3.3	Absence of Ocular Toxicity	190
6.4	STA-1474, a Phosphate Pro-drug of Ganetespib		191
6.5	Clinical Update		192
	6.5.1	Ganetespib in Combination with Docetaxel in Treatment of NSCLC Patients	192
	6.5.2	Efficacy in NSCLC Patients Harboring ALK Rearrangement	193
	6.5.3	Clinical Benefits in Patients of Various Cancers	193
6.6	Conclusion		194
Acknowledgment			195
References			195

Chapter 7 Discovery of the Serenex Hsp90 Inhibitor, SNX5422 198
Timothy Haystead and Philip Hughes

7.1	Introduction		198
7.2	Proteome Mining		199
	7.2.1	Discovery of SNX5422 by Proteome Mining	202
7.3	Lead Optimization Studies Leading to SNX2112		203
	7.3.1	Development of SNX5422, an Orally Bioavailable Hsp90 Inhibitor	205
	7.3.2	Synthesis of SNX5422	208

7.4	Pre-clinical Studies with SNX5422	208
7.5	Clinical Studies with SNX5422	209
Acknowledgements		211
References		211

Chapter 8 Discovery of NVP-AUY922   213
Paul A. Brough, Joseph Schoepfer, Andrew Massey and Michael Rugaard Jensen

8.1	Historical Context	213
8.2	Hit Identification	214
8.3	Lead Optimization	218
8.4	*In Vitro* Assays for Medicinal Chemistry Optimization	219
8.5	*In Vitro* Biomarker Discovery/Development	221
8.6	*In Vivo* Characterization	232
8.7	Clinical Development	236
References		236

Chapter 9 Discovery and Selection of NVP-HSP990 as a Clinical Candidate   241
Timothy D. Machajewski, Daniel Menezes and Zhenhai Gao

9.1	Introduction to the Target as a Cancer Therapeutic		241
9.2	Biochemical and Cellular Assay Development		242
9.3	Library Screening		243
	9.3.1	Selection of Lead Series	245
9.4	Structure and Modeling		245
	9.4.1	Structure Analysis – Comparison with Known Inhibitors	245
	9.4.2	Design of Inhibitors to Displace Structural Waters	246
	9.4.3	Attempted Optimization through Lys58	248
	9.4.4	Reorganization of the "Flexible Loop" Region	248
	9.4.5	Optimization of Binding Affinity	250
9.5	Lead Optimization		252
	9.5.1	Testing Cascade – Selection of GTL16 Cell Line for *in Vitro* and *in Vivo* Optimization	252
	9.5.2	Relationship of PK-PD and Efficacy for NVP-HSP990	254
	9.5.3	PK/PD Optimization	255
	9.5.4	*In Vivo* Efficacy of NVP-HSPP990 in Multiple Cancer Xenografts	255
9.6	Summary of Discovery of HSP990		256
References			257

Chapter 10 Inhibitors of the Hsp90 C-terminus   259
Huiping Zhao and Brian S. J. Blagg

    10.1   Introduction   259
          10.1.1 Hsp90-mediated Protein Folding Machinery   259
          10.1.2 Hsp90 Modulation for the Treatment of Cancer and Neurodegenerative Diseases   261
          10.1.3 The Hsp90 C-terminus: A Second Site to Modulate Hsp90 Function   264
    10.2   Hsp90 C-terminal Inhibitors Derived from Novobiocin   266
          10.2.1 Cytoprotective Hsp90 C-terminal Inhibitors   266
          10.2.2 Hsp90 C-terminal Inhibitors for the Treatment of Cancer   269
    10.3   Additional Hsp90 C-terminal Inhibitors that Induce the Heat-shock Response   287
          10.3.1 Novologues   287
          10.3.2 AEG3482   289
          10.3.3 ITZ-1   289
    10.4   Other Hsp90 C-terminal Inhibitors that Induce Client Protein Degradation   290
          10.4.1 EGCG   290
          10.4.2 Silybin   291
          10.4.3 Hybrids of Novobiocin and Silybin   292
          10.4.4 Cisplatin   294
    10.5   Conclusion   295
    References   296

Chapter 11 Hsp70 Inhibitors   302
Yaoyu Chen and Wenlai Zhou

    11.1   Hsp70 Chaperone Function   302
    11.2   How to Target Hsp70?   303
    11.3   Hsp70 Inhibitors   304
          11.3.1 Targeting the Nucleotide Binding Domain   304
          11.3.2 Targeting the Substrate Binding Domain   311
          11.3.3 Targeting the Co-chaperones   312
    11.4   The Prospectus of Hsp70 Inhibitor   314
    References   314

Chapter 12 The Cancer Super-chaperone Hsp90: Drug Targeting and Post-translational Regulation   318
Annerleim Walton-Diaz, Sahar Khan, Jane B. Trepel, Mehdi Mollapour and Len Neckers

    12.1   Background and Biology of Hsp90   318

	12.2	Post-translational Modifications of Hsp90	319
		12.2.1 Phosphorylation	320
		12.2.2 Nitrosylation/Oxidation	322
		12.2.3 Acetylation	322
		12.2.4 Client-dependent Regulation of Hsp90 PTMs: a Novel Approach to Therapy?	323
	12.3	Targeting Hsp90: Implications and Best Clinical Outcomes	324
		12.3.1 Targeting Sensitive Clients that are Also Tumor Drivers	325
		12.3.2 Hsp90 and Proteotoxic Stress	328
		12.3.3 Hsp90 Inhibitors and TKIs	329
	12.4	Concluding Remarks	330
	References		331

Chapter 13 Hsp90 Inhibitors in Clinic 336
Emin Avsar

13.1	Introduction		336
13.2	Geldanamycin Analogues		337
	13.2.1	17-AAG (Tanespimycin)	337
	13.2.2	17-DMAG (Alvespimycin)	339
	13.2.3	17-Allyamino-17-demethoxygeldanamycin Hydroquinone Hydrochloride (IPI-504; Retaspimycin)	341
13.3	Resorcinol Derivates		341
	13.3.1	NVP-AUY922	341
	13.3.2	NVP-HSP990	343
	13.3.3	STA9090 (Ganetespib)	344
	13.3.4	AT13387	344
13.4	Purine Scaffold-based Inhibitors		345
	13.4.1	Debio 0932/CUDC-305	345
	13.4.2	MPC-3100	346
	13.4.3	BIIB021/CNF 2024	347
13.5	Aminobenzamide Derivates		347
	13.5.1	SNX-5422/PF-04928473	347
13.6	Indications that Appear to Have Promise for Hsp90 Inhibitor Development		352
	13.6.1	Non-small-cell Lung Cancer (NSCLC)	352
	13.6.2	Breast Cancer	356
	13.6.3	Prostate Cancer	359
	13.6.4	Gastrointestinal Stromal Cancers (GIST)	361
	13.6.5	Gastric Cancer	362
	13.6.6	Multiple Myeloma	364
13.7	Conclusions		367
References			369

Chapter 14 Heat-shock Protein 90 as an Antimalarial Target 379
Ankit K. Rochani, Meetali Singh and Utpal Tatu

14.1	History of Malaria		379
14.2	Malaria and Prevention		380
14.3	Current Antimalarial Drugs and Drug Targets		381
14.4	Heat-shock Protein 90 as Antimalarial Drug Target		381
14.5	Hsp90 Targeted New Antimalarial Drug Discovery		385
	14.5.1	Repurposing for PfHsp90 Inhibitors	386
	14.5.2	High-throughput Screening of PfHsp90 Inhibitors	387
	14.5.3	Rational Approach to Hsp90 Targeted Drug Discovery	388
14.6	Future of PfHsp90 Inhibitors as Antimalarials		389
References			389

Chapter 15 Molecular Chaperones as Potential Therapeutic Targets for Neurological Disorders 392
Marion Delenclos and Pamela J. McLean

15.1	Molecular Chaperones and Neurodegenerative Disorders		392
	15.1.1	Protein Aggregation and Molecular Chaperones	392
	15.1.2	Molecular Chaperones in Neuronal Disorders	393
15.2	Modulation of Neurodegeneration by Molecular Chaperones		400
	15.2.1	Pharmacological Up-regulation of Molecular Chaperones	401
	15.2.2	Chemical Chaperones	403
	15.2.3	Viral Mediated Strategies	404
15.3	Concluding Remarks and Perspectives		405
References			406

Subject Index 414

CHAPTER 1

Overview of Molecular Chaperones in Health and Disease

TAI WANG, PABLO C. ECHEVERRÍA AND DIDIER PICARD*

Département de Biologie Cellulaire, Université de Genève, Sciences III, 30 Quai Ernest-Ansermet, 1211 Genève 4, Switzerland
*Email: didier.picard@unige.ch

1.1 Proteostasis and the Central Role of Molecular Chaperones

After being synthesized on ribosomes as linear amino acid chains, proteins need to be folded into their native states, a dynamic equilibrium of closely related three-dimensional structures. In addition to this initial process, cells also need protein quality control and the maintenance of proteome homeostasis (known as proteostasis), both of which are crucial for cellular and organismal health. That is why many diseases appear to be caused by misregulation of protein maintenance. Examples of this are the loss-of-function diseases such as cystic fibrosis and the gain-of-toxic-function diseases such as Alzheimer's, Parkinson's and Huntington's diseases. Proteostasis is maintained by a complex regulatory network, which comprises proteins, cofactors and processes that control protein synthesis, folding, trafficking, aggregation, disaggregation and degradation.[1] Molecular chaperones and their regulators are central players of the proteostatic network.[2]

Molecular chaperones interact with their targets to provide a temporary stabilization, which facilitates folding into a functionally active conformation, unfolding for degradation, or assembly or disassembly of multi-component complexes.[3] Typically, molecular chaperones are not associated with their target proteins once these have acquired their final functional conformation.[4] Recent findings show that there are exceptions. It was recently demonstrated that the glucorticoid receptor (GR) translocates into the nucleus as an Hsp90 heterocomplex upon stimulation by glucocorticoids before dissociating within the nucleus.[5] Moreover, the intranuclear dynamics of GR is still Hsp90-dependent,[6,7] which indicates that some molecular chaperones may escort their clients during their entire lifespan.

Molecular chaperones further ensure proteostasis and prevent proteotoxicity by promoting protein degradation and disposal through multiple pathways. For example, Hsp70, Hsp90 and their co-chaperones target unfolded proteins for degradation *via* the ubiquitin–proteasome system.[8–11] They also assist autophagy. Recent studies show the contribution of Hsp70 to this type of removal of pathogenic proteins. Hsp70 is involved in a certain type of autophagy involving late endosomes known as endosomal microautophagy,[12] and in a form of macroautophagy mediating the degradation of protein aggregates known as chaperone-assisted selective autophagy.[13] The more selective chaperone-mediated autophagy (CMA) requires that cytosolic proteins that contain the pentapeptide targeting motif KFERQ[14] are recognized by heat-shock cognate 70 (Hsc70) and delivered to the surface of lysosomes[15] to be translocated and degraded by lysosomal proteases.[16] A fraction of Hsp90 is present at lysosomes, bound to the luminal side of the lysosomal membrane, and it can either increase or decrease CMA activity depending on the cell type.[17,18]

1.2 The Major Classes of Molecular Chaperones

Molecular chaperones are classified into five families according to their molecular size, namely Hsp100, Hsp90, Hsp70 and J proteins, chaperonins and small heat-shock proteins (sHsp).[2]

1.2.1 Hsp100

Hsp100 chaperones are members of a large superfamily of AAA+ ATPases. They form oligomeric rings involved in protein refolding, disaggregation and degradation.[19] They are found in bacteria, yeast and plants but not in animal cells. Most members of this family use the energy derived from ATP to unfold substrates and to translocate them for degradation to a protease subunit that can be associated with them.[20] Other chaperone machines such as the Hsp70 and J proteins cooperate with the protein disaggregation activity of Hsp100 proteins.[21]

1.2.2 Hsp90

This molecular chaperone is highly abundant in the cytosol of bacterial and eukaryotic cells under physiological conditions, and can be further up-regulated by cellular stress. The Hsp90 family in mammalian cells is composed of four major homologs: Hsp90α (inducible form) and Hsp90β (constitutive form) are cytosolic isoforms; the 94 kDa glucose-regulated protein (GRP94) is localized in the endoplasmic reticulum,[22] and TRAP1 resides in the mitochondrial matrix.[23] The cytosolic forms of Hsp90 bind proteins in a metastable native-like state, which they may have acquired with the help of other chaperone machines. Hsp90 acts with a group of co-chaperones that modulate its client recognition, ATPase cycle and chaperone function. Due to the nature of its clients, its proteostatic functions affect several essential cellular activities such as development, transcription, cell cycle, intracellular signaling, apoptosis, protein degradation and innate and adaptive immunity.[24–29]

1.2.3 Hsp70 and J Proteins

Hsp70 proteins are highly conserved, present both as constitutively expressed and stress-inducible cytosolic isoforms, and isoforms localized to other cellular compartments such as the endoplasmic reticulum and mitochondria.[30] They are important for *de novo* folding, but also for other functions, including protein trafficking, unfolding and degradation of misfolded proteins. More generally speaking, Hsp70s together with a group of essential cofactors, the J proteins of the Hsp40 family and the large nucleotide-exchange factors (NEFs), are involved in ATP-regulated binding and release of non-native proteins including nascent polypeptide chains.[31] Binding and release by Hsp70 is achieved through the allosteric coupling of a conserved N-terminal ATPase domain with a separate substrate binding domain. Hydrolysis of ATP to ADP is strongly accelerated by Hsp40 proteins. Hsp40s also interact directly with unfolded polypeptides and can recruit Hsp70 to protein substrates.[32] Binding of Hsp70 to non-native substrates impedes aggregation by rapidly protecting exposed hydrophobic segments thereby reducing the presence of species tending to aggregate. Recently, it was described that the nucleotide exchange factor Hsp110, an Hsp70 homolog in eukaryotes, cooperates with the conventional Hsp70-Hsp40 machinery to disaggregate and to refold aggregated proteins.[33] Hsp70 machines act in concert with yet other molecular chaperone machines; for example, they often act upstream of chaperonins[34] and the Hsp90 chaperone machine.[35] Hsp70s are especially important under stress by preventing the aggregation of unfolded proteins and by refolding aggregated proteins.[36] This and different features described in Table 1.3 confer the capacity to Hsp70 to act as a survival factor, which is particularly relevant to provide resistance to apoptosis[37,38] and autophagy[39] in cancer cells.

1.2.4 Chaperonins

Chaperonins are ring-shaped multi-subunit chaperones that encapsulate non-native proteins in an ATP-dependent manner. In bacteria, the GroE machinery consists of a multi-subunit structure of the two proteins GroEL and GroES. The closely related proteins of eukaryotic mitochondria are called Hsp60 and

Hsp10, respectively. The non-native protein is trapped in the cavity of GroEL, which becomes a highly hydrophilic environment with a net negative charge where the protein is free to fold after the open structure is capped by GroES binding.[2] The eukaryotic cytosolic chaperonin TRiC is independent of Hsp10. Instead, it contains finger-like projections in its apical domain, which act as a lid and replace the Hsp10/GroES functions.[2]

1.2.5 Small Hsps

These are ATP-independent molecular chaperones that interact with large numbers of partially folded target proteins to prevent their aggregation upon stress. They act as a depository for unfolded proteins, which will later be refolded by other chaperone machines like Hsp70 and Hsp100. In their native state, they form ring-like oligomers of 12–32 subunits with internal spaces in symmetrically blocked dimeric subunits.[40]

1.3 Miscellaneous Other Molecular Chaperones

The following are molecular chaperones that are not part of the major chaperone families, but are nevertheless worth thinking about as potential drug targets.

- **ADCK3** (Chaperone activity of bc1 complex-like, mitochondrial): this chaperone-like protein kinase is essential for the proper conformation and functioning of protein complexes in the respiratory chain. Its absence produces a decrease of the Coenzyme Q10 (CoQ10), and increased ROS production and oxidation of lipids and proteins.[42] ADCK3 mutations were detected in patients with cerebellar ataxia.[43,44]
- **AHSP** (Alpha-hemoglobin-stabilizing protein): a molecular chaperone that is important to prevent the harmful aggregation of free α-hemoglobin during normal erythroid cell development and in β-thalassemic erythroid precursor cells.[45] Gene knockout studies in mice confirmed that AHSP is required for normal erythropoiesis. AHSP knockout mice exhibit anemia, decreased hematocrit and high levels of ROS, consistent with the presence of unstable α-globin.[46]
- **ANKRD13** (Ankyrin repeat domain-containing protein 13): acts as a molecular chaperone for G protein-coupled receptors, controlling their biogenesis and exit from the endoplasmic reticulum.[47] It also regulates the rapid internalization of ligand-activated EGFR.[48]
- **Histones chaperones:** a group of proteins interacting with histones from their synthesis, during import into the nucleus, and for association with target DNA throughout DNA replication, repair or transcription.[49]
- **CLU** (clusterin): functions as extracellular chaperone that prevents aggregation of non-native proteins. Maintains partially unfolded proteins in a state appropriate for subsequent refolding by other chaperones. In Alzheimer's disease, CLU contributes to limit amyloidogenic Aβ species misfolding and facilitates their clearance from the extracellular space.[50]

TOR1A (Dystonia 1 protein): TorsinA is a member of the AAA-ATPase family of molecular chaperones, assisting in the proper folding of secreted and/or membrane proteins. Defects in TOR1A are the cause of dystonia type 1.[51]

HYPK (Huntingtin-interacting protein K): it has a molecular chaperone activity that prevents polyglutamine (polyQ) aggregation of the huntingtin protein.[52]

1.4 Molecular Chaperones in Health and Disease

Molecular chaperones are sensitive hubs of the proteostasis network. As a consequence, genetic alterations of the expression or sequence of members of this network may cause disease. Table 1.1 summarizes the disorders associated with mutations of genes encoding members of the molecular chaperones families.

In addition, the phenotypes of the genetic ablation of different members of the molecular chaperone families in mouse models further emphasize the importance of these genes at the organismic level. Table 1.2 displays a complete compilation of the annotated data concerning the knockout of several molecular chaperones and their co-chaperones present in the PhenomicDB.[41]

Tables 1.1 and 1.2 clearly show that genetic polymorphisms, mutations or the complete ablation of a member of any molecular chaperone machine have an impact on organisms, often with catastrophic consequences. The more or less severely perturbed proteostasis can affect multiple organs and physiological processes, including aging. In accordance with their key hub function,

Table 1.1 Disease association with polymorphisms of molecular chaperones genes.

Molecular chaperone family	Associated diseases
Hsp90 and co-chaperones	Bipolar disorder (HSP90B1).[136] Depression in AD and unipolar depression (FKBP5).[137,138]
Hsp70, J proteins and co-chaperones	Ménière's disease (HSPA1A).[139] Schizophrenia and pulmonary edema (HSPA1A, HSPA1B and HSPA1L).[140,141] Alzheimer's disease (AD) (HSPA4, BAG, DNAJA-B-C, CHIP).[142,143] Coronary disease (HSPA1A/B).[144,145] Diabetes type 1 and 2 (HSPA1A/B).[146-148] Parkinson's disease (HSPA1A/B, DNAJC6).[149,150] Aging (HSPA1A, HSPA1L).[151,152] Crohn's disease (HSPA1B).[153] Myopathy myofibrillar type 6 (BAG3).[154]
Chaperonins	Spastic paraplegia autosomal dominant type 13 (Hsp60).[155]
Small HSPs	Charcot–Marie–Tooth disease type 2, distal hereditary motor neuropathy (Hsp27, Hsp22 and HspB3).[156-158] Different forms of cataracts (HspB3 and CRYAB).[159,160] Myopathy myofibrillar type 2 (HspB5).[161] Dilated cardiomyopathy (HspB5).[162]

Table 1.2 Survey of annotated mouse knockout data for members of the five major molecular chaperone families.

Molecular chaperone family	Gene ID	Official gene name	Phenotype	PubMedID
Hsp90 and co-chaperones	Hsp90aa1	Heat-shock protein Hsp 90-alpha	Male infertility, arrest of meiosis.	21209834
	Hsp90ab1	Heat-shock protein Hsp 90-beta	Embryonic lethality, fail to develop a placental labyrinth.	10654595
	Hsp90b1	Endoplasmin (=GRP94)	Abnormal cytokine secretion and inflammatory response, premature death, essential for mesoderm induction and muscle development. Also complete embryonic lethality between implantation and placentation was described. Spermatozoa deficient in Hsp90B1 could not naturally fertilize oocytes and exhibited large and globular heads with abnormal intermediate pieces (globozoospermia)	17275357, 17634284, 20520781 and 21208614
	PTGES3	Prostaglandin E synthase 3 (alias p23)	Perinatal death	17000766
	FKBP4	FK506 binding protein 4 (alias FKBP52)	Androgen and progesterone insensitivity, male and female infertility	15831525, 16176985, 17142810 and 17307907
	FKBP5	FK506 binding protein 5 (alias FKBP51)	Normal	17142810
	FKBP8	FK506 binding protein 8 (alias FKBP38)	Embryonically lethal, neural defects	15105374
	FKBP6	FK506 binding protein 6 (alias FKBP38)	Male infertility, arrest of male meiosis and azoospermia	12764197
	ITGB1BP2	Integrin beta 1 binding protein 2 (alias melusin)	Defective cardiac response to pressure overload	12496958
	CHORDC1	Cysteine and histidine-rich domain (CHORD)-containing, zinc-binding protein 1	Early embryonic lethal	20230755
Hsp70 and co-chaperones	Hspa1a	Heat-shock 70 kDa protein 1A	Cellular thermotolerance impaired	11713291
	Hspa1b	Heat-shock 70 kDa protein 1B	Impaired TNFα-induced hypothermia	12049720

	Gene	Name	Phenotype	PMID
	Hspa2	Heat-shock-related 70 kDa protein 2	Male infertility, abnormal meiosis	8622925
	Hspa4l	Heat-shock 70 kDa protein 4	Male infertility, abnormal spermiation, kidney morphology and hydronephrosis	16923965
	BAG1	BCL2-associated athanogene 1	Complete embryonic lethality during organogenesis	16116448
	BAG3	BCL2-associated athanogene 3	Increased cardiomyocyte apoptosis, postnatal lethality	16936253
	BAG4	BCL2-associated athanogene 4	Spleen hypoplasia, increased interleukin-6 secretion	12748303
	BAG6	BCL2-associated athanogene 6	Abnormal kidney morphology and brain development, aging	16287848
J proteins	DNAJA1	DnaJ protein homolog 2	Abnormal Sertoli cell, spermatid and spermatocyte morphology, oligozoospermia and reduced male fertility	15660130
	DNAJA3	DnaJ protein Tid-1	Critical for early embryonic development and cell survival	14993262
	DNAJB1	DnaJ protein homolog 1	Required for thermotolerance in early phase	17050614
	DNAJB6	Heat-shock protein J2	Embryonic lethality, fail to develop a placental labyrinth	10021343
	DNAJC17	DnaJ homolog subfamily C member 17	Complete embryonic lethality before implantation	20160132
	DNAJC3	Endoplasmic reticulum DnaJ protein 6	Decreased pancreatic beta cell mass and number, hyperglycemia, premature death	15793246
	DNAJC5	Cysteine string protein	Ataxia, blindness, conductive hearing impairment, premature death	15091340
	DNAJC6	DnaJ homolog subfamily C member 6, auxilin	Abnormal synaptic vesicle number and recycling, delayed female fertility, partial postnatal lethality	20160091
Chaperonins	Hspd1	60 kDa chaperonin	Early embryonic lethality	20393889
Small Hsps	Hspb1	28 kDa heat-shock protein	Normal	17661394
	CRYAB	Alpha-crystallin B chain	Abnormal skeletal muscle fiber and tongue muscle morphology; muscle dystrophic, premature death	11687538
	CRYAA	Alpha-crystallin A chain	Abnormal lens fiber morphology, cataracts	10493778

knockout mouse models are often embryonically lethal or have severe complications in development.

It is noteworthy that the presence (or disproportionate presence) of molecular chaperones is not always beneficial. Even when their participation in the protection against apoptosis could be favorable in the context of some disorders, it can unfortunately also promote the initiation and progression of cancer. In addition, several "clients" of these chaperone machines are oncoproteins. These relationships are summarized in Table 1.3.

This overview provides valuable information in view of the use of inhibitors of selected molecular chaperones for therapeutic interventions, for example against cancer or neurodegenerative diseases. It highlights the huge therapeutic potential, but it also gives a flavor of the extremes of the adverse effects that may have to be expected. To emphasize this point further, we have attempted to model the effect of removing, *i.e.* inhibiting, Hsp90 as one of the key hubs of proteostasis. Figure 1.1 illustrates the dramatic impact that such a treatment can have on part of the proteome.

Table 1.3 Potentially non-beneficial biological processes associated with the presence of the members of the five major molecular chaperone families at normal or over-expressed levels.

Molecular chaperone	Potentially non-beneficial biological process
Hsp90	**Cancer**
	• Hsp90 is "misused" by tumor cells to maintain a large set of oncoproteins (kinases, cell cycle proteins and transcription factors) in a competent state, despite often being mutated or over-expressed. It is also important to safeguard tumors from the stresses that they can be exposed to.[163] Hsp90 may even be present in an activated multi-chaperone complex in cancer cells.[164] Moreover, secreted Hsp90α is involved in promoting invasion of cancer cells.[165–167]
	Immune-related
	• Hsp90 assists the HIV-1 Tat protein (extracellular form) by promoting late steps of Tat translocation into the target cell cytosol. This is believed to play a key role in the development of the acquired immune-deficiency syndrome.[168]
	• Hsp90 is over-expressed in systemic lupus erythematosus patients, where it may contribute to inflammation and disease progression.[169]
	Neurodegeneration
	• Hsp90 increases the stability of mutant tau protein and p35 (an activator of the tau kinase cdk5) in Alzheimer's disease.[170]
	• Inhibition of Hsp90 can prevent the aggregation of huntingtin,[171] mutant androgen receptor proteins,[172] tau protein[170] and α-synuclein,[173] revealing the potential contribution of Hsp90 to several neurodegenerative diseases.

Table 1.3 (Continued)

Molecular chaperone	Potentially non-beneficial biological process
Hsp70 and J proteins	**Apoptosis** Hsp70 proteins regulate apoptosis at different levels affecting both extrinsic and extrinsic apoptotic pathways. • Mediate Bcr-Abl induced resistance to TNFα-related apoptosis-inducing ligand (TRAIL)-induced apoptosis.[118] • Inhibition of apoptotic signaling kinases.[174,175] • Stabilize mutated p53 and sequester wild-type in the cytoplasm.[176] • Hsp70 and Hsp40 proteins inhibit cisplatin-induced apoptosis by binding to Bax and impeding its translocation to mitochondria.[38] • Hsp70 binding to Apaf-1 prevents the recruitment of procaspase-9 to the apoptosome.[177] • Hsp70 can protect GATA-1 from caspase-3 cleavage thereby inhibiting apoptosis.[178] • Hsp70s promote cell survival by inhibiting lysosomal membrane permeabilization and efficient autophagy.[39] **Cancer** Hsp70 function is important at several stages of tumorigenesis. • Inhibits drug-induced Myc-mediated apoptosis of cancer cells.[179] • High levels of Hsp70 in tumors correlate with poor prognosis for cancer patients.[180] • Cisplatin resistance is associated with higher expression of Hsp70 in human ovarian cancer cells.[38] • J proteins were suggested to contribute to the development and spread of cancer as chaperones of various oncogenes.[181]
Chaperonins	**Apoptosis and cancer** • High levels of mitochondrial Hsp60 in tumor cells are associated with apoptosis resistance, lack of senescence and uncontrolled proliferation with neoplastic transformation.[182] • Extracellular Hsp60 is able to interact with many plasma membrane receptors like CD14, CD40 and Toll-like-receptors, with pro- and anti-inflammatory effects.[183–185] It can also stimulate the maturation of dendritic cells synergizing with the pro-inflammatory action of interferon-δ.[186] Extracellular Hsp60 can be a ligand of gamma delta T lymphocytes in oral cancer with antitumoral immunosuppression activity as a consequence.[187] • High levels of Hsp10 were reported in several tumors and pre-tumoral lesions. Hsp10 was also shown to be a pro- or anti-apoptotic factor in different physiological conditions.[188]
Small Hsps	**Apoptosis and cancer** • In several types of cancer cells, Hspb8 is over-expressed and possesses an anti-apoptotic activity.[189] • HspB1 can inactivate Bax and block the release of Smac and cytochrome C;[190,191] moreover, CRYAB protects cells against apoptotic effectors by inhibiting caspase-3 and PARP and by preventing the translocation of Bax and Bcl-2 from the cytosol.[192] • HspB1 is highly expressed in many cancers and is correlated with aggressive tumor behavior, metastasis and poor prognosis; however, in some cases it is associated with good prognosis.[193]

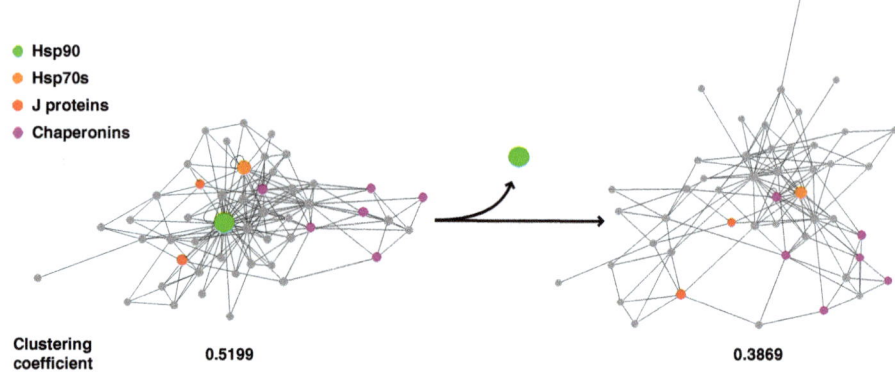

Figure 1.1 The topology of a subnetwork of the human molecular chaperone interactome is severely affected upon Hsp90 withdrawal. We built an up-to-date interactome of the fully annotated set of the members of the major molecular chaperone machines (Hsp90s, Hsp70s, J proteins, chaperonins and small Hsps; a total of 1048 proteins and 5127 interactions) using our previously described pipeline.[200] Using the Cytoscape plugin MCODE,[61] we extracted a molecular complex (49 proteins and 166 interactions) from the large interaction network that is involved in DNA damage response, apoptosis and protein ubiquitination. This network structure seems to be maintained by the different molecular chaperone machines. The ablation of Hsp90 destroys 35 interactions, which relaxes the network and results in a decrease of the clustering coefficient that is a measure of the density of a network.[200] To represent the complex, we used a spring-embedded layout (based on the Kamada–Kawai algorithm.[201] Using this layout, the graph simulates a physical system where interactions are "springs" and proteins are "electrically charged particles". The final conformation of the network gets established when the system comes to an equilibrium.

1.5 Strategies for Modulating Chaperone Activities

The next chapters present various strategies aimed at modulating chaperone activities. In the interest of space, we decided to focus on a few major chaperone machines. This is not a serious limitation since a limited number of molecular chaperones have received almost all of the attention of developers. Other molecular chaperones have yet to be explored as drug targets altogether. Most of the "strategies" reported to date are based on small organic molecules that inhibit molecular chaperones, but more diversity in terms of types of molecules, approaches and impact on molecular chaperones can be expected from future efforts. And to the best of our knowledge and despite very high hopes, "molecular chaperone drugs" have yet to make it into the clinic. Nevertheless, the emerging diversity attests to the huge interest that this class of proteins has attracted over the last few years.

1.5.1 Hsp90 Inhibitors

The functionality of Hsp90 requires highly complex conformational rearrangements regulated by a large spectrum of co-chaperones in which ATP hydrolysis plays a fundamental role in driving the chaperone machinery. Hsp90 features a conserved GHKL-type (gyrase, Hsp90, Histidine Kinase, MutL) ATPase domain at the N-terminus. Therefore, designing competitive inhibitors that target the N-terminal ATP binding pocket of Hsp90 has been the main strategy to block Hsp90.[53] The following paragraphs first present these inhibitors before giving an overview of inhibitors that target other Hsp90 domains and Hsp90 co-chaperones (see Table 1.4).

1.5.1.1 Inhibitors Targeting the N-Terminal ATP Binding Pocket

These types of inhibitors arrest the chaperone cycle by trapping Hsp90 in a conformation reminiscent of the ADP-bound conformation. This leads to an early release of immature client proteins followed by their degradation through the ubiquitin-proteasome pathway. Among the clients of Hsp90, many are oncogenic such as B-Raf, v-Src, HER2, Akt, mutant p53, HIF-1α and Bcr-Abl, which exemplifies the central role of Hsp90 in maintaining the homeostasis of cancer cells. As a result of Hsp90 inhibition and degradation of these and other clients, cancer cells stop proliferating and even undergo apoptosis. A corollary of the inhibition of Hsp90 is the induction of Hsp70 expression through the proteasomal stress response and by derepression of the heat-shock factor 1 (HSF-1).[54]

Several classes of N-terminally targeted Hsp90 inhibitors have emerged from both the pharmaceutical industry and the academic world. Based on the natural and prototypical inhibitor geldanamycin, tanespimycin (17-AAG) and 17-DMAG were identified as improved derivatives with less toxicity and better water solubility. However, the reduction of the benzoquinone core by NQO1/DT-diaphorase is required for these inhibitors to exert full efficacy. This prompted the development of IPI-504 (Infinity Pharmaceuticals), a tanespimycin derivative bearing a stabilized hydroquinone ring as an attempt to reduce or to abolish the dependence on the NQO1-mediated reduction. This inhibitor is currently in clinical trials and demonstrates promising potency against non-small-cell lung cancer. However, despite an excellent specificity for Hsp90, many benzoquinone ansamycin-based molecules have intrinsic hepatotoxicity.[54]

The discovery of the purine-based inhibitor PU3 brought a new impetus for the development of synthetic inhibitors. PU3 stabilizes the N-terminal ATPase pocket of Hsp90 in a conformation characterized by the formation of an α-helix between Leu107 and Gly114. In comparison with the ADP-bound form, this structural rearrangement creates a secondary hydrophobic binding pocket where appropriate hydrophobic decorations of ligands can be accommodated. Apart from the ansamycin family of inhibitors, most of the subsequently developed inhibitors take advantage of this secondary site.[55,56]

The purine-based inhibitors mimic the adenosine moiety of the natural ligand ATP, which ensures recognition by Hsp90. The more recently reported aminopyri(mi)dines, thienopyrimidines (from Astex Pharmaceuticals, Evotec,

Table 1.4 Hsp90 and Hsp90 co-chaperone inhibitors.

Chaperone name	Binding domain/ co-chaperone	Name of drugs/molecules	Type of molecule, other comments	Refs
Hsp90	N-terminal ATP binding pocket	Geldanamycin, Tanespimycin, 17-DMAG, IPI-504	– Benzoquinone ansamycin	54, 194
		PU3, PU-H71, BBIIB021, Debio-0932, CUDC-305	– Adenosine-scaffold or Purine-derivative (3-position NH_2 substituent)	54, 194
		Radicicol, NVP-AUY992/VER52296, KW2478, STA-9090/Ganetespib PF-3823863	– Resorcinol scaffold	for Pochonins/ Pochoximes 195, 196
		Pochonins/Pochoximes		
		SNX-5422, SNX-2112, XL888	– Benzamides	54, 194
		NVP-BEP800 and others	– Aminopyri(mi)dines/Thienopyrimidines	54, 194
		N-[4-(3H-Imidazo[4,5-c]pyridin-2-yl)-9H-fluoren-9(R)-yl]-1H-pyrrolo[2,3-b]pyridine-4-carboxamide	– Azaindole/Tricyclic Imidazo[4,5-c]pyridines	197
		CP9	– Competes with 17-AAG for binding – Disrupts p23/Hsp90α association – Inhibits Hsp90 and induces client degradation in cells	64
	Other N-terminal domain	Gambogic acid	– Likely binds to a site other than the ATPase pocket	58
		Shepherdin	– Peptide	59, 60
		Sansalvamide A	– Preferentially binds to closed conformation of yeast Hsc82	61–63

C-terminal domain	Novobiocin derivatives, e.g. KU-135, KU-32	–	Allosteric inhibitors of Hsp90	65
		–	Induce less proteasomal stress compared with N-terminal inhibitors	
	(–)-Epigallocatechin-3-gallate (EGCG)	–	Allosteric inhibitor of Hsp90	75, 198
		–	Degradation of clients such as B-Raf, pAKT and HER2 without inducing Hsp70	
	Withaferin A	–	Allosteric inhibitor	69
		–	Disrupts Cdc37-Hsp90 association (but not p23-Hsp90)	
		–	Induces Hsp70 up-regulation	
	Cisplatin and LA-12	–	C-terminal binder	71, 72
		–	Selective degradation of the clients AR, GR or ER	
		–	No Hsp70 up-regulation	
	Molybdate	–	Stabilizes closed conformation resembling ATP-bound form	73
		–	Competes with the binding of geldanamycin	
Hsp90 co-chaperone				
Cdc37-Hsp90	Celastrol	–	Disrupts Cdc37-Hsp90 complex without affecting ATP binding	66, 67
		–	Modifies cysteine residues in Cdc37 kinase-binding domain, which accounts for the impairment of the Cdc37-Hsp90 interaction	
		–	Binds to ADP-bound but not ATP-bound Hsp90	
		–	Inhibits Hsp90 ATPase activity	
		–	Induces Hsp70 up-regulation	
HOP	C9 (1,6-dimethyl-3-propylpyrimido[5,4-e][1,2,4]triazine-5,7-dione)	–	Disrupts interaction between Hsp90-MEEVD and HOP-TPR2A	82

Abbott or Vernalis), the azaindole (from Sanofi-Aventis) and the benzamide (SNX-5422, SNX-2112, XL888) derivatives actually share the same or very similar but more cryptic bioisosteres that were inspired by the adenosine moiety.[56]

Resorcinol derivatives constitute another important group of the N-terminally targeted Hsp90 inhibitors. Their dihydroxyl benzene group is present in radicicol, a natural antibiotic that binds tightly to Hsp90. The binding mode of the resorcinol motif appears to be very stable and conserved in Hsp90s and therefore provides an excellent "starting point" for further drug design by both structure-based and fragment-based approaches.[56]

1.5.1.2 Inhibitors Targeting Other Surfaces of the N-Terminal Domain of Hsp90

Gambogic acid

Gambogic acid binds to the N-terminal domain of Hsp90 and impedes the association with Hsp70 and Cdc37. This leads to the degradation of client proteins and up-regulation of Hsp70 and Hsp90.[57] It induces apoptosis through inactivation of the TNF-α/NFκB pathway. The binding of gambogic acid to the N-terminal domain of Hsp90 is not affected by the presence of geldanamycin, suggesting that the drug may bind to a site distinct from the ATP binding pocket.[58]

Shepherdin (Peptide)

Shepherdin is a fragment of a peptide from the Hsp90 client survivin. Initially designed to block its interaction with Hsp90, the peptide was shown to bind to the N-terminal domain of Hsp90 and to disrupt the Hsp90 ATPase activity.[59] It induces apoptosis in tumor cells following the degradation of Hsp90 clients such as Akt, CDK-4, CDK-6 and survivin without affecting the levels of Hsp70.[60]

Sansalvamide A Derivatives (Peptide)

Sansalvamide A is a cyclic pentapeptide that is extracted from a marine fungus of the genus *Fusarium sp*. The peptides were demonstrated to bind to the junction between the N-terminal and middle domains of Hsp90 and to inhibit the Hsp90 cycle. The mechanism is most likely an allosteric regulation, which is supported by the fact that the peptides preferentially bind to the closed conformation of yeast Hsc82 stabilized by the non-hydrolyzable ATP derivative AMPPNP.[61] Further biological characterization showed that Sansalvamide A derivatives[62] induce caspase-dependent apoptosis in cells. Some of them impair the recruitment of clients or co-chaperones including IP6K2, FKBP38, FKBP52 and HOP. Like 17-AAG, these compounds also elicit the up-regulation of Hsp70.[63] In contrast, one of the compounds causes also cytotoxic effects and caspase-mediated apoptosis while it does not affect the interaction between the co-chaperone/clients and Hsp90, suggesting that it may be used as

CP9

The 2-[6-(trifluoromethyl)pyrimidin-2-yl]thio-acetamide-based compound C9 was discovered as a molecule that disrupts the Hsp90-p23 interaction. It targets the N-terminus of Hsp90 and competes with 17-AAG for binding. The compound induces Hsp90 client degradation in cancer cell lines and selectively impairs the interaction of p23 with the Hsp90α isoform. Further chemical optimization is expected to deliver candidates with higher potency in live animals.[64]

1.5.1.3 Inhibitors Targeting the C-Terminal Domain of Hsp90

Derivatives of Coumarin Antibiotics (Novobiocin)

The coumarin antibiotic novobiocin binds to the C-terminal domain of Hsp90, but nevertheless inhibits the ATPase activity in an allosteric manner. A key feature of this family of molecules is that they seem to induce much less Hsp70 expression. Since the anti-apoptotic effects of Hsp70 are unwanted side effects of N-terminally targeted ATPase inhibitors, there is an interest in further developing those as alternatives. Indeed, certain derivatives of novobiocin such as KU-135 even have improved drug effectiveness and binding to Hsp90 compared to 17-AAG.[65]

Celastrol

The proteasomal inhibitor Celastrol disrupts the association of Hsp90-Cdc37. Even though it does not occupy the ATPase pocket of Hsp90, its binding is exclusive to the ADP bound or nucleotide-free form of Hsp90,[66] suggesting it is an allosteric inhibitor acting through a different site. Celastrol inhibits the ATPase activity of Hsp90 and induces the degradation of client proteins such as Akt and Cdk4. The action is accompanied by the simultaneous increase of Hsp70 expression through the activation of the HSF-1 pathway.[67] In addition, the quinone methide moiety of Celastrol confers a reactivity towards thiols, which brings about the direct modification of cysteine residues in many proteins including Hsp90, p23 and Cdc37. Presumably, thiol oxidation of cysteines accounts at least in part for the ability of Celastrol to inhibit the Hsp90 activity.[68]

Withaferin A

Withaferin A is a withanolide from the plant *Withania somnifera*. This steroidal lactone binds to the C-terminal domain of Hsp90 and inhibits its ATPase activity. Like Celastrol it disrupts the formation of the Hsp90-Cdc37 complex, and induces the degradation of Hsp90 clients (Akt, Cdk4, GR) and up-regulation of Hsp70 through the proteasomal stress response pathway.[69] Withaferin A shares the quinone methide moiety with Celastrol and therefore also induces direct thiol oxidation of cysteine residues on Hsp90, which causes the aggregation of Hsp90. Tubocapsenolide A inhibits Hsp90 using the same mechanism.[70]

Cisplatin and LA-12

In addition to binding to DNA, cisplatin has been reported to target the C-terminal domain of Hsp90 and to disrupt selectively the Hsp90 chaperoning function for transcriptional factor clients such as the androgen or glucocorticoid receptors. Similarly to the coumarin antibiotics, cisplatin does not induce Hsp70 expression.[71]

An optimized derivative of the platinum-based complex, LA-12, exhibits enhanced binding affinity for Hsp90, presumably due to the increased hydrophobicity, which favors the binding to the C-terminal domain. Compared to cisplatin, this compound also stimulates the degradation of more Hsp90 clients, such as cyclin D1 and the estrogen receptor.[72]

Molybdate

Molybdate stabilizes a closed conformation of Hsp90 that resembles the ATP-bound state. It may do this by replacing the cleaved γ-phosphate of ATP. It has been used extensively to stabilize Hsp90-client interactions. However, its poor cell permeability essentially limits its use to biochemical experiments.[73]

EGCG

(−)-Epigallocatechin-3-gallate (EGCG) is a polyphenolic cathechin found in green tea. Biochemical assays showed that the compound binds to the C-terminal domain of Hsp90 and induces the degradation of various client proteins such as Akt, Cdk4, Raf-1, HER2 and pERK.[74] EGCG also weakens the formation of Hsp90-p23 and Hsp90-Hsc70 complexes and suppresses Hsp70 expression induced by proteasomal stress.[75]

1.5.1.4 Inhibition of the Hsp90 Co-Chaperone Cdc37

Cdc37 is a co-chaperone that regulates the ATPase cycle of Hsp90. It appears to serve as an adapter that brings a wide spectrum of client kinases to Hsp90 and therefore acts as an important modulator of many signaling pathways. In comparison with Aha1 that promotes the Hsp90 ATPase activity, Cdc37 slows down the ATP turnover through an interaction with the lid segment of the N-terminal domain of Hsp90.[26,76]

Drugs (Celastrol, Withaferin A)

In addition to affecting Hsp90 directly, Celastrol was shown to target Cdc37 by covalently modifying the cysteine residues on the kinase-binding domain of Cdc37, which alters the global conformation of Cdc37 and leads to a disruption of the Hsp90-Cdc37 interaction.[77] The same appears to be true for Withaferin A.[68,69] Additional inhibitors that target Cdc37 directly and selectively would clearly constitute formidable tools and potential therapeutics since they might only affect kinases that are Hsp90 clients.

Cdc37 siRNA

Silencing Cdc37 by RNA interference results in a significant degradation of Hsp90 client kinases through the ubiquitin-proteasome pathway. Cdc37 silencing also sensitizes cancer cells against Hsp90 inhibitors. In addition, Cdc37 silencing does not derepress HSF-1 and therefore does not induce up-regulation of Hsp70.[76]

Phosphorylation/Dephosphorylation of Cdc37

As the phosphorylation of specific sites is required for Cdc37 function, inhibition of casein kinase II (with TBB), which decreases the phosphorylation of Cdc37, can be envisaged as an alternative to repress Cdc37.[78] Over-expression of the phosphatase PP5 results in the same outcome.[79]

1.5.1.5 Inhibition of the Hsp90 Co-chaperone HOP

By bridging Hsp70 and Hsp90, the Hsp70-Hsp90 Organizing Protein (HOP; also known as Sti1) helps to pass clients from Hsp70 to Hsp90. It contains several tetratricopeptide (TPR) domains. The TPR1 domain interacts with Hsp70 whereas the TPR2A domain enables the recognition of the C-terminal MEEVD motif of Hsp90 and the TPR2B domain interacts with both the N-terminal and middle domains of Hsp90. HOP prevents the N-terminal dimerization of Hsp90 and thereby inhibits the ATPase activity.[80]

C9

Using an *in vitro* high-throughput screen, a family of 7-azapteridine compounds was first identified as lead compounds that bind to the TPR2A domain of HOP and block its interaction with the MEEVD motif of Hsp90.[81] The derivative C9 was shown to be effective without inducing the HSF-1-mediated up-regulation of Hsp70. Interestingly, the cell death induced by C9 does not result from the activation of the caspase 3/7-dependent apoptotic pathway, which indicates a distinct mechanism in comparison with that of 17-AAG. C9 also suppresses the up-regulation of Hsp70 if applied in combination with 17-AAG or NVP-AUY922, although not with PU-H71, possibly due to the overwhelming inhibitory effect of PU-H71.[82]

Hybrid Antp-TPR Peptide

A hybrid TPR peptide was designed based on a structurally engineered TPR2A derivative that was rendered cell-permeable by fusion to a cell targeting peptide from the Antennapedia homeodomain protein. This hybrid peptide selectively induces apoptosis in cancer cell-lines and provokes Hsp90 client degradation without activating HSF-1 mediated Hsp70 expression.[83]

HOP siRNA

The application of siRNA against HOP expression in pancreatic cancer cells has been shown to decrease invasiveness through MMP-2 down-regulation. Decreased expression levels of Hsp90 clients such as HER2, Bcr-Abl, c-MET and v-Src were also observed.[84]

1.5.1.6 Inhibition of the Hsp90 Co-chaperone p23

Celastrol

It was reported that Celastrol can modify the cysteine residues of p23 through direct thiol oxidation. This alters the three-dimensional structure of p23, which drives p23 into amyloid-like fibrils.[85]

Genetic Interference

Deletion or silencing of p23 (Sba1 in yeast) results in a hypersensitivity to Hsp90 inhibitors (17-AAG) in the context of budding yeast and mammalian cells.[86]

1.5.1.7 Aha1

Aha1 is the only Hsp90 co-chaperone known to activate the Hsp90 ATPase. It does so by interacting with the middle and N-terminal domains of the Hsp90 dimer *in trans* and favors the dimerization of Hsp90.[87]

Genetic Interference

The knockdown of Aha1 expression by RNA interference leads to a higher sensitivity to Hsp90 inhibitors despite not affecting the levels of Hsp90 clients. However, a decreased phosphorylation of MEK1/2 and ERK1/2 was observed, suggesting that Aha1 may be more involved in regulating the activation of these client proteins rather than their stabilization.[88] Surprisingly, over-expression of Aha1 results in a lower rate of Hsp90-dependent refolding of denatured luciferase.[89] Moreover, mutations in the Hsp90 ATPase lid domain that result in an increased affinity for Aha1 are associated with an increased resistance to Hsp90 inhibitors, indicating a protective role of Aha1 through regulation of the ATPase activity of Hsp90.[90]

1.5.2 Hsp70 Inhibitors

The structure of Hsp70 can be functionally divided into two domains: an N-terminal nucleotide-binding domain and a C-terminal substrate-binding domain. As for Hsp90, the functionality of Hsp70 relies on the binding and hydrolysis of ATP, which drives conformational changes. The ATPase activity is stimulated by Hsp70 co-chaperones of the Hsp40/DnaJ family through an interaction that involves the J-domain of Hsp40. Other co-chaperones such as the nucleotide exchange factors Hsp110, Bag-1 or HspBP1 facilitate the release of ADP.[91]

Although both Hsp90 and Hsp70 are ATPase-dependent molecular chaperones, intrinsic tight ATP binding, more hydrophilic ligand interaction and flexible ligand-pocket accommodation present obstacles to the design of competitive small-molecule inhibitors targeting the ATP-binding pocket of Hsp70 compared with the situation for Hsp90.[53]

Nevertheless, given the significance of Hsp70 as a potential target for cancer therapy, there are many continued efforts for the development of Hsp70 targeting pharmacological agents in the form of small molecules, peptides or peptide aptamers. Among them, some directly compete with ATP for binding to the Hsp70 ATP binding pocket, some bind to the substrate-binding domain and inhibit the recruitment of client proteins, whereas some affect the association of Hsp40 with Hsp70, which impairs the activation of the Hsp70 ATPase. In the following paragraphs, we mention some examples of molecules that target different regions of Hsp70 (Table 1.5).

1.5.2.1 Inhibitors Targeting the Nucleotide Binding Domain

VER-155008

A small synthetic molecule mimicking adenosine, VER-155008, binds to the Hsp70 ATPase site. The compound also displays affinity for Hsc70 and the isoform GRP78 (Bip), which resides in the endoplasmic reticulum (ER). It induces the degradation of some Hsp90 client proteins such as Raf-1 and HER2, inhibits the proliferation and induces caspase 3/7-mediated apoptosis of several tumor cell lines.[92]

Apoptozole

Apoptozole (Az) was originally identified in an imidazole-based library as an agent that induces apoptosis in P19 and A549 lung cancer cells. Biochemical studies demonstrate that Az is able to bind to and inhibit the Hsp70/Hsc70 ATPase domain. Competitive binding between Az and ATP suggests that the compound may directly fit into the ATP-binding pocket. Moreover, Az has been shown to disrupt the association between the CFTR mutant ΔF508 and Hsp70, which promotes its degradation through the ubiquitin-proteasome pathway.[93]

MKT-007

MKT-007 (1-ethyl-2-[[3-ethyl-5-(3-methylbenzothiazolin-2-yliden)]-4-oxothiazolidin-2-ylidenemethyl] pyridinium chloride) is known to bind in the vicinity of the ATP-binding pocket of Hsc70. It favors the ADP-bound conformation of Hsc70, suggesting that the compound is an allosteric inhibitor of Hsc70. It has also been shown that MKT-007 promotes the ubiquitination and degradation of over-expressed tau.[94]

AdaSGC

The water-soluble glycosphingolipid mimic adamantly-sulfogalactosylceramide (AdaSGC) binds to the Hsc70 ATPase domain and disrupts the Hsp40-stimulated

Table 1.5 Hsp70 and Hsp70 co-chaperone inhibitors.

Chaperone name	Binding domain/co-chaperone	Name of drugs/molecules	Remarks	Refs
Hsp70	Nucleotide binding domain	VER-155008	– Purine-based inhibitor (Vernalis) binds ATPase domain of Hsp70	92
		MKT-007	– Allosteric inhibitor that binds to ADP-bound state of Hsp70	94
		Apoptozole	– Binds ATP-binding pocket and inhibits Hsc70	93
		AdaSGC	– Disrupts Hsp40-Hsp70 interaction and therefore inhibits Hsp70 ATPase activity – Inhibits peptide-binding regardless of the presence of Hsp40	95
		Aptamer A17 P17 (only the variable region of Aptamer A17)	– Inhibits tumor progression – Induces apoptosis (through caspase 3)	96
	Peptide-binding domain	2-phenylethynesulfonamide (PES)/Pifithrin-μ	– Induces client protein degradation through autophagy-lysosome pathway	97, 98
		ADD70 (AIF derivative)	– Engineered protein based on AIF – Binds to the substrate-binding domain of inducible Hsp70 – Synergizes with cisplatin and 17-AAG – Not naturally cell-permeable	99, 100
		Aptamers A8	– Discovered together with Aptamer A17 – Different portions of A8 alone do not work in tumor cells	96
Hsp70 co-chaperones	Hsp40/DnaJ	D-peptides	– Dissociate DnaJ from DnaK proteins	101
		Dihydropyrimidine NSC6340668-R/1 MAL3-101 116-9e	– Inhibits Hsp70 by disrupting Hsp40-Hsp70 association	199, 102, 103
		Dihydropyrimidine 115-7c	– Binds preferentially to the DnaK/DnaJ complex – Stimulates ATPase activity of DnaK	104
		AdaSGC	(See above)	95

ATPase activity, causing a decrease in client protein loading. It has been shown to promote the clearance of over-expressed CFTR ΔF508 through the ER-associated-degradation pathway.[95]

Peptide Aptamer A17

Yeast two-hybrid experiments led to the discovery of the peptide aptamers A8 and A17, which bind to the Hsp70 peptide-binding and nucleotide-binding domains (but not to those of the Hsc70s), respectively. Both aptamers, when expressed in mouse melanoma B16F10 cells, suppress tumor progression and synergize with the effects of cisplatin. The synthetic peptide P17, which contains only the variable region of aptamer A17, sensitizes tumor cells to cisplatin and induces apoptosis. Although it binds the nucleotide-binding domain, the exact binding site still remains to be elucidated.[96]

1.5.2.2 Inhibitors Targeting the Substrate Binding Domain

PES/Pifithrin-μ

2-phenylethyenesulfonamide (PES) is a small molecule that binds to both Hsc70 and Hsp70. It was first discovered as a molecule that impairs the mitochondrial localization of p53.[97] It was shown to induce apoptosis *via* activation of caspase-3 in leukemic cells, which also involves the degradation of client proteins such as Akt and ERK1/2. It thereby overcomes the protective effect afforded by the up-regulation of Hsp70 because of the proteasomal stress. This also contributes to enhanced anti-proliferative effects in combinatorial treatments with the Hsp90 inhibitor 17-AAG or with the histone deacetylase inhibitor SAHA.[98]

ADD70

ADD70 is an artificially engineered protein derived from the flavoprotein apoptosis-inducing factor (AIF). It only contains the Hsp70-interacting region and binds to the substrate-binding domain of the inducible Hsp70.[99] Tumor cells expressing ADD70 demonstrate increased sensitivity to a treatment with cisplatin or the Hsp90 inhibitor 17-AAG. Extended studies indicated that the induction of antitumorigenic immune responses may account for the cytotoxicity of ADD70. This is supported by the observation that the synergy of ADD70 and cisplatin disappears in immunodeficient animals.[100]

1.5.2.3 Inhibitors That Affect the Interaction between Hsp40 and Hsp70

D-peptides

D-peptides are derived from the N-terminal fragment of rhodanese, which was known to bind to the bacterial DnaJ. The peptides RI1-17 and RI1-10 bind to

DnaJ and disrupt the formation of the bacterial DnaJ/DnaK/GrpE chaperone complex.[101]

Dihydropyrimidine Family (NSC6340668-R/1, MAL3-101, 115-7c, 116-9e)

Based on the prototype molecule 15-deoxyspergualin (DSG) that demonstrates the ability to modulate the Hsc70 ATPase activity, NSC6340668-R/1 and MAL3-101 were identified as small molecules that inhibit Hsp70 by preventing the Hsp40 stimulation of the ATPase activity.[102] MAL3-101 was shown to be effective against multiple myeloma cell lines by inducing caspase-3-mediated apoptosis. It displays strong synergy with the Hsp90 inhibitor 17-AAG or with the proteasomal inhibitor MG-132.[103] In addition, other molecules in the family, such as MAL3-38, MAL3-90 or 115-7c, exert the opposite effect by stimulating the Hsp70 ATPase activity. Among them, 115-7c was demonstrated to compensate partially for the loss-of-function phenotype of yeast lacking the Hsp40 protein Ydj1. It preferentially binds to the bacterial DnaJ-DnaK complex rather than DnaK alone, suggesting that the compound regulates the DnaK ATPase activity through an allosteric mechanism. Compared to 115-7c, the derivative 116-9e toggles the effect back to a disruption of the DnaJ-DnaK interaction and inhibition of the DnaK ATPase activity. This illustrates the flexibility of the dihydropyrimidine family of compounds as allosteric regulators of the DnaK-DnaJ-GrpE chaperone complex.[104]

1.5.2.4 CHIP

CHIP Over-expression

The constitutive heat-shock cognate 70 (Hsc70)-interacting protein CHIP binds to Hsp70/Hsc70 through its TPR domain. It also possesses a U-box domain, which carries its E3 ubiquitin ligase activity. CHIP is responsible for the ubiquitin-mediated degradation of many Hsp70/Hsp90 client proteins. Overexpression of CHIP enhances client protein degradation *in vivo*.[105] Furthermore, it targets the molecular chaperones Hsp70 and Hsp90 themselves.[106]

1.5.3 Inhibitors Targeting Small Molecular Chaperones

1.5.3.1 Hsp27 Inhibitors

Hsp27 increases cell survival and enhances tumor migration and invasiveness. It is highly expressed in various cancers such as breast, ovarian or prostate cancer. There has been an increasing awareness that Hsp27 may be an interesting anticancer target.[107]

KRIBB3

KRIBB3 was identified from a pool of compounds that inhibit tumor cell migration. The compound binds to Hsp27 with high affinity and affects the protein kinase C-dependent phosphorylation of Hsp27, which is necessary for its activation.[108] Moreover, KRIBB3 was demonstrated to induce cancer cell apoptosis by activating the mitotic spindle checkpoint because of the inhibition of tubulin polymerization.[109]

Peptide Aptamers PA11 and PA50

Identified with a yeast two-hybrid screen, aptamers PA11 and PA50 have antiproliferative and apoptosis-inducing activities in HeLa cells, while the overall expression levels of Hsp60, Hsp70 and Hsp90 were not affected. In parallel, biochemical studies showed that PA50 affects Hsp27 dimerization while PA11 impairs the oligomerization of Hsp27, and both aptamers modified the phosphorylation state of Hsp27. Expressing PA50 and PA11 in cells appears to be more effective than an shRNA-mediated knockdown of Hsp27.[110]

OGX-042

OGX-042 (OncoGeneX Pharmaceuticals Inc.) is an antisense oligonucleotide (ASO) that blocks the translation initiation site of the Hsp27 mRNA. It has a 2′-O-methoxy-ethyl (MOE) backbone, which enhances its protection against nucleases and increases its half-life *in vivo*. OGX-042 inhibits cell growth, induces apoptosis and sensitizes cells to the chemotherapeutic agent gemcitabine.[111]

1.5.3.2 Clusterin Inhibitors

The expression of clusterin is known to be associated with higher grade, post-treatment stress and poor outcome in many cancers.[107]

OGX-011 (Custirsen)

Like OGX-042, OGX-011 is an antisense oligonucleotide that is designed to block the expression of human clusterin. The depletion of clusterin leads to an enhanced sensitivity to chemo- and radiation therapy. OGX-011 has demonstrated promising effects to postpone the recurrence of castration-resistant prostate cancer.[112] The combination of OGX-011 with the Hsp90 inhibitors 17-AAG or PF-04929113 potentiates the antitumor potency in xenografts by limiting the protective heat-shock response induced by the Hsp90 inhibitors.[113]

1.5.4 Modulators of the HSF-1 Pathway

In response to proteotoxic stress, heat-shock or pharmacological agents such as Hsp90 or proteasome inhibitors, HSF-1 initiates the transcription of several major protective proteins including Hsp70, Hsp90 and Hsp27. Since HSF-1 is required by cancer cells to maintain proteostasis, targeting the HSF-1 pathway represents a promising strategy to suppress the transformation and malignant progression of cancer cells and to potentiate other drugs that elicit stress responses. The mechanisms of activation of HSF-1 are incompletely characterized. Moreover, there is no druggable HSF-1 domain that is well defined. This makes the development of potent inhibitors very challenging.[114]

1.5.4.1 Inhibitors of the HSF-1 Pathway

QC

Some known antimalarial agents such as aminoacridine quinacrine (QC) are known to inhibit the HSF-1-mediated Hsp70 expression. The combination of QC with 17-AAG has shown enhanced potency in inducing caspase-mediated apoptosis. The pharmacological effects of QC on HSF-1 are fairly broad.[114] Despite its low potency and poor specificity, which prevents its clinical development, it serves as an encouraging model.[115]

KNK437

KNK437 (*N*-formyl-3,4-methylenedioxy-benzylidene-gamma-butyrolactam) suppresses the HSF-1-dependent transcription of Hsp70 and Hsp27.[116] The exact mechanism still remains to be established. As a result of lowering Hsp70 and Hsp27 levels and thus the proteasomal response, KNK437 potentiates the inhibitory effects of Hsp90 inhibitors in various cell lines.[117,118]

Triptolide

The diterpene triepoxide Triptolide is derived from the Chinese medicinal plant *Triptergium wilfordii*. It can effectively suppress the heat-shock response initiated by HSF-1 at nanomolar doses through direct interference with the transactivation domain of HSF-1, possibly at a post-transcriptional level.[119]

NZ28 and Emunin

NZ28 and Emunin were identified from 20,000 compounds by a screen for inhibitors of the induction of heat-shock proteins. Despite the analogy to emetine, which blocks translation, the compounds do not affect the overall level of transcription, translation or protein degradation. Instead they specifically suppress the up-regulation of Hsp70 and Hsp27 in response to Hsp90 or proteasome inhibitors. The inhibitory effect may be at the post-transcriptional level since synthesis and degradation of Hsp70 mRNA were unaffected by the treatment. Moreover, the molecules sensitize cancer cells to Hsp90 and proteasome inhibitors and induce Hsp90 client degradation.[120]

1.5.4.2 Activators of the HSF-1 Pathway

This avenue is pursued because HSF-1 activation may be beneficial to treat neurodegenerative diseases that are associated with the accumulation of aberrant proteins. An appropriate induction of the heat-shock response by the activation of HSF-1 enhances cytoprotection against detrimental consequences of misfolded or aggregated stressors.[121]

Hsp90 Inhibitors

The induction of Hsp70 through HSF-1 activation in response to Hsp90 inhibitors is well established. It is thought that the release of the inhibitory Hsp90 complex from HSF-1 contributes to its activation. In keeping with the beneficial effects of Hsp70 induction, geldanamycin is effective against the formation of α-synuclein oligomers in cells. Several benzamide derivatives (*e.g.* SNX-0723), which have the added advantage of being orally available and passing through the blood-brain barrier, have been reported to prevent α-synuclein oligomerization through HSF-1-stimulated expression of Hsp70.[122]

Celastrol

As described above in the section on Hsp90 inhibitors, Celastrol inhibits Hsp90 through various mechanisms and induces the HSF-1 pathway.[85,123] However, its intrinsic toxicity, in part to due to pleiotropic thiol oxidation of proteins, makes it difficult to develop it further for clinical applications.

Geranylgeranylacetone

Geranylgeranylacetone (GGA) has been widely used as an anti-ulcer drug. It is also known to induce the expression of Hsp70. Biochemical studies demonstrate that GGA competes with HSF-1 for the binding to the substrate-binding domain of Hsp70. Upon dissociating from Hsp70, HSF-1 is activated.[124] Moreover, GGA seems to trigger ER stress, which leads to the up-regulation of GRP78. GGA elicits both pro-apoptotic and anti-apoptotic unfolded protein responses (UPR) through modulating the ATF6-CHOP, ATF6-GRP78 and IRE1-XBP1 pathways.[125] The cytoprotective role of GGA may result from its induction of thioredoxin (TRX) expression, which enhances the ability to reduce the oxidized thiol groups of proteins and to protect cells against oxidative stresses.[126]

HSF1A

HSF1A is a benzyl pyrazole identified with a yeast-based high-throughput screen. The compound activates human HSF-1; it increases the expression of Hsp70 by facilitating the nuclear import and phosphorylation of HSF-1. HSF1A confers protection against cytotoxicity caused by polyQ aggregation. Interestingly, the TRiC/CCT chaperone complex was found to co-purify with biotinylated HSF1A, which is intriguing considering that TRiC/CCT acts as a negative regulator of polyQ aggregation.[127]

Bimoclomol/Arimoclomol

Bimoclomol alone does not activate HSF-1. Instead, it potentiates the expression of Hsps under a pre-existing stress condition. The mechanism seems to involve direct binding of bimoclomol to HSF-1, which leads to a prolonged HSF-1/DNA interaction accompanied by a mild increase of HSF-1 phosphorylation.[128] As co-inducers of HSF-1-mediated chaperone expression, the drugs may be well adapted for the treatment of neurodegenerative diseases in which a strengthened molecular chaperone system increases the tolerance to misfolded stressors. The derivative arimoclomol (BRX-220) is in a Phase II/III clinical trial for the treatment of patients with superoxide dismutase (SOD1) positive amyotrophic lateral sclerosis (ALS).[129]

Riluzole

Riluzole is an FDA-approved drug for the treatment of ALS. It has recently been shown that the drug inhibits chaperone-mediated autophagy of HSF-1. The increased availability of HSF-1 boosts heat-shock responses and ultimately improves cytoprotection against stresses.[130]

NSAIDs

Non-steroidal anti-inflammatory drugs (NSAIDs) have been known for a long time to promote HSF-1 activity.[131] However, only some (sulindac, cyclopentenone derivatives) are robust enough to elicit the HSF-1-induced Hsp70 expression by themselves.[132,133] Others (sodium salicylate, indomethacin) act as co-inducers, which facilitate Hsp70 expression when an additional stress pre-exists.[131,134] This differential ability may stem from the fact that while NSAIDs generally promote the binding of HSF-1 to the promoter of Hsp70, only some of the agents are capable of triggering its transactivation.

Small-molecule Proteostasis Regulators

With a cell-based high-throughput screen of over 900,000 compounds, the group of Morimoto[135] discovered several families of small-molecule proteostasis regulators that activate the heat-shock response. Some were demonstrated to promote the recruitment of HSF-1 to target promoters. Furthermore, they were shown to induce a whole panel of responses that are hallmarks of an activated stress response aimed at reestablishing proteostasis. One compound A1 was found to induce the degradation of Hsp90 client kinases without affecting Hsp90 activity in general, indicating that it may specifically affect the Cdc37-Hsp90 interaction.[135]

References

1. E. T. Powers, R. I. Morimoto, A. Dillin, J. W. Kelly and W. E. Balch, *Annu. Rev. Biochem.*, 2009, **78**, 959.
2. F. U. Hartl, A. Bracher and M. Hayer-Hartl, *Nature*, 2011, **475**, 324.
3. F. U. Hartl, *Nature*, 1996, **381**, 571.
4. F. U. Hartl and M. Hayer-Hartl, *Nat. Struct. Mol. Biol.*, 2009, **16**, 574.
5. P. C. Echeverría, G. Mazaira, A. Erlejman, C. Gomez-Sanchez, G. Piwien Pilipuk and M. D. Galigniana, *Mol. Cell. Biol.*, 2009, **29**, 4788.
6. J. Liu and D. B. DeFranco, *Mol. Endocrinol.*, 1999, **13**, 355.
7. D. A. Stavreva, M. Wiench, S. John, B. L. Conway-Campbell, M. A. McKenna, J. R. Pooley, T. A. Johnson, T. C. Voss, S. L. Lightman and G. L. Hager, *Nat. Cell Biol.*, 2009, **11**, 1093.
8. N. R. Jana, M. Tanaka, G. Wang and N. Nukina, *Hum. Mol. Genet.*, 2000, **9**, 2009.
9. G. C. Meacham, C. Patterson, W. Zhang, J. M. Younger and D. M. Cyr, *Nat. Cell Biol.*, 2001, **3**, 100.

10. M. A. Theodoraki, N. B. Nillegoda, J. Saini and A. J. Caplan, *J. Biol. Chem.*, 2012, **287**, 23911.
11. M. A. Theodoraki and A. J. Caplan, *Biochim. Biophys. Acta*, 2012, **1823**, 683.
12. R. Sahu, S. Kaushik, C. C. Clement, E. S. Cannizzo, B. Scharf, A. Follenzi, I. Potolicchio, E. Nieves, A. M. Cuervo and L. Santambrogio, *Dev. Cell*, 2011, **20**, 131.
13. V. Arndt, N. Dick, R. Tawo, M. Dreiseidler, D. Wenzel, M. Hesse, D. O. Furst, P. Saftig, R. Saint, B. K. Fleischmann, M. Hoch and J. Hohfeld, *Curr. Biol.*, 2010, **20**, 143.
14. J. F. Dice and S. R. Terlecky, *Crit. Rev. Ther. Drug Carrier Syst.*, 1990, **7**, 211.
15. S. R. Terlecky, H. L. Chiang, T. S. Olson and J. F. Dice, *J. Biol. Chem.*, 1992, **267**, 9202.
16. S. Kaushik and A. M. Cuervo, *Trends Cell Biol.*, 2012, **22**, 407.
17. U. Bandyopadhyay and A. M. Cuervo, *Autophagy*, 2008, **4**, 1101.
18. P. F. Finn, N. T. Mesires, M. Vine and J. F. Dice, *Autophagy*, 2005, **1**, 141.
19. M. Zolkiewski, T. Zhang and M. Nagy, *Arch. Biochem. Biophys.*, 2012, **520**, 1.
20. S. Hodson, J. J. Marshall and S. G. Burston, *J. Struct. Biol.*, 2012, **179**, 161.
21. A. Mogk, T. Tomoyasu, P. Goloubinoff, S. Rudiger, D. Roder, H. Langen and B. Bukau, *EMBO J.*, 1999, **18**, 6934.
22. P. K. Sorger and H. R. Pelham, *J. Mol. Biol.*, 1987, **194**, 341.
23. S. J. Felts, B. A. Owen, P. Nguyen, J. Trepel, D. B. Donner and D. O. Toft, *J. Biol. Chem.*, 2000, **275**, 3305.
24. P. L. Yeyati, R. M. Bancewicz, J. Maule and V. van Heyningen, *PLoS Genet.*, 2007, **3**, e43.
25. R. Sawarkar, C. Sievers and R. Paro, *Cell*, 2012, **149**, 807.
26. M. Taipale, I. Krykbaeva, M. Koeva, C. Kayatekin, K. D. Westover, G. I. Karras and S. Lindquist, *Cell*, 2012, **150**, 987.
27. S. C. Bishop, J. A. Burlison and B. S. Blagg, *Curr. Cancer Drug Targets*, 2007, **7**, 369.
28. N. Sato, T. Yamamoto, Y. Sekine, T. Yumioka, A. Junicho, H. Fuse and T. Matsuda, *Biochem. Biophys. Res. Commun.*, 2003, **300**, 847.
29. J. Kunisawa and N. Shastri, *Immunity*, 2006, **24**, 523.
30. H. H. Kampinga, J. Hageman, M. J. Vos, H. Kubota, R. M. Tanguay, E. A. Bruford, M. E. Cheetham, B. Chen and L. E. Hightower, *Cell Stress Chaperones*, 2009, **14**, 105.
31. J. C. Young, V. R. Agashe, K. Siegers and F. U. Hartl, *Nat. Rev. Mol. Cell. Biol.*, 2004, **5**, 781.
32. H. H. Kampinga and E. A. Craig, *Nat. Rev. Mol. Cell. Biol.*, 2010, **11**, 579.
33. H. Rampelt, J. Kirstein-Miles, N. B. Nillegoda, K. Chi, S. R. Scholz, R. I. Morimoto and B. Bukau, *EMBO J.*, 2012, **31**, 4221.

34. S. A. Teter, W. A. Houry, D. Ang, T. Tradler, D. Rockabrand, G. Fischer, P. Blum, C. Georgopoulos and F. U. Hartl, *Cell*, 1999, **97**, 755.
35. W. B. Pratt and D. O. Toft, *Exp. Biol. Med.*, 2003, **228**, 111.
36. M. P. Mayer and B. Bukau, *Cell. Mol. Life Sci.*, 2005, **62**, 670.
37. F. Guo, C. Sigua, P. Bali, P. George, W. Fiskus, A. Scuto, S. Annavarapu, A. Mouttaki, G. Sondarva, S. Wei, J. Wu, J. Djeu and K. Bhalla, *Blood*, 2005, **105**, 1246.
38. X. Yang, J. Wang, Y. Zhou, Y. Wang, S. Wang and W. Zhang, *Cancer Lett.*, 2012, **321**, 137.
39. J. Nylandsted, M. Gyrd-Hansen, A. Danielewicz, N. Fehrenbacher, U. Lademann, M. Hoyer-Hansen, E. Weber, G. Multhoff, M. Rohde and M. Jaattela, *J. Exp. Med.*, 2004, **200**, 425.
40. H. S. McHaourab, J. A. Godar and P. L. Stewart, *Biochem.*, 2009, **48**, 3828.
41. P. Groth, N. Pavlova, I. Kalev, S. Tonov, G. Georgiev, H. D. Pohlenz and B. Weiss, *Nucleic Acids Res.*, 2007, **35**, D696.
42. C. M. Quinzii, C. Garone, V. Emmanuele, S. Tadesse, S. Krishna, B. Dorado and M. Hirano, *FASEB J.*, 2013, **27**, 612.
43. J. Mollet, A. Delahodde, V. Serre, D. Chretien, D. Schlemmer, A. Lombes, N. Boddaert, I. Desguerre, P. de Lonlay, H. O. de Baulny, A. Munnich and A. Rotig, *Am. J. Hum. Genet.*, 2008, **82**, 623.
44. R. Horvath, B. Czermin, S. Gulati, S. Demuth, G. Houge, A. Pyle, C. Dineiger, E. L. Blakely, A. Hassani, C. Foley, M. Brodhun, K. Storm, J. Kirschner, G. S. Gorman, H. Lochmuller, E. Holinski-Feder, R. W. Taylor and P. F. Chinnery, *J. Neurol. Neurosurg. Psychiatry*, 2012, **83**, 174.
45. M. E. Favero and F. F. Costa, *Biochem. Res. Int.*, 2011, 373859.
46. A. J. Kihm, Y. Kong, W. Hong, J. E. Russell, S. Rouda, K. Adachi, M. C. Simon, G. A. Blobel and M. J. Weiss, *Nature*, 2002, **417**, 758.
47. A. Parent, S. J. Roy, C. Iorio-Morin, M. C. Lepine, P. Labrecque, M. A. Gallant, D. Slipetz and J. L. Parent, *J. Biol. Chem.*, 2010, **285**, 40838.
48. H. Tanno, T. Yamaguchi, E. Goto, S. Ishido and M. Komada, *Mol. Biol. Cell*, 2012, **23**, 1343.
49. N. Avvakumov, A. Nourani and J. Cote, *Mol. Cell*, 2011, **41**, 502.
50. Y. Charnay, A. Imhof, P. G. Vallet, E. Kovari, C. Bouras and P. Giannakopoulos, *Brain Res. Bull.*, 2012, **88**, 434.
51. A. Granata and T. T. Warner, *Eur. J. Neurol.*, 2010, **17**(1), 81.
52. S. Raychaudhuri, M. Sinha, D. Mukhopadhyay and N. P. Bhattacharyya, *Hum. Mol. Genet.*, 2008, **17**, 240.
53. A. J. Massey, *J. Med. Chem.*, 2010, **53**, 7280.
54. L. Neckers and P. Workman, *Clin. Cancer Res.*, 2012, **18**, 64.
55. L. Wright, X. Barril, B. Dymock, L. Sheridan, A. Surgenor, M. Beswick, M. Drysdale, A. Collier, A. Massey, N. Davies, A. Fink, C. Fromont, W. Aherne, K. Boxall, S. Sharp, P. Workman and R. E. Hubbard, *Chem. Biol.*, 2004, **11**, 775.

56. S. D. Roughley and R. E. Hubbard, *J. Med. Chem.*, 2011, **54**, 3989.
57. L. Zhang, Y. Yi, J. Chen, Y. Sun, Q. Guo, Z. Zheng and S. Song, *Biochem. Biophys. Res. Commun.*, 2010, **403**, 282.
58. J. Davenport, J. R. Manjarrez, L. Peterson, B. Krumm, B. S. Blagg and R. L. Matts, *J. Nat. Prod.*, 2011, **74**, 1085.
59. J. Plescia, W. Salz, F. Xia, M. Pennati, N. Zaffaroni, M. G. Daidone, M. Meli, T. Dohi, P. Fortugno, Y. Nefedova, D. I. Gabrilovich, G. Colombo and D. C. Altieri, *Cancer Cell*, 2005, **7**, 457.
60. B. Gyurkocza, J. Plescia, C. M. Raskett, D. S. Garlick, P. A. Lowry, B. Z. Carter, M. Andreeff, M. Meli, G. Colombo and D. C. Altieri, *J. Natl Cancer Inst.*, 2006, **98**, 1068.
61. G. D. Bader and C. W. Hogue, *BMC Bioinformatics*, 2003, **4**, 2.
62. D. M. Ramsey, J. R. McConnell, L. D. Alexander, K. W. Tanaka, C. M. Vera and S. R. McAlpine, *Bioorg. Med. Chem. Lett.*, 2012, **22**, 3287.
63. V. C. Ardi, L. D. Alexander, V. A. Johnson and S. R. McAlpine, *ACS Chem. Biol.*, 2011, **6**, 1357.
64. C. T. Chan, R. E. Reeves, R. Geller, S. S. Yaghoubi, A. Hoehne, D. E. Solow-Cordero, G. Chiosis, T. F. Massoud, R. Paulmurugan and S. S. Gambhir, *Proc. Natl Acad. Sci. USA*, 2012, **109**, E2476.
65. A. K. Samadi, X. Zhang, R. Mukerji, A. C. Donnelly, B. S. Blagg and M. S. Cohen, *Cancer Lett.*, 2011, **312**, 158.
66. T. Zhang, Y. Li, Y. Yu, P. Zou, Y. Jiang and D. Sun, *J. Biol. Chem.*, 2009, **284**, 35381.
67. T. Zhang, A. Hamza, X. Cao, B. Wang, S. Yu, C. G. Zhan and D. Sun, *Mol. Cancer Ther.*, 2008, **7**, 162.
68. A. Trott, J. D. West, L. Klaic, S. D. Westerheide, R. B. Silverman, R. I. Morimoto and K. A. Morano, *Mol. Biol. Cell*, 2008, **19**, 1104.
69. Y. Yu, A. Hamza, T. Zhang, M. Gu, P. Zou, B. Newman, Y. Li, A. A. Gunatilaka, C. G. Zhan and D. Sun, *Biochem. Pharmacol.*, 2010, **79**, 542.
70. W. Y. Chen, F. R. Chang, Z. Y. Huang, J. H. Chen, Y. C. Wu and C. C. Wu, *J. Biol. Chem.*, 2008, **283**, 17184.
71. M. C. Rosenhagen, C. Soti, U. Schmidt, G. M. Wochnik, F. U. Hartl, F. Holsboer, J. C. Young and T. Rein, *Mol. Endocrinol.*, 2003, **17**, 1991.
72. V. Kvardova, R. Hrstka, D. Walerych, P. Muller, E. Matoulkova, V. Hruskova, D. Stelclova, P. Sova and B. Vojtesek, *Mol. Cancer*, 2010, **9**, 147.
73. W. P. Sullivan, B. A. Owen and D. O. Toft, *J. Biol. Chem.*, 2002, **277**, 45942.
74. Z. Yin, E. C. Henry and T. A. Gasiewicz, *Biochem.*, 2009, **48**, 336.
75. P. L. Tran, S. A. Kim, H. S. Choi, J. H. Yoon and S. G. Ahn, *BMC Cancer*, 2010, **10**, 276.
76. J. R. Smith, P. A. Clarke, E. de Billy and P. Workman, *Oncogene*, 2009, **28**, 157.
77. S. Sreeramulu, S. L. Gande, M. Gobel and H. Schwalbe, *Angew. Chem. Intl Ed.*, 2009, **48**, 5853.

78. Y. Miyata, *Cell. Mol. Life Sci.*, 2009, **66**, 1840.
79. C. K. Vaughan, M. Mollapour, J. R. Smith, A. Truman, B. Hu, V. M. Good, B. Panaretou, L. Neckers, P. A. Clarke, P. Workman, P. W. Piper, C. Prodromou and L. H. Pearl, *Mol. Cell*, 2008, **31**, 886.
80. C. T. Lee, C. Graf, F. J. Mayer, S. M. Richter and M. P. Mayer, *EMBO J.*, 2012, **31**, 1518.
81. F. Yi and L. Regan, *ACS Chem. Biol.*, 2008, **3**, 645.
82. G. Pimienta, K. M. Herbert and L. Regan, *Mol. Pharmaceutics*, 2011, **8**, 2252.
83. T. Horibe, M. Kohno, M. Haramoto, K. Ohara and K. Kawakami, *J. Transl. Med.*, 2011, **9**, 8.
84. N. Walsh, A. Larkin, N. Swan, K. Conlon, P. Dowling, R. McDermott and M. Clynes, *Cancer Lett.*, 2011, **306**, 180.
85. A. Chadli, S. J. Felts, Q. Wang, W. P. Sullivan, M. V. Botuyan, A. Fauq, M. Ramirez-Alvarado and G. Mer, *J. Biol. Chem.*, 2010, **285**, 4224.
86. F. Forafonov, O. A. Toogun, I. Grad, E. Suslova, B. C. Freeman and D. Picard, *Mol. Cell. Biol.*, 2008, **28**, 3446.
87. M. Retzlaff, F. Hagn, L. Mitschke, M. Hessling, F. Gugel, H. Kessler, K. Richter and J. Buchner, *Mol. Cell*, 2010, **37**, 344.
88. J. L. Holmes, S. Y. Sharp, S. Hobbs and P. Workman, *Cancer Res.*, 2008, **68**, 1188.
89. L. Sun, T. Prince, J. R. Manjarrez, B. T. Scroggins and R. L. Matts, *Biochim. Biophys. Acta*, 2012, **1823**, 1092.
90. A. Zurawska, J. Urbanski, J. Matuliene, J. Baraniak, M. P. Klejman, S. Filipek, D. Matulis and P. Bieganowski, *Biochim. Biophys. Acta*, 2010, **1803**, 575.
91. A. R. Goloudina, O. N. Demidov and C. Garrido, *Cancer Lett.*, 2012, **325**, 117.
92. A. J. Massey, D. S. Williamson, H. Browne, J. B. Murray, P. Dokurno, T. Shaw, A. T. Macias, Z. Daniels, S. Geoffroy, M. Dopson, P. Lavan, N. Matassova, G. L. Francis, C. J. Graham, R. Parsons, Y. Wang, A. Padfield, M. Comer, M. J. Drysdale and M. Wood, *Cancer Chemother. Pharmacol.*, 2010, **66**, 535.
93. H. J. Cho, H. Y. Gee, K. H. Baek, S. K. Ko, J. M. Park, H. Lee, N. D. Kim, M. G. Lee and I. Shin, *J. Am. Chem. Soc.*, 2011, **133**, 20267.
94. A. Rousaki, Y. Miyata, U. K. Jinwal, C. A. Dickey, J. E. Gestwicki and E. R. Zuiderweg, *J. Mol. Biol.*, 2011, **411**, 614.
95. H. J. Park, M. Mylvaganum, A. McPherson, S. W. Fewell, J. L. Brodsky and C. A. Lingwood, *Chem. Biol.*, 2009, **16**, 461.
96. A. L. Rerole, J. Gobbo, A. De Thonel, E. Schmitt, J. P. Pais de Barros, A. Hammann, D. Lanneau, E. Fourmaux, O. Deminov, O. Micheau, L. Lagrost, P. Colas, G. Kroemer and C. Garrido, *Cancer Res.*, 2011, **71**, 484.
97. J. I. Leu, J. Pimkina, P. Pandey, M. E. Murphy and D. L. George, *Mol. Cancer Res.*, 2011, **9**, 936.

98. M. Kaiser, A. Kuhnl, J. Reins, S. Fischer, J. Ortiz-Tanchez, C. Schlee, L. H. Mochmann, S. Heesch, O. Benlasfer, W. K. Hofmann, E. Thiel and C. D. Baldus, *Blood Cancer J.*, 2011, **1**, e28.
99. E. Schmitt, A. Parcellier, S. Gurbuxani, C. Cande, A. Hammann, M. C. Morales, C. R. Hunt, D. J. Dix, R. T. Kroemer, F. Giordanetto, M. Jaattela, J. M. Penninger, A. Pance, G. Kroemer and C. Garrido, *Cancer Res.*, 2003, **63**, 8233.
100. E. Schmitt, L. Maingret, P. E. Puig, A. L. Rerole, F. Ghiringhelli, A. Hammann, E. Solary, G. Kroemer and C. Garrido, *Cancer Res.*, 2006, **66**, 4191.
101. P. Bischofberger, W. Han, B. Feifel, H. J. Schonfeld and P. Christen, *J. Biol. Chem.*, 2003, **278**, 19044.
102. S. W. Fewell, B. W. Day and J. L. Brodsky, *J. Biol. Chem.*, 2001, **276**, 910.
103. M. J. Braunstein, S. S. Scott, C. M. Scott, S. Behrman, P. Walter, P. Wipf, J. D. Coplan, W. Chrico, D. Joseph, J. L. Brodsky and O. Batuman, *J. Oncol.*, 2011, 232037.
104. S. Wisen, E. B. Bertelsen, A. D. Thompson, S. Patury, P. Ung, L. Chang, C. G. Evans, G. M. Walter, P. Wipf, H. A. Carlson, J. L. Brodsky, E. R. Zuiderweg and J. E. Gestwicki, *ACS Chem. Biol.*, 2010, **5**, 611.
105. W. Xu, M. Marcu, X. Yuan, E. Mimnaugh, C. Patterson and L. Neckers, *Proc. Natl Acad. Sci. USA*, 2002, **99**, 12847.
106. L. Kundrat and L. Regan, *J. Mol. Biol.*, 2010, **395**, 587.
107. A. Zoubeidi and M. Gleave, *Int. J. Biochem. Cell Biol.*, 2012, **44**, 1646.
108. K. D. Shin, M. Y. Lee, D. S. Shin, S. Lee, K. H. Son, S. Koh, Y. K. Paik, B. M. Kwon and D. C. Han, *J. Biol. Chem.*, 2005, **280**, 41439.
109. K. D. Shin, Y. J. Yoon, Y. R. Kang, K. H. Son, H. M. Kim, B. M. Kwon and D. C. Han, *Biochem. Pharmacol.*, 2008, **75**, 383.
110. B. Gibert, E. Hadchity, A. Czekalla, M. T. Aloy, P. Colas, C. Rodriguez-Lafrasse, A. P. Arrigo and C. Diaz-Latoud, *Oncogene*, 2011, **30**, 3672.
111. V. Baylot, C. Andrieu, M. Katsogiannou, D. Taieb, S. Garcia, S. Giusiano, J. Acunzo, J. Iovanna, M. Gleave, C. Garrido and P. Rocchi, *Cell Death Dis.*, 2011, **2**, e221.
112. H. Miyake, I. Hara and M. E. Gleave, *Int. J. Urol.*, 2005, **12**, 785.
113. F. Lamoureux, C. Thomas, M. J. Yin, H. Kuruma, E. Beraldi, L. Fazli, A. Zoubeidi and M. E. Gleave, *Cancer Res.*, 2011, **71**, 5838.
114. E. de Billy, M. V. Powers, J. R. Smith and P. Workman, *Cell Cycle*, 2009, **8**, 3806.
115. N. Neznanov, A. V. Gorbachev, L. Neznanova, A. P. Komarov, K. V. Gurova, A. V. Gasparian, A. K. Banerjee, A. Almasan, R. L. Fairchild and A. V. Gudkov, *Cell Cycle*, 2009, **8**, 3960.
116. S. Yokota, M. Kitahara and K. Nagata, *Cancer Res.*, 2000, **60**, 2942.
117. C. H. Lee, H. M. Hong, Y. Y. Chang and W. W. Chang, *Biochimie*, 2012, **94**, 1382.
118. F. Guo, K. Rocha, P. Bali, M. Pranpat, W. Fiskus, S. Boyapalle, S. Kumaraswamy, M. Balasis, B. Greedy, E. S. Armitage, N. Lawrence and K. Bhalla, *Cancer Res.*, 2005, **65**, 10536.

119. S. D. Westerheide, T. L. Kawahara, K. Orton and R. I. Morimoto, *J. Biol. Chem.*, 2006, **281**, 9616.
120. N. Zaarur, V. L. Gabai, J. A. Porco, Jr., S. Calderwood and M. Y. Sherman, *Cancer Res.*, 2006, **66**, 1783.
121. D. W. Neef, A. M. Jaeger and D. J. Thiele, *Nat. Rev. Drug Discov.*, 2011, **10**, 930.
122. P. Putcha, K. M. Danzer, L. R. Kranich, A. Scott, M. Silinski, S. Mabbett, C. D. Hicks, J. M. Veal, P. M. Steed, B. T. Hyman and P. J. McLean, *J. Pharmacol. Exp. Ther.*, 2010, **332**, 849.
123. B. Peng, L. Xu, F. Cao, T. Wei, C. Yang, G. Uzan and D. Zhang, *Mol. Cancer*, 2010, **9**, 79.
124. M. Otaka, S. Yamamoto, K. Ogasawara, Y. Takaoka, S. Noguchi, T. Miyazaki, A. Nakai, M. Odashima, T. Matsuhashi, S. Watanabe and H. Itoh, *Biochem. Biophys. Res. Commun.*, 2007, **353**, 399.
125. S. Endo, N. Hiramatsu, K. Hayakawa, M. Okamura, A. Kasai, Y. Tagawa, N. Sawada, J. Yao and M. Kitamura, *Mol. Pharmacol.*, 2007, **72**, 1337.
126. K. Hirota, H. Nakamura, T. Arai, H. Ishii, J. Bai, T. Itoh, K. Fukuda and J. Yodoi, *Biochem. Biophys. Res. Commun.*, 2000, **275**, 825.
127. D. W. Neef, M. L. Turski and D. J. Thiele, *PLoS Biol.*, 2010, **8**, e1000291.
128. J. Hargitai, H. Lewis, I. Boros, T. Racz, A. Fiser, I. Kurucz, I. Benjamin, L. Vigh, Z. Penzes, P. Csermely and D. S. Latchman, *Biochem. Biophys. Res. Commun.*, 2003, **307**, 689.
129. V. Lanka, S. Wieland, J. Barber and M. Cudkowicz, *Expert Opin. Investig. Drugs*, 2009, **18**, 1907.
130. J. Yang, K. Bridges, K. Y. Chen and A. Y. Liu, *PLoS ONE*, 2008, **3**, e2864.
131. B. S. Lee, J. Chen, C. Angelidis, D. A. Jurivich and R. I. Morimoto, *Proc. Natl Acad. Sci. USA*, 1995, **92**, 7207.
132. J. N. Housby, C. M. Cahill, B. Chu, R. Prevelige, K. Bickford, M. A. Stevenson and S. K. Calderwood, *Cytokine*, 1999, **11**, 347.
133. A. Ianaro, A. Ialenti, P. Maffia, P. Di Meglio, M. Di Rosa and M. G. Santoro, *Mol. Pharmacol.*, 2003, **64**, 85.
134. K. Ishihara, N. Yamagishi and T. Hatayama, *Eur. J. Biochem.*, 2004, **271**, 4552.
135. B. Calamini, M. C. Silva, F. Madoux, D. M. Hutt, S. Khanna, M. A. Chalfant, S. A. Saldanha, P. Hodder, B. D. Tait, D. Garza, W. E. Balch and R. I. Morimoto, *Nat. Chem. Biol.*, 2012, **8**, 185.
136. C. Kakiuchi, M. Ishiwata, S. Nanko, H. Kunugi, Y. Minabe, K. Nakamura, N. Mori, K. Fujii, T. Umekage, M. Tochigi, K. Kohda, T. Sasaki, K. Yamada, T. Yoshikawa and T. Kato, *J. Hum. Genet.*, 2007, **52**, 794.
137. S. Arlt, C. Demiralay, B. Tharun, O. Geisel, N. Storm, M. Eichenlaub, J. T. Lehmbeck, K. Wiedemann, B. Leuenberger and H. Jahn, *Curr. Alzheimer Res.*, 2013, **10**, 72.
138. A. Zobel, A. Schuhmacher, F. Jessen, S. Hofels, O. von Widdern, M. Metten, U. Pfeiffer, C. Hanses, T. Becker, M. Rietschel, L. Scheef,

W. Block, H. H. Schild, W. Maier and S. G. Schwab, *Int. J. Neuropsychopharmacol.*, 2010, **13**, 649.
139. S. Kawaguchi, A. Hagiwara and M. Suzuki, *Acta Otolaryngol.*, 2008, **128**, 1173.
140. C. U. Pae, T. S. Kim, O. J. Kwon, P. Artioli, A. Serretti, C. U. Lee, S. J. Lee, C. Lee, I. H. Paik and J. J. Kim, *Neurosci. Res.*, 2005, **53**, 8.
141. Y. Qi, W. Q. Niu, T. C. Zhu, J. L. Liu, W. Y. Dong, Y. Xu, S. Q. Ding, C. B. Cui, Y. J. Pan, G. S. Yu, W. Y. Zhou and C. C. Qiu, *Clin. Chim. Acta*, 2009, **405**, 17.
142. L. Broer, M. A. Ikram, M. Schuur, A. L. DeStefano, J. C. Bis, F. Liu, F. Rivadeneira, A. G. Uitterlinden, A. S. Beiser, W. T. Longstreth, A. Hofman, Y. Aulchenko, S. Seshadri, A. L. Fitzpatrick, B. A. Oostra, M. M. Breteler and C. M. van Duijn, *J. Alzheimers Dis.*, 2011, **25**, 93.
143. P. Tsvetkov, Y. Adamovich, E. Elliott and Y. Shaul, *J. Biol. Chem.*, 2011, **286**, 8839.
144. M. He, H. Guo, X. Yang, X. Zhang, L. Zhou, L. Cheng, H. Zeng, F. B. Hu, R. M. Tanguay and T. Wu, *PLoS ONE*, 2009, **4**, e4851.
145. R. Giacconi, C. Cipriano, E. Muti, L. Costarelli, M. Malavolta, C. Caruso, D. Lio and E. Mocchegiani, *Biogerontology*, 2006, **7**, 347.
146. S. S. Mir, D. Fiedler and A. G. Cashikar, *Mol. Cell. Biol.*, 2009, **29**, 187.
147. M. Buraczynska, A. Swatowski, K. Buraczynska, M. Dragan and A. Ksiazek, *Clin. Sci.*, 2009, **116**, 81.
148. K. Zouari Bouassida, L. Chouchane, K. Jellouli, S. Cherif, S. Haddad, S. Gabbouj and J. Danguir, *Diabetes Metab.*, 2004, **30**, 175.
149. Y. R. Wu, C. M. Chen, L. S. Ro, R. K. Lyu and L. M. Tang, *J. Formos. Med. Assoc.*, 2004, **103**, 727.
150. S. Edvardson, Y. Cinnamon, A. Ta-Shma, A. Shaag, Y. I. Yim, S. Zenvirt, C. Jalas, S. Lesage, A. Brice, A. Taraboulos, K. H. Kaestner, L. E. Greene and O. Elpeleg, *PLoS ONE*, 2012, **7**, e36458.
151. K. Altomare, V. Greco, D. Bellizzi, M. Berardelli, S. Dato, F. DeRango, S. Garasto, G. Rose, E. Feraco, V. Mari, G. Passarino, C. Franceschi and G. De Benedictis, *Biogerontology*, 2003, **4**, 215.
152. O. A. Ross, M. D. Curran, K. A. Crum, I. M. Rea, Y. A. Barnett and D. Middleton, *Exp. Gerontol.*, 2003, **38**, 561.
153. L. Zouiten-Mekki, S. Karoui, M. Kharrat, M. Fekih, S. Matri, J. Boubaker, A. Filali and H. Chaabouni, *Eur. J. Gastroenterol. Hepatol.*, 2007, **19**, 225.
154. D. Selcen, F. Muntoni, B. K. Burton, E. Pegoraro, C. Sewry, A. V. Bite and A. G. Engel, *Ann. Neurol.*, 2009, **65**, 83.
155. J. J. Hansen, A. Durr, I. Cournu-Rebeix, C. Georgopoulos, D. Ang, M. N. Nielsen, C. S. Davoine, A. Brice, B. Fontaine, N. Gregersen and P. Bross, *Am. J. Hum. Genet.*, 2002, **70**, 1328.
156. O. V. Evgrafov, I. Mersiyanova, J. Irobi, L. Van Den Bosch, I. Dierick, C. L. Leung, O. Schagina, N. Verpoorten, K. Van Impe, V. Fedotov, E. Dadali, M. Auer-Grumbach, C. Windpassinger, K. Wagner, Z. Mitrovic, D. Hilton-Jones, K. Talbot, J. J. Martin, N. Vasserman,

S. Tverskaya, A. Polyakov, R. K. Liem, J. Gettemans, W. Robberecht, P. De Jonghe and V. Timmerman, *Nat. Genet.*, 2004, **36**, 602.
157. J. Irobi, K. Van Impe, P. Seeman, A. Jordanova, I. Dierick, N. Verpoorten, A. Michalik, E. De Vriendt, A. Jacobs, V. Van Gerwen, K. Vennekens, R. Mazanec, I. Tournev, D. Hilton-Jones, K. Talbot, I. Kremensky, L. Van Den Bosch, W. Robberecht, J. Van Vandekerckhove, C. Van Broeckhoven, J. Gettemans, P. De Jonghe and V. Timmerman, *Nat. Genet.*, 2004, **36**, 597.
158. S. J. Kolb, P. J. Snyder, E. J. Poi, E. A. Renard, A. Bartlett, S. Gu, S. Sutton, W. D. Arnold, M. L. Freimer, V. H. Lawson, J. T. Kissel and T. W. Prior, *Neurology*, 2010, **74**, 502.
159. M. Litt, P. Kramer, D. M. LaMorticella, W. Murphey, E. W. Lovrien and R. G. Weleber, *Hum. Mol. Genet.*, 1998, **7**, 471.
160. V. Berry, P. Francis, M. A. Reddy, D. Collyer, E. Vithana, I. MacKay, G. Dawson, A. H. Carey, A. Moore, S. S. Bhattacharya and R. A. Quinlan, *Am. J. Hum. Genet.*, 2001, **69**, 1141.
161. P. Vicart, A. Caron, P. Guicheney, Z. Li, M. C. Prevost, A. Faure, D. Chateau, F. Chapon, F. Tome, J. M. Dupret, D. Paulin and M. Fardeau, *Nat. Genet.*, 1998, **20**, 92.
162. N. Inagaki, T. Hayashi, T. Arimura, Y. Koga, M. Takahashi, H. Shibata, K. Teraoka, T. Chikamori, A. Yamashina and A. Kimura, *Biochem. Biophys. Res. Commun.*, 2006, **342**, 379.
163. J. Trepel, M. Mollapour, G. Giaccone and L. Neckers, *Nat. Rev. Cancer*, 2010, **10**, 537.
164. K. Moulick, J. H. Ahn, H. Zong, A. Rodina, L. Cerchietti, E. M. Gomes Dagama, E. Caldas-Lopes, K. Beebe, F. Perna, K. Hatzi, L. P. Vu, X. Zhao, D. Zatorska, T. Taldone, P. Smith-Jones, M. Alpaugh, S. S. Gross, N. Pillarsetty, T. Ku, J. S. Lewis, S. M. Larson, R. Levine, H. Erdjument-Bromage, M. L. Guzman, S. D. Nimer, A. Melnick, L. Neckers and G. Chiosis, *Nat. Chem. Biol.*, 2011, **7**, 818.
165. D. Picard, *Nat. Cell Biol.*, 2004, **6**, 479.
166. W. Li, Y. Li, S. Guan, J. Fan, C. F. Cheng, A. M. Bright, C. Chinn, M. Chen and D. T. Woodley, *EMBO J.*, 2007, **26**, 1221.
167. Y. Yang, R. Rao, J. Shen, Y. Tang, W. Fiskus, J. Nechtman, P. Atadja and K. Bhalla, *Cancer Res.*, 2008, **68**, 4833.
168. S. Debaisieux, F. Rayne, H. Yezid and B. Beaumelle, *Traffic*, 2012, **13**, 355.
169. H. D. Shukla and P. M. Pitha, *Autoimmune Dis.*, 2012, 728605.
170. W. Luo, F. Dou, A. Rodina, S. Chip, J. Kim, Q. Zhao, K. Moulick, J. Aguirre, N. Wu, P. Greengard and G. Chiosis, *Proc. Natl Acad. Sci. USA*, 2007, **104**, 9511.
171. A. Sittler, R. Lurz, G. Lueder, J. Priller, H. Lehrach, M. K. Hayer-Hartl, F. U. Hartl and E. E. Wanker, *Hum. Mol. Genet.*, 2001, **10**, 1307.
172. M. Waza, H. Adachi, M. Katsuno, M. Minamiyama, F. Tanaka, M. Doyu and G. Sobue, *J. Mol. Med.*, 2006, **84**, 635.

173. M. Riedel, O. Goldbaum, L. Schwarz, S. Schmitt and C. Richter-Landsberg, *PLoS ONE*, 2010, **5**, e8753.
174. H. S. Park, S. G. Cho, C. K. Kim, H. S. Hwang, K. T. Noh, M. S. Kim, S. H. Huh, M. J. Kim, K. Ryoo, E. K. Kim, W. J. Kang, J. S. Lee, J. S. Seo, Y. G. Ko, S. Kim and E. J. Choi, *Mol. Cell. Biol.*, 2002, **22**, 7721.
175. V. L. Gabai, A. B. Meriin, D. D. Mosser, A. W. Caron, S. Rits, V. I. Shifrin and M. Y. Sherman, *J. Biol. Chem.*, 1997, **272**, 18033.
176. M. Zylicz, F. W. King and A. Wawrzynow, *EMBO J.*, 2001, **20**, 4634.
177. H. M. Beere, B. B. Wolf, K. Cain, D. D. Mosser, A. Mahboubi, T. Kuwana, P. Tailor, R. I. Morimoto, G. M. Cohen and D. R. Green, *Nat. Cell Biol.*, 2000, **2**, 469.
178. J. A. Ribeil, Y. Zermati, J. Vandekerckhove, S. Cathelin, J. Kersual, M. Dussiot, S. Coulon, I. C. Moura, A. Zeuner, T. Kirkegaard-Sorensen, B. Varet, E. Solary, C. Garrido and O. Hermine, *Nature*, 2007, **445**, 102.
179. E. A. Afanasyeva, E. Y. Komarova, L. G. Larsson, F. Bahram, B. A. Margulis and I. V. Guzhova, *Int. J. Cancer*, 2007, **121**, 2615.
180. D. R. Ciocca and S. K. Calderwood, *Cell Stress Chaperones*, 2005, **10**, 86.
181. J. N. Sterrenberg, G. L. Blatch and A. L. Edkins, *Cancer Lett.*, 2011, **312**, 129.
182. F. Cappello, E. Conway de Macario, L. Marasa, G. Zummo and A. J. Macario, *Cancer Biol. Ther.*, 2008, **7**, 801.
183. U. Steinhoff, V. Brinkmann, U. Klemm, P. Aichele, P. Seiler, U. Brandt, P. W. Bland, I. Prinz, U. Zugel and S. H. Kaufmann, *Immunity*, 1999, **11**, 349.
184. A. G. Pockley, B. Fairburn, S. Mirza, L. K. Slack, K. Hopkinson and M. Muthana, *Methods*, 2007, **43**, 238.
185. A. G. Pockley, M. Muthana and S. K. Calderwood, *Trends Biochem. Sci.*, 2008, **33**, 71.
186. W. Chen, U. Syldath, K. Bellmann, V. Burkart and H. Kolb, *J. Immunol.*, 1999, **162**, 3212.
187. I. Kaur, S. D. Voss, R. S. Gupta, K. Schell, P. Fisch and P. M. Sondel, *J. Immunol.*, 1993, **150**, 2046.
188. S. Corrao, C. Campanella, R. Anzalone, F. Farina, G. Zummo, E. Conway de Macario, A. J. Macario, F. Cappello and G. La Rocca, *Life Sci.*, 2010, **86**, 145.
189. M. D. Gober, S. Q. Wales and L. Aurelian, *Front. Biosci.*, 2005, **10**, 2788.
190. C. Garrido, M. Brunet, C. Didelot, Y. Zermati, E. Schmitt and G. Kroemer, *Cell Cycle*, 2006, **5**, 2592.
191. A. P. Arrigo, *Adv. Exp. Med. Biol.*, 2007, **594**, 14.
192. J. Acunzo, M. Katsogiannou and P. Rocchi, *Int. J. Biochem. Cell Biol.*, 2012, **44**, 1622.
193. A. Zoubeidi and M. Gleave, *Int. J. Biochem. Cell Biol.*, 2012, **44**, 1646.
194. S. D. Roughley and R. E. Hubbard, *J. Med. Chem.*, 2011, **54**, 3989.

195. G. Karthikeyan, C. Zambaldo, S. Barluenga, V. Zoete, M. Karplus and N. Winssinger, *Chemistry*, 2012, **18**, 8978.
196. S. Barluenga, C. Wang, J. G. Fontaine, K. Aouadi, K. Beebe, S. Tsutsumi, L. Neckers and N. Winssinger, *Angew. Chem. Int. Ed.*, 2008, **47**, 4432.
197. F. Vallee, C. Carrez, F. Pilorge, A. Dupuy, A. Parent, L. Bertin, F. Thompson, P. Ferrari, F. Fassy, A. Lamberton, A. Thomas, R. Arrebola, S. Guerif, A. Rohaut, V. Certal, J. M. Ruxer, T. Gouyon, C. Delorme, A. Jouanen, J. Dumas, C. Grepin, C. Combeau, H. Goulaouic, N. Dereu, V. Mikol, P. Mailliet and H. Minoux, *J. Med. Chem.*, 2011, **54**, 7206.
198. Y. Li, T. Zhang, Y. Jiang, H. F. Lee, S. J. Schwartz and D. Sun, *Mol. Pharm.*, 2009, **6**, 1152.
199. S. W. Fewell, C. M. Smith, M. A. Lyon, T. P. Dumitrescu, P. Wipf, B. W. Day and J. L. Brodsky, *J. Biol. Chem.*, 2004, **279**, 51131.
200. P. C. Echeverria, A. Bernthaler, P. Dupuis, B. Mayer and D. Picard, *PLoS ONE*, 2011, **6**, e26044.
201. T. Kamada and S. Kawai, *Inform. Process. Lett.*, 1989, **31**, 7.

CHAPTER 2
Structural Basis of Hsp90 Function

CHRISOSTOMOS PRODROMOU* AND
LAURENCE H. PEARL

Genome Damage and Stability Centre, School of Life Sciences, Science Park Road, Falmer, Brighton, East Sussex BN1 9RQ, UK
*Email: chris.prodromou@sussex.ac.uk

2.1 Introduction

Hsp90 is a remarkable molecular chaperone that is responsible not only for the maintenance, maturation or activation of an ever-increasing cohort of client proteins,[1] but also for the assembly and disassembly of protein complexes[2] and can buffer mutations and thus suppress phenotypic variation.[3–6] Hsp90 is a highly conserved protein and represents around 1–3% of the soluble protein in eukaryotic cells.[7] Client proteins that are dependent on Hsp90 include many that are involved in signal transduction, such as ERBB2, BRAF and CDK4, making Hsp90 an extremely attractive anticancer target.[8–13] In higher eukaryotes there are typically four paralogues; two cytoplasmic (Hsp90α and Hsp90β), an endoplasmic reticulum-specific (Grp94) and a mitochondrial (TRAP1). In contrast bacteria contain typically one homologue, HtpG, while Archaea lack an equivalent protein.

The ATPase activity of Hsp90 is at the heart of this chaperone machine and is coupled to a conformational cycle consisting of a complex set of structural changes. We have only recently understood these conformational rearrangements, but how these relate to the maturation and activation of client protein still remains one of the most pressing issues in this field. It is now universally accepted that all Hsp90s are ATPases and that the rate-limiting step of the Hsp90 cycle is structural rearrangement,[14–18] although the precise kinetic details of the reaction varies between cytosolic, ER, mitochondrial and bacterial proteins. It has been suggested that species-dependent ensembles[19–24] of a conserved conformational cycle are the result of evolutionary fine-tuning to meet specific requirements imposed by their own client proteins and metabolic status.[19]

2.2 Domain Structure of Hsp90

Hsp90 consists of three globular domains, a highly flexible charged linker and a C-terminal extension that possesses the conserved MEEVD TPR-domain (tetratricopeptide) binding motif (ref. 25, Figure 2.1). The N-terminal domain has an approximate molecular mass of 25 kD and provides the binding site for ATP, as well as the natural antibiotics geldanamycin and radicicol and a growing number of synthetic inhibitors (see review[26]). The N-terminal domain contains the catalytic glutamate (Glu 33 in yeast: from here on yeast numbering will be used unless otherwise stated) that is responsible for the hydrolysis of ATP[27] and is separated from the middle domain by a highly charged linker that is significantly shorter in TRAP1 and the bacterial homologs. The charge linker

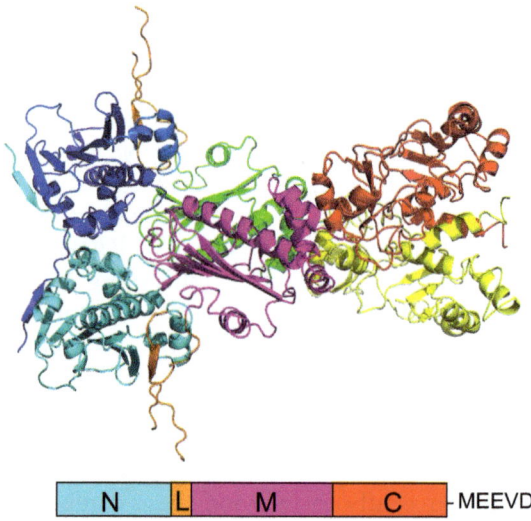

Figure 2.1 PyMol cartoon showing the domain structure of yeast Hsp90. The full-length structure of yeast Hsp90 is shown in the closed N-terminally dimerized state (PDB 2CG9). The N-terminal domains (N) are shown in blue and cyan, the charged linker (L) in orange, the middle domains (M) in green and magenta and the C-terminal domains (C) in red and yellow. The conserved MEEVD motif is shown in the schematic below the structure, but was not visible in the full-length structure.

has been implicated in the interaction with co-chaperones and client proteins.[28–30] The middle domain has an approximate molecular mass of around 33 kD and consists of two subdomains that have an α-β-α- sandwich fold separated by three small helices arranged in a right-handed coil. The longest α-helix of the first subdomain connects to the catalytic loop that possesses Arg 380, which interacts with the γ-phosphate of the ATP bound to the N-terminal domain. Its interaction with the γ-phosphate of ATP is essential for hydrolysis.[31,32] The C-terminal domain has a molecular mass of around 13 kD and is responsible for the constitutive dimerization of Hsp90. At the extreme C-terminus there is a conserved MEEVD motif that is responsible for mediating interaction with a variety of proteins containing TPR domains. These interactions are important for recruiting a variety of different enzymatic activities into Hsp90 complexes. These include phosphatase (Ppp5/Pptlp), peptidylprolyl isomerase (CYP40/Cpr6/FKBP51 *etc*.) and ubiquitin ligase activities (CHIP). The MEEVD motif also acts as a site by which specific co-chaperones, such as Hop/Sti1, bind and ultimately regulate the ATPase activity of Hsp90 but also mediate client protein loading.

The domain structure of the endoplasmic reticulum-specific form, glucose-regulated protein 94 (GRP94, also known as gp96 and endoplasmin), is very similar. However, GRP94 lacks the conserved MEEVD motif at the extreme C terminus and instead possesses the C-terminal ER-retention sequence, KDEL. The first ∼50 amino acids residues, following the signal peptide sequence, of GRP94 are distinct from those of the cytosolic Hsp90s. How this affects the β-strand exchange reaction and dimerization of the N-terminal domains is currently unknown. Unlike other Hsp90s the charged linker of GRP94 appears to be required for its ATPase activity[33,34] and is also known to contain 1–2 calcium binding sites. Occupation of these sites by calcium is thought to impact on the N-terminal domain.[35]

The mitochondrial TRAP1 is perhaps the simplest in terms of its domain structure. It has a shorter linker and lacks the conserved MEEVD motif and, in its processed form (it possesses a signal sequence), any N-terminal extensions. The clientele of GRP94 and TRAP1 are not as well characterized as those of the cytosolic Hsp90s.[36,37]

2.3 Determining the Rate-limiting Step of the Hsp90 Cycle

A thorough comprehension of the ATPase-coupled conformational cycle of Hsp90 is crucial if its mechanism of action is to be fully understood. Direct evidence for ATP binding and hydrolysis initially came from structural studies showing ATP bound to the N-terminal domain of yeast Hsp90 and the presence of a conserved catalytic residue (Glu 33) required for ATP hydrolysis.[38] However, understanding the precise mechanism by which ATP was hydrolyzed took a considerable amount of time following this initial observation. Support for a cycle that was rate limited by conformational movements was initially suggested.[39] Thus a mutant, A107N, originally isolated as a temperature-sensitive mutation,[40] caused a substantial increase in the catalytic activity of Hsp90. It was suggested that this mutation stabilized a loop, named the ATP "lid", in a closed state over the bound ATP. This closed state was considered a prerequisite for ATP hydrolysis and reaching this conformation represented the rate-limiting step of the cycle.[41] However, this contrasted with reports that the rate-limiting step was the hydrolysis of the ATP.[42] It then emerged, when the structure was determined of the middle domain of Hsp90, and in complex with the N-terminal domain of Aha1,[32] that reaching the catalytically active state was conformationally more complex than originally conceived. These structures showed that Aha1 could modulate the catalytic loop of the middle domain of Hsp90 and increase the ATPase activity of Hsp90. Thus, conformational changes in Hsp90 were directly linked to catalytic activity. With the publication of the full-length structure of yeast Hsp90 in complex with Sba1 and the ATP-analogue AMP-PNP the full extent of the structural changes required to establish a catalytically active state was finally defined.[25]

2.4 Nucleotide Interactions of the ADP and ATP/AMP-PNP Bound States of Hsp90

Structural studies showed that ADP binds deep in a cleft in the N-terminal domain of Hsp90 (Figure 2.2A) making few direct hydrogen bonds to the protein (Figure 2.2B). Most of the interactions are mediated through water molecules or the bound Mg^{2+} ion. Direct interactions to the bound ADP include a direct hydrogen bond from the exocyclic N6 of adenine to the carboxylate side-chain of Asp 79. The side-chain of Asn 92 also makes a direct hydrogen bond to the 2′-hydroxyl of the ribose sugar. The remaining direct hydrogen bonds are to the phosphate oxygens of ADP. These include bonds from the main chain carbonyl of Gly121, the main-chain amide of Phe 124, the side-chain amide of Lys 98 and the side-chain amide of Asn 37 (Figure 2.2B). The interactions that AMP-PNP makes are similar to those made by ADP. The structure of the isolated N-terminal domain of human Hsp90 bound to ATP and AMP-PNP has recently been determined.[43] ATP/AMP-PNP bind in a similar way to ADP except for some additional interactions mediated through the γ-phosphate of these nucleotides, which are essentially the same as those seen in the structure of full-length yeast Hsp90.[25] However, in switching from the open state to the closed state a small number of hydrogen bonds are disrupted and a few new interactions are formed (Figure 2.2C). This is mainly due to the movement of the ATP lid. Thus, for the open ATP-lid human structure the main-chain carbonyl of Gly 132 forms two direct bonds to two oxygens of the γ-phosphate of ATP. However, in the closed yeast conformation the equivalent bonds are broken and two new bonds are formed to the same oxygen atoms of ATP by the main-chain carbonyl of Gln 119. In addition, a further hydrogen bond is established between the same main-chain carbonyl of Gln 119

Structural Basis of Hsp90 Function 41

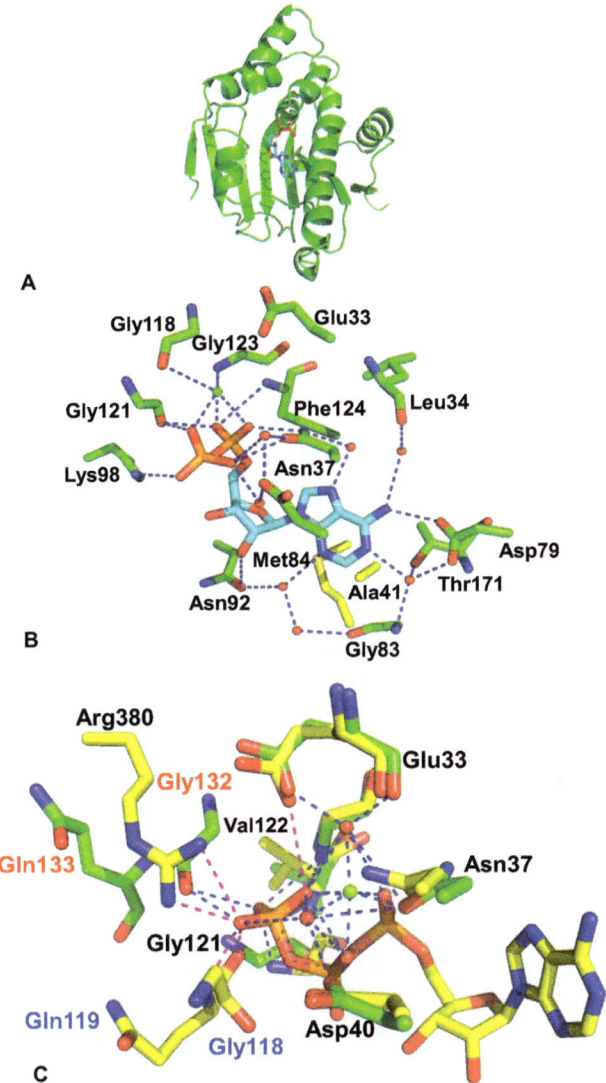

Figure 2.2 PyMol cartoon showing the nucleotide interactions of Hsp90. A) ADP bound to the N-terminal domain of Hsp90 (PDB 1AM1). B) Interactions between ADP and the open conformation of yeast N-terminal domain. Amino acid residues involved in polar interactions are shown in green while those in hydrophobic contact are shown in yellow. ADP is shown in cyan, water molecules as red spheres and hydrogen bonds as dotted blue lines. The green sphere represents the Mg^{2+} ion. C) Comparison of the interactions of the open conformation of human N-terminal domain (yellow) bound with ATP (PDB 3T0H) and the closed conformation of yeast Hsp90 (green, PDB 2CG9). Water molecules are shown as red spheres, while the Mg^{2+} ion is shown as a green sphere. Hydrogen bonds are shown as blue dotted lines. Those that represent the new bonds formed in the closed yeast structure are, however, shown as magenta coloured dashed lines.

and the oxygen of the phosphodiester bond between the β- and γ-phosphate atoms of ATP. It also appears that the side-chain of Glu 33 of the yeast protein moves closer (from 3.9 to 3.4 Å) and interacts with an oxygen atom of the γ-phosphate of ATP. In the yeast full-length structure interactions between Arg 380 of the middle domain with two oxygens of the γ-phosphate of ATP are also seen. The greater number of hydrogen bonds that are formed in the closed ATP bound state may help shift the equilibrium towards a catalytically active state.

2.5 Structural Changes in Cytosolic Hsp90 upon ATP Binding

In addition to the closed state of the Hsp90 dimer stabilized by binding of ATP/AMP-PNP, numerous open conformations of Hsp90 have been described,[19,44] but their functional significance in the chaperone cycle, if any, remains unclear. However, once nucleotide triphosphate is bound it helps to stabilize the closed ATP lid state (Figure 2.3A), in which the γ-phosphate of ATP provides the hydrogen bonding that promotes a stable association of the ATP lid with the main body of the N-terminal domain. It is this state that is promoted by the A107N mutation (Figure 2.3A) and it is noteworthy that an additional hydrogen bond formed between the amine nitrogen of the side-chain of A107N and the hydroxyl of the side-chain of Tyr 47 has such a profound effect on the stability of the lid, by way of stabilizing the closed state, and consequently on the ATPase activity of Hsp90.[25,39] Conversely, the T101I mutation has a completely opposing effect on the conformation of the ATP lid and inhibits the ATPase activity of Hsp90. T101I stabilizes the ATP lid in an open state by enhancing the hydrophobic interactions that keep the lid in its open state (Figure 2.3A).

Closure of the lid following the binding of ATP allows Hsp90 to form a stable N-terminally dimerized state. In the open state the ATP lid is packed against residues Ile 12 to Asn 21 in α-helix 1 of the N-terminal domain (Figure 2.3B). Thus, closure of the ATP lid exposes hydrophobic residues (Ile 12, Leu 15 and Lue 18) to solvent but also increases the mobility of residues 1–25 due to the lack of a stable association with the now closed ATP lid. The hydrophobic surface (Ile 12, Leu 15 and Leu 18), which was responsible for maintaining the ATP lid equilibrium towards an open conformation (Figure 2.3B), now provides the platform for N-terminal dimerization. Thus, exposure to solvent likely provides a favorable entropic force for the burial of these residues into the dimerization interface. Consequently, α-helix 1 associates with the equivalent helix from the N-terminal domain in the other monomer (Figure 2.3C). Dimerization is further stabilized by a β-strand exchange (Glu4 to Glu 7) that occurs between the N-terminal ends of the N-terminal domains (Figure 2.3C). This β-strand exchange provides further stability for the N-terminally dimerized state of Hsp90. The stable dimerization of these domains is a prerequisite for the hydrolysis of ATP, although as we will see later this may not be the only state in which Hsp90 is able to hydrolyze ATP.

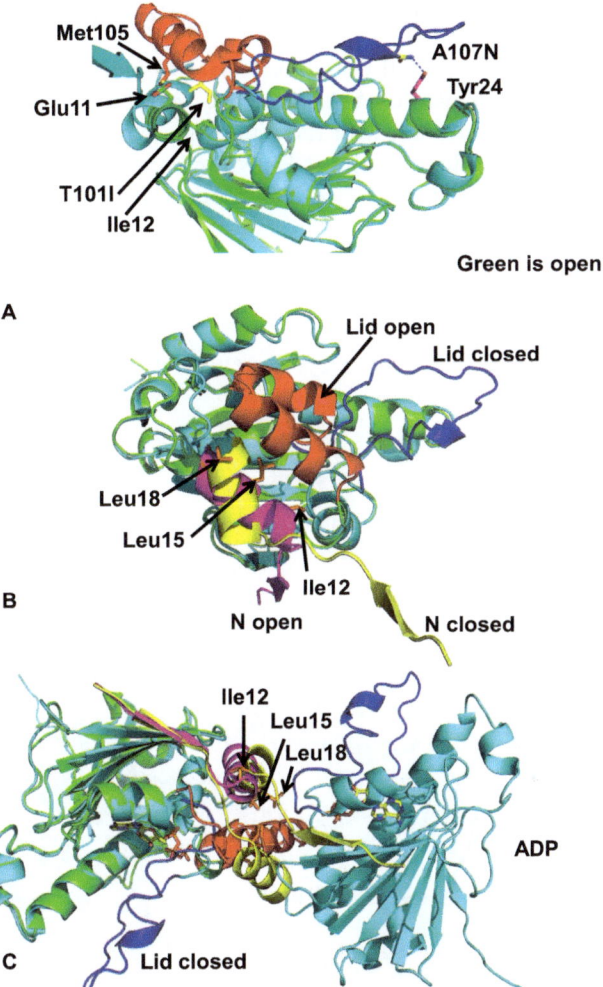

Figure 2.3 PyMol cartoon showing the structural changes in the N-terminal domains of Hsp90. A) Superimposition of the open (PDB 1AM1) and closed (PDB 2CG9) states of the N-terminal domain of yeast Hsp90. Green, open ATP lid structure (red ATP lid); cyan, closed ATP lid structure (blue ATP lid). The hydrogen bond between A107N and Tyr 24 is shown as is the inactivating mutation T101I. B) Superimposition of the open and closed states of the N-terminal domain of yeast Hsp90 showing the hydrophobic residues of α-helix 1 that interact with the open ATP lid. Green, open ATP lid structure (red ATP lid); cyan, closed ATP lid structure (blue ATP lid). C) N-terminal dimerization of yeast Hsp90. Green, open ATP lid structure (red, ATP lid; magenta, α-helix 1) showing that the ATP lid (only one red ATP lid and one magenta coloured α-helix 1 are shown in the dimer) is in an incompatible conformation for dimerization; cyan, closed ATP lid structure (blue ATP lid) showing dimerization between residues found on α-helix 1 (yellow helices) of the N-terminal domains.

2.6 Structural Changes in the Middle Domain of Hsp90

Dimerization of the N-terminal domains of Hsp90 alone is not sufficient for ATPase activity. To achieve this Arg 380 in the "catalytic loop" of the middle domain of Hsp90 must interact with the γ-phosphate of ATP[25,32] (Figure 2.4). The structure of the middle domain of Hsp90 in complex with the N-terminal domain of Aha1 showed how the co-chaperone remodels the catalytic loop

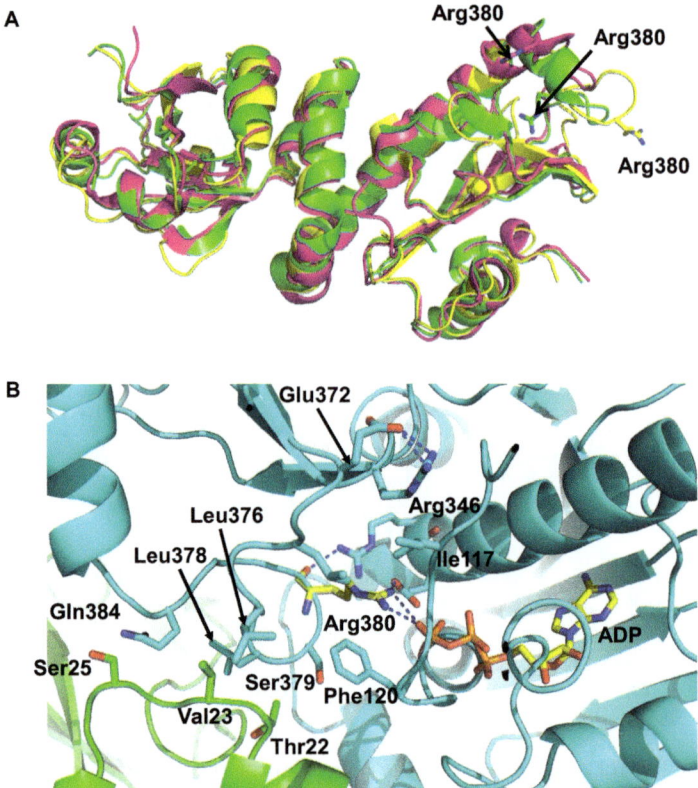

Figure 2.4 PyMol cartoon showing the conformational flexibility and stabilization of the catalytic loop of the middle domain of yeast Hsp90. A) Superimposition of three structures showing multiple conformations of the catalytic loop. Green, structure of the middle domain of yeast Hsp90 showing the closed inactive conformation of the catalytic loop (PDB 1HK7). Yellow, structure of the middle domain of Hsp90 as seen in the N-terminally dimerized crystal structure of full-length Hsp90 showing the open active state of the catalytic loop (PDB 2CG9). Magenta, structure of the middle domain of Hsp90 in complex with the N-terminal domain of Aha1 showing an intermediate catalytic loop conformation (PDB 1USU). B) Stabilization of the catalytic loop of yeast Hsp90 in its open active state. The N-terminal domains of Hsp90 are shown in cyan and green. Amino acid residues contributing to the stabilization of the catalytic loop are shown. AMPPNP is shown in yellow.

such that the loop extends towards the N-terminal domain and allows Arg 380 to directly interact with the γ-phosphate of AMPPNP.[25,32] A comparison of the different conformational states observed for the catalytic loop is shown in Figure 2.4A.

The active conformation of the catalytic loop is stabilized by a series of interactions with the N-terminal domains of Hsp90. Firstly, the closed ATP lid of Hsp90 forms a series of hydrophobic interactions with the catalytic loop of the same Hsp90 monomer. Thus, ATP lid residues, especially Ile117 and Phe120, form interactions with Leu 374 and Leu 376 of the catalytic loop (Figure 2.4B). Secondly, the N-terminal end of the N-terminal domain in the neighboring Hsp90 monomer stabilizes the conformation of the catalytic loop *via* an interaction involving Thr 22, Val 23 and Ser 25 of the N-terminal domain and Leu 376, Leu 378 and Gln 384 of the catalytic loop. Thus, a convergence of mainly hydrophobic residues, driven by movements in both N- and middle domains, is required to stabilize the catalytically active state of Hsp90.[25] In addition to the interactions with the N-terminal domain, the active state of the catalytic loop naturally makes a series of interactions with the middle domain (Figure 2.4B). The hydrogen bonds between the side-chain carboxyl group of Glu 372 of the catalytic loop and the side-chain amines of Arg 346 of the middle domain are of particular importance. In support of this model it has been shown that removal of the first 24 amino acid residues of Hsp90 prevents the transactivation required to stimulate the ATPase activity of the neighboring monomer and that removal of the ATP lid also disrupts the ATPase activity of Hsp90.[45]

2.7 Function of the Charged Linker

Although eukaryotic Hsp90s can retain biological function with a shortened linker[38,46] the evolutionary prevalence of a long, highly charged linker suggests that it serves some mechanistic role. It has been suggested that the long linker may help provide flexibility between domains to accommodate Hsp90's large repertoire of client proteins.[15,23,47,48] An F349A mutation weakens the interface between the N- and middle domains of Hsp90 leading to the significantly lower ATPase activity.[31,41] However, the Aha1 co-chaperone suppresses this mutation.[41] Thus the N- and middle domains of Hsp90 might become fully disassociated under some circumstances and Aha1 may serve to stabilize their association. Evidence that the N-terminal domains can dissociate and rotate relative to the middle domains of Hsp90 was seen in the structure of the *E. coli* Hsp90, HtpG.[49] In this structure the N-terminal domains were rotated 180° relative to the middle domains (Figure 2.5) when compared to the closed conformation of yeast Hsp90.[25] The charged linker that separates the domains might be of significance in relation to this.

Further evidence for N- and middle-domain rotations has been observed in EM and SAXS structural studies and with molecular and thermodynamic simulations.[19,22,50] Mechanistically, the significance of such a motion between the N-terminal and middle domains of Hsp90 remains unknown, but intriguing. Motion between the N- and middle domains may help transfer

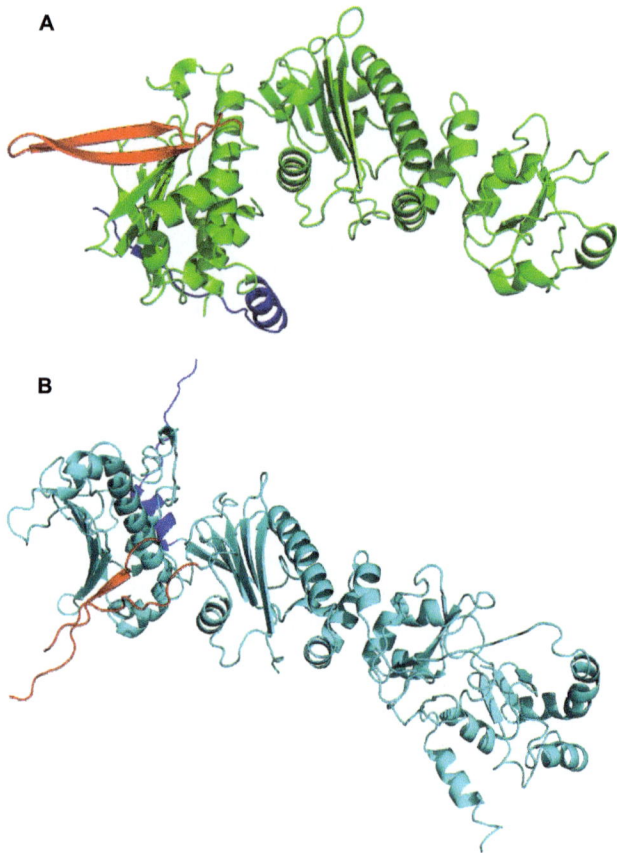

Figure 2.5 PyMol cartoon comparing the structure of full-length yeast Hsp90 and the N- and middle-domain structure of HtpG. A) HtpG in the open inactive state (PDB 1Y4U). B) Yeast Hsp90 in the closed active state (PDB 2CG9) in which the middle domain is in the same orientation and view as that of HtpG in panel A. The ATP lids are shown in blue and the linkers in red. The N-terminal domains of HtpG are rotated 180° relative to the equivalent domain of the yeast structure.

movement to client proteins and bring about structural changes in them that lead to their maturation or activation.

Recent biochemical data suggest that the charged linker can modulate nucleotide-dependent chaperone activity.[51] Another study has shown that reducing the length of the charged linker can suppress specific mutations (yeast I205A and Leu207, Human Ile 218 and Leu 220) that occur at the interface of the N-terminal and middle domains of Hsp90.[28] Yet another study has shown that the linker sequence is also important and provides regulatory sites that modulate Hsp90's activity.[29] These data, together with those for the F349A mutation,[41] support the idea that the linker provides flexibility required for Hsp90's conformational dynamics. Thus, it appears that the linker not only provides a means by which the N-terminal and middle domains of Hsp90 can undergo motions independently, which might be mechanistically important, but may also act as a regulatory site for its chaperone activity.

2.8 Structural Changes of the C-terminal Domain of Hsp90

Although formation of the C-terminal dimerization interface appears to be constitutive, there is some evidence to suggest that it may nonetheless undergo cycles of opening and closing, while the N-terminal domains themselves remain associated; the functional significance of these observations remains unclear.[52] Cys598 in human Hsp90 (equivalent to Ala 577 in yeast) has been shown to be a regulatory switch point in Hsp90 that affects its ATPase activity and chaperoning of client proteins.[53] It was suggested that this switch point alters the properties of both C- and N-terminal association and shifts the conformational equilibrium of the ATPase cycle.

2.9 Long-range Communication between Domains

The structure of full-length Hsp90 in complex with Sba1 and AMP-PNP showed that the closed Hsp90 dimer adopts a left-handed helical twist with inter-monomer interactions across the full length of the protein, indicating that the nucleotide state is communicated throughout the whole molecule. We have also seen that Cys 598 (Human) acts as a regulatory switch point in Hsp90 that affects its ATPase activity by altering the properties of both C- and N-terminal association and ultimately impacting on the chaperoning of client proteins.[52,53] The structure of HtpG also shows that its N-terminal domains can rotate 180° relative to the middle domains, an observation supported by hydrogen-exchange mass spectroscopy and EM experiments.[23,44] Client proteins and co-chaperones (Hop) also influence rotation of the N-terminal domains relative to the middle domains. Thus, there appears to be a rotation of the N-terminal domains from a position observed in the HtpG structure[49] to one in which the domains now face each other as observed in the N-terminally dimerized state. It has been suggested that this might represent a state in which client protein binding sites are now accessible.[50,54] While these conformational changes are themselves complex it has also been seen that opening and closing of the N- and C-terminal domains appear to be co-operative. Collectively, these results suggest that the nucleotide and client protein state of Hsp90 can be communicated throughout the molecule.

2.10 Co-chaperone Regulation and Modulation of Structural Elements of the Hsp90 ATPase Machinery

The set of complex structural changes that occur in Hsp90 are ideal for the evolution of a variety of regulatory mechanisms that are able to modulate the ATP-coupled conformational cycle of this chaperone. Thus, a whole array of co-chaperones have taken advantage of these potential regulatory points to suit the wide range of client proteins that are dependent on Hsp90 for their activation and maturation. Whatever the specific requirements of an individual client protein, certain aspects will undoubtedly be common to all client protein activation mechanisms; the client protein needs to be loaded, its structure modified or stabilized, and finally released. Consequently, different co-chaperones that are specific to subsets of client proteins may have similar effects on the Hsp90 function, but achieve this mechanistically in very different ways.

2.11 Co-chaperones that Deliver Client Proteins to Hsp90 Inhibit its ATPase Activity

A subset of co-chaperones has evolved to deliver their own specific client proteins to Hsp90. These include Sti1/Hop, Cdc37/p50, Sgt1 and perhaps AIP. Hop in higher eukaryotes is responsible for delivering steroid-hormone receptors to the Hsp90 complex. It does this by recruiting Hsp70, with bound receptor, *via* a set of TPR domains that bind specifically to Hsp90 or Hsp70. Thus, Hop acts as a scaffold protein physically coupling the Hsp70 and Hsp90 chaperone complexes. Following formation of this "super chaperone" complex the steroid hormone is passed from Hsp70 to Hsp90. This requires that Hsp70 be in its low-affinity state for substrate-protein binding,[55] but it is not clear whether association with Hsp70 nucleotide-exchange factors, such as Bag1,[55] is required to achieve this, or whether this is promoted by Hop itself. The reported effects of Hop on the ATPase activity of Hsp70 are contradictory.[56,57] The Hsp70 co-chaperone Ydj1, which stimulates high-affinity client binding by Hsp70, can also stimulate the refolding of luciferase in *in vitro* systems containing Hsp70 and Hsp90.[58] Thus the mechanism of transfer of client protein from Hsp70 to Hsp90 remains poorly understood.

Loading of client proteins into a stable complex with Hsp90 appears to need "arrest" of the ATPase-coupled conformational cycle, and co-chaperones such as Hop and Cdc37 possess the ability to inhibit ATP turnover and to lock the conformation of the chaperone in to a fixed state. Hop primarily binds Hsp90 *via* its middle TPR domain by binding to the C-terminal MEEVD motif of Hsp90, but also makes interactions in several other parts of the chaperone. Thus it may interact with the N-terminal domain of Hsp90 and prevent its dimerization and ATPase activity and a single monomeric molecule of Hop may be able to achieve this.[59] This is an attractive proposal, as it provides a means by which other TPR-domain co-chaperones such as the immunophilin FKBP51 might enter the complex during progression of the chaperone cycle. However, structural and biochemical insights into how Hop arrests the ATPase cycle and how immunophilins relieve this arrest are completely lacking.

Cdc37 is the co-chaperone responsible for delivering protein kinases to the Hsp90 system, and like Hop is able to arrest the ATPase cycle of the chaperone.[60] Unlike Hop, Cdc37 delivers its client proteins by interacting with the N-terminal domains of Hsp90.[61,62] Why loading factors for different client proteins interact with different regions of Hsp90 is a matter of intense debate. One possibility is that different clients need to interact with different binding sites on the chaperone, and undergo different activation mechanisms; all nonetheless coupled to the basic chaperone cycle. This idea at least helps explain how Hsp90 deals with such a variety of client proteins that require different forms of activation. For example, steroid-hormone receptors need to bind their appropriate small-molecule hormone. Kinases might require phosphorylation and modulation of their structure into a catalytically active state. However, whatever the reason the current understanding of this key aspect of Hsp90 function remains very poor.

Cdc37 achieves inhibition of the Hsp90 ATPase activity at a number of levels. Firstly, it binds to the N-terminal ATP lids of Hsp90 (Figure 2.6A). This prevents them from moving to a closed state. Secondly, Cdc37 is bound between the N-terminal domains so that N-terminal dimerization becomes impossible. Finally, a hydrogen bond between Arg 167 of Cdc37 and the catalytic Glu 33 of Hsp90 is formed, thus inactivating Glu 33 and preventing ATP hydrolysis (Figure 2.6A). This interaction does not appear to inhibit ATP binding to the chaperone, which might be required for the subsequent step in the activation of the kinase client.

The 19 Å negative stain electron microscopy reconstruction of the complex of Cdk4 with Cdc37 and Hsp90 remains the only three-dimensional model for client protein complex currently available.[47] Despite the limited resolution, this complex has provided some very important insights into the nature of kinase complexes with the Hsp90 system, although how general these are to other classes of client proteins is uncertain. Within the ternary complex a single Cdc37 molecule is bound between the N-terminal domains of an Hsp90 dimer, with the two monomers adopting substantially different conformations at the junction of the middle and N-terminal domains. Similarly a single Cdk4 kinase molecule is bound in the complex, contacting Cdc37 and the N- and middle domains of only one of the two Hsp90 molecules (Figure 2.6B). The apparent conformation of the kinase suggests that one lobe of the kinase (tentatively identified as the N-terminal lobe) contacts the N-domain of the chaperone while the other lobe of the kinase contacts the middle domain of Hsp90. Given the substantial change in relative orientation that the N- and middle domains of Hsp90 can undergo, this mode of binding by the kinase suggests that structural changes of Hsp90 coupled to its ATPase cycle would bring about conformational movement of one lobe of the bound kinase relative to the other, and this might enable whatever conformational changes are required to facilitate activation and cellular stabilization of the kinase client.

Sgt1 is an Hsp90 co-chaperone involved in a number of Hsp90-dependent processes including kinetochore assembly[63] and the activation of NOD-like receptors involved in the cytosolic innate immune response in plants and animals.[64] The structure of the CS domain of Sgt1 in complex with the N-terminal domain of Hsp90 has been determined.[65] Although the CS domain of Sgt1 is structurally homologous to another Hsp90 co-chaperone, p23/Sba1 that also interacts with the N-terminal domain of Hsp90, Sgt1-CS binds to a different

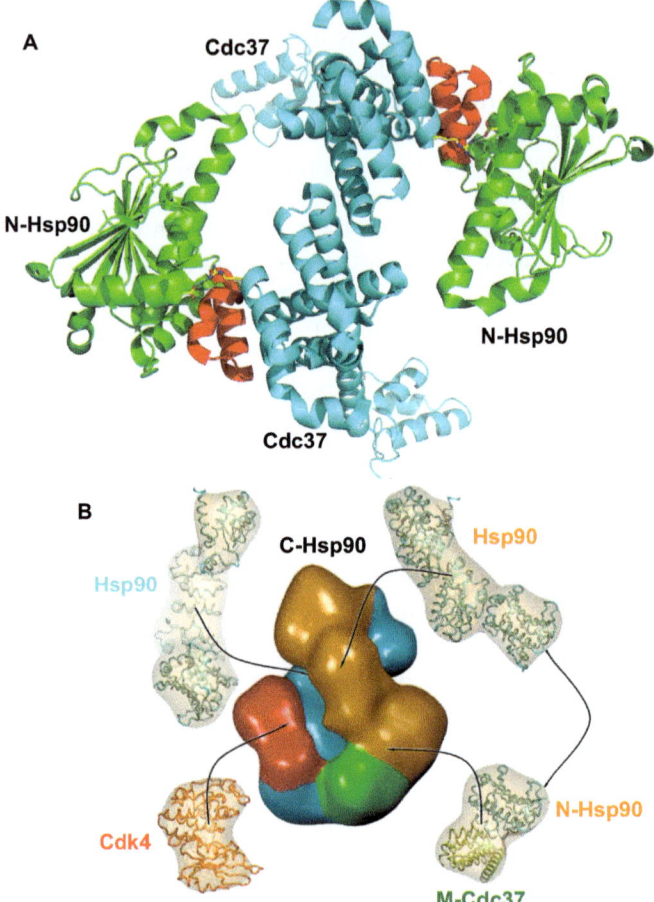

Figure 2.6 PyMol cartoon showing interactions of Hsp90 with Cdc37 and Cdk4. A) Crystal structure of the N-terminal domain of Hsp90 in complex with the C-terminal domain of Cdc37 (PDB 1US7). B) EM structure of the yeast Hsp90 (gold and cyan) in complex with Cdc37 (green) and Cdk4 (red).

surface and does not interact directly with the ATP-lid (Figure 2.7). Unlike p23/Sb1, Sgt1-binding generates a steric clash in the ATP-bound conformation of an Hsp90 dimer, explaining the observation that Sgt1 binding is favored by the open ADP-bound or apo conformation of the chaperone.[66]

Recently, the structures of the N-terminal immunophilin and C-terminal TPR domains of AIP (also known as XAP2 and ARA9) have been determined.[67,68] The structure of the TPR domain showed that AIP interacted with the conserved extreme C-terminal TPR binding motifs of Hsp90 (MEEVD), Hsp70 (IEEVD) and TOMM20 (EDDVE). AIP was found to bind these conserved motifs in a similar way to that seen by the TPR domain of CHIP in binding to Hsp70 and Hsp90 (Figure 2.8A,B). Thus, CHIP and AIP bind these conserved motifs in a *trans*-configuration in which the hydrophobic pockets

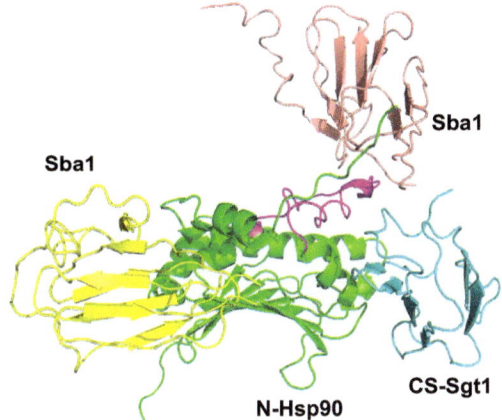

Figure 2.7 PyMol cartoon showing the interaction of Sba1 and the CS domain of Sgt1 with the N-terminal domain of yeast Hsp90. The interaction sites of Sba1 (salmon and yellow) are shown on a single monomer of the dimerized N-terminal domains of Hsp90 (green, PDB 2CG9). One of these interaction sites is on the exposed surface of the ATP lid (magenta) in the closed state. The CS domain of Sgt1 (cyan) does not interact with the ATP lids (PDB 2JKI).

that accept the conserved methionine and valine, in the case of the MEEVD of Hsp90, are on opposite sides of the walls of the binding cleft of the TPR domain.[68] In contrast, the hydrophobic pockets of Hop are found on one side of the TPR cleft (*cis*-mode of binding, Figure 2.8A,B). It appears that TPR domains, such as those of AIP and CHIP, which bind a variety of conserved motifs use a *trans*-mode of binding. Thus, for AIP and CHIP this results in sequences upstream of the conserved binding-motif being directed up and out of the binding cleft of the TPR domain. Consequently, these sequences, which are variable between Hsp90, Hsp70 and TOMM20, are avoided. In contrast, Hop utilizes different TPR domains to simultaneously bind Hsp90 and Hsp70 (Figure 2.8A,B). AIP is known to interact with many different client proteins, such as AHR, ERα, Survivin, G proteins, RET, EBNA3 and PDE4A5.[69] It appears that the C-7 terminal α-helix of the AIP domain may act as a client protein interaction site and that AIP may be responsible for delivering its client proteins to Hsp90 (Figure 2.8C).[68] However, unlike Hop/Sti1 and Cdc37 it appears that human AIP does not inhibit the ATPase activity of yeast Hsp90. Whether another as yet unknown factor achieves such inhibition remains an open question.

2.12 Co-chaperones that Activate the ATPase Activity of Hsp90

Aha1 remains the most potent protein activator of the ATPase activity of Hsp90[70] so far identified. Structural studies have revealed the mechanism of

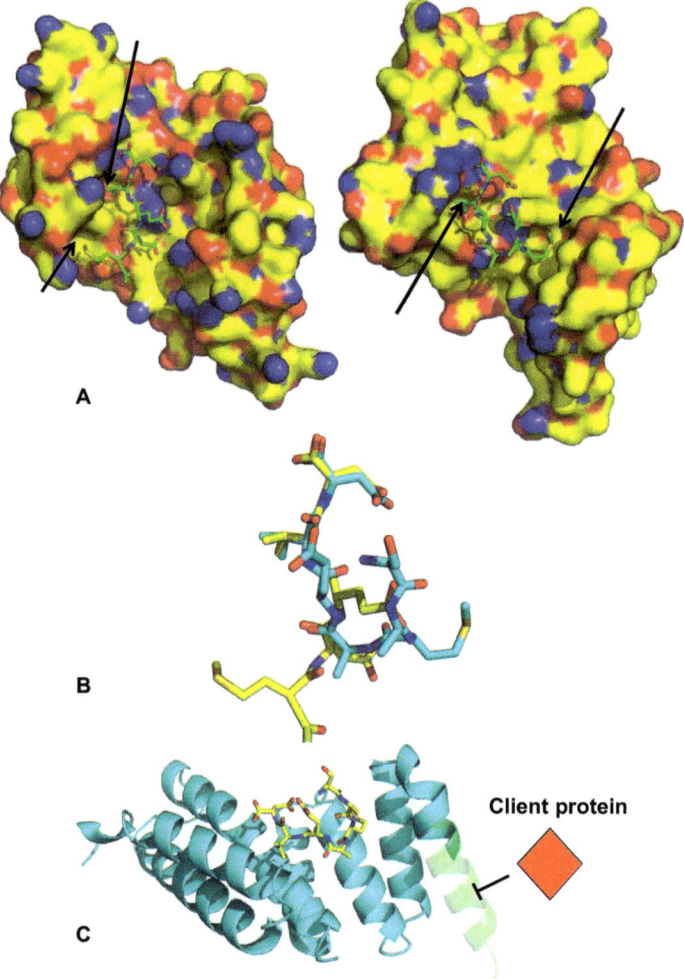

Figure 2.8 PyMol diagrams showing the interaction of Hsp90 peptide sequences. A) *Cis*-binding of MEEVD to the TPR 2A domain of Hop (left, PDB 1ELR) and *trans*-binding to the TPR domain of AIP (right, PDB 4AIF). The two hydrophobic pockets that accept methionine and valine (in the case of Hsp90) are shown by arrows. B) Superimposition of the bound peptides bound to Hop (bound with MEEVD) and AIP (bound with EDASRMEEVD, only SRMEEVD is visible). C) The structure of AIP showing loss of the client protein binding site (light green) of the C-7 α-helix in the R304* mutant (nonsense mutant).

Hsp90 ATPase activation by the N-terminal domain of Aha1, which modulates the conformation of the catalytic loop in the middle domain of the chaperone. This facilitates engagement of Arg380 with the γ-phosphate of ATP bound in the N-terminal domain (Figure 2.9).[32] However, the C-terminal domain of Aha1 is also required for full activation of Hsp90 ATPase activity. This domain

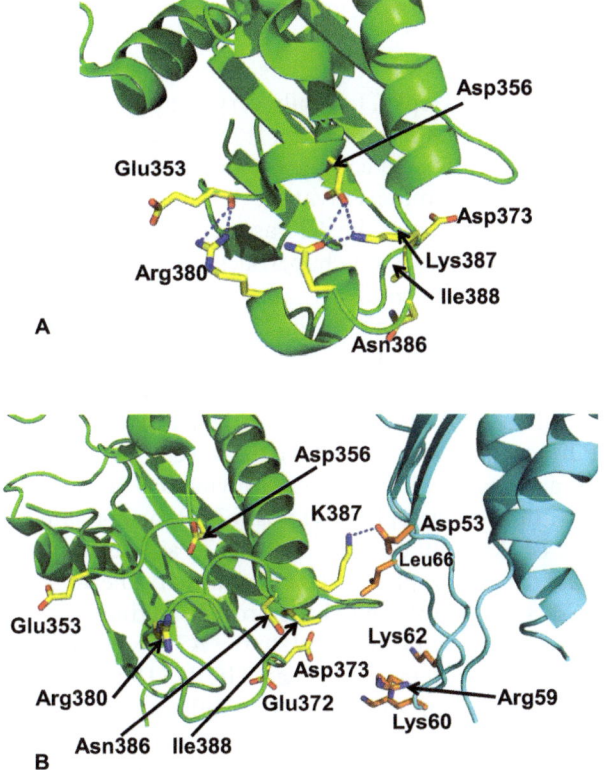

Figure 2.9 PyMol cartoon showing the interaction of Aha1 with yeast Hsp90. A) The middle domain of Hsp90 with the catalytic loop in the closed inactive state (PDB 1HK7). B) Interaction of Aha1 (cyan) with the middle domain of Hsp90 (green, PDB 1USU). The molecular interaction releases the catalytic loop (Arg 380) into an open state.

has been shown to bind between the N-terminal domains of Hsp90, overlapping with Sba1 binding, and thus might provide further stabilization of the N-terminally dimerized state.[71]

Unlike Sti1/Hop and Cdc37, Aha1 is not involved in client protein delivery and most likely partakes in a role downstream of client protein binding. Aha1 has been implicated in the activation of protein kinases, steroid-hormone receptors, CFTR, MC4R and ANT1.[70,72–81] Thus Aha1 appears to be a regulatory co-chaperone of Hsp90 activation of client proteins in general. However, an alternative view is that Aha1 is specifically required for the regulation and assembly of oligomeric Hsp90 complexes.[82]

Budding yeast possesses a small Aha1 paralogue, Hch1,[83] which resembles the N-terminal domain of Aha1 and can also activate Hsp90 ATPase activity.[70] Recently, it was shown that the biological function of Hch1 is very different from that of Aha1.[84] Deletion of Hch1, but not Aha1, was shown to mitigate the temperature-sensitive phenotype of two Hsp90 mutations, G313S and

A587T, while its over-expression greatly increased sensitivity to the Hsp90 inhibitor NVP-AUY922.

The TPR-domain cyclophilin Cpr6 has also been shown to stimulate the Hsp90 ATPase activity of Hsp90, but only very weakly relative to that of Aha1. The significance of this remains unknown. It might be that such immunophilins take the place of Aha1 in Hsp90 complexes and in the context of client protein may show stronger effects on the activation of the ATPase activity of Hsp90. It may be that weak stimulation of the ATPase activity of Hsp90 might act as the trigger for release of inhibitory co-chaperones such as Hop/Sti1p and Cdc37. However, this remains to be tested.

The TPR-domain protein Tah1, which is recruited to Hsp90 *via* the C-terminal conserved EEVD residues,[85] is a weak activator of the ATPase activity of Hsp90.[86,87] However, in the context of a ternary complex also involving its partner protein Pih1, the ATPase activity of Hsp90 is instead inhibited.[87] The Hsp90-Tah1-Pih complex is involved in the assembly of a number of large complexes such as SnoRNAP and RNA polymerase II, as well as the stabilization and activation of phosphatidylinositol-3 kinase-related protein kinase complexes.[2,88] It is likely that arrest of the ATPase activity of Hsp90 in the assembly of these large complexes serves a similar role to ATPase arrest by Hop/Sti1 and Cdc37 in facilitating client protein loading.

Another co-chaperone that can stimulate the ATPase activity of Hsp90 is the plant CHORD domain protein Rar1. The CHORD II domain of Rar1 binds in complex with the CS domain of Sgt1, and stabilizes an open conformation of the Hsp90 dimer in which the N-terminal domains are prevented from coming into direct contact, as is required for activity in the ATP-bound conformation stabilized by Sba1.[25] Despite this, binding of Rar1 not only permits ATPase activity, but also stimulates it. Structural analysis suggests that the CHORD II domain of Rar1 bound to the N-terminal domain of the chaperone facilitates the displacement of the ATP-lid, stabilizes the active conformation of the catalytic loop from the middle domain and facilitates ATP-hydrolysis by direct interaction of a conserved and functionally essential histidine with the β-phosphate of the bound ATP[89] (Figure 2.10).

2.13 Co-chaperones that Modulate the ATPase Activity of Hsp90

Yeast Sba1 is unusual in that it does not completely inhibit the ATPase activity of Hsp90[70] but modulates it to a lower level. Thus, although Sba1 binds the ATP lids in the closed ATP-bound conformation and inhibits their opening in the next phase of the ATPase cycle, it also stabilizes the catalytic loop from the middle domain in an active conformation (Figure 2.11). Consequently, Sba1 promotes hydrolysis of ATP, but inhibits ADP release as long as it remains bound. The biological role this may play remains obscure, but it may contribute to stabilization of specific client protein conformations. For example, steroid-hormone receptors need to bind their appropriate hormone, but access to the

Structural Basis of Hsp90 Function

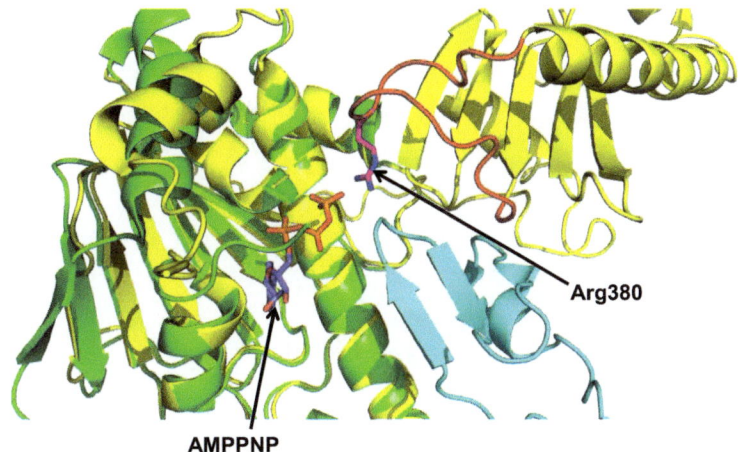

Figure 2.10 PyMol Model of the activation of the ATPase activity of Hsp90 by the CHORD II domain of Rar1. The CHORD II domain of Rar1 (cyan) binds to a region on the N-terminal domain of Hsp90 (green) that is normally occupied by the closed ATP lid (PDB 2XCM). Superimposition of the full-length Hsp90 in its closed active state (yellow, PDB 2CG9) shows that the catalytic loop comes close to the CHORD II domain of Rar1 and could therefore be subject to conformational modulation by Rar1.

steroid-binding pocket is guarded by a small helix.[90,91] Clearly, this helix must move to allow hormone binding and thus a stable state in which the helix has been physically moved allowing access for the hormone could be achieved by slowing the release phase in the ATPase cycle of Hsp90. However, this so far has not been observed directly and remains speculative. In contrast to Sba1, human p23 has been shown to completely inhibit the ATPase activity of human Hsp90.[92] Whether this is a fundamental difference between the mechanisms of Human and yeast Hsp90 remains to be confirmed.

2.14 Co-chaperones that Target Hsp90 Client Proteins for Proteasome Degradation

The crystal structure of CHIP bound to the conserved C-terminal MEEVD motifs of Hsp90 has been determined.[93] CHIP is a E3 ubiquitin ligase that can associate with the E2 ubiquitin-conjugating enzyme Ubc13-Uev1 and direct the formation of Lys 63 linkages.[94] CHIP can also associate with Hsp90 and Hsp70 and apparently contribute to the proteasomal destruction of some bound client proteins *via* recruitment of the Lys 48-specific E2 UbcH5.[95] However, it is far from certain that CHIP provides the route by which the majority of Hsp90 clients become ubiquitinated and degraded when the ATPase cycle is blocked by inhibitors. The effect of CHIP on the ATPase activity of Hsp90 has not, as far as we are aware, been determined. The structure of dimeric CHIP shows an

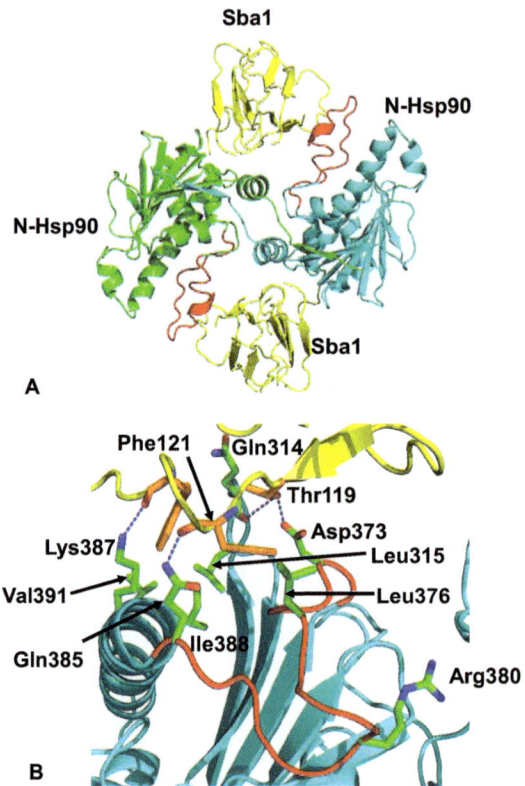

Figure 2.11 PyMol cartoon showing the interaction of Sba1 with yeast Hsp90. A) Sba1 bound between the N-terminal domains of Hsp90 in a dimerized state (PDB 2CG9). B) Interaction of the N-terminal tail of Sba1 (yellow with gold amino acid residues) with the middle domain of Hsp90 (cyan with green amino acid residues). The molecular interaction releases the catalytic loop (red) into its open active state. This is similar to the activation of Hsp90 by the N-terminal domain of Aha1.

asymmetric arrangement of the TPR domains such that one of the Ubc sites is precluded, thus providing a means by which a dimeric chaperone is coupled to a single ubiquitin system (Figure 2.12).

2.15 Client Proteins and Activation of Hsp90

The interaction of the ligand-binding domain of glucocorticoid receptor stimulates the ATPase activity of Hsp90.[96] Recently, it has been shown that a fragment of a staphylococcus nuclease stimulates conformational change in the bacterial Hsp90 homologue HtpG. It appears that this protein fragment stimulates an unfavorable 180° rotation of the N-terminal domains (relative to the middle domains) so that they are now posed to undergo dimerization. This stimulation is coupled to activation of the ATPase activity of Hsp90.[50]

Structural Basis of Hsp90 Function

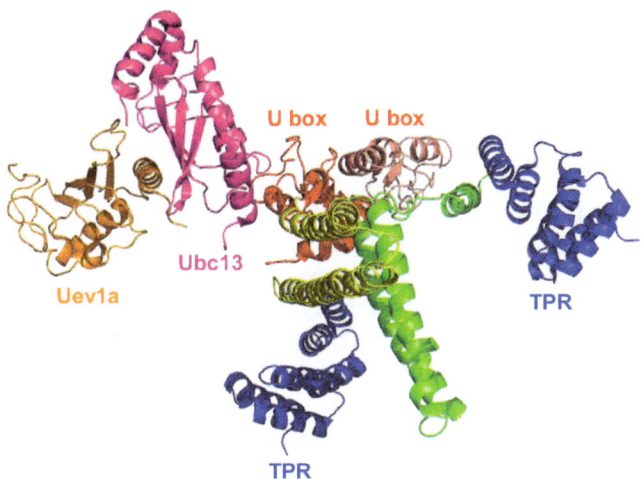

Figure 2.12 PyMol model of the CHIP-Ubc13-Uev1a structure. The asymmetric structure of CHIP (PDB 2C2L) prevents the binding of a second Ubc13-Uev1a module (PDB 2C2V).

Furthermore, it appears that this conformational change, which primes chaperone activity, is dependent on cross monomer contacts between the client protein and Hsp90.[97] The significance of these observations is difficult to assess, as staphylococcus nuclease is not known to be a client protein of HtpG.

The EM structure of Hsp90 in complex with Cdc37 and Cdk4 showed that the N-terminal and the middle domain of Hsp90 were probably involved in interactions with protein kinases.[47] Trp 300, identified as functionally important in a mutagenesis screen of the middle domain,[31] was identified as a key residue in client protein interaction. Bridging between the N- and middle domains might provide a mechanism for transmitting conformational changes driven by the ATPase cycle to the bound client. The interaction of a staphylococcal nuclease fragment with HtpG appears to involve a separate patch on the chaperone's middle domain, formed by residues Phe 390, Trp 467 and Asp 476.[50] Other regions including two amphipathic helices in the C-terminal dimerization domain, the unstructured extreme C-terminal extension and a patch centered on Glu 412 and Glu 415 have also been implicated in client protein binding,[98,99] however, there are no direct structural observations to support these. Whether the range of different sites implicated reflects the different classes of interacting client protein involved remains unknown.

2.16 Phosphorylations and other Modifications

Post-translational modification of Hsp90 and its impact on the ATPase cycle and client protein interactions is comprehensively covered by Neckers and

Mollapour.[100] Thus we will only briefly cover modifications that impact on the main aspect of the chaperone cycle.

Phosphorylation of Tyr 24 (human Tyr 38) modulates the chaperone cycle of Hsp90, with a phosphomimetic Y24E mutation resulting in the complete inhibition of Hsp90 ATPase activity.[101] It appears that this site influences the productive chaperoning of protein kinase client-proteins. Another site of phosphorylation is Thr 22 (Human Thr 36). Like Tyr 24 this is part of the nexus of residues that come together to form the catalytically active closed-state of Hsp90. As with Y24E, a phosphomimetic T22E mutation also affected client protein activation depending on the precise type of client protein investigated. Whether this is because some clients are more sensitive to changes in the overall rate of the cycle or whether this reflects mechanistic differences in their activation is not known.

Away from the core of the N-terminal ATP-dependent conformational switch, another series of tyrosine phosphorylations has also been implicated in regulating the chaperone cycle of Hsp90. For example, phosphorylation of Tyr 197 (human) was shown to dissociate p50^{Cdc37} from Hsp90. Hsp90 phosphorylation on Tyr 313 (human) promotes recruitment of Aha1, which appears to stimulate the ATPase activity of Hsp90. Finally, at completion of the chaperone cycle, Hsp90 Tyr 627 (Human) phosphorylation appears to induce dissociation of the client and remaining co-chaperones.

S-nitrosylation of Cys 598 (human) has been reported to inhibit the ATPase activity of Hsp90α.[53,102] In yeast similar inhibition is seen with the nitrosylated A577C mutant. Interestingly, it appears that mutation of Ala 577 to isoleucine or asparagine alters the dimerization of the N- and C-terminal domains. Finally, acetylation of Hsp90 has also been shown.[103–105] However, no effects on the ATPase activity of Hsp90 have to our knowledge been demonstrated.

2.17 Natural Resistance Mechanisms

The N-terminal nucleotide-binding site of Hsp90 is the target for the natural antibiotics, geldanamycin and radicicol,[106] and a vast array of synthetically synthesized inhibitors (see review[26]). Although the nucleotide-binding site of Hsp90 is highly conserved, a number of resistance mechanisms to natural antibiotics have been discovered.[107,108] A single mutation, L34I, results in a pocket opening up in the ATP-binding site of yeast Hsp90. Consequently, a number of water molecules enter and form a stable network of hydrogen bonds. One of these waters impacts on the chlorine atom of radicicol and lowers its affinity for Hsp90[108] (Figure 2.13A). In contrast, a pair of mutations, E88G and N992L, decreases the affinity for geldanamycin.[107] These mutations disrupt hydrogen bonding to the C-12 methoxy group of geldanamycin. Furthermore, there is also a loss of Van-der-Waals interactions between the same methoxy group and E88G and N92L. Thus, the C-12 methoxy group is less ordered and binding affinity is compromised (Figure 2.13B).

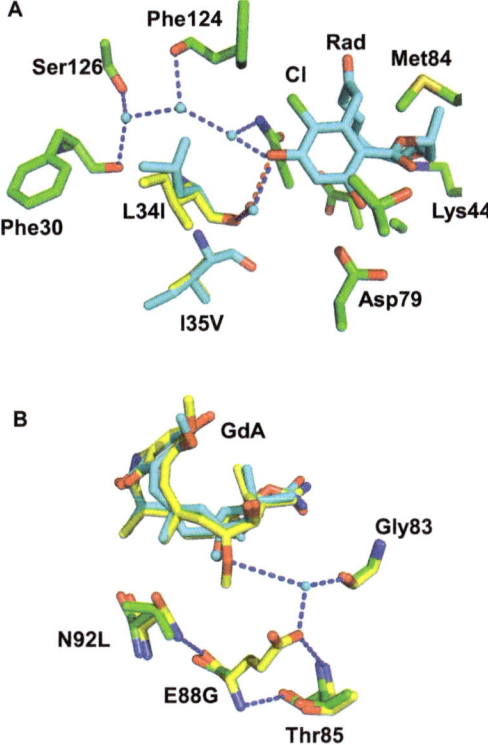

Figure 2.13 Pymol cartoons showing resistance to inhibitor binding in Hsp90. A) The resistance mechanism of radicicol binding. The L34I mutation opens a pocket that allows a network of water-mediated hydrogen bonds to form that impact on the chlorine atom (Cl) of radicicol (Rad) and lowers its affinity for Hsp90 (PDB 2WEP). B) The E88G-N92L double mutation results in the loss of hydrogen bonding to the C-12 methoxy of geldanamycin and loss of Van-der-Waals interactions and thus reduces the affinity of geldanamycin (GdA) for Hsp90 (PDB 2YGE).

2.18 Concluding Remarks

The complex conformational cycle of Hsp90 is central to its chaperoning activity and significant progress has been made in understanding the detailed molecular events that are involved. We have also made significant progress on understanding how the vast array of co-chaperones influence the cycle and are now just beginning to understand how client proteins and post-translational modifications can also contribute to this process. However, exactly what these conformational changes do to or for client proteins still remains enigmatic. Nonetheless, the structure of Hsp90 has driven the development of a vast array of small molecule inhibitors that are showing real promise as anticancer drugs.

References

1. S. E. Jackson, *Topics Curr. Chem.*, 2012, **328**, 155–240.
2. T. Makhnevych and W. A. Houry, *Biochim. Biophys. Acta*, 2012, **1823**, 674–682.
3. S. L. Rutherford and S. Lindquist, *Nature*, 1998, **396**, 336–342.
4. S. Lindquist, *Cold Spring Harbor Symposia on Quantitative Biology*, 2009, **74**, 103–108.
5. T. A. Williams and M. A. Fares, *Genome Biol. Evol.*, 2010, **2**, 609–619.
6. I. Yahara, *Genes Cells*, 1999, **4**, 375–379.
7. K. A. Borkovich, F. W. Farrelly, D. B. Finkelstein, J. Taulien and S. Lindquist, *Mol. Cell Biol.*, 1989, **9**, 3919–3930.
8. Y. Miyata, H. Nakamoto and L. Neckers, *Curr. Pharm. Des.*, 2012, **19**, 347–365.
9. L. Whitesell, S. Santagata and N. U. Lin, *Curr. Mol. Med.*, 2012, **12**, 1108–1124.
10. L. Whitesell and N. U. Lin, *Biochim. Biophys. Acta*, 2012, **1823**, 756–766.
11. L. Neckers and P. Workman, *Clin. Cancer Res.*, 2012, **18**, 64–76.
12. L. H. Pearl, C. Prodromou and P. Workman, *Biochem. J.*, 2008, **410**, 439–453.
13. J. Travers, S. Sharp and P. Workman, *Drug Discov. Today*, 2012, **17**, 242–252.
14. C. K. Vaughan, P. W. Piper, L. H. Pearl and C. Prodromou, *FEBS*, 2009, **276**, 199–209.
15. A. K. Shiau, S. F. Harris, D. R. Southworth and D. A. Agard, *Cell*, 2006, **127**, 329–340.
16. S. Frey, A. Leskovar, J. Reinstein and J. Buchner, *J. Biol. Chem.*, 2007, **282**, 35612–35620.
17. A. Leskovar, H. Wegele, N. D. Werbeck, J. Buchner and J. Reinstein, *J. Biol. Chem.*, 2008, **283**, 11677–11688.
18. C. Graf, M. Stankiewicz, G. Kramer and M. P. Mayer, *EMBO J.*, 2009, **28**, 602–613.
19. D. R. Southworth and D. A. Agard, *Mol. Cell*, 2008, **32**, 631–640.
20. T. O. Street, K. A. Krukenberg, J. Rosgen, D. W. Bolen and D. A. Agard, *Protein Sci.*, 2010, **19**, 57–65.
21. K. A. Krukenberg, D. R. Southworth, T. O. Street and D. A. Agard, *J. Mol. Biol.*, 2009, **390**, 278–291.
22. K. A. Krukenberg, F. Forster, L. M. Rice, A. Sali and D. A. Agard, *Structure*, 2008, **16**, 755–765.
23. P. Bron, E. Giudice, J. P. Rolland, R. M. Buey, P. Barbier, J. F. Diaz, V. Peyrot, D. Thomas and C. Garnier, *Biol. Cell*, 2008, **100**, 413–425.
24. M. Simunovic and G. A. Voth, *Biophys. J.*, 2012, **103**, 284–292.
25. M. M. Ali, S. M. Roe, C. K. Vaughan, P. Meyer, B. Panaretou, P. W. Piper, C. Prodromou and L. H. Pearl, *Nature*, 2006, **440**, 1013–1017.
26. D. S. Hong, U. Banerji, B. Tavana, G. C. George, J. Aaron and R. Kurzrock, *Cancer Treat. Rev.*, 2013, **39**, 375–387.

27. B. Panaretou, C. Prodromou, S. M. Roe, R. O'Brien, J. E. Ladbury, P. W. Piper and L. H. Pearl, *EMBO J.*, 1998, **17**, 4829–4836.
28. S. Tsutsumi, M. Mollapour, C. Graf, C. T. Lee, B. T. Scroggins, W. Xu, L. Haslerova, M. Hessling, A. A. Konstantinova, J. B. Trepel, B. Panaretou, J. Buchner, M. P. Mayer, C. Prodromou and L. Neckers, *Nat. Struct. Mol. Biol.*, 2009, **16**, 1141–1147.
29. S. Tsutsumi, M. Mollapour, C. Prodromou, C. T. Lee, B. Panaretou, S. Yoshida, M. P. Mayer and L. M. Neckers, *Proc. Natl Acad. Sci. USA*, 2012, **109**, 2937–2942.
30. O. Hainzl, M. C. Lapina, J. Buchner and K. Richter, *J. Biol. Chem.*, 2009, **284**, 22559–22567.
31. P. Meyer, C. Prodromou, B. Hu, C. Vaughan, S. M. Roe, B. Panaretou, P. W. Piper and L. H. Pearl, *Mol. Cell*, 2003, **11**, 647–658.
32. P. Meyer, C. Prodromou, C. Liao, B. Hu, S. M. Roe, C. K. Vaughan, I. Vlasic, B. Panaretou, P. W. Piper and L. H. Pearl, *EMBO J.*, 2004, **23**, 1402–1410.
33. T. W. Schulte, S. Akinaga, T. Murakata, T. Agatsuma, S. Sugimoto, H. Nakano, Y. S. Lee, B. B. Simen, Y. Argon, S. Felts, D. O. Toft, L. M. Neckers and S. V. Sharma, *Mol. Endocrinol.*, 1999, **13**, 1435–1448.
34. S. Vogen, T. Gidalevitz, C. Biswas, B. B. Simen, E. Stein, F. Gulmen and Y. Argon, *J. Biol. Chem.*, 2002, **277**, 40742–40750.
35. C. Biswas, O. Ostrovsky, C. A. Makarewich, S. Wanderling, T. Gidalevitz and Y. Argon, *Biochem. J.*, 2007, **405**, 233–241.
36. M. Marzec, D. Eletto and Y. Argon, *Biochim. Biophys. Acta*, 2012, **1823**, 774–787.
37. D. C. Altieri, G. S. Stein, J. B. Lian and L. R. Languino, *Biochim. Biophys. Acta*, 2012, **1823**, 767–773.
38. C. Prodromou, S. M. Roe, R. O'Brien, J. E. Ladbury, P. W. Piper and L. H. Pearl, *Cell*, 1997, **90**, 65–75.
39. C. Prodromou, B. Panaretou, S. Chohan, G. Siligardi, R. O'Brien, J. E. Ladbury, S. M. Roe, P. W. Piper and L. H. Pearl, *EMBO J.*, 2000, **19**, 4383–4392.
40. Y. Kimura, S. Matsumoto and I. Yahara, *Mol. Gen. Genet.*, 1994, **242**, 517–527.
41. G. Siligardi, B. Hu, B. Panaretou, P. W. Piper, L. H. Pearl and C. Prodromou, *J. Biol. Chem.*, 2004, **279**, 51989–51998.
42. T. Weikl, P. Muschler, K. Richter, T. Veit, J. Reinstein and J. Buchner, *J. Mol. Biol.*, 2000, **303**, 583–592.
43. J. Li, L. Sun, C. Xu, F. Yu, H. Zhou, Y. Zhao, J. Zhang, J. Cai, C. Mao, L. Tang, Y. Xu and J. He, *Acta Biochim. Biophys. Sin.*, 2012, **44**, 300–306.
44. J. J. Phillips, Z. P. Yao, W. Zhang, S. McLaughlin, E. D. Laue, C. V. Robinson and S. E. Jackson, *J. Mol. Biol.*, 2007, **372**, 1189–1203.
45. K. Richter, S. Moser, F. Hagn, R. Friedrich, O. Hainzl, M. Heller, S. Schlee, H. Kessler, J. Reinstein and J. Buchner, *J. Biol. Chem.*, 2006, **281**, 11301–11311.

46. J. F. Louvion, R. Warth and D. Picard, *Proc. Natl Acad. Sci. USA*, 1996, **93**, 13937–13942.
47. C. K. Vaughan, U. Gohlke, F. Sobott, V. M. Good, M. M. Ali, C. Prodromou, C. V. Robinson, H. R. Saibil and L. H. Pearl, *Mol. Cell*, 2006, **23**, 697–707.
48. S. J. Felts, B. A. Owen, P. Nguyen, J. Trepel, D. B. Donner and D. O. Toft, *J. Biol. Chem.*, 2000, **275**, 3305–3312.
49. Q. Huai, H. Wang, Y. Liu, H. Y. Kim, D. Toft and H. Ke, *Structure*, 2005, **13**, 579–590.
50. T. O. Street, L. A. Lavery and D. A. Agard, *Mol. Cell*, 2011, **42**, 96–105.
51. T. Scheibel, H. I. Siegmund, R. Jaenicke, P. Ganz, H. Lilie and J. Buchner, *Proc. Natl Acad. Sci. USA*, 1999, **96**, 1297–1302.
52. C. Ratzke, M. Mickler, B. Hellenkamp, J. Buchner and T. Hugel, *Proc. Natl Acad. Sci. USA*, 2010, **107**, 16101–16106.
53. M. Retzlaff, M. Stahl, H. C. Eberl, S. Lagleder, J. Beck, H. Kessler and J. Buchner, *EMBO Rep.*, 2009, **10**, 1147–1153.
54. D. R. Southworth and D. A. Agard, *Mol. Cell*, 2011, **42**, 771–781.
55. M. P. Mayer and B. Bukau, *Cell. Mol. Life Sci.*, 2005, **62**, 670–684.
56. M. Gross and S. Hessefort, *J. Biol. Chem.*, 1996, **271**, 16833–16841.
57. B. D. Johnson, R. J. Schumacher, E. D. Ross and D. O. Toft, *J. Biol. Chem.*, 1998, **273**, 3679–3686.
58. H. Wegele, S. K. Wandinger, A. B. Schmid, J. Reinstein and J. Buchner, *J. Mol. Biol.*, 2006, **356**, 802–811.
59. K. Richter, P. Muschler, O. Hainzl, J. Reinstein and J. Buchner, *J. Biol. Chem.*, 2003, **278**, 10328–10333.
60. G. Siligardi, B. Panaretou, P. Meyer, S. Singh, D. N. Woolfson, P. W. Piper, L. H. Pearl and C. Prodromou, *J. Biol. Chem.*, 2002, **277**, 20151–20159.
61. W. Zhang, M. Hirshberg, S. H. McLaughlin, G. A. Lazar, J. G. Grossmann, P. R. Nielsen, F. Sobott, C. V. Robinson, S. E. Jackson and E. D. Laue, *J. Mol. Biol.*, 2004, **340**, 891–907.
62. S. M. Roe, M. M. Ali, P. Meyer, C. K. Vaughan, B. Panaretou, P. W. Piper, C. Prodromou and L. H. Pearl, *Cell*, 2004, **116**, 87–98.
63. L. B. Lingelbach and K. B. Kaplan, *Mol. Cell Biol.*, 2004, **24**, 8938–8950.
64. Y. Kadota, K. Shirasu and R. Guerois, *Trends Biochem. Sci.*, 2010, **35**, 199–207.
65. M. Zhang, M. Boter, K. Li, Y. Kadota, B. Panaretou, C. Prodromou, K. Shirasu and L. H. Pearl, *EMBO J.*, 2008, **27**, 2789–2798.
66. M. G. Catlett and K. B. Kaplan, *J. Biol. Chem.*, 2006, **281**, 33739–33748.
67. M. Linnert, K. Haupt, Y. J. Lin, S. Kissing, A. K. Paschke, G. Fischer, M. Weiwad and C. Lucke, *Biomol. NMR Assignments*, 2012, **6**, 209–212.
68. R. M. Morgan, L. C. Hernandez-Ramirez, G. Trivellin, L. Zhou, S. M. Roe, M. Korbonits and C. Prodromou, *PLoS One*, 2012, **7**, e53339.
69. G. Trivellin and M. Korbonits, *J. Endocrinol.*, 2011, **210**, 137–155.
70. B. Panaretou, G. Siligardi, P. Meyer, A. Maloney, J. K. Sullivan, S. Singh, S. H. Millson, P. A. Clarke, S. Naaby-Hansen, R. Stein,

R. Cramer, M. Mollapour, P. Workman, P. W. Piper, L. H. Pearl and C. Prodromou, *Mol. Cell*, 2002, **10**, 1307–1318.
71. M. Retzlaff, F. Hagn, L. Mitschke, M. Hessling, F. Gugel, H. Kessler, K. Richter and J. Buchner, *Mol. Cell*, 2010, **37**, 344–354.
72. G. P. Lotz, H. Lin, A. Harst and W. M. Obermann, *J. Biol. Chem.*, 2003, **278**, 17228–17235.
73. A. Harst, H. Lin and W. M. Obermann, *Biochem. J.*, 2005, **387**, 789–796.
74. A. V. Koulov, P. LaPointe, B. Lu, A. Razvi, J. Coppinger, M. Q. Dong, J. Matteson, R. Laister, C. Arrowsmith, J. R. Yates, 3rd and W. E. Balch, *Mol. Biol. Cell*, 2010, **21**, 871–884.
75. X. Wang, J. Venable, P. LaPointe, D. M. Hutt, A. V. Koulov, J. Coppinger, C. Gurkan, W. Kellner, J. Matteson, H. Plutner, J. R. Riordan, J. W. Kelly, J. R. Yates, 3rd and W. E. Balch, *Cell*, 2006, **127**, 803–815.
76. E. Meimaridou, S. B. Gooljar, N. Ramnarace, L. Anthonypillai, A. J. Clark and J. P. Chapple, *Mol. Endocrinol.*, 2011, **25**, 1650–1660.
77. J. L. Holmes, S. Y. Sharp, S. Hobbs and P. Workman, *Cancer Res.*, 2008, **68**, 1188–1197.
78. L. Swick and G. Kapatos, *J. Neurochem.*, 2006, **97**, 1447–1455.
79. F. Ran, N. Gadura and C. A. Michels, *J. Biol. Chem.*, 2010, **285**, 13850–13862.
80. F. Sun, Z. Mi, S. B. Condliffe, C. A. Bertrand, X. Gong, X. Lu, R. Zhang, J. D. Latoche, J. M. Pilewski, P. D. Robbins and R. A. Frizzell, *FASEB J.*, 2008, **22**, 3255–3263.
81. X. Y. Zhong, J. H. Ding, J. A. Adams, G. Ghosh and X. D. Fu, *Genes Dev.*, 2009, **23**, 482–495.
82. L. Sun, T. Prince, J. R. Manjarrez, B. T. Scroggins and R. L. Matts, *Biochim. Biophys. Acta*, 2012, **1823**, 1092–1101.
83. D. F. Nathan, M. H. Vos and S. Lindquist, *Proc. Natl Acad. Sci. USA*, 1999, **96**, 1409–1414.
84. H. Armstrong, A. Wolmarans, R. Mercier, B. Mai and P. Lapointe, *PLoS One*, 2012, **7**, e49322.
85. B. Jimenez, F. Ugwu, R. Zhao, L. Orti, T. Makhnevych, A. Pineda-Lucena and W. A. Houry, *J. Biol. Chem.*, 2012, **287**, 5698–5709.
86. S. H. Millson, C. K. Vaughan, C. Zhai, M. M. Ali, B. Panaretou, P. W. Piper, L. H. Pearl and C. Prodromou, *Biochem. J.*, 2008, **413**, 261–268.
87. K. Eckert, J. M. Saliou, L. Monlezun, A. Vigouroux, N. Atmane, C. Caillat, S. Quevillon-Cheruel, K. Madiona, M. Nicaise, S. Lazereg, A. Van Dorsselaer, S. Sanglier-Cianferani, P. Meyer and S. Morera, *J. Biol. Chem.*, 2010, **285**, 31304–31312.
88. Y. Kakihara and W. A. Houry, *Biochim. Biophys. Acta*, 2012, **1823**, 101–107.
89. M. Zhang, Y. Kadota, C. Prodromou, K. Shirasu and L. H. Pearl, *Mol. Cell*, 2010, **39**, 269–281.
90. M. Ruff, M. Gangloff, J. M. Wurtz and D. Moras, *Breast Cancer Res.*, 2000, **2**, 353–359.

91. D. J. Kojetin, T. P. Burris, E. V. Jensen and S. A. Khan, *Endocr. Relat. Cancer*, 2008, **15**, 851–870.
92. S. H. McLaughlin, F. Sobott, Z. P. Yao, W. Zhang, P. R. Nielsen, J. G. Grossmann, E. D. Laue, C. V. Robinson and S. E. Jackson, *J. Mol. Biol.*, 2006, **356**, 746–758.
93. M. Zhang, M. Windheim, S. M. Roe, M. Peggie, P. Cohen, C. Prodromou and L. H. Pearl, *Mol. Cell*, 2005, **20**, 525–538.
94. R. M. Hofmann and C. M. Pickart, *Cell*, 1999, **96**, 645–653.
95. P. Connell, C. A. Ballinger, J. Jiang, Y. Wu, L. J. Thompson, J. Hohfeld and C. Patterson, *Nat. Cell Biol.*, 2001, **3**, 93–96.
96. S. H. McLaughlin, H. W. Smith and S. E. Jackson, *J. Mol. Biol.*, 2002, **315**, 787–798.
97. T. O. Street, L. A. Lavery, K. A. Verba, C. T. Lee, M. P. Mayer and D. A. Agard, *J. Mol. Biol.*, 2012, **415**, 3–15.
98. S. F. Harris, A. K. Shiau and D. A. Agard, *Structure*, 2004, **12**, 1087–1097.
99. F. Hagn, S. Lagleder, M. Retzlaff, J. Rohrberg, O. Demmer, K. Richter, J. Buchner and H. Kessler, *Nat. Struct. Mol. Biol.*, 2011, **18**, 1086–1093.
100. M. Mollapour and L. Neckers, *Biochim. Biophys. Acta*, 2012, **1823**, 648–655.
101. M. Mollapour, S. Tsutsumi, A. C. Donnelly, K. Beebe, M. J. Tokita, M. J. Lee, S. Lee, G. Morra, D. Bourboulia, B. T. Scroggins, G. Colombo, B. S. Blagg, B. Panaretou, W. G. Stetler-Stevenson, J. B. Trepel, P. W. Piper, C. Prodromou, L. H. Pearl and L. Neckers, *Mol. Cell*, 2010, **37**, 333–343.
102. A. Martinez-Ruiz, L. Villanueva, C. Gonzalez de Orduna, D. Lopez-Ferrer, M. A. Higueras, C. Tarin, I. Rodriguez-Crespo, J. Vazquez and S. Lamas, *Proc. Natl Acad. Sci. USA*, 2005, **102**, 8525–8530.
103. X. Yu, Z. S. Guo, M. G. Marcu, L. Neckers, D. M. Nguyen, G. A. Chen and D. S. Schrump, *J. Natl Cancer Inst.*, 2002, **94**, 504–513.
104. J. J. Kovacs, P. J. Murphy, S. Gaillard, X. Zhao, J. T. Wu, C. V. Nicchitta, M. Yoshida, D. O. Toft, W. B. Pratt and T. P. Yao, *Mol. Cell*, 2005, **18**, 601–607.
105. B. T. Scroggins, K. Robzyk, D. Wang, M. G. Marcu, S. Tsutsumi, K. Beebe, R. J. Cotter, S. Felts, D. Toft, L. Karnitz, N. Rosen and L. Neckers, *Mol. Cell*, 2007, **25**, 151–159.
106. S. M. Roe, C. Prodromou, R. O'Brien, J. E. Ladbury, P. W. Piper and L. H. Pearl, *J. Med. Chem.*, 1999, **42**, 260–266.
107. S. H. Millson, C. S. Chua, S. M. Roe, S. Polier, S. Solovieva, L. H. Pearl, T. S. Sim, C. Prodromou and P. W. Piper, *FASEB J.*, 2011, **25**, 3828–3837.
108. C. Prodromou, J. M. Nuttall, S. H. Millson, S. M. Roe, T. S. Sim, D. Tan, P. Workman, L. H. Pearl and P. W. Piper, *ACS Chem. Biol.*, 2009, **4**, 289–297.

CHAPTER 3

Structure and Function of Hsp70 Molecular Chaperones

EUGENIA M. CLERICO*[a] AND LILA M. GIERASCH*[a,b]

[a] Department of Biochemistry and Molecular Biology, University of Massachusetts, Amherst, MA 01003, USA; [b] Department of Chemistry, University of Massachusetts, Amherst, MA 01003, USA
*Email: eclerico@biochem.umass.edu; gierasch@biochem.umass.edu

3.1 Introduction

The 70-kDa heat-shock proteins (Hsp70s) constitute a family of ubiquitous and conserved molecular chaperones that perform a wide-ranging group of cellular functions and play crucial roles in cellular homeostasis.[1,2] They are found in most kingdoms of life from bacteria to humans,[3] and sequence and structural similarities among Hsp70s make them one of the most conserved families in Nature.[4,5] Hsp70s, with the help of their co-chaperones, are involved in numerous cellular processes where they function as "holdases" by binding non-native proteins and transiently protecting them from aggregation, subsequently releasing them to fold or participate in downstream interactions.

In eukaryotes some Hsp70s are expressed at relatively high levels under normal physiological conditions and perform housekeeping functions, whereas others become highly over-expressed only under many forms of stress, when they have irreplaceable functions.[6] Interestingly, tumor cells are exposed to diverse unfavorable metabolic and environmental conditions like oxidative stress, low nutrient and oxygen availability, unregulated growth, *etc.*, and concomitantly they express constant high levels of inducible Hsp70s.[7–10]

Hsp70s help tumor cells by protecting their cellular proteome under conditions that would otherwise cause many proteins to unfold, by overcoming high mutational loads and helping mutant proteins to reach their native state, and by inhibiting cell death pathways.[11] Hsp70s are therefore indispensable for tumor cells to thrive, and in recent years there has been an increase in possible strategies to specifically inhibit inducible Hsp70s in cancer. The protective roles of Hsp70 are also implicated in neuropathology. Many neurodegenerative processes are associated with the deposit of insoluble misfolded proteins inside or outside the cells, and it is believed that these aggregates interfere with intracellular transport and metabolism, and may activate apoptosis.[12,13] Increased expression of Hsp70 has been shown to lead to the reduction of the size and number of certain aggregates as well as improved cell viability (reviewed in[14,15]).

In this chapter we first discuss the structures of representative Hsp70s and what is known about their mechanisms. The architecture and mechanism of action of these proteins are highly conserved among members of the Hsp70 family. The *E. coli* Hsp70 DnaK shares 47% sequence identity with the human cytosolic HspA1.[5] This striking sequence conservation of these proteins allows us to extend to other Hsp70s the enormous progress made in structural and biochemical studies of DnaK.[2,4] A comprehensive review on the allosteric mechanism of DnaK has recently been published.[16] Here we will focus on advances in understanding the structure and mechanism of Hsp70s made since this latter review was published: for example, the structure of the ATP-bound state of DnaK[17,392] and the characterization of the "allosterically active" state in the DnaK cycle.[18] We will also provide an overview of current knowledge about the Hsp70 co-chaperones: J proteins and nucleotide exchange factors ("NEF") and how they co-operate with Hsp70s to recruit them to specific cellular locations and confer substrate specificity. Because greater detail is beyond the scope of this review, we will only describe the major co-chaperone groups and their main characteristics and refer the reader to more comprehensive recent reviews.

Finally, we will describe the wide array of cellular functions of Hsp70s. As mentioned above, Hsp70s are involved in numerous critical cellular processes, and with each passing day new functions are attributed to members of the Hsp70 family. Consistent with the participation of Hsp70s in this wide variety of functions, this family of proteins is implicated in many diseases including neurodegeneration, cancer and infection, as previously reviewed.[14] We do not attempt here to provide an encyclopedic compilation of the vast literature on cellular functions of Hsp70s; instead we focus our discussion on the general categories of key Hsp70 functions in cellular processes, in particular those that are most relevant to the potential use of these chaperones as therapeutic targets. More detailed information can be found in the references cited herein and in subsequent chapters of this book.

3.2 General Background on Hsp70 Chaperones

Hsp70s are molecular chaperones of approximately 70 kDa that are involved in numerous cellular processes where they function by binding hydrophobic sequences that are exposed to the solvent in non-native proteins and transiently protect them from aggregation by binding and release. Once freed from the chaperone, the substrate protein may refold with or without the help of a downstream chaperone or, if unable to fold in a timely fashion, may re-bind Hsp70, be transferred to a downstream chaperone or be targeted for degradation. Substrates of Hsp70 include proteins that are newly synthesized, stress-denatured or aggregated, targeted for degradation, or in complexes to be assembled, as well as pre-proteins to be transported to subcellular compartments.[19,20] Remarkably, once proteins have established non-native interactions, Hsp70 can, also by binding and release, locally undo the offending interactions in the protein substrate and give it a new opportunity to fold.[19,21] In all cases, Hsp70s do not provide structural information to the folding mechanism but improve client folding yield by preventing aggregation.[21–23] Additionally, Hsp70s, in co-operation with Hsp90, Hsp60s and specific co-chaperones, play key roles in signal transduction and cell cycle regulation by modulating the biological activity of regulatory proteins.[19,24]

Hsp70 proteins are composed of two domains and function by allosterically passing a signal from one domain to the other through a conserved interdomain linker. The C-terminal Hsp70 substrate-binding domain ("SBD") transiently binds short, solvent-accessible hydrophobic sequences.[6,16,25] At the N-terminal nucleotide-binding domain ("NBD"), regulated cycles of ATP hydrolysis alter the SBD affinity for substrates, causing them to be alternately bound and released from the SBD. At their extreme C-terminus, most Hsp70s contain a structurally disordered domain that incorporates a conserved motif for interaction with tetratricopeptide repeat (TPR) domains. This motif is responsible for binding a variety of co-chaperones including carboxy terminus of Hsc70-interacting protein (CHIP) and Hsc70-/Hsp90-organizing protein (Hop), thus mediating the formation of Hsp70 complexes.[19]

Although Hsp70s can hydrolyze ATP and bind/release substrates, the efficiency of the allosteric cycle is greatly enhanced by co-chaperones of the J proteins family, which accelerate the Hsp70 ATPase rate, and diverse NEFs, which facilitate the exit of ADP to allow ATP re-binding.[20] Other cofactors partner with Hsp70s to further confer specificity for substrates or couple their functions with other cellular systems like Hsp90, the proteasome *etc.* (see below). The diversity of functions that Hsp70s perform is accomplished by amplification and evolution of the *hsp70* genes, diversity of the co-chaperones that enhance the specificity of the Hsp70 functions and partnership with other chaperone systems.[19]

As mentioned above, eukaryotes contain multiple Hsp70s. While some are constitutively expressed at relatively high levels under physiological conditions for housekeeping functions, others are highly over-expressed under cellular stress.[6] Increased expression of the inducible Hsp70s is achieved by triggering a "heat-shock response". The heat-shock response is an essential and very conserved mechanism used by a wide range of organisms to up-regulate stress proteins (Hsp70 among them) that help the cell survive under harsh conditions.[26,27] In eukaryotes, induced expression of Hsp70 is mediated at the transcriptional level by a complex regulatory mechanism that involves DNA sequences called "heat-shock elements" that are present upstream of the Hsp genes, and by the heat stress factor ("Hsf") proteins. In the absence of stress, Hsf1 is constitutively expressed but maintained in an inactivate state by binding to Hsp90, Hsp70 and Hsp40. Upon stress, denatured proteins compete Hsp70 away from Hsf1, which then becomes activated by release from the chaperones and also by post-translational modifications (such as phosphorylation and sumoylation[26]) that convert inactive monomeric Hsf1 to high-affinity DNA-binding trimers; Hsf1 enters the nucleus where it binds to the heat-shock promoter elements and promotes transcription of a monocistronic HspA1 mRNA[5,28] and other heat-shock genes.[29] Hsf1 transcriptional activity is eliminated by negative feedback from Hsps, which repress the transactivation of DNA-bound Hsf1. Also, Hsf1 modulates the activity of histone deacetylases, which in turn govern gene expression.[30] In addition to stress-induced transcription of heat-shock genes, the heat-shock response is regulated by other mechanisms, such as mRNA stability, stress-induced translational control and effects on the activity and subcellular localization of chaperones. Accordingly, the heat-shock response also decreases numerous basal functions during stress and recovery phases.[31–34] For example, early ribosomal elongation pausing is induced by the sequestration of chaperone molecules (specifically Hsc70) by misfolded proteins, and this effect is induced by direct inhibition of Hsc/HspA1, whereas an increase in Hsc70 availability restores normal translation efficiency.[35] At the organismic level, these cellular regulations have profound effects on growth and development.[33] Tumor cells are exposed to diverse unfavorable metabolic and environmental conditions like oxidative stress, low nutrient and oxygen availability and unregulated growth. Cancer cells have activated Hsf1 and they maintain the constant high levels of inducible Hsp70s,[7,36,37] which are essential for tumor cell survival.

3.3 Classes of Human Hsp70s

The human genome contains more than 17 *hsp70* genes and about 30 *hsp70*-related gene sequences that contain frame shifts, in-frame stops and other features that make them pseudo genes[38,39] The nomenclature and prominent features of the human Hsp70 genes have been extensively described.[38,39] In this section, we will briefly summarize what has been learned from *hsp70* sequence analysis.

Twelve distinct proteins were identified as true Hsp70s: HspA1A and HspA1B (encoded by separate genes but the same protein), Hsp1L, HspA2, HspA5, HspA6, HspA7 (sometimes considered a pseudo gene), HspA8, HspA9, HspA12A, HspA12B, HspA13 and HspA14 (Table 3.1). Some Hsp70s are expressed in the cytosol and nucleus, while others are restricted to cellular compartments (Table 3.1). Also, some Hsp70s are differentially expressed in some tissues and at specific developmental stages.[38] Hsp70-encoding genes and pseudo genes are distributed over all chromosomes except for 15, 16, 17, 19 and 22. HspA1A, HspA1B and HspA1L are encoded in close proximity on chromosome 6, while *hspA6* and *hspA7* are close to one another on

Table 3.1 Names and major properties of the human Hsp70 proteins. Based on Kampinga et al.[39] and Brocchieri et al.[38]

Gene	Protein	Other names	Cellular compartment	Expression
hspA1A	HspA1A	Hsp70-1; Hsp72; HspA1	Cyt/Nuc	S I
hspA1B	HspA1B	Hsp70-2	Cyt/Nuc	S I
hspA1L	Hsp1L	hHsp70t; Hum70t; Hsp-hom	Cyt/Nuc Spermatida	C
hspA2	HspA2	Heat-shock 70 kDa protein-2	Nuc	C
hspA5	HspA5	BiP; Grp78; Mif2	ER	S I (ER)
hspA6	HspA6	Hsp70B'	Cyt/Nuc	S I
hspA7*	HspA7	Heat-shock 70 kDa protein-7	U	U
hspA8	HspA8	Hsc70; Hsc71; Hsp71; Hsp73	Cyt/Nuc/Surface of embryonic stem cells	I S
hspA9	HspA9	Grp75; HspA9B; Mot; Mot2; PbP74; mot-2; mortalin; mtHsp70	Mit, ER, Cyt, vesicles, membrane surface	C
hspA12A	HspA12A	FLJ13874; KIAA0417	U	C and S I
hspA12B	HspA12B	RP23-32L15.1; 2700081N06Rik	U	C and S I
hspA13	HspA13	Stch	U	U
hspA14	HspA14	Hsp70-4; Hsp70L1; MGC131990	Cyt, ribosome associated	U

*This gene is sometimes annotated as a gene and a pseudogene. Cyt: Cytosol; Nuc: nucleus; U: Undetermined; C: Constitutive expression; S I: Stress Inducible; ER: Endoplasmic reticulum. Mit: Mitochondria.

chromosome 1; all other Hsp70 genes are distributed in the genome irrespective of their evolutionary relations.[38]

Brocchieri et al.[38] constructed a phylogenetic tree of the human hsp70 genes based on a multiple sequence alignment of the Hsp70 proteins. NBD is strongly evolutionarily conserved and thus constitutes a parameter to recognize members of the Hsp70 family, while the SBD is significantly less conserved.

It is not certain whether HspA12A and HspA12B are related to other Hsp70s. They have a more divergent NBD than other Hsp70s, and their SBDs do not share any similarities with other Hs70s. They are constitutively expressed and inducible by stress. The mitochondrial protein mtHsp70 (HspA9B) constitutes a different evolutionary group. It is found primarily in mitochondria and is constitutively expressed. HspA14 (aka HspAL1, as the only member of a different evolutionary group) is part of the mammalian ribosome-associated complex (mRAC). mRAC consist of Hsp70L1 and a ribosome-interacting J domain protein (MPP11).[40] This complex participates in avoiding aggregation of ribosome nascent chains (see below). HspA13 or Stch is the only member of another separate group. Consistent with the fact that Stch has a signal peptide for translocation to the ER, it has been isolated from microsomes, but unlike HspA5 (aka BiP), the main ER-resident Hsp7, Stch does not possess an ER retention sequence. Stch contains an insertion in the NBD that is associated with interactions with the ubiquitin-system.

Cytosolic Hsp70 proteins include three subgroups, one consisting of HspA1A ("HspA1"), HspA1B (identical proteins, two different genes) and HspA1L (only expressed in spermatids), a second including HspA8 (Hsc70) and HspA2, and the last including HspA6 and HspA7. HspA1, HspA1B, HspA1L, Hsc70 and HspA2, along with the mitochondrial and the ER-resident Hsp70 proteins, are considered "typical" Hsp70 proteins because they contain conserved NBD and SBDs[38] and have been the best characterized proteins of the human Hsp70 family.

While some of the Hsp70 gene expansion may have arisen to meet the need for Hsp70s in all cellular compartments and at different developmental stages, many Hsp70 proteins co-exist in the cytosol and nucleus at the same time, contributing to the functional specificities among the cytosolic isoforms.[41]

Some of the human Hsp70 family members are separated by differential spatio-temporal expression, while others are present in the same compartment at the same time, often showing overlapping functions.[41] However, it has been proposed that a dynamic network of chaperones and co-chaperones evolved to take advantage of specific functions of the many Hsp70 members, implying that different Hsp70s could have distinct substrate and/or co-chaperone preferences.[41,42]

The most intensively studied members of the Hsp70 family in the cytosol of human cells are HspA1 (from now on "HspA1" refers to HspA1A) and Hsc70 (for heat-shock cognate protein). Hsc70 is the most abundant Hsp70 in the cell and is constitutively expressed. On the other hand, HspA1 expression levels are very low under physiological conditions, and its expression is highly augmented under stress, as also observed for HspA6, although these two proteins have different functions in the cell.[42] HspA1 and Hsc70 are 86% identical, with their disordered C-termini and SBD α-helical lid (see below) showing the greatest variation.

Both similarities and differences have been noted between the mechanisms of the two cytoplasmic Hsp70s. For example, both are able to promote refolding of the model substrate luciferase *in vitro* using the same set of co-chaperones, Hdj1 and Bag-1.[43] Also, Hsp70 and Hsc70 exert opposite effects on the trafficking of an epithelial sodium chloride channel in *Xenopus* oocytes: HspA1 promotes its maturation whereas Hsc70 interferes with it.[44] The NEF HspBP1 binds more tightly *in vivo* to HspA1 than to Hsc70,[45] while other co-chaperones, like Bag-1M, act as a substrate-discharging factor for both Hsc70 and HspA1. In the clathrin-uncoating system, a specific peptide sequence in the clathrin substrate and another in the co-chaperone auxilin seem to be exclusively used by Hsc70.[46] Importantly, methylene blue abrogates HspA1 ATPase activity by oxidation-induced conformation changes that block ATP binding, while Hsc70 is entirely resistant to this inhibition, as Hsc70 lacks one of the two Cys that methylene blue can modify in HspA1.[47]

If Hsp70 inhibition were chosen as an anticancer strategy, it would be crucial to understand the mechanistic differences between HspA1 and Hsc70, as it would be desirable to interfere with the pro-survival functions of the stress-induced HspA1 without affecting the housekeeping functions of Hsc70.

3.4 Hsp70 Structure

As we pointed out above, the architecture and mechanism of action of Hsp70s proteins are conserved among members of the Hsp70 family. Consequently, we can apply to other members of the Hsp70 family the extensive information gathered over the last several years about the structure and mechanism of action of the *E. coli* Hsp70 DnaK.[2–4]

Crystal and NMR based structures have been solved for several members of the Hsp70 family in different nucleotide states, primarily as individual NBDs. Two important two-domain structures are available for the *E. coli* DnaK: an NMR model that depicts the ADP/substrate-bound structure (PDB: 2KHO[48]) and the long-awaited crystal structures of the ATP-bound state (PDB: 4B9Q[17] and 4JN4[392]). Previously, a structure of bovine apo-Hsc70 had revealed comparable domain architecture.[49] Hsp70 proteins are organized in three distinct domains: NBD, SBD and the disordered C-terminal domain, which does not appear in the available crystal or NMR structures. Both the SBD and NBD can be further divided into subdomains (Figure 3.1).

3.4.1 The Hsp70 NBD

The N-terminal ~45-kDa NBD of Hsp70s belongs to the actin/hexokinase/Hsp70 superfamily.[50] The NBD is comprised of two lobes, I and II, and each lobe consists, in turn, of two subdomains: IA and IB for lobe I, and IIA and IIB for lobe II. Nucleotides bind at the bottom of the central, deep interlobe cleft between subdomains IB and IIB, and all four subdomains are involved in nucleotide coordination. The α-phosphate and β-phosphate of the nucleotide interact with subdomain IIA, and the γ-phosphate and Mg^{2+} are coordinated (with help of a K^+ ion) primarily by contacts with subdomain IA (Figure 3.1C and 3.2).

The crystal structures of the NBD in different nucleotide states and in complex with NEFs, as well as results of NMR analysis[51–54] and molecular dynamics (MD) calculations,[55] revealed conformational flexibility that arises largely from subdomain reorientations in different nucleotide-bound states of the NBD. These conformational rearrangements are responsible for the allosteric signals that are transmitted to the SBD to cause higher or lower affinity for substrate, and consequent substrate binding or release (see below). For example, the structures of NBDs in complex with NEFs (shown in Figure 3.5: PDB entries 3FZF[56] and 3D2F[57]) show opening of the nucleotide-binding pocket relative to other NBD structures, which facilitates nucleotide exchange[57–60] (Figure 3.2).

The NBD exists as an ensemble of conformations that interconvert *via* subdomain motions.[53] NMR chemical shift perturbation analysis enabled the identification of residues within the NBD that together constitute an allosteric network, which transmits the signal from nucleotide binding to the SBD (Figure 3.2).[54] The residues in the network link the nucleotide-binding site to a hydrophobic pocket below the crossing helices that is a conserved feature of the actin-fold family and plays a role in interaction with partner proteins.[61]

Here, this pocket is key to communicating the nucleotide state of the NBD, and is the site of docking of the interdomain linker, as well as serving as a binding site for some co-chaperones.

3.4.2 The Hsp70 Interdomain Linker

The NBD is connected to the SBD *via* an interdomain linker about ten residues in length that includes the consensus, largely hydrophobic sequence D/E V/I/ LLLDV*P[16,62] (in *E. coli* DnaK, "VLLL"). This sequence is highly conserved and intolerant to mutation; substitution particularly of the hydrophobic residues results in loss of DnaK chaperone function.[63,64] The interdomain linker is implicated in the transmission of the allosteric signal between the Hsp70

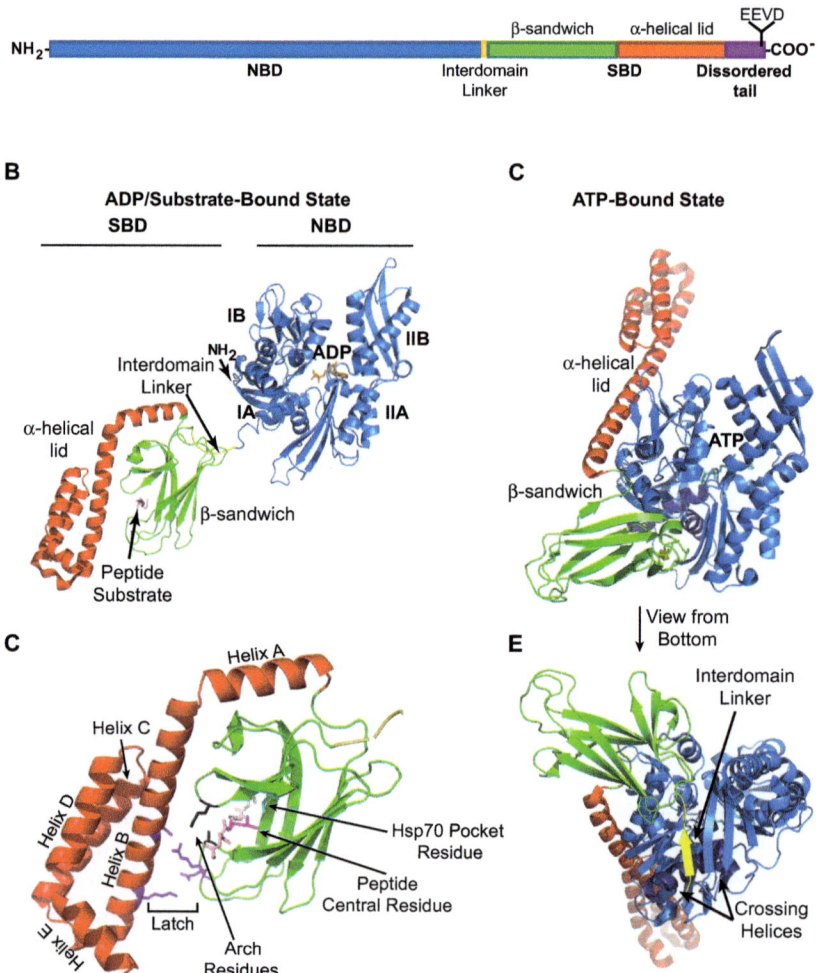

domains. Indeed, a DnaK NBD construct that includes the hydrophobic sequence of the linker (1-392) displays ATPase activity elevated to the same extent as seen in the full-length protein in the presence of substrate (8 fold), whereas one without this portion of the linker (1-388) has ATPase activity comparable to the basal, substrate-free full-length protein.[52,62] Importantly, the linker changes both conformation and environment depending on the nucleotide/substrate state of the protein: it is flexibly exposed to the solvent and connects the relatively free-moving NBD and SBD in the absence of ATP, and is protected from the solvent, specifically docked into the hydrophobic pocket beneath the crossing helices, in the ATP-bound NBD (as shown by proteolysis,[65] hydrogen exchange,[66] NMR techniques[18,48,52,54] and X-ray crystallography,[17] Figure 3.1 B and C).

3.4.3 The Hsp70 SBD

The ~27 kDa SBD of Hsp70s is composed of a β-sandwich subdomain, an α-helical subdomain and the unstructured C-terminal subdomain (see below). The β-sandwich is organized in two sheets, each containing four antiparallel β-strands.[25,67] This array forms the hydrophobic pocket where substrate binds.[25] The bottom sheet (strands 3, 6, 7 and 8) makes up a twisted β-sheet with hairpin turns connecting the strands. The top sheet (strands 5, 4, 1 and 3)

Figure 3.1 General structure of Hsp70. All structures were plotted using Pymol (Molecular Graphics System, Schrödinger, LLC). The color code for the domains and subdomains is the same for all panels of the figure: blue: NBD; yellow: interdomain linker; red: α-helical subdomain; green: β-sandwich subdomain; pink: peptide substrate; orange: ADP; cyan: ATP; purple: disordered C-terminal domain (only depicted in panel A). In panels B and C, the SBD is shown on the left, and NBD on the right for best visualization of the structural features. **A.** Cartoon representation of the Hsp70 sequence (*E. coli* DnaK). The three Hsp70 domains are labeled, along with the SBD subdomains. The EEVD sequence is the conserved site for binding of TPR-containing proteins in eukaryotes. **B.** Illustration of DnaK structure in the ADP and substrate bound form (PDB: 2KHO[48]). The NBD subdomains are labeled. Substrate was added into the 2KHO structure from the PDB entry 1DKX[25] and ADP from the structure of the HspA1 NBD (PDB ID: 3ATU[393]). **C.** ATP-bound state of DnaK (PDB: 4B9Q[17]). **D.** Structure of the SBD of *E. coli* DnaK with bound peptide (PDB entry 1DKX[25]). The central residue in the peptide that contacts the pocket residue on Hsp70 (shown in light blue) is depicted in magenta sticks; in black sticks are shown the residues that surround the peptide forming the "hydrophobic arch" (Met 404 and Ala 429 in DnaK);[68] the residues that form the "latch" between the Helix B of the α-helical lid and the β-sandwich are shown in purple sticks. Helices A to E in the α-helical subdomain are labeled. D. Structure of ATP-bound DnaK (same as in panel C) seen from the bottom of the NBD, where the structure adopted by the interdomain linker in the NBD is shown. Behind the docked linker, the crossing helices that lie at the base of the pocket that harbors the interdomain linker are showed in dark blue.

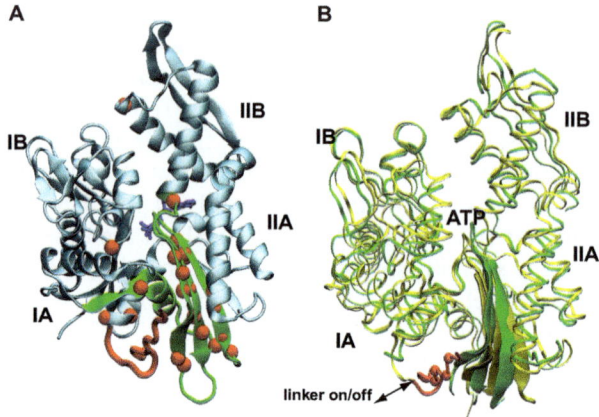

Figure 3.2 ATP-induced conformational change in the NBD and linker binding. **A.** Model structure of the ATP-Bound NBD of *E. coli* where the regions marked in green depict the residues that show the chemical-shift differences between the ADP and ATP-bound states. The residues that showed large chemical-shift perturbations are shown as red spheres. **B.** Superimposed models of the ADP- (green) and ATP- (yellow) bound conformations. For the ATP-bound conformation, the C-terminal 12 residues are shown in red, and the interdomain linker bound to the hydrophobic cleft is shown as red spheres.
© 2011 National Academy of Sciences. Reproduced from Zhuravleva and Gierasch, 2011.[54]

is rather irregular, and its irregularity is related to the loops that protrude from the structure and help to enclose the substrate in its binding site (see below and Figure 3.1 B and D).

Connected to the β-sandwich structure of the SBD by a short linker is the 15 kDa α-helical lid. The helical lid subdomain comprises five α-helices (from A to E) in a bundle that constitutes a stable entity with a defined hydrophobic core (ref. 25 and PDB entry 3LOF) (Figure 3.1B and D). Helices A and B act as a true lid for the β-sandwich subdomain, with Helix B spanning over the β-subdomain and acting as a shield that stabilizes the substrate in its binding pocket.[25] But the helices do not contact the substrate directly; instead, Helix B establishes latch-like salt-bridge interactions with the distal part of the β-sandwich to sequester the bound substrate (Figure 3.1D). The α-helical lid shows a lower degree of conservation among members of the Hsp70 family than the NBD.

3.4.4 Hsp70 Substrate Binding

The crystal structure of the DnaK SBD bound to a model substrate peptide was solved in 1996 and provided insight into the mechanism by which Hsp70 chaperones interact with their substrates.[25] Specifically, strands 3 and 4 (that flank loop $L_{3,4}$) and strands 5 and 6 (flanking loop $L_{5,6}$) of the β-sandwich participate in substrate binding (Figure 3.1 D). The peptide-binding cavity is formed by several key structural elements: (a) the deep central binding hydrophobic pocket (called "position 0") that binds the single essential hydrophobic residue of the substrate. In DnaK, Ile 401 makes contact with the central position of the bound substrate (in the case of the model peptide NRLLLTG used for the crystal structure, the central Leu 4) in a pocket well protected from the solvent[25] (Figure 3.1D). This position is very conserved in the Hsp70 family and provides the main contribution to the binding affinity for substrates; (b) a hydrophobic "arch" formed by two hydrophobic residues of DnaK (Met 404 and Ala 429, see Figure 3.1D) that contact the peptide substrate. The hydrophobic arch contributes to substrate specificity (and, to a lesser extent, affinity), and the arch-forming residues are the only residues found both to make direct contact with the substrate and also to show substantial evolutionary variation in the Hsp70 family. Also, these residues were proposed to contribute to functional specialization of Hsp70s;[68] (c) the α-helical lid, which does not contact the substrate directly but helps cover the pocket and hence slows the peptide dissociation rate;[25] Helix B of the lid subdomain closes the cavity by establishing a salt bridge and two hydrogen bonds to outer loops $L_{3,4}$ and $L_{5,6}$ (Figure 3.1D). Consistent with this analysis, truncated SBD constructs lacking the helical domain show a highly dynamic β-sandwich domain and a significant decrease in substrate-binding affinity, primarily due to higher substrate dissociation rates.[69–71]

Substrate peptides bind to Hsp70 in an extended conformation in the β-sandwich (Figure 3.1 B and D).[25,72,73] DnaK binds sequences that normally would be buried in the protein interior, but are exposed to the solvent in their non-native or partially folded substrates. For DnaK and BiP, the preferred binding motifs were determined using the peptide array method for DnaK peptides[74] and phage display[75] and synthetic labeled peptides for BiP.[76] The DnaK-binding peptide is composed of a hydrophobic core of four to five amino acids enriched particularly in L, but also in I, V, F and Y, flanked by two regions rich in basic residues and poor in negatively charged residues.[74] Preferred DnaK-binding sequences are predicted to happen on average every 36 residues.[77] The interaction between DnaK and peptide requires that peptide substrate be positioned with the peptide N-terminus close to strand 3 and the C-terminus close to strand 4. By contrast, the highly specialized bacterial Hsp70 HscA, which plays a role in iron-sulfur center assembly, binds a substrate peptide in the reverse orientation from that observed for DnaK and rat Hsc70.[78] For BiP binding, the peptide-binding motif shows a significant enrichment in hydrophobic amino acids alternating with aromatic amino acids (Trp and Phe).[75] Although the binding motif is argued to be similar for all Hsp70s,[74] there are differences that may be important for specificity and/or affinity. For example, the peptides NRLLLTG and NDLLLTG are both well bound by DnaK (albeit with different affinities), but they bind BiP only weakly.[79] A recent report showed that DnaK and BiP bound to the same sequences with comparable affinities but different kinetics of binding.[73]

3.4.5 The Disordered Hsp70 C-terminal Domain

The extreme C-terminal region of Hsp70 proteins is dynamic and comprises about 30 unstructured residues.[48] In eukaryotes, the conserved EEVD at the C-terminus is essential for interaction with TPR-containing proteins, like CHIP to target substrates for degradation, or Hop to mediate co-operation with Hsp90s. This interaction site is conserved in the sequences of most human Hsp70s, but it is not present in BiP, mtHsp70, Hsp70L1, HspA12A or HspA12B.

This disordered C-terminal domain was speculated to be involved in the association of Hsc70 with protein substrates.[80,81] Interestingly, this domain is also present in most bacterial Hsp70s, even though bacteria lack homologs of Hsp70-interacting TPR domain proteins. In *E. coli*, specific elements in this disordered C-terminus were shown to enhance protein refolding activity and physiological chaperone function in the cell by providing a weak and non-essential secondary substrate-binding site.[82]

3.5 Hsp70 Allosteric Mechanism

In order for Hsp70s to function, the signals from nucleotide binding or hydrolysis in the NBD must be transmitted to the SBD, where they favor substrate binding or release, and, in turn, the ATP hydrolysis rate of the protein dramatically increases in the presence of substrate. As the main features of the Hsp70 allosteric mechanism have been recently reviewed by Zuiderweg *et al.*,[16] we will briefly describe the mechanism and add the major contributions to the field that were reported in the last year.

The conformational transitions that Hsp70s undergo during their allosteric cycle allow this chaperone to bind substrates and release them in an ATP-dependent fashion. Allosteric changes are reversible and regulated by the availability and affinity of substrates, the ATP/ADP balance and co-chaperone interactions (Figure 3.3).

When ATP is bound to the NBD, the SBD is intimately packed against the NDB and adopts an "open" conformation with very low to no affinity for substrates; the interdomain linker is buried into the NBD at the hydrophobic pocket beneath the crossing helices. This state constitutes the ATP-bound or "docked" conformation, and structural details for this state were recently revealed in two crystal structures.[17,392] In the ATP-bound state, the SBD and NBD establish an intimate interdomain contact interface: The α-helical lid of the SBD lifts off the SBD β-sandwich, and both of these subdomains form contacts with different sides of the NBD. The β-sandwich of the SBD contacts NBD subdomains IA, IB and IIA, whereas Helix A (which forms a continuous helix with Helix B) is packed against subdomain IB of the NBD. Not surprisingly, many of the residues that make up the interdomain interfaces are important for allosteric regulation.[17,18,392] The interdomain linker is "inserted" into the NBD, forming hydrophobic interactions and making hydrogen bonds with strand 13, which extends the central β sheet of subdomain IIA[17,18] (Figure 3.1E).

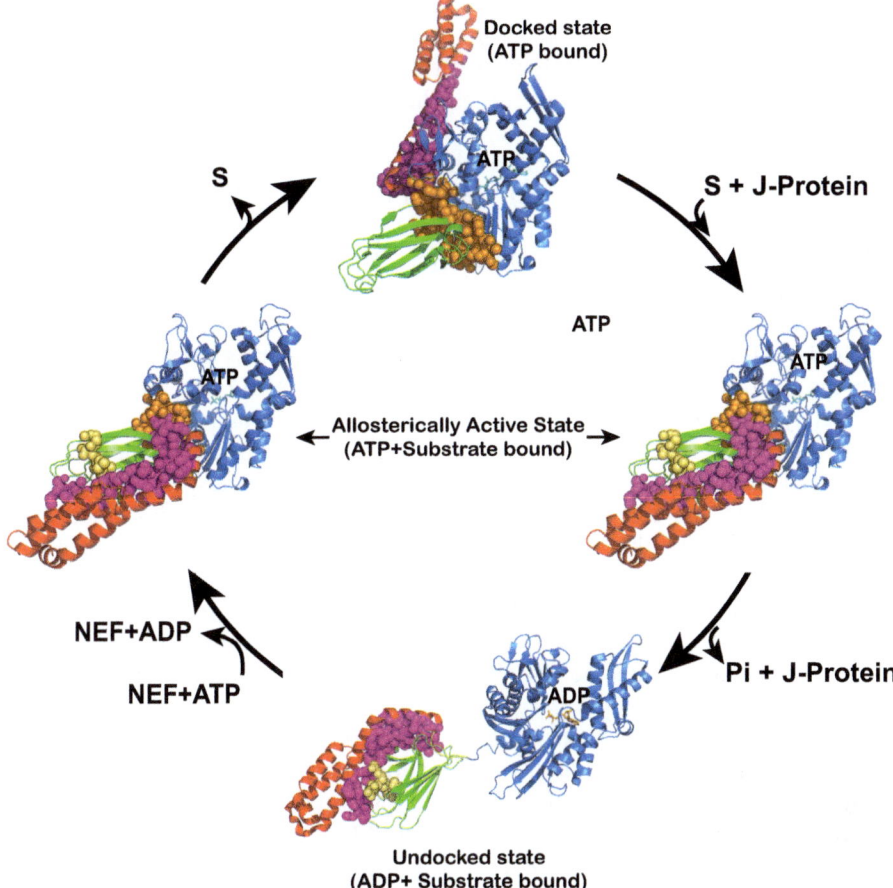

Figure 3.3 The allosteric cycle of *E. coli* Hsp70. The ADP and substrate-bound form is based on the PDB entry 2KHO,[48] where substrate was added to the structure from the PDB entry 1DKX[25] and ADP from the structure of the HspA1 NBD (PDB ID: 3ATU[393]). The intermediate state (allosterically active) is represented using a modeled structure[18] with the substrate added to the structure from the PDB entry 1DKX[25] and ATP added from the ATP-bound form of DnaK (PDB: 4B9Q[17]). The residues that form the interfaces critical for allostery are shown as spheres, in orange the interdomain interfaces and in purple the interface formed between the α-helical domain and β-sandwich of the SBD; NEF: nucleotide exchange factor and J protein: co-chaperone of the J protein family.

The presence of ATP in the nucleotide-binding site alters the orientation of the turn in subdomain IIA that is responsible for coordination of the ATP γ-phosphate, and this causes a conformational change that exposes a surface competent to bind the interdomain linker at the edge of the IIA β-sheet; linker binding to this site results in rotation of subdomain IIA relative to subdomain IA and the crossing α-helices of NBD (Figure 3.2). Thus, the presence of the

ATP γ-phosphate along with the bound interdomain linker are essential for reaching an ATPase active structure.[54] Intriguingly, the hydrophobic cleft where the interdomain linker docks overlaps with at least part of the binding site for J domain co-chaperones, which themselves modulate the rate of ATP hydrolysis.[43,83–85]

The steps in the allosteric cycle of the *E. coli* Hsp70 can be summarized as follows (Figure 3.3): when a substrate with the proper affinity is presented to the substrate-binding site of an ATP-bound DnaK the ATPase activity of the NBD is stimulated. Upon ATP hydrolysis, the Hsp70 undergoes a massive conformational change (domain uncoupling) that leads to the domain-undocked, high substrate affinity, ADP/substrate bound conformation,[48,52] in which Helix B is closed over the bound substrate and forms several salt bridges with the β-sandwich. In this ADP-bound form, the NBD and SBD are now completely undocked and independent of each other, behaving as "beads on a string". They are connected through the interdomain linker, which has been discharged from its buried position inside the NBD and is now highly dynamic and exposed to the solvent.[48,52,53] The interdomain linker is described as a "switch",[54] because linker binding to its docking site in a nucleotide-dependent fashion modulates the NBD conformation and mediates interdomain communication. The cycle completes when ADP is exchanged for ATP, with the help of an NEF, which stabilizes the "open" NBD conformation to facilitate nucleotide exchange, and the two domains dock onto one another and release the substrate.

We now know that the cycle occurs because DnaK exists as a conformational ensemble, and it is the result of an energetic competition between conformations:[18] The "undocked" and "docked" conformations of Hsp70s represent the end points of the cycle (see Figure 3.1B and C and Figure 3.3). As substrate binding enhances the rate of ATP hydrolysis by the NBD, there is a point in the cycle where both substrate and ATP must bind to an intermediate, allosterically active state. This state (named "allosterically active", "ATP/substrate bound") constitutes the tipping point between the two "end point" states, and recent work from our laboratory exploited NMR methods to reveal three conformations sampled by DnaK during the allosteric cycle.[18] The conformational ensemble populated by the allosterically active state of DnaK was obtained by promoting simultaneous binding of ATP and peptide substrate to a hydrolysis-impaired mutant (Thr 199 replaced by Ala).[54] The concomitant binding of ATP, which stabilizes linker binding to NBD and domain docking, and substrate, which favors domain dissociation and stabilizes the "closed" conformation of the SBD, creates a conformational "tug-of-war" in DnaK.[18] The activation of the ATP hydrolysis activity of DnaK by substrate binding is then explained by the action of linker binding on the NBD in the allosterically active state (without domain docking), which is necessary and sufficient to enhance ATPase activity.[18,52,62] Progressing towards ATP-hydrolysis during the cycle will depend on the affinity of the Hsp70 for the client protein.

Two orthogonal interfaces in DnaK characterize its allosteric cycle: the NBD/β-sandwich interface, and the β-sandwich/α-helical lid interface. These interfaces constitute "tunable" elements in the allosteric landscape.[18] For instance, amino acid substitutions on either the NBD/β-sandwich or the β-sandwich/α-helical lid interface will influence the allosteric properties of different Hsp70s. Despite the high conservation of these regions, there are some residue changes among different Hsp70 classes, and the conformational distribution for a particular Hsp70 will be modulated by these sequence changes.[18] In addition, the energetic balance of these orthogonal interfaces can be modulated by co-chaperone binding.

3.6 Co-chaperones

The functions of some Hsp70s are redundant, but it is known that functional specificity exists among isoforms. Importantly, in the cell Hsp70s never work alone. Hsp70s and client proteins function within a complex arrangement of protein folding networks regulated by specialized co-chaperones. The astonishing functional diversity of the Hsp70s is accomplished by the amplification and diversification of the *hsp70* genes, but also by the numerous co-chaperones and other partners that recruit and modulate Hsp70 activities.

ATP hydrolysis and nucleotide exchange reactions cause Hsp70 to cycle among conformations that dictate substrate binding and release. Even though Hsp70 is able to undergo such conformational transitions by itself, the intrinsic rate of such changes is very slow, and thus requires greater efficiency for efficient allosteric cycling. This is achieved through Hsp70 interaction with co-chaperones.

3.6.1 J proteins

"J proteins" (also called heat-shock proteins 40 or Hsp40s in eukaryotic systems, or DnaJs in bacteria) are a group of Hsp0 co-chaperones orthologues of *E. coli* DnaJ, with some highly conserved features in the family. Even though they are named as Hsp40s, most members do not have a molecular weight of 40 kDa. J proteins partner with Hsp70s and Hsp90s to perform several functions. The canonical Hsp70 co-chaperone functions of the Hsp40/DnaJ family are to accelerate ATPase activity of Hsp70, to recruit the chaperone system to specific cellular locations and to deliver substrates to their Hsp70 partners (for recent reviews see refs. 86–88). However, an increasing number of biological activities of J proteins have been found to be independent of their chaperone partners.[88] Genomic analysis uncovered 41 members of the Hsp40 family in humans, localized in many compartments and organelles (Table 3.2).[86]

All Hsp40s are by definition characterized by the presence of a "J domain", which mediates their interaction with Hsp70s: the J domain alone is sufficient to bind and stimulate the ATPase activity of partner Hsp70s: mutation of the HPD motif (see below) eliminates functional interactions;[89,90] some Hsp40 proteins only contain J domains[86] or have a J domain within an organelle, like sec63 (see below); and all Hsp40s and J proteins show high conservation only in their J domains, pointing to a conserved mechanism of interaction between

Table 3.2 Human Hsp40 proteins.[86]

Hsp40	Other names	Localization	Function	Client Binding	Features
DnaJA1	HSJ2;Dj-2;DJA1	C/N/M?	Horm. receptor maturation	P	
DnaJA2	DNAJ;DNAJ3	C/N/M	G-Protein signaling	P	
DnaJA3	TID1;hTid-1	Mit-i/C?	Signaling	P	
DnaJA4	DJ4;Hsj4	C/N/M	Folding	P	
DNAJB1	Hsp40;Hdj-1	C/N	(Re)folding	P	CTDI no ZFLR
DNAJB2a,b	Hsj1a,b	C/ER-a	Proteasomal degradation	P	Ubiquitin ID
DNAJB3	HCG3;Hsj3;Msj1	C	Folding in sperm	U	
DNAJB4	HLJ1;Hsc40	C/N	?	P	CTDI no ZFLR
DNAJB5	Hsc40;Hsp40-3	C/N	HDAC shuttling	P	CTDI no ZFLR
DNAJB6a,b	HSJ2;MRJ;MSJ1	C/N	Anti-aggregation	P	CTDI w/ HDAC BD
DNAJB7	Dj-5;mDj5;Hsc3	C/N	U	P	CTDI w/ HDAC BD
DNAJB8	mDj6	C/N	Anti-aggregation	P	CTDI w/ HDAC BD
DNAJB9	ERdj-4;Mdg1	ER-i	ERAD; folding	P	TMD
DNAJB11	ERdj-3;HEDJ	ER-i	Folding	P	CTDI no ZFLR
DNAJB12a,b	DJ10;mDJ10	ER-a	Erad	P	TMD
DNAJB13	TSARG6;RSPH16A	C?	No HPD!	U	No canonical HPD
DNAJB14a,b	FLJ14281	?	U	P	TMD
DNAJC1	ERdj1;Mtj1	ER-i	Translation	No	3 TMD/ 2 SANT D
DNAJC2	MPP11;zuotin	C	Translation	No	RBS bindin/ SANT D
DNAJC3	PRKRI;P58IPK	ER-i	Translation under stress	S	8 TPR D
DNAJC4	Hspf2;MCG18	?	U	No	TMD
DNAJC5,5b,g	CSP	Exosomes?	Exocytosis	S	Cys Rich D
DNAJC6	auxilin	C	Clathrin uncoating	S	Clathrin BD
DNAJC7	TTC2;TPR2;mDJ11	C	Hormone maturation	S	8 TPR D
DNAJC8	SPF31	C	Phosphorylation?	U	
DNAJC9	JDD1;HDJC9	N	Nuclear exit upon stress	U	
DNAJC10	ERdj-5;JDI	ER-i	Erad	S	4 Trx Box
DNAJC11	FLJ10737	Mit-i?	U	U	Coiled Coil D
DNAJC12	JPD1	?	U	No	
DNAJC13	RME-8;KIAA0678	C/M	Endosome trafficking	U	
DNAJC14	DRIP78;HDJ3;LIP6	C/ER/M	Cell surface export	S	3 TMD
DNAJC15	DNAJD1;MCJ	Golgi	Protein degradation	No	TMD
DNAJC16	KIAA0962	?	U	S	Trx Box/ Extracell Frag

Table 3.2 (*Continued*)

Hsp40	Other names	Localization	Function	Client Binding	Features
DNAJC17	FLJ10634	?	U	S	Spliceosome ID
DNAJC18	MGC29463	?	U	P	TMD
DNAJC19	TIMM14;Tim14	Mit-i	Protein import	No	TMD
DNAJC20	Hscb;Hsc20	Mit-i	Fe-S cluster biogenesis	S	Isu1 BD
DNAJC21	DNAJA5;JJJ1	?	U	S	3 Coiled-Coil D
DNAJC22	wus;FLJ13236	C	Endocytosis?	S	3 TMD
DNAJC23	Sec63L;ERdj2	ER-M	Protein import	No	2 Sec23 D/2 TMD
DNAJC24	DPH4;JJJ3	C	Dipthamide synthesis	S	Zn Finger D
DNAJC25	bA16L21.2.1	?	U	U	3 TMD
DNAJC26	GAK	C/N	Clathrin uncoating	S	Clathrin BD/Tensin BD/PKD
DNAJC27	RBJ;RabJS	?	U	S	3 GTP B sites
DNAJC28	Orf28;C21orf55	?	U	U	Coiled Coil D
DNAJC29	ARSACS;sacsin	C/Mit?	Protein degradation?	S	TPR D/HEPN D
DNAJC30	WBSCR18	?	U	No	

C: Cytosol; ER: Endoplasmic Reticulum; N: Nucleus; Mit: Mitochondria; M: Membrane; a: Associated; i: Inside; ?: Uncertain. P: Promiscuous; S: Specific; U: Unknown; D: Domain; BD: Bindong Domain; ID: Interaction Domain; Trx: Thiorexoxin; TMD: Transmembrane Domain; PKD: Protein Kinase Domain; CTDI: C-Terminal Domain I; RBS: Ribosome; HPD: His/Pro/Asp Conserved Motif; TPR: Tetratricopeptide Repeat; SANT: Swi3, Ada2, N-Cor And TFIIIB Domain; HDAC: Histone Deacetylases; ZFLR: Zinc Finger-Like Region.

Hsp40s and Hsp70s. Although interactions between NBDs of Hsp70s and J domains have been identified, more sites of contact between Hsp70s and J proteins have been proposed (reviewed in ref. 16).

J domains are comprised of about 70 amino acids arranged in four α-helices; helices 2 and 3 are antiparallel and appear as a protruding "finger" stabilized by helices 1 and 4 (Figure 3.4). The loop between helices 2 and 3 contains the conserved tripeptide sequence His-Pro-Asp (HPD) crucial for the ATPase stimulatory function on Hsp70. The interaction surface of J domains on the NBD of Hsp70 is proximal to the hydrophobic pocket where the interdomain linker binds in the ATP-bound state. The ability of J domains to stimulate the Hsp70 ATPase activity is augmented by the presence of substrates bound to the Hsp70 SBD, and the *in vitro* interaction between J domains and Hsp70s seems to be transient and highly dynamic.[85] Significant progress has been made towards understanding how J domains stimulate Hsp70 ATPase activity, which in turn results in the stabilization of the bound client protein,[84,91–94] but the detailed mechanism is not yet clear. The enhancement of the Hsp70 ATPase activities by Hsp40s was recently explained in terms of the ability of DnaJ to shift the equilibrium between the linker-bound and linker-unbound

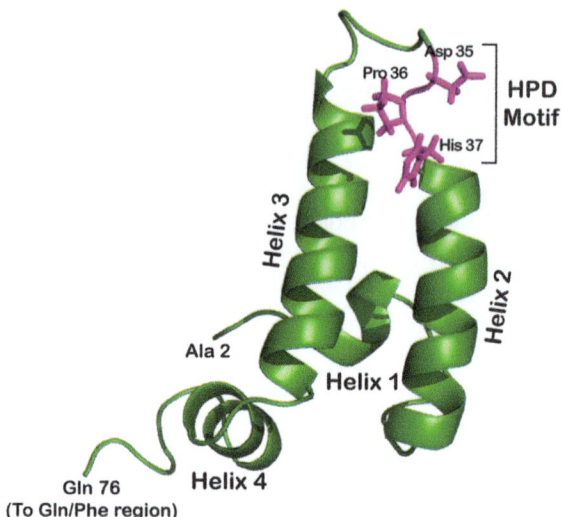

Figure 3.4 Structure of the J domain of Hsp40 proteins. The J domain of *E. coli* DnaJ is represented. The HPD motif important for interaction with Hsp70 is colored in magenta, and the side-chains of these residues are shown as sticks (PDB entry 1XBL[394]).

conformations of Hsp70,[54] and this model is supported by the identification of a dynamic interface between DnaJ and DnaK that overlaps with the NBD-SBD interdomain interfaces of DnaK,[85] thus providing another way to tune the Hsp70 allosteric cycle.

Unlike Hsp70s, J proteins show sequence and structural divergence, according to their role in affecting the versatility of the Hsp70 system. J proteins are usually divided into three classes based on their domain similarity to the *E. coli* DnaJ but while this classification is helpful for nomenclature purposes it is not informative regarding the proteins' functionalities (Table 3.2):[86]

(a) Class I: The first class (DnaJA, with four members identified in humans) includes the Hsp40s that have domain structures identical to *E. coli* DnaJ. They have an N-terminal J domain, followed by the Gly/Phe rich domain (G/F), which, together with the J domain, is essential for maximal stimulation of the Hsp70 ATPase activity. Hsp40 Class I have four repeats of the conserved CXXCXGXG zinc finger motif that form two zinc binding centers[95] followed by an extension that contains two barrel topology domains (C-terminal domain I and C-terminal domain II, CTDI and CTDII). CTDI includes a hydrophobic patch where client proteins are proposed to bind.[96,97] The extreme C-terminal extension is responsible for Hsp40 dimerization.[98] Co-chaperone members of this group (*i.e.* DnaJA1) are able to bind substrates independently of Hsp70 and protect them from aggregation.

(b) Class II: Hs40 members of this group (DnaJB, with 13 members found in humans) possess the N-terminal J domain with the G/F region but lack the zinc-finger domain. Some members of this group can also bind and prevent aggregation of client proteins independently of Hsp70 (like DnaJB1, DnaJB6b and DnaJB8). This class includes members with non-standard domains like the ubiquitin-interacting motifs.[99]

Client binding by Class I and II Hsp40 proteins is important not only to prevent aggregation but also for protein degradation, remodeling of folded proteins *etc*.[86] For example, DnaJB11 (ERdj3) is an ER Hsp40 protein that even in the absence of a functional J domain can bind directly to many nascent or unfolded proteins prior to binding to BiP.[100] Other Hsp40 proteins contain ubiquitin-interacting motifs like DnaJB2 (HSJ1), which interacts with ubiquitinylated substrates making the system specific for clients destined for degradation in an Hsp70-dependent function.[99]

(c) Class III: Members of this class (DnaJC, with 32 members found in humans) contain the J domain anywhere in the protein, and this group includes proteins that cannot be classified as Class I or Class II. They often include dedicated domains that function in the recognition of specific substrates for delivery to Hsp70s; Class III Hsp40s require interaction with Hsp70 to function.[101] In some cases, the presence of only a J domain is sufficient for some cellular functions, for example J domains in specific compartments target Hsp70 to client proteins at these locations. In another example, Hsp40 protein at the cytosolic face of the ER membrane attached by a single *trans* membrane domain recruits cytosolic Hsp70s to assist in the degradation of proteins dislocated from the ER (see below). Another Class III Hsp40, DnaJC2, is tethered near the polypeptide exit of the ribosome and recruits cytosolic Hsc70.[91,102]

Even though all Hsp40 proteins contain the J domain, which determines their interaction with Hsp70, the Hsp40 family is very diverse, and multiple forms occur within the same cell and compartment. However, specific Hsp40-Hsp70 pairs have been identified for certain client proteins.[103] As mentioned before, Hsp40s interact with Hsp70s through their J domains, but there are additional determinants important in specific Hsp40–Hsp70 interactions. Different Hsp40 proteins are not completely interchangeable for different Hsp70s,[103] and the level of stimulation of the ATPase activity of Hsp70s by different Hsp40s varies. For example, MmDjC7 (a murine Hsp40) is able to stimulate the ATPase activity of DnaK, human Hsc70 and murine BiP to different extents.[104] *E. coli* DnaJ can enhance the ATPase activity of mammalian Hsc70, but mammalian Hdj1 cannot stimulate the ATPase activity of DnaK.[105] Also, only the J domains from the ER Hsp40 proteins can bind and stimulate the ER-resident Hsp70, BiP.[103,106] Clearly, this inability of Hsp40s to interact with all Hsp70s implies binding discrimination that could arise from the non-conserved regions of Hsp40s or indeed from changes within their J domains, which translate into differences in specificity and/or in affinity.[103] On the other hand, regulation of specific interactions could arise from cellular co-localization or expression of partners in eukaryotes, or at the level of co-expression under certain conditions.

There is increased interest in investigating the roles of Hsp40 proteins in human diseases. For example, TID1 (DnaJA3) is currently the only DnaJ-like protein identified as a tumor suppressor; the functions of several members of the Hsp40 family in tumor growth (such as hTid I (class DnaJA3), or HLJ1 (class DnaJB4)) are under study and constitute a fertile emerging field.[87,88]

3.6.2 NEFs

As part of the allosteric cycle of Hsp70s, ADP and P_i have to be released from the NBD to allow binding of a new ATP molecule, with the subsequent release of substrate. ADP and P_i dissociation, which becomes rate-limiting for the cycle in conditions of high ATP availability, requires the opening of the nucleotide-binding cleft. Variations among Hsp70 proteins exist in a loop in subdomain IIB around the nucleotide-binding pocket and in the cleft itself: some NBDs contain a hydrophobic patch and potential salt bridges, whereas others (like the specialized bacterial Hsp70, HscA) lack all elements that help to lock the nucleotide in the pocket. Not surprisingly, most Hsp70s require NEFs to accelerate the rate of ADP dissociation and ATP binding, while HscA does not. X-ray structures of the NDBs of several Hsp70s in complex with their NEFs show an outward rotation of subdomain IIB of about 10–20° compared to the structures of the isolated NBDs, reflecting an "open" conformation of the domain (see Figure 3.5).[58–60,107,108]

3.6.2.1 GrpE

The DnaK/DnaJ/GrpE system from *E. coli* constitutes the paradigm for the Hsp70 system and the DnaK NEF, GrpE, is the best studied of all Hsp70 NEFs. Recent NMR studies from our laboratory on the NDB of *E. coli* DnaK showed that the "open" conformation is present in the conformational ensemble of an isolated NBD, and is selected by binding to its NEF.[18] This result is consistent with previous elastic network modeling and molecular dynamics simulations.[55,109] The mechanism by which GrpE accelerates ADP release from Hsp70 then seems to rely on stabilization of the open conformation of the NBD, which also facilitates ATP binding[43,110] and thus determines the lifetime of the Hsp70-substrate complex.[111]

There is a GrpE homologue in mitochondria (GrpEL1/2), but none in the cytosol of eukaryotes,[112] where the other three different families of NEFs play analogous roles to GrpE in modulating the activity of Hsp70s.[113]

3.6.2.2 Bags

Human cells contain six members of the Bag (Bcl-2-associated athanogene) family of NEFs, named Bag-1 to Bag-6, all of which act as HspA1/Hsc70 co-chaperones. All contain at least one copy of a conserved C-terminal Bag-domain (Bag-5 has five Bag domains), which is itself sufficient to stimulate

Structure and Function of Hsp70 Molecular Chaperones 85

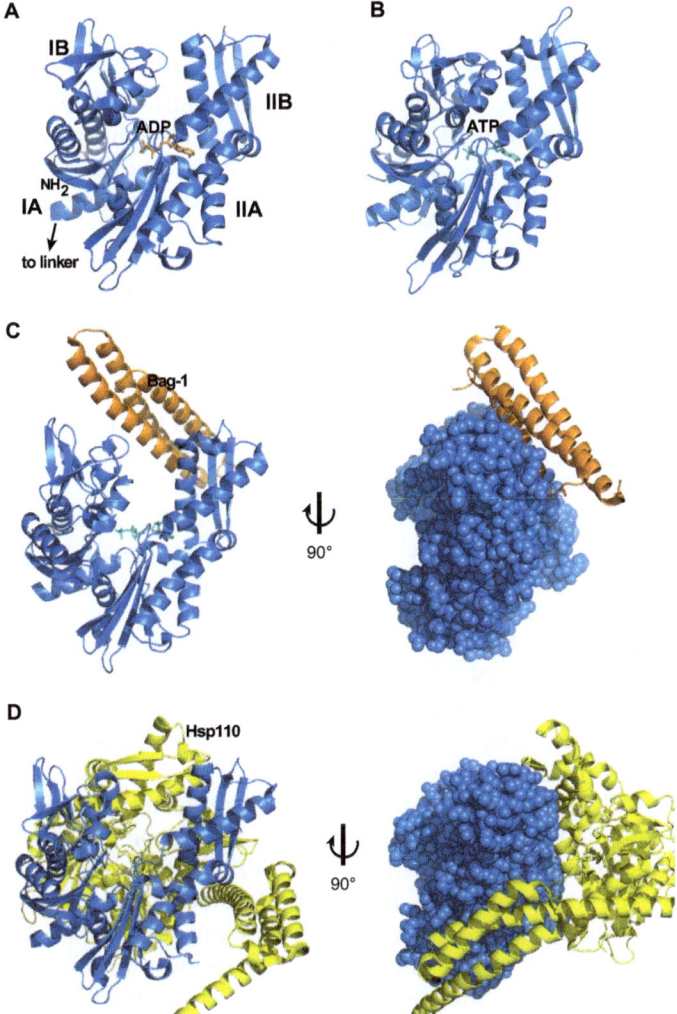

Figure 3.5 Structures of the nucleotide-binding domain of Hsp70s (in blue) in different nucleotide-bound states and in the presence of NEFs. All figures contain the NBD oriented in the same fashion unless indicated; labels in A are the same for all subsequent panels. ADP is shown in orange and ATP in cyan. **A**. ADP-bound form of the human NBD domain of HspA1 (PDB entry 3 ATU[393]). **B**. ATP-bound form of DnaK from PDB entry 4B9Q.[17] **C**. Crystal structure of human Hsc70 in complex with the Hsp70-interacting domain of Bag-1 (in orange) and ATP (PDB entry 3FZF[56]). Right panel shows the rotated complex where the Hsc70 NBD is depicted in space fill for visual simplification. **D**. Crystal structure of the complex between human HspA1 NBD and yeast Hsp110 homolog Sse1p (PDB entry 3D2F[57]). Right panel shows the rotated complex where the Hsp70 NBD is depicted in space fill for visual simplification.

the dissociation of ADP from Hsc70 and HspA1. Structures are available for the NBD of Hsc70 interacting with Bag-1[56,60] or Bag-2,[114] and for the human HspA1 NBD interacting with the Bag domain 5 of Bag-5[58] (Figure 3.5). The Bag domain consists of a three-helix bundle that binds to subdomains IB and IIB of the NBD, thus keeping the cleft open. Even though Bag-1 binds to the NBD of Hsc70/HspA1 and causes analogous conformational changes to those caused in DnaK by GrpE, these NEFs operate through different mechanisms and have different functions.[19,60,115,116] For example, GrpE can catalyze the release of ADP and ATP from DnaK while Bag-1 can only accelerate the release of ADP; ADP dissociation from Hsc70 can be stimulated 100- to 600-fold by Bag-1 depending on the presence of inorganic phosphate,[43] whereas the function of GrpE on DnaK is not affected by P_i. Also, the selectivity of these NEFs for different Hsp70 partners is stringent.[43,110]

In addition to the Bag domain, the other domains in Bag proteins are diverse, and they have been proposed to function as targeting factors for Hsp70s by recruiting them to certain locations for specific tasks.

Bag-1 (described later in the chapter) was first identified as an interaction partner of the anti-apoptotic Bcl-2 protein and the activated glucocorticoid receptor.[117,118] Bag-1 participates in numerous functions in normal and cancerous cells, including apoptosis, proliferation, transcription and metastasis.[119] It is produced in the cell as four translation initiation variants, Bag-1L, Bag-1M, Bag-1p33 and Bag-1S of 50, 46, 33 and 29 kDa, respectively. All the translation products are in frame, so the Bag-1 isoforms contain the same C-terminal end and different N-terminal ends; they also differ in their levels of expression and cellular localization.[119]

It was recently discovered that in malignant prostate tissue Bag-5 is over-expressed compared to the benign samples.[120] Also, upon ER-stress induction Bag-5 translocates to the ER and interacts with BiP, and its levels are increased. Moreover, over-expression of Bag-5 in prostate cancer cells inhibited ER-stress-induced apoptosis in the unfolded protein response (UPR), whereas Bag-5 down-regulation resulted in increased stress-induced cell death.[120]

Bag-1 and Bag-2 have opposite effects on the Tau protein, which makes up the neurofibrillary tangles involved in Alzheimer's disease.[121] The Hsc70/Bag-1 complex inhibits Tau turnover whereas a complex of Hsc70 and Bag-2 forms a microtubule-attached complex that can capture insoluble and phosphorylated tau and deliver it to the proteasome for degradation.[122,123]

3.6.2.3 HspBP1 and BAP

HspBP1 (yeast orthologue, Fes1p) acts as an Hsc70 NEF in the mammalian cytosol.[124] Again, this protein is structurally and mechanistically different from the GrpE and Bag proteins[110,113,124] as is clear from the structure of the complex between a fragment of the Hsc70 NBD and HspB1.[124,125] HspBp1 consists of four repeats of an α-helical "armadillo-like" fold that wraps around subdomain IIB of the Hsp70 NBD, thus causing a distortion of the lobes of the nucleotide-binding cleft and favoring nucleotide dissociation.

BAP (yeast orthologue, Sls1p) promotes co-translational translocation of proteins across the ER membrane by specifically associating with the ADP-bound state of BiP and boosting ADP nucleotide release.[126] Mutations of BAP have been implicated in the development of neurodegenerative diseases in mouse models and in humans: certain homozygous mice with BAP mutations develop an accumulation of misfolded proteins in the ER with subsequent UPR induction, apoptosis and autophagy.[127]

3.6.2.4 Hsp110

The eukaryotic cytosol and ER contain a class of NEFs in the Hsp110 family (also called Grp170 when resident in the ER). These proteins belong to a branch of the Hsp70 super family. They resemble Hsp70s in domain arrangement but are larger, containing many insertions, primarily in the SBD and at the C-terminus), and they do not perform the same functions as the Hsp70 proteins. Hsp110 can themselves bind ATP and act as client "holdases", but they do not perform a canonical Hsp70 allosteric cycle,[128–130] and they are strong NEFs of Hsp70s. The structure of the yeast Hsp110 interacting with the NBD of yeast Hsp70 and bovine Hsc70 has been solved.[57,59]

Again, Hsp110s act as NEFs of Hsp70s in a mechanism that is distinct from those used by GrpE and Bag-1. Moreover, yeast Hsp110 must first bind ATP, then adopt an active, more compact conformation, which is able to bind Hsp70, and act as a NEF.[131,132] The Hsp110:Hsp70 complex is stable and comes apart when ATP binds to the Hsp70. It has been proposed that direct interactions of Hsp110 with substrates assist Hsp70 function in a co-operative process.[57]

3.6.3 Other Co-chaperones

3.6.3.1 Hip

The cytosolic Hsc70-interacting protein or "Hip" (also known as p48 or ST13) was identified using a yeast two-hybrid screen that aimed to identify interacting partners of the rat Hsc70.[133] Hip plays a role in the maturation of the progesterone receptor by facilitating the co-operation between Hsp70 and Hsp90.[134,135] Hip is an Hsc70 regulatory protein that interacts with the Hsc70 NBD in a reaction that depends on stimulation of ATP hydrolysis by Hsp40; it is unable to bind the chaperone in the presence of Mg^{2+} and ATP. Hip is proposed to bind the ADP-bound conformation of Hsp70s, preventing ADP dissociation and thus stabilizing the complex, which helps avoid premature substrate release. Also, Hip can protect unfolded substrates from aggregation in vitro.[133] Even though oligomers of Hip were found to bind to Hsc70, monomeric Hip fragments are still able to bind the NBD of Hsc70, thus homo-oligomerization is not essential.[136]

Hip is composed of an N-terminal oligomerization domain, a central TPR module that interacts with Hsc70, followed by a highly charged region, a GGMP-rich module (involved in substrate binding[135–137]), and a C-terminal DP-containing region.[138] These domains are conserved, arguing for an important role in the regulation of the Hsp70 and Hsp90 proteins.[136]

Some Bag-independent functions of Hip have been reported. For instance a Hip protein that has the TPR motif deleted and a point mutation, and consequently has reduced low ability to bind Hsc70, was still able to stimulate maturation of the glucocorticoid receptor, pointing to the existence of an Hsp70-independent function of Hip in this process, at least in yeast.[139]

Hip was shown to be down-regulated in tumors, and was proposed to be a putative tumor suppressor gene.[140,141] Hip contains a loop with a sequence homologous to the counterpart in BID (Bcl-2 Interacting Protein), which serves as a cleavage site for caspase-8 and granzyme-B in the cytosol. The resulting truncated form of BID mediates mitochondrial-mediated apoptosis (see Section 3.7.5). Hip is cleaved by granzyme B but not by caspase-8 *in vitro* and *in vivo*, and the loss of Hip upon cleavage by granzyme B is correlated with the loss of BID. Modulation of the Hsp70 complexes by proteolytic cleavage of its co-chaperones was proposed to be a pathway to regulate granzyme B-mediated apoptosis.[140]

3.6.3.2 Hop

Hop (Hsp70/Hsp90 organizing protein, or Stl1) was first identified in yeast as a protein involved in the heat-shock response,[142] and it has been well studied for its role in the maturation of protein structures (*i.e.* the progesterone receptor complex).[143] Hop is a co-chaperone that can simultaneously bind Hsp70 and Hsp90; it efficiently targets Hsp90 to a pre-formed complex containing Hsp70 and client substrates. It preferentially associates with the ADP-bound form of both chaperones,[144] thus it coordinates the actions of the Hsps in folding protein at the right nucleotide-binding states.[145] Partially folded Hsp90-dependent proteins are transferred from Hsp70 to Hsp90, where they complete the acquisition of their final structure.[146–148]

Hop contains three TPR repeat domains that together form TPR-binding pockets (carboxylate clamps) and bind the conserved C-terminal residues GPTIEEVD of Hsp70s and MEEVD of Hsp90. The TPR domains are formed by three tandem repeats of a 34 amino acid sequence motif,[149] and the core TPR domain comprises six α-helices that form a saddle-like structure with a curved surface where the interacting peptides bind.[150] Hop binds to Hsp70 and Hsp90 *via* TPR1 and TPR2A, respectively.[151] Crystal structures of TPR1 and TPR2A in complex with peptides corresponding to the C-terminal sequences of the chaperone show that the substrates are bound as extended peptides, and affinity is primarily contributed by electrostatic interactions.[150]

Interactions between Hop and Hsp70 appear to be more complex than those with Hsp90: in addition to the C-terminal part of Hsp70, additional sites in the Hsp70 appear to be involved in binding.[152]

Hop is over-expressed in invasive cancers (such as pancreatic cancer).[153] Moreover, Hop knockdown by small interfering (si)RNA reduces the invasiveness of pancreatic cancer cells, with a subsequent decreased expression of matrix metalloproteinases-2 (MMP-2), which is highly important for metastasis, and with which Hop directly immunoprecipitates. Also, a diminution of Hop expression also led to reduced expression levels of Hsp90 client proteins, HER2, Bcr-Abl, c-MET and v-Src. Hop displays differential localization in invasive ductal pancreatic cancer, indicating that its localization constitutes a key factor in pancreatic tumors. Moreover, attenuation of Hop expression inactivates signal transduction proteins, which may be a result of the modulation of Hsp90 activity, making Hop an interesting target for cancer therapy.[153]

3.7 Cellular Functions of Hsp70s

3.7.1 *De Novo* Protein Folding

A nascent protein exits the ribosomal tunnel after the first 35–40 amino acids have been added to the chain.[154] Depending on the protein, folding can commence co-translationally. It is speculated that some proteins do not acquire stable structure until all elements of the sequence are outside of the ribosome, making the nascent chain vulnerable to misfolding and aggregation. To protect the polypeptide chain, multiple cellular factors comprise a "greeting committee" for the translating protein and influence its early steps of folding; the greeting committee includes chaperones as well as targeting factors that deliver the secreted protein to the right cellular location. This chaperone network in eukaryotes that is dedicated to protein biogenesis has been dubbed "CLIPS" (chaperones linked to protein synthesis).[155] In mammals this ribosome-bound "reception committee" includes the heterodimeric nascent polypeptide-associated complex (NAC) (see ref. 156 for a review) and the Hsp70-containing mammalian ribosome associated complex (mRAC).

NAC protects nascent chains from deleterious interactions as they emerge from the ribosome.[157] NAC is a ribosome-anchored heterodimer of an α and a β subunit, very conserved throughout Nature.[158] Besides prevention of misfolding and aggregation, multiple functions have been assigned to NAC, including roles in transport to mitochondria and ER.[159]

RAC is a stable complex that in yeast consists of the Hsp70 proteins Ssz and the Hsp40 homolog Zuotin. Yeast RAC contains an extra Hsp70 homolog protein (Ssb-RAC) also bound to the ribosome. The mammalian counterpart contains a homologue of yeast Ssz-like Hsp70 protein, Hsp70L1 (although Hsp70L1 and Ssz1p contain fairly dissimilar NBD and SBD)[102] and the ribosome-associated DnaJC2. DnaJC2 (MMPP11) is the human counterpart of yeast Zuotin, which contains an unusual N-terminal domain only present in a minor group of eukaryotic Hsp40 homologs.[102,160] Mammals do not possess the ribosome-bound Hsp70 corresponding to the yeast Ssb protein; instead, mRAC recruits the cytoplasmic HspA1 (but not Hsc70) to nascent chains.[40] Recruitment of cytosolic HspA1 to the ribosome may be accomplished by the enhancement of its ATPase activity by mRAC, or mRAC may bind to nascent chains first and hand them over to Hsp70. On the other hand, cytosolic Hsc70 can be recruited to the ribosome in an mRAC-independent fashion.[40,102]

After a nascent chain is shepherded by the CLIPS, it can spontaneously fold or be handed to downstream chaperones for assisted folding, including the Hsp70/Hsp40/NEF and the chaperonins, and subsequently Hsp90 for the final structural acquisition step. Provocatively, a recent report by Willmund et al.[161] reveals the essential contribution of the yeast Hsp70s during the early events of the chaperone-sorting process at the ribosome: Ssb binds approximately 70% of newly synthesized polypeptides in yeast, and 45% of nascent polypeptides were strongly enriched in the Ssb-binding set *versus* the whole translatome. This latter group includes the more slowly translated and aggregation-prone nascent polypeptides; Ssb deletion led to extensive aggregation of this class of protein. Ssb was suggested to stabilize aggregation-prone nascent chains and to serve as an initial chaperone for nascent chains, before they were handled by more specialized chaperone complexes.[161]

3.7.2 Protein Re-folding

Many forms of stress, including changes in the cellular conditions due to aging or mutations, subject cellular proteins to denaturation. Thus, hydrophobic regions in proteins or protein complexes become exposed to the solvent and may kinetically partition between aggregating and binding to chaperones that inhibit aggregation. Concurrently, acute increases in the amount of proteins that require chaperone binding deplete the cellular pool of available chaperones causing Hsf1 to be activated, in turn increasing the amount of chaperones made until the stress challenge ceases.[23] Hsp70 proteins (along with their co-chaperones) passively protect these denatured proteins from aggregation by their "holdase" activity, *i.e.* by transient binding and release, giving the protein more chances to fold. Along with stress-induced unfolding and chaperone depletion, the pool of available ATP in the cell decreases; concomitantly, the ADP-bound form of Hsp70 that has high affinity for substrates predominates and is able to hold substrates longer.

The depletion of chaperone capacity may lead to serious health risks associated with what are known as "protein misfolding" or "protein conformational" diseases.[12,162,163] Along with other chaperones and components of the ubiquitin-proteasome system, Hsp70s are found associated with aggregates typical of misfolding diseases.[15] This is due to the fact that once proteins are already aggregated or misfolded, Hsp70s are also able to act as "unfoldase" machines fueled by ATP hydrolysis; Hsp70s locally unfold and remove proteins from the surface of some aggregates and allow them to fold. Likewise, non-native interactions formed within protein monomers can be locally unfolded by Hsp70 allowing the protein to reach its native conformation after release.[21,128,164–166] As this activity of Hsp70s requires that a binding site within the substrate be accessible for interaction with the SBD of Hsp70, the cores of amyloid-type fibrils cannot generally be reached by Hsp70s and cannot be remodeled/rescued, whereas Hsp70 can chaperone misfolded proteins while they are part of pre-fibrillar species[12,167–170] (reviewed in ref. 12).

Neurons are very sensitive to the accumulation of toxic misfolded proteins as a result of aging or other stresses. This sensitivity underlies several neurodegenerative disorders including Huntington's, Alzheimer's and Parkinson's diseases.[14] Because of their abilities to rescue certain aggregation prone proteins from forming toxic aggregates, Hsp70s have been linked to these diseases and are now emerging as potential drug targets (reviewed in ref. 14). For example, polyglutamine (polyQ) diseases (which include Huntington's disease, spinocerebellar ataxia and spinal and bulbar muscular atrophy) arise from the expansion of CAG repeats in the disease genes, leading to extended stretches of Gln in the resulting aggregation-prone proteins. *In vitro*, Hsp70 and Hdj1 are able to reduce the aggregation of the polyQ-expanded exon 1 of huntingtin. The addition of Hsp70/Hdj1 is only effective during the lag phase of the aggregation reaction, consistent with a mechanism whereby the chaperones interact with pre-fibrillar species.[14,171] *In vivo*, the rescuing effect of Hsp70s seems to follow a more complicated mechanism, as the action of Hsp70s on the polyQ substrates may be attributed to more complex effects on pro-survival pathways.

3.7.3 Protein Degradation

The cellular proteome is highly dynamic, and even under non-stressed conditions proteins have to be turned over. One of the functions of Hsp70s in cellular proteostasis is to facilitate degradation of misfolded or otherwise defective proteins by coordinating chaperoning activities with the two major protein degradation systems: the autophagy-lysosome and the proteasome pathways.[172,173] Both pathways are important as "housekeeping" mechanisms, but they are highly up-regulated under stress. Hsc70 plays a crucial role in both catabolic pathways. As described in the sections below, linkage between Hsp70s functions and the degradation routes involves the participation of its co-chaperones CHIP and Bag-1.

3.7.3.1 Protein Degradation via the 26S Proteasome

Short-lived cytosolic proteins are covalently tagged for proteolytic degradation with a chain of ubiquitins, and transfer of ubiquitin molecules from ubiquitin-conjugating enzyme (E2) to the substrate protein is ultimately mediated by the ubiquitin ligase (E3).[174,175] The specificity of this targeting process is dictated by E3. Ultimately, the substrate protein carrying the "degradation tag" is targeted to the 26S proteasome: a large 26S complex composed of the 20S particle (four stacked rings around a central pore that harbors the protease activity), and the 19S regulatory "cap". Usually, a chain of four or more ubiquitin tags attached *via* their K48 residues binds to the proteasomal 19S subunits through ubiquitin-interacting motifs, resulting in unfolding of the substrate and removal of the ubiquitin. Subsequently, the substrate is unfolded and proteolyzed to short peptides. All these mechanisms are complex and tightly regulated.[176,177]

3.7.3.1.1 Role of CHIP.
Hsp70s participate in the degradation process by binding the substrate protein and recruiting the co-chaperone CHIP. CHIP is the first recognized chaperone-dependent E3 ligase and links Hsp70s functions with the proteasome. By binding to Hsp70/Hsc70, CHIP shifts the chaperone/substrate complex from the protein-folding pathway to the degradation pathway, and accelerates proteasomal degradation. CHIP contains three domains: a TPR domain, which binds to the conserved C-terminal EEVD motif of Hsp70, a U-box domain that contains the ubiquitin ligase function and a middle, charged domain.[178–180] Interaction with Hsp70/Hsc70 via the TPR domain and ubiquitylation of the bound substrate by the E3 domain results in substrate degradation by the proteasome. CHIP inhibits the ATP hydrolysis activity of Hsp70, thus decreasing its substrate refolding activity. It is speculated that this mechanism would slow the Hsc70 reaction cycle under stress conditions, or it may direct misfolded proteins into the ubiquitin-proteasome machinery.[181] It is proposed that CHIP has a role in delivering substrates to the proteasome by interaction with the proteasomal protein S5a.[179,182]

As part of the complete protein clearance mechanism, CHIP can ubiquitylate Hsc70[180] and Bag-1 in a non-canonical way to mediate binding to the proteasome in a degradation-independent manner.[183]

CHIP-mediated proteasomal degradation has been implicated in many diseases. For instance, degradation of the tumor suppressor p53 is regulated by CHIP as well as by another ubiquitin ligase, Mdm2.[184] CHIP is also involved in the degradation of the ErbB family of receptor tyrosine kinases, the overexpression and activation of which is implicated in the development and progression of a variety of solid tumors.[185] Other substrates of CHIP include glucocorticoid receptors[179] and the misfolded cystic fibrosis transmembrane-conductance regulator (CFTR).[182]

The Hop and CHIP co-chaperones bind to the same Hsp70 site. Thus, the balance of Hsp70s bound to one or the other co-chaperone will regulate the balance between client proteins being folded or degraded.[186] Consequently, Hsp70 binding to these co-chaperones is a regulated process. Phosphorylation of the C-terminal region of both Hsp70 and Hsp90 prevents binding to CHIP and consequently promotes binding to Hop favoring folding over degradation.[187] Not surprisingly, cancer cells show an increased level of chaperone phosphorylation and an increased level of Hop, adding to the survival role of Hsp70s in cancer.[187]

3.7.3.1.2 Role of Bag-1.
Bag proteins comprise a class of Hsp70 co-chaperone that represents another point of connection between these chaperones and the proteasome degradation pathway: in addition to the Bag domain, Bag-1 groups of proteins are characterized by the presence of a ubiquitin-like (UBL) domain, proposed to mediate interaction with the 20S

core and the 19S subunit of the proteasome. Binding of the 26S proteasome in human cells is not mediated directly by Hsp70s:[188] instead, Bag-1 simultaneously binds Hsp70 and the proteasome, promoting the association of Hsp70s with the protease system *in vitro* and *in vivo*. This simultaneous binding causes substrate release from the Hsp70 proteins, thus directly coupling nucleotide-exchange activity with proteasomal targeting functions.

As an example, Bag-1 has been intensively studied in breast cancer cells where its over-expression impairs apoptosis and growth inhibition induced by some stresses or serum deprivation.[189–191] This activity of Bag-1 is dependent on residues responsible for its interaction with Hsp70/Hsc70 and requires the UBL domain. Therefore, the role of Bag-1 in breast cancer has been linked to its function as adaptor for Hsp70 chaperone functions with the proteasome.[119]

Many activities of CHIP require the presence of Bag-1 in the complex along with the substrate protein and Hsp70s and *vice versa*. A model has been proposed where these two co-chaperones promote the formation of a complex that couples the functions of the Hsp70 chaperone system with the ubiquitin/proteasome degradation machinery to tune proteostasis.[119]

CHIP and Bag both have an impact on human diseases. For example, CHIP ubiquitylates polyQ-expanded exon 1 of huntingtin and promotes its degradation;[192,193] also, CHIP over-expression rescues the symptoms of spinocerebellar ataxia, spinal and bulbar muscular atrophy and Huntington's diseases in animal models, whereas CHIP double negative mice have exacerbated disease progression in models of Huntington's diseases and spinocerebellar ataxia.[14,193,194]

3.7.3.2 Protein Degradation via the Lysosome

Protein degradation occurs in the lysosome, and three different lysosomal pathways are involved: macro-autophagy, micro-autophagy and chaperone-mediated autophagy (CMA). The latter is a selective clearance pathway characterized by the receptor-mediated delivery of a set of cytosolic chaperone-substrates into the lysosome,[195] and it constitutes an important player in the quality control system of the cell to maintain proteostasis. Degradation by the lysosome results in processing of misfolded proteins by the lysosomal hydrolases, yielding amino acids that can be re-used by the cell. In order to be targeted to this pathway, cytosolic substrate proteins must contain a KFERQ consensus tag. Upon partial unfolding, hydrophobic sequences and the KFERQ tag become exposed to the solvent and thus become substrates for Hsc70. The substrate/chaperone complex is then targeted to the lysosomal receptor Lamp2a. Lamp2a is a membrane protein heavily glycosylated in the luminal region and positively charged in the cytosolic region. This charged region is involved in substrate binding.[196] Once the substrate protein/Hsc70 complex is attached to Lamp2a, the receptor helps to further unfold the protein and facilitate its transport into the lysosomal lumen. Lamp2a has been recognized to be important both for binding and for substrate translocation.[197,198] Upon translocation, intralysosomal Hsc70 (required for the functioning of the

CMA pathway) binds the substrate prior to hydrolysis by lysosomal proteases. It has been speculated that lysosomal Hsc70 is required to pull the substrate into the lumen,[196,199] but the detailed mechanism by which Hsc70 enters the lysosome remain elusive.

Other chaperones and co-chaperones are involved in the lysosomal degradation process, like specific Hsp40s and a Hip co-chaperone that stimulates assembly of the Hsp70 chaperone/CHIP/lysosomal Hsc70/Bag-1 complex.

Although they constitute distinct mechanisms, the CMA and the proteasome pathways are linked: some ubiquitylated proteins are also targeted for lysosomal removal *via* autophagy,[197,200] and the two mechanisms share substrates and protein factors that in some cases compensate for each other. Inhibition of the proteasome stimulates the activity of the CMA, possibly through regulatory processes,[201,202] although the mechanisms of crosstalk continue to be unclear. Conversely, inhibition of autophagy slows down the degradation of the proteasomal substrates.[203] Several ubiquitin-binding proteins have been implicated in the selective autophagy of ubiquitylated proteins, and this constitutes a topic of increasing interest.[204] Although ubiquitin is also involved in the CMA, it is speculated that the accumulation of CMA substrates inhibits proteasome activity by blocking the ubiquitin receptors at the 19S particles or the unfolding steps.[204] When the function of the proteasome is compromised, up-regulation of the CMA pathway assumes the task of removing unwanted proteins.[205]

3.7.4 Intracellular Protein Traffic

3.7.4.1 *Endocytosis*

Many endocytic pathways function at the plasma membrane of mammalian cells.[206] Protein transport inside the cell, both between the plasma membrane and the endosome, and between the endosome and the Golgi compartment, is mediated by clathrin-coated vesicles, and the process requires a number of accessory proteins and adaptors. After a vesicle is formed, the assembly team is recycled back to the cytoplasm and re-used.[207] Clathrin-mediated endocytosis controls basal as well as stimulated internalization of several receptors with or without their cargos,[206] like immunoglobulin G,[208] low-density lipoproteins,[209,210] epidermal growth factor receptor,[211] insulin,[212] G-protein-coupled receptors[213] and transferrin.[214] Internalized cargos are transferred into endosomes where they get trafficked back to membranes or into compartments for degradation. Not surprisingly, this endocytic pathway is exploited by toxins, viruses and bacteria as a gateway into cells.

The endocytic pathway proceeds through many stages,[215] and Hsc70 is essential for the last step that involves uncoating the vesicle and recycling the clathrin.

Hsc70 and its J protein auxilin[216,217] utilize ATP energy to disassemble clathrin cages into "triskelia". This results in the "stripped" vesicle that is now able to fuse with the target endosome. The uncoating reaction can be recapitulated *in vitro*, where the C-terminal fragment of auxilin that includes the clathrin-binding site and J domain is sufficient to dissociate clathrin baskets in the presence of Hsc70 and ATP.[218] Moreover, the structures of the proteins that participate in this process have been determined.[219–224]

Auxilin is recruited after a coated vesicle buds. This specialized J protein binds directly to the triskelia[46,219,220,225,226] and recruits Hsc70, which binds near the C-terminus of the triskelion at the base of the tripod. Hsc70 causes a distortion of the clathrin lattice, which is proposed to trap a transient fluctuation, and accumulation of this local strain at multiple vertices of the cage leads to disassembly.[226] The uncoating reaction discharges the clathrin and chaperone machinery back into the cytoplasm to be re-used in another round of clathrin-coated vesicle formation.

The QLMLT sequence near the C-terminus of a clathrin heavy chain is the required binding site for Hsc70-mediated uncoating. Three sites in the clathrin trimer are exposed under each vertex of a coat in a narrow space. Auxilin binding causes local displacements of the ankle segments leading to expansion of the structure allowing a single Hsc70 to access its target. ATP hydrolysis clamps this Hsc70 locking the locally distorted structure. By using sensitive single-particle fluorescence experiments with high time resolution the Hsc70/Auxilin-dependent disassembly of clathrin coats was followed.[227] A mechanism whereby Hsc70 traps a conformational fluctuation at one vertex for long enough to allow more fluctuations at other vertices that can reinforce each other in destabilizing the coat was proposed. Also, it was observed that the sequence and location of the specific Hsc70 binding site is critical for the reaction as Hsc70 binding elsewhere on the coat fails to facilitate the coat disassembly.[46,227]

3.7.4.2 Exocytosis

Hsc70 exploits its chaperone functions to mediate cellular exocytosis of synaptic vesicles and pancreatic zymogen, insulin and mucin granules.[228–233] In particular, the role of Hsp70s proteins has been studied extensively in calcium-dependent exocytosis of synaptic vesicles, where localization and specificity is dictated by the J protein: Hsc70 gets recruited to the vesicles by the J domain-containing Cys string protein (CSP or DnaJC5). Deletion mutants of CSP in *Drosophila* undergoes neurodegeneration and synaptic dysfunction.[234] Interestingly, CSP mutants in *Drosophila* show the same phenotype as upon mutation of the Hsc70 gene.[235] CSP acts as an organizer of protein–protein interactions, and with Hsc70 it stabilizes the proteins involved in exocytosis and their complexes.[236] In these complexes, CSP and Hsc70 have been found associated with another co-chaperone, a small glutamine-rich, TPR-containing protein (SGT). SGT has the ability to bind Hsc70 through its TPR domain and CSP through its N-terminus. Over-expression of SGT in cultured neurons inhibits neurotransmitter release.[237]

SNARE ("soluble *N*-ethylmaleimide sensitive factor attachment protein receptor") proteins bring into proximity the lipid bilayers of membranes of the cell and secretory vesicle and induce their fusion. SNARES are partially folded and prone to non-specific interactions.[229] The vesicle-membrane associated Hsc70/CSP/SGT complex binds directly to the Synaptosomal-associated protein 25 (a SNARE protein on the target membrane) and maintains its functional competence to participate in complex formation.

Furthermore, the Hsc70/SCP/SGT chaperone system acts at many stages of the vesicle secretion process: for example, during uptake of GABA into synaptic vesicles, CSP and Hsc70 form a complex with glutamate decarboxylase (which catalyzes the decarboxylation of Glu to GABA), the vesicular GABA transporter and the Ca^{2+}-calmodulin dependent kinase.[228,238] During targeting and docking of synaptic vesicles, CSP and Hsc70 form a complex with the α-GDP-dissociation inhibitor, which regulates the cycling of Rab3a, involved in the targeting and docking of vesicles.

3.7.4.3 Pre-protein Transport to Organelles

3.7.4.3.1 Mitochondria. Mitochondria are organelles of α-proteobacterial origin acquired by eukaryotic cells. However, because most mitochondrial genes were transferred to the cellular nucleus during evolution, precursors of most mitochondrial proteins are synthesized in the cytosol and have to be imported into the organelle. Mitochondrial protein precursors generally carry a targeting sequence, which insures their productive interaction with the import apparatuses of the outer and inner mitochondrial membranes, and most are imported post-translationally (reviewed in ref. 239). To be competent for mitochondrial import, precursor proteins must be unfolded.[240] While there may be some examples of co-translational import or localization of mRNA for mitochondrial precursors near the import apparatus,[239,241] in general there is not direct coupling of translation with translocation into mitochondria as there is for entry into the endoplasmic reticulum (ER) lumen (see below). Instead, the cell utilizes cytoplasmic Hsp70s to keep mitochondrial pre-proteins from aggregating or being stably folded, thus maintaining them in a "translocation-competent state".[242,243] In mammals, mitochondrial outer membrane receptors Tom70 and Tom20 are responsible for translocation of cytosolic Hsc70- and Hsp90-bound proteins to the mitochondria. Tom70 contains seven cytosolic TPR motifs[240,244] that mediate the interaction between Tom70 and cytosolic Hsp70 and Hsp90. The recognition of the pre-protein by Tom70 is dependent on its interaction with these cytosolic chaperones. Disruption of this interaction impairs client translocation. Tom70 preferentially binds to precursors with internal targeting sequences, like members of the mitochondrial carrier family.[245]

Tom20, which also contains TPR domains, interacts with a peptidyl-prolyl isomerase-type chaperone (AIP), which acts as an adaptor to mediate the interaction of the pre-protein with Tom20.[246] AIP interacts with Hsc70 as well, and recognizes positively charged N-terminal amphipathic targeting helices of the precursors for mitochondrial matrix proteins. Although they have different specificities, mutagenesis in yeast revealed that the functions of Tom70 and Tom20 overlap.[240] By using human proteins, and employing cross-linking, co-precipitation and NMR techniques, it was shown that the TPR-binding motif at the C-terminus of Tom20 is necessary to interact with the TPR clamp of Tom70. Tom20 competes with the chaperones for interaction with Tom70, but blocking of Tom20 disfavored the import of both Tom20- and Tom70-dependent pre-protein. This interaction between the receptors was proposed to represent a mechanism to displace the molecular chaperones so that pre-proteins can be released for translocation.[247]

Mammalian cells utilize both Hsp90 and Hsc70 to transfer pre-proteins to the TOM complex. Several studies showed the essentiality of association between Hsc70 and mitochondrial precursors.[240,248] For example, depletion of Hsc70 from reticulocytes lysate renders precursors to ornithine transcarbamylase incapable of being imported.[249] This impairment can only be overcome when Hsc70 is supplied during translation (and not during import), supporting the notion that Hsc70 associates with the nascent protein during translation and maintains it in a translocation-competent conformation.[249] The same authors co-immunoprecipitated mitochondrial precursor proteins with Hsc70.[250] A hybrid precursor protein constructed by fusing the mitochondrial matrix-targeting signal of rat pre-ornithine carbamyl transferase to murine cytosolic dihydrofolate reductase could be translocated to mitochondria if diluted from denaturing conditions into import medium with purified mitochondria lacking cytosolic extracts. Purified Hsp70 prevented aggregation of the precursor and retarded its folding.[251]

3.7.4.3.2 Peroxisomes. Found in virtually all eukaryotic cells, peroxisomes are organelles that are involved in the catabolism of long chain fatty acids, branched chain fatty acids, D-amino acids, polyamines and biosynthesis of plasmalogens.[252] Proteins destined for the peroxisomal matrix are synthesized in the cytosol on free ribosomes and imported post-translationally as stably folded proteins. They are targeted through two types of peroxisomal targeting signals (PTS): PTS1 is a carboxy-terminal sequence, ending in the consensus sequence SKL, which is not cleaved after import; PTS2 sequences are found at the N-terminus of a small group of peroxisomal matrix proteins, and the tag undergoes cleavage upon import. Each type of targeting sequence has distinct cellular receptors.[240,253] Proteins bearing PTS1 targeting sequence form complexes with peroxisomal receptor Pex5p in the cytosol and then travel to peroxisomes where they shuttle on and off the membrane for targeting instead of forming a dynamic or stable complex at the membrane. Pex5P belongs to the TPR protein family, and the TPR domain is involved in the recognition of PTS1.

Peroxisomes do not contain chaperones and although it is recognized that Hsp70s are not required for import,[240] Hsc70 was proposed to enhance the translocation efficiency:[254,255] for example, when using a rat system to study the involvement of Hsc70 in the process, import of acyl-CoA oxidase to peroxisomes was synergistically enhanced by Pex5p and Hsp70;[254] also, it was shown that PTS2-containing reporter proteins were imported to the peroxisome in a process that required Hsc70 and Hsp40 in addition to the signal receptors.[256]

Thus, the involvement of Hsp70s in the process of peroxisome protein import is essentially different from the role they play in protein import to mitochondria. Here, Hsp70 seems to enhance protein import through a process other than the unfolding of cargo proteins.

3.7.4.4 Protein Translocation into Organelles

3.7.4.4.1 mtHsp70. mtHsp70 is the mitochondrial Hsp70, generally considered to be the major mitochondrial chaperone and import motor,[257,258] although this Hsp70 has been reported in other cellular localizations like ER, vesicles and cytosol.[259,260] Intriguingly, mtHsp70 folds and keeps its structural integrity in the mitochondrial matrix with the aid of a conserved and specialized partner Hep1 (also named Tim15 and Zim17).[261]

As described previously, most mitochondrial proteins are nuclear encoded, synthesized as precursor proteins on cytosolic ribosomes, and targeted for mitochondrial import by the presence of N-terminal pre-sequences. They cross the mitochondrial membranes as unfolded proteins. These precursors translocate across the Tom complex, which constitutes the general entry gate for all mitochondrial proteins.[262] After that, sorting pathways diverge to send precursor proteins to their mitochondrial locations. Proteins destined for the outer membrane are sorted by SAM, also named TOB. Inner-membrane and mitochondrial matrix proteins are sorted through the Tim complex, where Tim23 constitutes the central unit, while specific components regulate the two processes.

Before they undergo proteolytic processing and folding inside the mitochondrial matrix (usually with the help of chaperones), matrix mitochondrial pre-proteins are translocated by the concerted action of the electric potential across the inner membrane, $\Delta\psi$, and the motor system of which mtHsp70 is the only ATPase.[263] mtHsp70 is the human Hsp70 closest in sequence to DnaK. mtHsp70 is tethered to the Tim23 complex on the matrix side by Tim44 and the pre-sequence translocase-associated motor (PAM); PAM is a complex of J proteins: Pam17, Pam18 (Tim14 or DnaJC19 in humans) and Pam16. mtHsp70 works with the nucleotide exchanger GrpE-L1/2.[264] It appears only Pam18 can stimulate the ATPase activity of mtHsp70.[265,266]

mtHsp70 meets the pre-protein as it enters the mitochondrial matrix and, through interaction with Pam16/Pam18, binds stably to the pre-protein; then the mtHsp70/pre-protein complex leaves the Tim machinery at the membrane.[267–269] To explain how mtHsp70 provides a driving force for translocation across the inner membrane, two different models have been proposed:[270] the biased diffusion (or Brownian ratchet) model basically considers that the protein can go in either direction, but mtHsp70 binding prevents it from moving backwards into the cytosol, which eventually leads to translocation of the entire protein into the matrix without active pulling.[271] The power stroke model[272,273] proposed that Hsp70 applies an unfolding force on the pre-protein, where ATP hydrolysis induces a conformational change within the Tim44–mtHsp70 complex, using Tim44 as a fulcrum, and mechanically pulling the protein into the matrix while unfolding the non-translocated protein segment in the cytoplasm. To reconcile the evidence compiled to support each model, the "entropic pulling" model was proposed,[270] in which mtHsp70 binding to the substrate and release from the Tim complex translates into an entropic force that can pull the pre-protein into the matrix.[274] In this model, force comes from the limited freedom of movement of the locked Hsp70 caused by the physical presence of the membrane and the pore.

mtHsp70 expression is not induced by heat shock, but it is induced by other forms of stress like glucose deprivation, low-level radiation and some cytotoxins.[275] mtHsp70 undergoes a cellular redistribution in immortal cells (where it is perinuclear) compared to normal cells, and this redistribution is reversible upon induction of senescence.[276–278] mtHsp70 expression is elevated in many human tumor tissues and cells, and such up-regulation grants malignant characteristics to the transformed cells,[279] where, not surprisingly, mtHsp70 has a pro-survival role.[257] For example, p53 translocates to mitochondria where it induces apoptosis;[280–282] mtHsp70 can remove p53 from the p53-BclxL/2 complex and target it to proteasomal degradation.[257] Also, mot-2 (a mutant form of mtHsp70) can interact in the cytoplasm with p53, resulting in its cytoplasmic retention and transcriptional inactivation; the exclusion of p53 from the nucleus has been proposed as a possible mechanism of its inactivation in some tumors.[283]

mtHsp70 is multi-functional. For protein translocation mtHsp70 chaperone partners with Tim14, but it partners with the J protein DnaJA3 (hTid1) for the folding/refolding of mitochondrial proteins.[284,285] mtHsp70 also facilitates degradation of misfolded proteins by interacting with matrix-localized proteases of the AAA+ family.[258,286,287]

Additionally, mtHsp70 is involved in the process of producing Fe-S containing proteins.[288] Mitochondria assemble their own Fe/S proteins, and also participate in the biogenesis of cytoplasmic and nuclear counterparts.[289] Biogenesis of Fe-S proteins by the "ISC assembly machinery" serves as an essential regulator of cellular iron homeostasis. Biogenesis of mitochondrial Fe/S proteins, which is a conserved mechanism, commences with a complex process by which the [2Fe-2S] cluster is synthesized *de novo* on a Cys-containing scaffold protein IscU. Once the cluster is assembled, mtHsp70 (along with at least ten other proteins) participates in a complex process by which the Fe/S cluster gets dislocated from the IscU scaffold protein, to be subsequently inserted into the target apoproteins by specific targeting factors. The J protein Hsc20 helps recruit IscU to the ATP-bound mtHsp70. IscU acts as a native substrate for both the J protein and mtHsp70. Both IscU and Hsc20 stimulate mtHsp70 ATPase activity, and the ATPase cycle is completed by GrpEL1/2-assisted ADP-ATP exchange[289] with concomitant substrate release.

Mitochondria are often closely associated with the ER, thereby establishing dynamic and advantageous communication, and mtHsp70 has been proposed to play a role in calcium regulation in the ER.[290,291] mtHsp70 can provide scaffolding contacts between ER and mitochondria, thus determining the number of places where mitochondria are exposed to the high Ca^{2+}-micro-domains generated by IP$_3$R. Also, mtHsp70 can control the interaction of ER and mitochondrial proteins, thus facilitating communication between ion channels.[257,291]

3.7.4.4.2 BiP. BiP (or Grp78) is the dominant Hsp70 protein found in the ER lumen. It is expressed under all growth conditions, but its expression is stimulated by accumulation of unfolding polypeptides under ER stress. The ER is implicated in many crucial cellular functions like secretory protein biogenesis, modification, export and insertion in membranes, calcium homeostasis and signal transduction. The versatile player BiP is involved in each of those processes.[292] While BiP is the most abundant Hsp70 in the ER, distant relatives of Hsp70s (the glucose-regulated protein Grp170 or Hyou1 and Stch, with only the ATPase domain) are also found in the ER. Grp170 can be found in a complex with BiP in the absence of ATP.[293–295]

BiP functions rely heavily on partnerships with co-chaperones. Around ten different Hsp40-type co-chaperones have been identified in the human ER. The characterized members are named Erj1 to Erj7; some of these are soluble proteins whereas others are membrane proteins with a lumenal J domain (see ref. 292 and Table 3.2). Because of the presence of additional domains besides the J domain, Erj3 and Erj4 have the ability to deliver substrates to BiP.[292] Erj4, Erj7, Erj2 (Sec63) and Erj1 possess different types of transmembrane domains, whereas Erj3, Erj5 and Erj6 (p58^{IPK}) are luminal soluble proteins. Erj5 also contains four thioredoxin domains, and Erj6 (p58IPK) contains a TPR domain. ErJ5 and ErJ6 can also play a role in substrate binding.[296–299]

Sil1 (or BAP) and Grp170 interact with BiP as nucleotide exchange factors, thus promoting efficient exchange of ADP for ATP.[126,300] Sil1 is related to cytosolic HspBP1, and Grp170 is related to Hsp110 (an alternative NEF of Hsc70). Other partners of BiP are other chaperones/lectins like Grp94, calreticulin, calnexin; folding catalysts like protein disulfide isomerases (PDIs) and peptidylprolyl-*cis*/*trans*-isomerases (PPIases); the Sec61 complex; UDP-glucose-glycoprotein-glycosyltransferase (UGGT); stromal cell-derived factor-2-like 1 (SDF2L1); calcium calumenin, reticulocalbin; and the mediators of the UPR: IRE1, ATF6, PERK, Sig-1R. Importantly, some BiP interaction partners are not present in every cell type.

Proteins targeting to the ER starts when the pre-protein (bearing a signal sequence for either insertion in the membrane or secretion) is targeted to the ER membrane by the signal recognition particle and its receptor at the membrane; simultaneously, the ribosome engages the translocon machinery at the ER membrane. The translocon is a dynamic proteinaceous aqueous complex through which nascent chains travel to the ER lumen or get inserted into the ER membrane.[301] It is assembled from four core proteins (the heterotrimeric Sec61α, Sec61β, Sec61γ, and the translocating nascent chain-associated membrane protein (TRAM)) plus various accessory proteins.[302] After insertion and prior to translocation, the signal peptide on the pre-protein gets proteolytically removed by a signal peptidase located at the ER membrane, and the chain undergoes modifications, like modification by oligosaccharides. In the translocation process, BiP is involved in the opening of the Sec61 channel; also, it binds to the nascent polypeptide and drives translocation. BiP has been implicated (directly or indirectly) in maintaining the membrane permeability barrier in the translocon at the ER luminal side while the cytosolic end is sealed by the translating ribosome.[303] While translating a secretory protein into the ER lumen, the channel broadens and the seal is maintained on the cytoplasmic side by the translating ribosome.[303,304] When a multi-spanning membrane protein is translated, the closures provided by the ribosome and BiP alternate depending on whether the nascent chain segment is directed to the endoplasmic reticulum lumen or the cytosol.[305] After translocation, protein folding occurs in the ER, and BiP is a key player in promoting folding and inhibiting aggregation. For example, the C_H1 domain of immunoglobulins binds to BiP for efficient assembly of immunoglobulins[306] along with Erj3 and Sil1.[297,307]

Some secretory proteins can translocate post-translationally into the ER. This process occurs essentially as described for the process of importing proteins into mitochondria. For post-translational import, the channel partners are the Sec62 and Sec63 at the membrane and BiP at the ER lumen.[301] Basically, the substrate (kept in a translocation-competent state by cytosolic chaperones) binds to the translocon where it is stripped of all chaperone interactions.[308] Once translocation begins, it is helped by BiP, most likely through the entropic pulling mechanism.[270] Pre-protein binding and binding to the J domain of Sec63 stimulate BiP ATP hydrolysis, allowing the SBD to close and interact tightly with the substrate. As the protein enters the ER lumen, the next BiP molecule can bind. This mechanism proceeds until the complete polypeptide has crossed the translocon,[302] when BiP exchanges ATP for ADP for ATP and releases the substrate into the lumen.

Proteins undergo strict quality control in the ER. Misfolded or incompletely processed proteins (as a consequence of mutation or stress) are translocated back to the cytoplasm, where they are ubiquitylated and degraded by the 26S proteasome.[309] This process is called ER-associated degradation (ERAD), and numerous human diseases have been associated with it.[310] The Sec61 complex, Sec63 and BiP may be involved in at least one (but not the only) pathway for the retro-translocation of some substrates to the cytosol.[311] Erj3 through Erj5 have been implicated in ERAD in mammalian cells.[296,299,312] Some ERAD substrates can be exported to the cytoplasm still containing folded domains,[313] while others have to fully unfold to be dislocated, likely requiring disulfide bond breakage. The main candidate for this task seems to be ERdj5, which, in addition to the J domain, possesses four thioredoxin motifs, and binds BiP and EDEM1 (the ER stress-induced mannose-binding lectin that regulates the removal of folding-defective glycoproteins[314]). It was proposed that the substrate protein transfers from calnexin to the EDEM1-ERdj5 complex and then to the retro-translocation channel, while being chaperoned by BiP.[315]

The ER plays a central role in calcium homeostasis and signal transduction by releasing Ca^{2+} to the cytoplasm upon plasma membrane receptor stimulation using the ER membrane ryanodine-receptors (RyR) or inositol-1,4,5-trisphosphate-receptors (IP_3R), or by refilling the ER-calcium stores by using the sarcoplasmic ER calcium (SERCA) pumps.[292] Certain secretory proteins may require high calcium concentrations during their folding and assembly. BiP is a chaperone with low affinity for Ca^{2+} (CaBPs proteins), and its activity is modulated by Ca^{2+} concentration.[316]

As described above, BiP participates in the signaling process between ER and mitochondria, and it is associated with the protein sigma-1 receptor (Sig-1R, which is also a chaperone). Upon Ca^{2+} depletion, Sig-1R dissociates from BiP and associates with IP$_3$R, protecting it from ERAD and extending the Ca^{2+} efflux to the mitochondria. The role of BiP in initiation and transmission of the UPR will be described later in this chapter.

Notably, BiP is a key player in promoting cell survival, as it is involved in protecting cancer cells from ER stress-induced apoptosis.[317]

3.7.5 ER Stress

Secretory and membrane proteins are synthesized by ER-bound ribosomes and inserted into the ER membrane or delivered into the ER lumen where a complex network of chaperones helps proteins fold and undergo modification, and performs quality control surveillance.[222,318] In the ER, an oxidizing environment is maintained as well as a delicate balance of Ca^{2+} and ATP. Proteins that fail quality control are retained in the ER until properly folded, or alternatively they may be subjected to degradation either by the proteasome-dependent ER-associated protein degradation (ERAD) pathway or by autophagy. Certain conditions like heat shock, mutations, perturbation on the Ca^{2+} levels, inadequate energy sources or imbalances in the redox properties of the ER create ER stress, leading to accumulation and/or aggregation of ER proteins. These conditions turn on the unfolded protein response (UPR) in order to protect the cell and restore normal ER function.[32,319] If chronic or severe stress persist, this protective mechanism ensures that the cell dies by inducing a pro-apoptotic response.[31]

The ER-resident chaperone Hsp70, BiP plays a central role in the UPR and in triggering ER stress-induced apoptosis.[318,320,321] This Hsp70 chaperone binds to folding intermediates of ER proteins, preventing aggregation and facilitating folding (described in Section 3.7.4.4.2).

When stress occurs, the pool of unfolded proteins increases, triggering activation of the receptors that mediate the UPR: pancreatic ER kinase or PKR-like ER kinase (PERK), activating transcription factor-6 (ATF6) and inositol-requiring enzyme 1 (IRE1).[318,320,321] In addition to the ATF6, PERK and IRE1 systems, mammals have some tissue-specific UPR sensors, most of which are transmembrane bZIP transcription factors that are activated by regulated intramembrane proteolysis similar to ATF6.[318]

IRE1 is the most ancient of the three ER-signaling receptors. This transmembrane receptor possesses a Ser-Thr kinase domain facing the cytoplasm and a C-terminal endoribonuclease domain; it is activated by accumulation of unfolded proteins in the ER lumen, most likely by direct binding,[322] although interaction with BiP may play a role as well. In the UPR, IRE1 is auto-transphosphorylated, and activated IRE1 removes a 26-nucleotide from the XBP1 mRNA transcript, turning it into a strong transcription factor[323] with different targets including ER chaperones and p58^{IPK}, an Hsp40 protein[324] that functions as a BiP co-chaperone.[325] The resulting response helps cells cope with stress on the ER folding capacity. Under prolonged ER stress, IRE1 becomes pro-apoptotic. IRE1 activates nuclease activity that turns over mRNAs for membrane and secreted proteins, through the regulated IRE1-dependent decay (RIDD) pathway.[326] Additionally, IRE1 kinase binds to the TNF receptor associated factor-2 (TRAF2), promoting a cascade of phosphorylation events and eventual activation of Jun amino-terminal kinase, JNK, which is pro-apoptotic.[222,327]

PERK possesses a luminal ER dimerization domain and a cytosolic kinase domain. Upon accumulation of unfolded proteins, PERK oligomerizes and auto-phosphorylates. PERK then phosphorylates the α subunit of the eukaryotic initiation factor 2 (eIF2α), which inhibits global protein translation[33] thus preventing further accumulation of unfolded proteins. mRNA for transcription factors (like "ATF4") that promote transcription of genes involved in amino acid metabolism, redox reactions, stress response and protein secretion[328] can circumvent the translational arrest and be translated at higher rates[32] to promote cell survival. Other genes induced by ATF4, like the transcription factor C/EBP homologous protein (CHOP), promote apoptosis by up-regulating apoptosis-related genes.[329–331] Thus, activation of PERK leads to a delicate balance between pro-survival and pro-death responses.

Accumulation of unfolded proteins causes ATF6 (an ER-associated type 2 *trans* membrane basic Leu zipper (bZIP) transcription factor) to be packaged into vesicles for delivery to the Golgi, where it undergoes specific proteolysis; the resulting N-terminal fragment is transported to the nucleus and induces expression of ER-resident genes including BiP, Grp94, protein disulfide isomerase and XBP1, the transcription factor important in IRE1 signaling, which help the cell cope with the ER stress.[332] Intriguingly, the expression of CHOP is also elevated, linking the ATF6 activation pathway directly to pro-apoptotic signaling, and pointing to a highly integrated response by the three different branches of the UPR.

Hsp70s and its co-chaperones are implicated in ER stress signaling at both the cytoplasmic and the ER-luminal side. As described above, BiP is a central player in the luminal compartment by participating directly in protein homeostasis. ATF6 associates with BiP, and BiP release under conditions of ER stress may contribute to its activation.[320] On the cytoplasmic side, overexpression of HspA1 has been reported to enhance cell survival under ER stress conditions by a direct interaction with the cytoplasmic domain of Ire1 and resulting enhancement of XBP1 mRNA splicing activity.[333] This kind of interaction among ER stress mediators and cytoplasmic Hsp70s would provide means to integrate cellular stress pathways and UPR signaling.[318,320,321]

The UPR adapts the ER capacity to the load of unfolded proteins in response to the physiological conditions by up-regulation of BiP expression in the ER. However, when the load of unfolded substrates is reduced, high amounts of BiP compete with folding thus promoting degradation and reducing yield of folded protein folding.[334,335] To avoid this outcome of the high luminal BiP concentration, the cell quickly and reversibly inactivates BiP by adenosine diphosphate (ADP)-ribosylation.[336] This modification (likely enzymatic) at two conserved positions of the SBD of BiP adds specific charges that hamper the allosteric communication of the domains, interfering with substrate binding. This rapid inactivation allows proteins to be released to fold, thus protecting the ER from excess unfolded protein, possible aggregation and energetically expensive degradation of unfolded proteins bound to BiP.[336]

3.7.6 Apoptosis

3.7.6.1 General Roles of Hsp70s in Apoptosis

Apoptosis or programmed cell death is a tightly regulated mechanism by which cells die in multi-cellular organisms.[337] Apoptosis generally occurs as an advantageous protective mechanism to remove damaged or unhealthy cells that may be detrimental for the organism. Failure of a cell to undergo apoptosis if needed can cause a myriad of diseases, including uncontrolled cell proliferation and cancer. On the other hand, unnecessary apoptosis can cause atrophy, neurodegenerative diseases and tissue damage.[337]

Apoptosis is executed through extrinsic and intrinsic pathways, depending on where the "death" signal comes from and with crosstalk between them. Both pathways ultimately may culminate with the activation of the family of Cys protease-effector proteins caspases that work in a proteolytic cascade to disassemble and destroy the cell.[337] Also, apoptosis can be carried out in a caspase-independent way (see below). Hsp70s inhibit apoptosis and promote cell survival by acting on the caspase-dependent pathway at several steps both upstream and downstream of caspase activation, and also on the caspase-independent pathway.[338] Hsp70 regulate apoptosis at different levels affecting intrinsic and extrinsic pathways.

Extrinsic pathways begin outside a cell with binding of death ligands of the tumor necrosis factor (TNF) super family to their cell surface death receptors. This signal is transmitted to the cytosol where a cellular death cascade is triggered by cleavage and activation of procaspase-8. Caspase-8 can induce mitochondrial outer membrane permeabilization (MOMP) by cleaving BID (Bcl-2 interacting protein), which subsequently translocates its C-terminal portion into mitochondria where it triggers cyt c (Cytochrome-c) release. Mitochondrial factors released to the cytoplasm (including cyt c, Apaf-1 and EndoG) promote apoptosis through caspase-dependent and independent pathways (see below). Also, caspase-8 can trigger the protease cascade of effector caspases-3,-6 and/or -7 that will lead to cell nuclear signals and cell death.[339]

Over-expression of Hsp70 blocks TNF-induced apoptosis. For example, Hsp70 interferes with the BID-dependent apoptotic pathway *via* inhibition of the c-Jun N-terminal kinases.[340] At the level of the death receptors, Hsp70 can bind the receptors (for example DR4 and DR5) therefore inhibiting the assembly of the complex that propagates the signal for apoptosis.[341] TNFα also induces the activity of the pro-apoptotic double-stranded RNA-dependent protein kinase (PKR). Hsp70 inhibits PKR by interacting with its inhibitor FANCC. Thus, the ternary complex Hsp70-Hsp40/PKR/FANCC inhibits apoptosis.[342,343] Even at very late stages of apoptosis Hsp70 is able to rescue the cell from death: the DNA-digestion activity and structural integrity of CAD, a caspase-activated DNAse, is regulated by Hsp70, Hsp40 and a specific CAD inhibitor (ICAD) downstream of caspase-3.[344] Remarkably, Hsp70 is able to inhibit late caspase-dependent events like activation of cytosolic phospholipase A2 (which causes a collapse in the plasma membrane potential) and changes in nuclear morphology.[36]

Conditions that alter intracellular homeostasis, like DNA damage, oxidative stress, starvation[345] and mitochondrial danger signals (for example reactive oxygen species (ROS) or mitochondrial-specific UPR[223]) also trigger MOMP. Cyt c discharge from the mitochondrion induces oligomerization of Apaf-1 (Apoptotic protease activating factor 1) and activation of the initiator caspase-9 forming a complex known as "apoptosome".[346] Activation of caspase-9 cleaves and activates the execution caspases -3, -6 and/or -7, which in turn cut cellular targets that will dismantle the cell.[328] At the mitochondrion, Hsp70 blocks heat-induced apoptosis by impeding Bax (a direct executor of MOMP) translocation and insertion in the mitochondrial membrane, thus preventing exit of the mitochondrial factors that initiate apoptosis. Direct interaction between HspA1 and Bax was shown by immunoprecipitation,[347] and this function requires the presence of both ATPase and SBD domains of Hsp70.[348]

Inhibition of caspase-9 by Hsp70 has an impact in Alzheimer's disease. This disease is characterized by the accumulation of senile plaques made out of β-amyloid (Aβ) and neurofibrillary tangles assembled from tau protein into β-containing oligomers likely responsible for cell death in Alzheimer's disease. The role of HspA1 in the progression of this disease has been widely studied (reviewed in ref. 14). Despite blocking *in vitro* the early steps of Aβ aggregation, HspA1 protects from Aβ-induced cytotoxicity by inhibiting caspase-9, which accelerates the elimination of Aβ.[349]

After mitochondrial permeabilization, Hsp70 inhibits apoptosis downstream of the release of cyt c but upstream of activation of caspase-3. Importantly, HspA1 interacts with procaspase-3 and -7 hampering their maturation thus inhibiting all the subsequent apoptotic stages.[350] At the post-mitochondrial level, HspA1 interacts with Apaf-1 preventing its recruitment with procaspase-9 to the apoptosome,[351] although it has been proposed that the inhibition of apoptosis by interference with the caspase cascade is the result of the inhibition of factors released from the mitochondrion and not because of a direct effect on Apaf-1.[352]

In the caspase-independent pathway the flavoprotein apoptosis inducing factor (AIF, a mitochondrial inter-membrane protein released after an apoptotic stimulus) translocates to the nucleus where it activates and initiates chromatin condensation and large-scale DNA fragmentation.[353] Additionally, poly(ADP-ribose) polymerase activation could also contribute to caspase-independent apoptosis: during incomplete energy failure poly(ADP-ribose) polymerase activation contributes to mitochondrial depolarization, AIF release and apoptosis.[221] Endonuclease-G is a mitochondrial nuclease that moves to the nucleus and digests DNA during apoptosis independently of DNA-Fragmentation Factor and Caspase-Activated DNAse (CAD). Hsp70 proteins inhibit caspase-independent apoptosis.[354,355] For example, Hsp70 can prevent serum withdrawal-induced apoptosis in an Apaf-1-/- cell.[9,355] Moreover, Hsp70 was found to directly bind to AIF inhibiting its translocation to the nucleus and the subsequent chromatin condensation.[47,355–357]

p53 (protein 53) is primarily a transcriptional activator implicated in the regulation of the cell cycle that functions as a "tumor suppressor", and its mutant forms are selected in almost all types of cancer.[358] Activation of p53 involves a sharp increase in its cellular amount and stabilization of the short-lived p53, making it able to modulate gene expression that results in growth arrest, to prevent replication of the damaged DNA, transcription of genes that encode proteins involved in DNA repair and cell suicide by apoptosis as a last resource to avoid propagation of cells with damaged DNA. p53 initiates apoptosis by transcriptional activation of genes that encode death receptors and by transcriptional repression of the anti-apoptotic genes *bcl-2*, and also acts through a transcription-independent mechanism by direct interaction with the apoptosis machinery at the mitochondria,[280,282,359,360] Stress-induced accumulation of p53 in the cytosol or mitochondria leads to direct activation of Bax and/or Bak (an inductor of MOMP similar to Bax). Oncogenic mutations of p53 are known to additionally destabilize p53 structure and/or directly disrupt its DNA contact resulting in oncogenic gain-of-function.[361] It was observed that transgenic *Hsf1*-knockout carrying p53 oncogenic mutation mice do not develop cancer, proposing that heat-shock proteins contribute to mutant p53 stabilization thus promoting tumor progression.[362] Hsp90 and Hsp70s partner together to neutralize the effects of *p53* missense mutations and to suppress the unstable phenotype of the WT protein.[363] Mdm2-dependent ubiquitination and degradation of p53 R175H mutant protein in mouse embryonic fibroblasts can be inhibited by HspA1 and it was proposed that interaction leads to an increase in the half-life of the mutant p53.[364]

3.7.6.2 Hypoxia

Hypoxia-inducible factors (HIFs) are conserved transcription factors responsible for mediating adaptation to hypoxia, leading to the transcriptional induction of genes that participate in angiogenesis, iron metabolism, glucose metabolism and cell proliferation/survival.[360] Up-regulation of heat-shock proteins plays a role in adaptation to hypoxic (and hyperosmotic) stress.[365] A central mechanism in this adaptation process is the impact of Hsp70 and co-chaperones on HIF-1, which is composed of a constitutively expressed subunit HIF-1β and an oxygen-regulated subunit HIF-1α (and its paralogs HIF-2α and HIF-3α). HIFα subunit is readily degraded in normoxia, but is stable under hypoxia. During hypoxia, HIF-1α is resistant to prolyl hydroxylation and translocated to the nucleus,[366,367] where it dimerizes with HIF-1β, and leads to transcriptional activation of genes that promotes adaptation.[368,369] HIF-1α constitutes a client protein of Hsp70, which promotes HIF-1α degradation through the proteasomal pathway,[370,371] and such oxygen-independent degradation depends on the recruitment of CHIP. Hsp70- and CHIP-dependent proteasomal targeting constitutes a molecular mechanism by which the levels of HIF-1α but not HIF-2α protein are selectively reduced under prolonged hypoxia.[372] In this scenario, disruption of Hsp70-CHIP interaction impairs chaperone-mediated degradation of HIF-1α. Moreover, inhibition of Hsp70 or CHIP synthesis increases the levels of HIF-1α protein but not HIF-2α, thus attenuating the decay of HIF-1α levels during prolonged hypoxia.

As HIFs are responsible for the induction of genes that facilitate adaptation and survival of cells and organisms from normoxia to hypoxia,[360,366,373] they play a crucial role in cancer. For example, in renal cell carcinoma defective function of the VHL tumor suppressor ablates proteolytic regulation of hypoxia-inducible factor α subunits (HIF-1α and HIF-2α, which have contrasting properties in this tumor), leading to constitutive activation of hypoxia pathways.[374]

3.7.6.3 Autophagy

As described in Section 3.7.3.2, three forms of autophagy exist: macro autophagy, micro autophagy and chaperone-mediated autophagy. The most studied form of autophagy is macro autophagy, which gives the cell the ability to "eat itself" and plays an essential role under conditions of starvation thus helping cancerous cells to survive.[375,376]

Autophagy is tightly regulated and starts with the formation of a double-membrane-bound vesicle that engulfs cytoplasmic components directed for recycling. The protein LC3-B is a cytoplasmic autophagosome marker that becomes membrane-attached by lipidation to the nascent structure. LC3-B acts as interaction partner to adaptor proteins that target the phagosome to the lysosome, where fusion ends with degradation of the cargo components.[198] A role for autophagy as an alternative mechanism for cell death has been proposed,[375] but it is a matter of debate as evidence that shows direct involvement of autophagy in cell death does not seem to be conclusive. Autophagy could play a tumor-promoting role by providing nutrients to starved tumor cells, thus being both a pro-survival and a pro-death mechanism.[198,375]

As mentioned in Section 3.7.3.2, protein degradation occurs in the lysosomal compartment, where lysosomal enzymes degrade the substrates to amino acids that can be re-used by the cell. Release of the lysosomal enzymes into the cytosol promotes apoptosis. Intralysosomal Hsc70 plays a role in apoptosis by promoting the stability of the lysosome membrane, therefore preventing its permeabilization with the concomitant release of apoptotic factors.[377,378] Hsp70 is localized to the membranes of the endosomes and lysosomes compartment of tumor cells and it inhibits lysosomal permeabilization induced by cytokines, anticancer drugs, γ-irradiation *etc.*[377] Depletion of Hsp70 from tumor cells triggers lysosome membrane permeabilization and cathepsin-mediated apoptosis.

Stressed and cancer cells translocate a small fraction of Hsp70 to the lysosome, where it binds to bis (monoacylglycero)phosphate (BMP), an endolysosomal anionic phospholipid essential cofactor for lysosomal sphingomyelin metabolism.[377,379] Both A and B forms of Niemann–Pick disease are severe lysosomal storage disorders caused by mutations in the sphingomyelinase gene. Cells of patients with these diseases have an accumulation of sphingomyelin within lysosomes and a decreased lysosomal stability. It has been proposed that this phenotype can be corrected by treatment with recombinant Hsp70.[379]

3.7.7 Senescence

Cellular senescence (the "limited" number of divisions that a normal cell can sustain) can be triggered by some DNA damage that results in build-up of the cell cycle inhibitors proteins p16 and p21.[380–382] It constitutes a complex cellular program with end points including growth arrest, cell enlargement, vacuolization, repression and de-repression of certain genes, secretion of various signaling molecules, inhibition of the heat-shock response *etc*. Not surprisingly, tumor cells can overcome the mechanisms of senescence and become "immortalized". These cells reactivate the telomerase and inactivate p16 by mutation or by methylation of the promoter,[383,384] and many anticancer drugs and therapies function by activation of the senescence program.[385,386]

It has been shown that Hsp70 proteins contribute to tumor development in part by controlling senescence programs.[380] Premature senescence in cancer cells depends on activation of p53 and its target, p21. Furthermore, by p53-independent pathway senescence can be activated by oncogenes like Ras or Raf and activation of the ER stress response.[387] It is not then surprising that depletion of Hsp70s (specifically Hsp72, mtHsp70 and Hsp70-2) leads to senescence of cancer cells[388–391] associated with up-regulation of p53. For example, depletion of Hsp72 in cancer cells promotes activation of p53, stimulation of p21 and cell cycle arrest.[380]

3.8 Perspectives

The vast array of cellular functions in which Hsp70s are involved presents a two-edged sword for those contemplating using it as a therapeutic target. On the one hand, it is abundantly clear that inhibition or activation of specific Hsp70-mediated processes offers tremendous promise for therapeutic intervention in cancer or protein misfolding diseases, such as neurodegenerative diseases. But on the other hand, the intricate network of functions that rely on Hsp70s challenges their use as drug targets. How will a desired therapeutic end point be achieved when so many processes will be affected by inhibition or activation of Hsp70s? In addition, the many highly conserved isoforms present in mammalian cells present a daunting challenge for selective targeting of a unique Hsp70. Despite these obstacles, efforts to design therapeutic modalities that interfere in Hsp70-mediated processes are likely to be pursued and could lead to desired outcomes. There are several reasons to believe this: 1) The enhanced understanding of the biology of Hsp70s as a result of expansion of basic research on these chaperones will elucidate the cellular balances under normal and stress states and hone expectations about the consequence of inhibition or activation of particular Hsp70s. For this statement to come true, we need to continue to push hard on basic research on Hsp70s and other protein homeostasis players. 2) Enhanced structural and mechanistic descriptions of Hsp70 functions will yield new insights into how they work. From this information will emerge new potential sites to target with predictable effects on Hsp70 functions. 3) As in the case of proteasome inhibitors, the sensitivity of cancerous cells to modulation of several of the Hsp70-mediated functions will be much greater than that of healthy, less rapidly growing and less stressed cells. Thus, there may be therapeutic advantages realized with acceptable effects on non-cancerous cells.

This review compiles many representative functions ascribed to Hsp70s. Not only were we unable to cover every suggested function, but also in those we did describe we have only been able to give quite superficial treatment. Nonetheless, we hope that the reader has been struck as dramatically as we have by the incredibly widespread and pervasive influence of this family of chaperones. There are some functional specializations observed for particular Hsp70s, but overall these chaperones are highly conserved and functionally redundant. One gains the impression that the cell has networks of proteins involved in different pathways, and that the Hsp70 family of molecular chaperones provides a background medium in which these processes occur. Specificity is realized by the evolution of co-chaperones dedicated to particular processes. This general principle suggests strongly that efforts to target drugs to the Hsp70-mediated processes will require manipulation of the partnership with the specialized co-chaperones.

Acknowledgements

We thank Anastasia Zhuravleva for contributions throughout the process of writing this chapter, Robert Smock for his participation at the early stages of our Hsp70 work and Mandy Blackburn for critical reading of the manuscript. The writing of this review was carried out with the support of NIH grant GM027616 to L.M.G.

References

1. R. J. Ellis, *Trends Biochem. Sci.*, 2006, **31**, 395–401.
2. F. U. Hartl, A. Bracher and M. Hayer-Hartl, *Nature*, 2011, **475**, 324–332.
3. S. Gribaldo, V. Lumia, R. Creti, E. Conway de Macario, A. Sanangelantoni and P. Cammarano, *J. Bacteriol.*, 1999, **181**, 434–443.

4. R. S. Gupta and G. B. Golding, *J. Mol. Evol.*, 1993, **37**, 573–582.
5. C. Hunt and R. I. Morimoto, *Proc. Natl Acad. Sci. USA*, 1985, **82**, 6455–6459.
6. B. Bukau, J. Weissman and A. Horwich, *Cell*, 2006, **125**, 443–451.
7. M. Jaattela, *Exp. Cell Res.*, 1999, **248**, 30–43.
8. J. L. Brodsky and G. Chiosis, *Curr. Top. Med. Chem.*, 2006, **6**, 1215–1225.
9. C. Garrido, M. Brunet, C. Didelot, Y. Zermati, E. Schmitt and G. Kroemer, *Cell Cycle*, 2006, **5**, 2592–2601.
10. S. K. Calderwood, M. A. Khaleque, D. B. Sawyer and D. R. Ciocca, *Trends Biochem. Sci.*, 2006, **31**, 164–172.
11. J. Nylandsted, M. Rohde, K. Brand, L. Bastholm, F. Elling and M. Jaattela, *Proc. Natl Acad. Sci. USA*, 2000, **97**, 7871–7876.
12. S. A. Broadley and F. U. Hartl, *FEBS Lett.*, 2009, **583**, 2647–2653.
13. I. Guzhova and B. Margulis, in *Int. Rev. Cyt.*, ed. W. J. Kwang, Academic Press, 2006, vol. 254, pp. 101–149.
14. C. G. Evans, L. Chang and J. E. Gestwicki, *J. Med. Chem.*, 2010, **53**, 4585–4602.
15. J. M. Barral, S. A. Broadley, G. Schaffar and F. U. Hartl, *Semin. Cell Dev. Biol.*, 2004, **15**, 17–29.
16. E. R. Zuiderweg, E. B. Bertelsen, A. Rousaki, M. P. Mayer, J. E. Gestwicki and A. Ahmad, *Top. Curr. Chem.*, 2013, **328**, 99–153.
17. R. Kityk, J. Kopp, I. Sinning and M. P. Mayer, *Mol. Cell*, 2012, **48**, 863–874.
18. A. Zhuravleva, E. M. Clerico and L. M. Gierasch, *Cell*, 2012, **151**, 1296–1307.
19. M. P. Mayer and B. Bukau, *Cell. Mol. Life Sci.*, 2005, **62**, 670–684.
20. F. U. Hartl, *Nature*, 1996, **381**, 571–579.
21. S. K. Sharma, P. De los Rios, P. Christen, A. Lustig and P. Goloubinoff, *Nat. Chem. Biol.*, 2010, **6**, 914–920.
22. D. R. Palleros, L. Shi, K. L. Reid and A. L. Fink, *J. Biol. Chem.*, 1994, **269**, 13107–13114.
23. R. M. Vabulas, S. Raychaudhuri, M. Hayer-Hartl and F. U. Hartl, *Cold Spring Harbor Perspect. Biol.*, 2010, **2**, a004390.
24. W. B. Pratt, *Annu. Rev. Pharmacol. Toxicol.*, 1997, **37**, 297–326.
25. X. Zhu, X. Zhao, W. F. Burkholder, A. Gragerov, C. M. Ogata, M. E. Gottesman and W. A. Hendrickson, *Science*, 1996, **272**, 1606–1614.
26. M. Akerfelt, R. I. Morimoto and L. Sistonen, *Nat. Rev. Mol. Cell Biol.*, 2010, **11**, 545–555.
27. J. T. Silver and E. G. Noble, *Cell Stress Chaperones*, 2012, **17**, 1–9.
28. C. M. Milner and R. D. Campbell, *Immunogenetics*, 1990, **32**, 242–251.
29. S. D. Westerheide, J. Anckar, S. M. Stevens, Jr., L. Sistonen and R. I. Morimoto, *Science*, 2009, **323**, 1063–1066.
30. S. Fritah, E. Col, C. Boyault, J. Govin, K. Sadoul, S. Chiocca, E. Christians, S. Khochbin, C. Jolly and C. Vourc'h, *Mol. Biol. Cell*, 2009, **20**, 4976–4984.

31. E. Szegezdi, S. E. Logue, A. M. Gorman and A. Samali, *EMBO Rep.*, 2006, **7**, 880–885.
32. M. Schroder and R. J. Kaufman, *Annu. Rev. Biochem.*, 2005, **74**, 739–789.
33. H. P. Harding, Y. Zhang and D. Ron, *Nature*, 1999, **397**, 271–274.
34. N. G. Theodorakis and R. I. Morimoto, *Mol. Cell. Biol.*, 1987, **7**, 4357–4368.
35. B. Liu, Y. Han and S.-B. Qian, *Mol. Cell*, 2013, **49**, 453–463.
36. M. Jaattela, D. Wissing, K. Kokholm, T. Kallunki and M. Egeblad, *EMBO J.*, 1998, **17**, 6124–6134.
37. J. Nylandsted, K. Brand and M. Jaattela, *Ann. NY Acad. Sci.*, 2000, **926**, 122–125.
38. L. Brocchieri, E. Conway de Macario and A. Macario, *BMC Evol. Biol.*, 2008, **8**, 19.
39. H. Kampinga, J. Hageman, M. Vos, H. Kubota, R. Tanguay, E. Bruford, M. Cheetham, B. Chen and L. Hightower, *Cell Stress Chaperones*, 2009, **14**, 105–111.
40. H. Jaiswal, C. Conz, H. Otto, T. Wolfle, E. Fitzke, M. P. Mayer and S. Rospert, *Mol. Cell. Biol.*, 2011, **31**, 1160–1173.
41. M. Kabani and C. N. Martineau, *Curr. Genomics*, 2008, **9**, 338–248.
42. J. Hageman, M. A. van Waarde, A. Zylicz, D. Walerych and H. H. Kampinga, *Biochem. J.*, 2011, **435**, 127–142.
43. C. S. Gassler, T. Wiederkehr, D. Brehmer, B. Bukau and M. P. Mayer, *J. Biol. Chem.*, 2001, **276**, 32538–32544.
44. S. B. Goldfarb, O. B. Kashlan, J. N. Watkins, L. Suaud, W. Yan, T. R. Kleyman and R. C. Rubenstein, *Proc. Natl Acad. Sci. USA*, 2006, **103**, 5817–5822.
45. S. Tanimura, A. I. Hirano, J. Hashizume, M. Yasunaga, T. Kawabata, K. Ozaki and M. Kohno, *J. Biol. Chem.*, 2007, **282**, 35430–35439.
46. I. Rapoport, W. Boll, A. Yu, T. Bocking and T. Kirchhausen, *Mol. Biol. Cell*, 2008, **19**, 405–413.
47. Y. Miyata, J. N. Rauch, U. K. Jinwal, A. D. Thompson, S. Srinivasan, C. A. Dickey and J. E. Gestwicki, *Chem. Biol.*, 2012, **19**, 1391–1399.
48. E. B. Bertelsen, L. Chang, J. E. Gestwicki and E. R. Zuiderweg, *Proc. Natl Acad. Sci. USA*, 2009, **106**, 8471–8476.
49. J. Jiang, K. Prasad, E. M. Lafer and R. Sousa, *Mol. Cell.*, 2005, **20**, 513–524.
50. K. M. Flaherty, C. DeLuca-Flaherty and D. B. McKay, *Nature*, 1990, **16**, 623–628.
51. A. Bhattacharya, A. V. Kurochkin, G. N. Yip, Y. Zhang, E. B. Bertelsen and E. R. Zuiderweg, *J. Mol. Biol.*, 2009, **388**, 475–490.
52. J. F. Swain, G. Dinler, R. Sivendran, D. L. Montgomery, M. Stotz and L. M. Gierasch, *Mol. Cell*, 2007, **26**, 27–39.
53. Y. Zhang and E. R. Zuiderweg, *Proc. Natl Acad. Sci. USA*, 2004, **101**, 10272–10277.
54. A. Zhuravleva and L. M. Gierasch, *Proc. Natl Acad. Sci. USA*, 2011, **108**, 6987–6992.

55. H. J. Woo, J. Jiang, E. M. Lafer and R. Sousa, *Biochemistry*, 2009, **48**, 11470–11477.
56. D. S. Williamson, J. Borgognoni, A. Clay, Z. Daniels, P. Dokurno, M. J. Drysdale, N. Foloppe, G. L. Francis, C. J. Graham, R. Howes, A. T. Macias, J. B. Murray, R. Parsons, T. Shaw, A. E. Surgenor, L. Terry, Y. Wang, M. Wood and A. J. Massey, *J. Med. Chem.*, 2009, **52**, 1510–1513.
57. S. Polier, Z. Dragovic, F. U. Hartl and A. Bracher, *Cell*, 2008, **133**, 1068–1079.
58. A. Arakawa, N. Handa, N. Ohsawa, M. Shida, T. Kigawa, F. Hayashi, M. Shirouzu and S. Yokoyama, *Structure*, 2010, **18**, 309–319.
59. J. P. Schuermann, J. Jiang, J. Cuellar, O. Llorca, L. Wang, L. E. Gimenez, S. Jin, A. B. Taylor, B. Demeler, K. A. Morano, P. J. Hart, J. M. Valpuesta, E. M. Lafer and R. Sousa, *Mol. Cell*, 2008, **31**, 232–243.
60. H. Sondermann, C. Scheufler, C. Schneider, J. Hohfeld, F. U. Hartl and I. Moarefi, *Science*, 2001, **291**, 1553–1557.
61. R. Dominguez, *Trends Biochem. Sci.*, 2004, **29**, 572–578.
62. M. Vogel, M. P. Mayer and B. Bukau, *J. Biol. Chem.*, 2006, **281**, 38705–38711.
63. W. Han and P. Christen, *FEBS Lett.*, 2001, **497**, 55–58.
64. T. Laufen, M. P. Mayer, C. Beisel, D. Klostermeier, A. Mogk, J. Reinstein and B. Bukau, *Proc. Natl Acad. Sci. USA*, 1999, **96**, 5452–5457.
65. A. Buchberger, A. Valencia, R. McMacken, C. Sander and B. Bukau, *EMBO J.*, 1994, **13**, 1687–1695.
66. W. Rist, C. Graf, B. Bukau and M. P. Mayer, *J. Biol. Chem.*, 2006, **281**, 16493–16501.
67. R. C. Morshauser, H. Wang, G. C. Flynn and E. R. Zuiderweg, *Biochemistry*, 1995, **34**, 6261–6266.
68. S. Rudiger, M. P. Mayer, J. Schneider-Mergener and B. Bukau, *J. Mol. Biol.*, 2000, **304**, 245–251.
69. M. Pellecchia, D. L. Montgomery, S. Y. Stevens, C. W. Vander Kooi, H. P. Feng, L. M. Gierasch and E. R. Zuiderweg, *Nat. Struct. Biol.*, 2000, **7**, 298–303.
70. G. Buczynski, S. V. Slepenkov, M. G. Sehorn and S. N. Witt, *J. Biol. Chem.*, 2001, **276**, 27231–27236.
71. S. V. Slepenkov and S. N. Witt, *Biochemistry*, 2002, **41**, 12224–12235.
72. S. J. Landry, R. Jordan, R. McMacken and L. M. Gierasch, *Nature*, 1992, **355**, 455–457.
73. M. Marcinowski, M. Rosam, C. Seitz, J. Elferich, J. Behnke, C. Bello, M. J. Feige, C. F. W. Becker, I. Antes and J. Buchner, *J. Mol. Biol.*, 2013, **425**, 466–474.
74. S. Rudiger, L. Germeroth, J. Schneider-Mergener and B. Bukau, *EMBO J.*, 1997, **16**, 1501–1507.
75. S. Blond-Elguindi, S. E. Cwirla, W. J. Dower, R. J. Lipshutz, S. R. Sprang, J. F. Sambrook and M. J. Gething, *Cell*, 1993, **75**, 717–728.

76. G. C. Flynn, J. Pohl, M. T. Flocco and J. E. Rothman, *Nature*, 1991, **353**, 726–730.
77. S. Rudiger, J. Schneider-Mergener and B. Bukau, *EMBO J.*, 2001, **20**, 1042–1050.
78. T. L. Tapley and L. E. Vickery, *J. Biol. Chem.*, 2004, **279**, 28435–28442.
79. A. Gragerov and M. E. Gottesman, *J. Mol. Biol.*, 1994, **241**, 133–135.
80. S. M. Hu and C. Wang, *Arch. Biochem. Biophys.*, 1996, **332**, 163–169.
81. M. Y. Tsai and C. Wang, *J. Biol. Chem.*, 1994, **269**, 5958–5962.
82. R. G. Smock, M. E. Blackburn and L. M. Gierasch, *J. Biol. Chem.*, 2011, **286**, 31821–31829.
83. J. Jiang, E. G. Maes, A. B. Taylor, L. Wang, A. P. Hinck, E. M. Lafer and R. Sousa, *Mol. Cell*, 2007, **28**, 422–433.
84. W. C. Suh, W. F. Burkholder, C. Z. Lu, X. Zhao, M. E. Gottesman and C. A. Gross, *Proc. Natl Acad. Sci. USA*, 1998, **95**, 15223–15228.
85. A. Ahmad, A. Bhattacharya, R. A. McDonald, M. Cordes, B. Ellington, E. B. Bertelsen and E. R. P. Zuiderweg, *Proc. Natl Acad. Sci. USA*, 2011, **108**, 18966–18971.
86. H. H. Kampinga and E. A. Craig, *Nat. Rev. Mol. Cell. Biol.*, 2010, **11**, 579–592.
87. A. Mitra, L. A. Shevde and R. S. Samant, *Clin. Exp. Metastasis*, 2009, **26**, 559–567.
88. J. N. Sterrenberg, G. L. Blatch and A. L. Edkins, *Cancer Lett.*, 2011, **312**, 129–142.
89. M. P. Mayer, T. Laufen, K. Paal, J. S. McCarty and B. Bukau, *J. Mol. Biol.*, 1999, **289**, 1131–1144.
90. J. Tsai and M. G. Douglas, *J. Biol. Chem.*, 1996, **271**, 9347–9354.
91. H. A. Hundley, W. Walter, S. Bairstow and E. A. Craig, *Science*, 2005, **308**, 1032–1034.
92. L. Chang, Y. Miyata, P. M. U. Ung, E. B. Bertelsen, T. J. McQuade, H. A. Carlson, E. R. P. Zuiderweg and J. E. Gestwicki, *Chem. Biol.*, 2011, **18**, 210–221.
93. C. S. Gassler, A. Buchberger, T. Laufen, M. P. Mayer, H. Schroder, A. Valencia and B. Bukau, *Proc. Natl Acad. Sci. USA*, 1998, **95**, 15229–15234.
94. S. J. Landry, *Structure*, 2003, **11**, 1465–1466.
95. C. Y. Fan, H. Y. Ren, P. Lee, A. J. Caplan and D. M. Cyr, *J. Biol. Chem.*, 2005, **280**, 695–702.
96. P. Kota, D. W. Summers, H. Y. Ren, D. M. Cyr and N. V. Dokholyan, *Proc. Natl Acad. Sci. USA*, 2009, **106**, 11073–11078.
97. K. Linke, T. Wolfram, J. Bussemer and U. Jakob, *J. Biol. Chem.*, 2003, **278**, 44457–44466.
98. Y. Wu, J. Li, Z. Jin, Z. Fu and B. Sha, *J. Mol. Biol.*, 2005, **346**, 1005–1011.
99. J. P. Chapple, J. van der Spuy, S. Poopalasundaram and M. E. Cheetham, *Biochem. Soc. Trans.*, 2004, **32**, 640–642.
100. Y. Shen and L. M. Hendershot, *Mol. Biol. Cell*, 2005, **16**, 40–50.

101. C. Y. Fan, S. Lee and D. M. Cyr, *Cell Stress Chaperones*, 2003, **8**, 309–316.
102. H. Otto, C. Conz, P. Maier, T. Wolfle, C. K. Suzuki, P. Jeno, P. Rucknagel, J. Stahl and S. Rospert, *Proc. Natl Acad. Sci. USA*, 2005, **102**, 10064–10069.
103. F. Hennessy, W. S. Nicoll, R. Zimmermann, M. E. Cheetham and G. L. Blatch, *Protein Sci.*, 2005, **14**, 1697–1709.
104. B. Kroczynska and S. Y. Blond, *Gene*, 2001, **273**, 267–274.
105. Y. Minami, J. Hohfeld, K. Ohtsuka and F.-U. Hartl, *J. Biol. Chem.*, 1996, **271**, 19617–19624.
106. F. Hennessy, M. E. Cheetham, H. W. Dirr and G. L. Blatch, *Cell Stress Chaperones*, 2000, **5**, 347–358.
107. S. Polier, F. U. Hartl and A. Bracher, *J. Mol. Biol.*, 2010, **401**, 696–707.
108. C. J. Harrison, M. Hayer-Hartl, M. Di Liberto, F. Hartl and J. Kuriyan, *Science*, 1997, **276**, 431–435.
109. Y. Liu, L. M. Gierasch and I. Bahar, *PLoS Comput. Biol.*, 2010, **6**, e1000931.
110. D. Brehmer, S. Rudiger, C. S. Gassler, D. Klostermeier, L. Packschies, J. Reinstein, M. P. Mayer and B. Bukau, *Nat. Struct. Mol. Biol.*, 2001, **8**, 427–432.
111. D. Brehmer, C. Gassler, W. Rist, M. P. Mayer and B. Bukau, *J. Biol. Chem.*, 2004, **279**, 27957–27964.
112. R. J. Schumacher, W. J. Hansen, B. C. Freeman, E. Alnemri, G. Litwack and D. O. Toft, *Biochemistry*, 1996, **35**, 14889–14898.
113. M. Kabani, *Protein Pept. Lett.*, 2009, **16**, 623–660.
114. Z. Xu, R. C. Page, M. M. Gomes, E. Kohli, J. C. Nix, A. B. Herr, C. Patterson and S. Misra, *Nat. Struct. Biol.*, 2008, **15**, 1309–1317.
115. L. Brive, S. Takayama, K. Briknarova, S. Homma, S. K. Ishida, J. C. Reed and K. R. Ely, *Biochem. Biophys. Res. Commun.*, 2001, **289**, 1099–1105.
116. K. Briknarova, S. Takayama, L. Brive, M. L. Havert, D. A. Knee, J. Velasco, S. Homma, E. Cabezas, J. Stuart, D. W. Hoyt, A. C. Satterthwait, M. Llinas, J. C. Reed and K. R. Ely, *Nat. Struct. Biol.*, 2001, **8**, 349–352.
117. S. Takayama, T. Sato, S. Krajewski, K. Kochel, S. Irie, J. A. Millan and J. C. Reed, *Cell*, 1995, **80**, 279–284.
118. M. Zeiner and U. Gehring, *Proc. Natl Acad. Sci. USA*, 1995, **92**, 11465–11469.
119. A. Sharp, S. J. Crabb, P. A. Townsend, R. I. Cutress, M. Brimmell, X. H. Wang and G. Packham, *Exp. Rev. Mol. Med.*, 2004, **6**, 1–15.
120. A. Bruchmann, C. Roller, T. V. Walther, G. Schafer, S. Lehmusvaara, T. Visakorpi, H. Klocker, A. C. Cato and D. Maddalo, *BMC Cancer*, 2013, **13**, 96.
121. C. Ballatore, V. M. Lee and J. Q. Trojanowski, *Nat. Rev. Neurosci.*, 2007, **8**, 663–672.
122. E. Elliott, P. Tsvetkov and I. Ginzburg, *J. Biol. Chem.*, 2007, **282**, 37276–37284.

123. D. C. Carrettiero, I. Hernandez, P. Neveu, T. Papagiannakopoulos and K. S. Kosik, *J. Neurosci.*, 2009, **29**, 2151–2161.
124. M. Kabani, C. McLellan, D. A. Raynes, V. Guerriero and J. L. Brodsky, *FEBS Lett.*, 2002, **531**, 339–342.
125. Y. Shomura, Z. Dragovic, H. C. Chang, N. Tzvetkov, J. C. Young, J. L. Brodsky, V. Guerriero, F. U. Hartl and A. Bracher, *Mol. Cell*, 2005, **17**, 367–379.
126. K. T. Chung, Y. Shen and L. M. Hendershot, *J. Biol. Chem.*, 2002, **277**, 47557–47563.
127. L. Zhao, C. Longo-Guess, B. S. Harris, J. W. Lee and S. L. Ackerman, *Nat. Genet.*, 2005, **37**, 974–979.
128. Z. Dragovic, S. A. Broadley, Y. Shomura, A. Bracher and F. U. Hartl, *EMBO J.*, 2006, **25**, 2519–2528.
129. H. Raviol, H. Sadlish, F. Rodriguez, M. P. Mayer and B. Bukau, *EMBO J.*, 2006, **25**, 2510–2518.
130. L. Shaner, A. Trott, J. L. Goeckeler, J. L. Brodsky and K. A. Morano, *J. Biol. Chem.*, 2004, **279**, 21992–22001.
131. C. Andreasson, J. Fiaux, H. Rampelt, M. P. Mayer and B. Bukau, *J. Biol. Chem.*, 2008, **283**, 8877–8884.
132. L. Shaner, R. Sousa and K. A. Morano, *Biochemistry*, 2006, **45**, 15075–15084.
133. J. Hohfeld, Y. Minami and F. U. Hartl, *Cell*, 1995, **83**, 589–598.
134. V. Prapapanich, S. Chen, S. C. Nair, R. A. Rimerman and D. F. Smith, *Mol. Endocrinol.*, 1996, **10**, 420–431.
135. V. Prapapanich, S. Chen, E. J. Toran, R. A. Rimerman and D. F. Smith, *Mol. Cell. Biol.*, 1996, **16**, 6200–6207.
136. H. Irmer and J. Hohfeld, *J. Biol. Chem.*, 1997, **272**, 2230–2235.
137. M. Velten, N. Gomez-Vrielynck, A. Chaffotte and M. M. Ladjimi, *J. Biol. Chem.*, 2002, **277**, 259–266.
138. V. Prapapanich, S. Chen and D. F. Smith, *Mol. Cell. Biol.*, 1998, **18**, 944–952.
139. G. M. Nelson, V. Prapapanich, P. E. Carrigan, P. J. Roberts, D. L. Riggs and D. F. Smith, *Mol. Endocrinol.*, 2004, **18**, 1620–1630.
140. J. A. Caruso and J. J. Reiners, Jr., *Apoptosis*, 2006, **11**, 1877–1885.
141. Y. Zhang, X. Cai, B. Schlegelberger and S. Zheng, *Cytogenet. Cell Genet.*, 1998, **83**, 56–57.
142. C. M. Nicolet and E. A. Craig, *Mol. Cell. Biol.*, 1989, **9**, 3638–3646.
143. Y. Morishima, K. C. Kanelakis, A. M. Silverstein, K. D. Dittmar, L. Estrada and W. B. Pratt, *J. Biol. Chem.*, 2000, **275**, 6894–6900.
144. S. Chen and D. F. Smith, *J. Biol. Chem.*, 1998, **273**, 35194–35200.
145. H. Wegele, L. Muller and J. Buchner, *Rev. Physiol. Biochem. Pharmacol.*, 2004, **151**, 1–44.
146. I. Grad and D. Picard, *Mol. Cell. Endocrinol.*, 2007, **275**, 2–12.
147. L. H. Pearl and C. Prodromou, *Annu. Rev. Biochem.*, 2006, **75**, 271–294.
148. K. A. Krukenberg, T. O. Street, L. A. Lavery and D. A. Agard, *Q. Rev. Biophys.*, 2011, **44**, 229–255.

149. D. F. Smith, *Cell Stress Chaperones*, 2004, **9**, 109–121.
150. C. Scheufler, A. Brinker, G. Bourenkov, S. Pegoraro, L. Moroder, H. Bartunik, F. U. Hartl and I. Moarefi, *Cell*, 2000, **101**, 199–210.
151. T. Kajander, J. N. Sachs, A. Goldman and L. Regan, *J. Biol. Chem.*, 2009, **284**, 25364–25374.
152. J. Demand, J. Luders and J. Hohfeld, *Mol. Cell. Biol.*, 1998, **18**, 2023–2028.
153. N. Walsh, A. Larkin, N. Swan, K. Conlon, P. Dowling, R. McDermott and M. Clynes, *Cancer Lett.*, 2011, **306**, 180–189.
154. S. Jha and A. A. Komar, *Biotechnol. J.*, 2011, **6**, 623–640.
155. V. Albanese, A. Y.-W. Yam, J. Baughman, C. Parnot and J. Frydman, *Cell*, 2006, **124**, 75–88.
156. S. Preissler and E. Deuerling, *Trends Biochem. Sci.*, 2012, **37**, 274–283.
157. B. Wiedmann, H. Sakai, T. A. Davis and M. Wiedmann, *Nature*, 1994, **370**, 434–440.
158. M. Pech, T. Spreter, R. Beckmann and B. Beatrix, *J. Biol. Chem.*, 2010, **285**, 19679–19687.
159. S. Rospert, Y. Dubaquie and M. Gautschi, *Cell. Mol. Life Sci.*, 2002, **59**, 1632–1639.
160. M. Rakwalska and S. Rospert, *Mol. Cell. Biol.*, 2004, **24**, 9186–9197.
161. F. Willmund, M. Del Alamo, S. Pechmann, T. Chen, V. Albanese, E. B. Dammer, J. Peng and J. Frydman, *Cell*, 2013, **152**, 196–209.
162. F. Chiti and C. M. Dobson, *Annu. Rev. Biochem.*, 2006, **75**, 333–366.
163. S. Baglioni, F. Casamenti, M. Bucciantini, L. M. Luheshi, N. Taddei, F. Chiti, C. M. Dobson and M. Stefani, *J. Neurosci.*, 2006, **26**, 8160–8167.
164. S. K. Sharma, P. Christen and P. Goloubinoff, *Curr. Protein Pept. Sci.*, 2009, **10**, 432–446.
165. S. Diamant, A. P. Ben-Zvi, B. Bukau and P. Goloubinoff, *J. Biol. Chem.*, 2000, **275**, 21107–21113.
166. M. P. Hinault, A. Ben-Zvi and P. Goloubinoff, *J. Mol. Neurosci.*, 2006, **30**, 249–265.
167. M. M. Dedmon, J. Christodoulou, M. R. Wilson and C. M. Dobson, *J. Biol. Chem.*, 2005, **280**, 14733–14740.
168. C. G. Evans, S. Wisen and J. E. Gestwicki, *J. Biol. Chem.*, 2006, **281**, 33182–33191.
169. K. C. Luk, I. P. Mills, J. Q. Trojanowski and V. M. Lee, *Biochemistry*, 2008, **47**, 12614–12625.
170. J. L. Wacker, M. H. Zareie, H. Fong, M. Sarikaya and P. J. Muchowski, *Nat. Struct. Biol.*, 2004, **11**, 1215–1222.
171. P. J. Muchowski, G. Schaffar, A. Sittler, E. E. Wanker, M. K. Hayer-Hartl and F. U. Hartl, *Proc. Natl Acad. Sci. USA*, 2000, **97**, 7841–7846.
172. A. Ciechanover, *Nat. Rev. Mol. Cell. Biol.*, 2005, **6**, 79–87.
173. E. Wong and A. M. Cuervo, *Cold Spring Harbor Perspect. Biol.*, 2010, **2**, 1–19.
174. A. Hershko and A. Ciechanover, *Annu Rev Biochem*, 1998, **67**, 425–479.

175. P. K. Jackson, A. G. Eldridge, E. Freed, L. Furstenthal, J. Y. Hsu, B. K. Kaiser and J. D. Reimann, *Trends Cell Biol.*, 2000, **10**, 429–439.
176. D. Lanneau, G. Wettstein, P. Bonniaud and C. Garrido, *Sci. World J.*, 2010, **10**, 1543–1552.
177. S. Elsasser and D. Finley, *Nat. Cell Biol.*, 2005, **7**, 742–749.
178. C. A. Ballinger, P. Connell, Y. Wu, Z. Hu, L. J. Thompson, L. Y. Yin and C. Patterson, *Mol. Cell. Biol.*, 1999, **19**, 4535–4545.
179. P. Connell, C. A. Ballinger, J. Jiang, Y. Wu, L. J. Thompson, J. Hohfeld and C. Patterson, *Nat. Cell Biol.*, 2001, **3**, 93–96.
180. J. Jiang, C. A. Ballinger, Y. Wu, Q. Dai, D. M. Cyr, J. R. Höhfeld and C. Patterson, *J. Biol. Chem.*, 2001, **276**, 42938–42944.
181. H. McDonough and C. Patterson, *Cell Stress Chaperones*, 2003, **8**, 303–308.
182. G. C. Meacham, C. Patterson, W. Zhang, J. M. Younger and D. M. Cyr, *Nat. Cell Biol.*, 2001, **3**, 100–105.
183. S. Alberti, J. Demand, C. Esser, N. Emmerich, H. Schild and J. Hohfeld, *J. Biol. Chem.*, 2003, **278**, 18702–18703.
184. C. Esser, M. Scheffner and J. Hohfeld, *J. Biol. Chem.*, 2005, **280**, 27443–27448.
185. K. L. Carraway III, *Semin. Cell Dev. Biol.*, 2010, **21**, 936–943.
186. D. Hutt and W. E. Balch, *Science*, 2010, **329**, 766–767.
187. P. Muller, E. Ruckova, P. Halada, P. J. Coates, R. Hrstka, D. P. Lane and B. Vojtesek, *Oncogene*, 2013, **32**, 3101–3110.
188. J. Lüders, J. Demand and J. Höhfeld, *J. Biol. Chem.*, 2000, **275**, 4613–4617.
189. M. Kudoh, D. A. Knee, S. Takayama and J. C. Reed, *Cancer Res.*, 2002, **62**, 1904–1909.
190. P. A. Townsend, R. I. Cutress, A. Sharp, M. Brimmell and G. Packham, *Cancer Res.*, 2003, **63**, 4150–4157.
191. P. A. Townsend, R. I. Cutress, A. Sharp, M. Brimmell and G. Packham, *Biochim. Biophys. Acta*, 2003, **1603**, 83–98.
192. N. R. Jana, P. Dikshit, A. Goswami, S. Kotliarova, S. Murata, K. Tanaka and N. Nukina, *J. Biol. Chem.*, 2005, **280**, 11635–11640.
193. A. J. Williams, T. M. Knutson, V. F. Colomer Gould and H. L. Paulson, *Neurobiol. Dis.*, 2009, **33**, 342–353.
194. I. Al-Ramahi, Y. C. Lam, H. K. Chen, B. de Gouyon, M. Zhang, A. M. Perez, J. Branco, M. de Haro, C. Patterson, H. Y. Zoghbi and J. Botas, *J. Biol. Chem.*, 2006, **281**, 26714–26724.
195. W. Li, Q. Yang and Z. Mao, *Cell. Mol. Life Sci.*, 2011, **68**, 749–763.
196. A. E. Majeski and J. F. Dice, *Int. J. Biochem. Cell Biol.*, 2004, **36**, 2435–2444.
197. V. Arndt, N. Dick, R. Tawo, M. Dreiseidler, D. Wenzel, M. Hesse, D. O. Furst, P. Saftig, R. Saint, B. K. Fleischmann, M. Hoch and J. Hohfeld, *Curr. Biol.*, 2010, **20**, 143–148.
198. D. Glick, S. Barth and K. F. Macleod, *J. Pathol.*, 2010, **221**, 3–12.
199. F. A. Agarraberes and J. F. Dice, *J. Cell Sci.*, 2001, **114**, 2491–2499.

200. V. Kirkin, D. G. McEwan, I. Novak and I. Dikic, *Mol. Cell*, 2009, **34**, 259–269.
201. U. B. Pandey, Y. Batlevi, E. H. Baehrecke and J. P. Taylor, *Autophagy*, 2007, **3**, 643–645.
202. U. B. Pandey, Z. Nie, Y. Batlevi, B. A. McCray, G. P. Ritson, N. B. Nedelsky, S. L. Schwartz, N. A. DiProspero, M. A. Knight, O. Schuldiner, R. Padmanabhan, M. Hild, D. L. Berry, D. Garza, C. C. Hubbert, T. P. Yao, E. H. Baehrecke and J. P. Taylor, *Nature*, 2007, **447**, 859–863.
203. V. I. Korolchuk, A. Mansilla, F. M. Menzies and D. C. Rubinsztein, *Mol. Cell*, 2009, **33**, 517–527.
204. C. Kraft, M. Peter and K. Hofmann, *Nat. Cell Biol.*, 2010, **12**, 836–841.
205. S. J. Orenstein and A. M. Cuervo, *Semin. Cell Dev. Biol.*, 2010, **21**, 719–726.
206. A. Benmerah and C. Lamaze, *Traffic*, 2007, **8**, 970–982.
207. J. Z. Rappoport, S. M. Simon and A. Benmerah, *Traffic*, 2004, **5**, 327–337.
208. R. Rodewald, *J. Cell Biol.*, 1973, **58**, 189–211.
209. R. G. Anderson, M. S. Brown, U. Beisiegel and J. L. Goldstein, *J. Cell Biol.*, 1982, **93**, 523–531.
210. J. L. Goldstein, R. G. Anderson and M. S. Brown, *Ciba Found. Symp.*, 1982, 77–95.
211. P. Gorden, J. L. Carpentier, S. Cohen and L. Orci, *Proc. Natl Acad. Sci. USA*, 1978, **75**, 5025–5029.
212. J. Y. Fan, J. L. Carpentier, P. Gorden, E. Van Obberghen, N. M. Blackett, C. Grunfeld and L. Orci, *Proc. Natl Acad. Sci. USA*, 1982, **79**, 7788–7791.
213. B. L. Wolfe and J. Trejo, *Traffic*, 2007, **8**, 462–470.
214. C. Watts, *J. Cell Biol.*, 1985, **100**, 633–637.
215. H. T. McMahon and E. Boucrot, *Nat. Rev. Mol. Cell Biol.*, 2011, **12**, 517–533.
216. E. Ungewickell, H. Ungewickell, S. E. H. Holstein, R. Lindner, K. Prasad, W. Barouch, B. Martini, L. E. Greene and E. Eisenberg, *Nature*, 1995, **378**, 632–635.
217. M. J. Taylor, D. Perrais and C. J. Merrifield, *PLoS Biol.*, 2011, **9**, e1000604.
218. S. E. Holstein, H. Ungewickell and E. Ungewickell, *J. Cell Biol.*, 1996, **135**, 925–937.
219. A. Fotin, Y. Cheng, N. Grigorieff, T. Walz, S. C. Harrison and T. Kirchhausen, *Nature*, 2004, **432**, 649–653.
220. J. M. Gruschus, C. J. Han, T. Greener, J. A. Ferretti, L. E. Greene and E. Eisenberg, *Biochemistry*, 2004, **43**, 3111–3119.
221. X. Zhang, Y. Chen, L. W. Jenkins, P. M. Kochanek and R. S. Clark, *Crit. Care*, 2005, **9**, 66–75.
222. S. S. Cao and R. J. Kaufman, *Curr. Biol.*, 2012, **22**, R622–626.
223. L. Galluzzi, O. Kepp and G. Kroemer, *Nat. Rev. Mol. Biol.*, 2012, **13**, 780–788.

224. J. Jiang, A. B. Taylor, K. Prasad, Y. Ishikawa-Brush, P. J. Hart, E. M. Lafer and R. Sousa, *Biochemistry*, 2003, **42**, 5748–5753.
225. E. Eisenberg and L. E. Greene, *Traffic*, 2007, **8**, 640–646.
226. Y. Xing, T. Bocking, M. Wolf, N. Grigorieff, T. Kirchhausen and S. C. Harrison, *EMBO J.*, 2010, **29**, 655–665.
227. T. Bocking, F. Aguet, S. C. Harrison and T. Kirchhausen, *Nat. Struct. Biol.*, 2011, **18**, 295–301.
228. E. Meimaridou, S. B. Gooljar and J. P. Chapple, *J. Mol. Endoc.*, 2009, **42**, 1–9.
229. T. C. Sudhof and J. Rizo, *Cold Spring Harbor Perspect. Biol.*, 2011, **3**, a005637.
230. J. E. Braun and R. H. Scheller, *Neuropharmacology*, 1995, **34**, 1361–1369.
231. H. Brown, O. Larsson, R. Branstrom, S. N. Yang, B. Leibiger, I. Leibiger, G. Fried, T. Moede, J. T. Deeney, G. R. Brown, G. Jacobsson, C. J. Rhodes, J. E. Braun, R. H. Scheller, B. E. Corkey, P. O. Berggren and B. Meister, *EMBO J.*, 1998, **17**, 5048–5058.
232. L. H. Chamberlain, J. Henry and R. D. Burgoyne, *J. Biol. Chem.*, 1996, **271**, 19514–19517.
233. A. Mastrogiacomo, S. M. Parsons, G. A. Zampighi, D. J. Jenden, J. A. Umbach and C. B. Gundersen, *Science*, 1994, **263**, 981–982.
234. J. A. Umbach, K. E. Zinsmaier, K. K. Eberle, E. Buchner, S. Benzer and C. B. Gundersen, *Neuron*, 1994, **13**, 899–907.
235. P. Bronk, J. J. Wenniger, K. Dawson-Scully, X. Guo, S. Hong, H. L. Atwood and K. E. Zinsmaier, *Neuron*, 2001, **30**, 475–488.
236. G. J. O. Evans, A. Morgan and R. D. Burgoyne, *Traffic*, 2003, **4**, 653–659.
237. S. Tobaben, P. Thakur, R. Fernandez-Chacon, T. C. Sudhof, J. Rettig and B. Stahl, *Neuron*, 2001, **31**, 987–999.
238. M. Zylicz and A. Wawrzynow, *IUBMB Life*, 2001, **51**, 283–287.
239. C. M. Koehler, *Annu. Rev. Cell Dev. Biol.*, 2004, **20**, 309–335.
240. V. Kriechbaumer, O. von Löffelholz and B. Abell, *Protoplasma*, 2012, **249**, 21–30.
241. A. U. Ahmed and P. R. Fisher, in *Int. Rev. Cell Mol. Biol.*, ed. W. J. Kwang, Academic Press, 2009, vol. 273, pp. 49–68.
242. H. Murakami, D. Pain and G. Blobel, *J. Cell Biol.*, 1988, **107**, 2051–2057.
243. J. C. Young, N. J. Hoogenraad and F. U. Hartl, *Cell*, 2003, **112**, 41–50.
244. H. F. Steger, T. Sollner, M. Kiebler, K. A. Dietmeier, R. Pfaller, K. S. Trulzsch, M. Tropschug, W. Neupert and N. Pfanner, *J. Cell Biol.*, 1990, **111**, 2353–2363.
245. J. Brix, S. Rudiger, B. Bukau, J. Schneider-Mergener and N. Pfanner, *J. Biol. Chem.*, 1999, **274**, 16522–16530.
246. M. Yano, K. Terada and M. Mori, *J. Cell Biol.*, 2003, **163**, 45–56.
247. A. C. Fan, G. Kozlov, A. Hoegl, R. C. Marcellus, M. J. Wong, K. Gehring and J. C. Young, *J. Biol. Chem.*, 2011, **286**, 32208–32219.
248. A. Ferramosca and V. Zara, *Biochim. Biophys. Acta*, 2013, **1833**, 494–502.
249. K. Terada, K. Ohtsuka, N. Imamoto, Y. Yoneda and M. Mori, *Mol. Cell. Biol.*, 1995, **15**, 3708–3713.

250. K. Terada, I. Ueda, K. Ohtsuka, T. Oda, A. Ichiyama and M. Mori, *Mol. Cell. Biol.*, 1996, **16**, 6103–6109.
251. W. P. Sheffield, G. C. Shore and S. K. Randall, *J. Biol. Chem.*, 1990, **265**, 11069–11076.
252. L. Pieuchot and G. Jedd, *Annu. Rev. Microbiol.*, 2012, **66**, 237–263.
253. H. W. Platta and R. Erdmann, *FEBS Lett.*, 2007, **581**, 2811–2819.
254. T. Harano, S. Nose, R. Uezu, N. Shimizu and Y. Fujiki, *Biochem. J.*, 2001, **357**, 157–165.
255. P. A. Walton, M. Wendland, S. Subramani, R. A. Rachubinski and W. J. Welch, *J. Cell Biol.*, 1994, **125**, 1037–1046.
256. J. E. Legakis and S. R. Terlecky, *Traffic*, 2001, **2**, 252–260.
257. C. C. Deocaris, S. C. Kaul and R. Wadhwa, *Biogerontology*, 2008, **9**, 391–403.
258. W. Voos, *Res. Microbiol.*, 2009, **160**, 718–725.
259. T. Bhattacharyya, A. N. Karnezis, S. P. Murphy, T. Hoang, B. C. Freeman, B. Phillips and R. I. Morimoto, *J. Biol. Chem.*, 1995, **270**, 1705–1710.
260. D. Pilzer and Z. Fishelson, *Int. Immunol.*, 2005, **17**, 1239–1248.
261. M. Blamowska, W. Neupert and K. Hell, *J. Cell Biol.*, 2012, **199**, 125–135.
262. M. J. Baker, A. E. Frazier, J. M. Gulbis and M. T. Ryan, *Trends Cell Biol.*, 2007, **17**, 456–464.
263. A. Geissler, J. Rassow, N. Pfanner and W. Voos, *Mol. Cell. Biol.*, 2001, **21**, 7097–7104.
264. A. A. Choglay, J. P. Chapple, G. L. Blatch and M. E. Cheetham, *Gene*, 2001, **267**, 125–134.
265. M. van der Laan, A. Chacinska, M. Lind, I. Perschil, A. Sickmann, H. E. Meyer, B. Guiard, C. Meisinger, N. Pfanner and P. Rehling, *Mol. Cell. Biol.*, 2005, **25**, 7449–7458.
266. D. Mokranjac, A. Berg, A. Adam, W. Neupert and K. Hell, *J. Biol. Chem.*, 2007, **282**, 18037–18045.
267. Q. Liu, P. D'Silva, W. Walter, J. Marszalek and E. A. Craig, *Science*, 2003, **300**, 139–141.
268. P. R. D'Silva, B. Schilke, W. Walter and E. A. Craig, *Proc. Natl Acad. Sci. USA*, 2005, **102**, 12419–12424.
269. P. D. D'Silva, B. Schilke, W. Walter, A. Andrew and E. A. Craig, *Proc. Natl Acad. Sci. USA*, 2003, **100**, 13839–13844.
270. P. Goloubinoff and P. D. L. Rios, *Trends Biochem. Sci.*, 2007, **32**, 372–380.
271. W. Neupert and M. Brunner, *Nat. Rev. Mol. Cell Biol.*, 2002, **3**, 555–565.
272. G. Schatz, *J. Biol. Chem.*, 1996, **271**, 31763–31766.
273. B. S. Glick, *Cell*, 1995, **80**, 11–14.
274. P. De Los Rios, A. Ben-Zvi, O. Slutsky, A. Azem and P. Goloubinoff, *Proc. Natl Acad. Sci. USA*, 2006, **103**, 6166–6171.
275. S. C. Kaul, C. C. Deocaris and R. Wadhwa, *Exp. Gerontol.*, 2007, **42**, 263–274.
276. R. Wadhwa, O. M. Pereira-Smith, R. R. Reddel, Y. Sugimoto, Y. Mitsui and S. C. Kaul, *Exp. Cell Res.*, 1995, **216**, 101–106.

277. R. Wadhwa, T. Sugihara, A. Yoshida, H. Nomura, R. R. Reddel, R. Simpson, H. Maruta and S. C. Kaul, *Cancer Res.*, 2000, **60**, 6818–6821.
278. N. Widodo, C. C. Deocaris, K. Kaur, K. Hasan, T. Yaguchi, K. Yamasaki, T. Sugihara, T. Ishii, R. Wadhwa and S. C. Kaul, *J. Gerontol. A Biol. Med. Sci.*, 2007, **62**, 246–255.
279. R. Wadhwa, S. Takano, K. Kaur, C. C. Deocaris, O. M. Pereira-Smith, R. R. Reddel and S. C. Kaul, *Int. J. Cancer*, 2006, **118**, 2973–2980.
280. M. Mihara, S. Erster, A. Zaika, O. Petrenko, T. Chittenden, P. Pancoska and U. M. Moll, *Mol. Cell*, 2003, **11**, 577–590.
281. J. E. Chipuk and D. R. Green, *Cell Cycle*, 2004, **3**, 429–431.
282. J. E. Chipuk, T. Kuwana, L. Bouchier-Hayes, N. M. Droin, D. D. Newmeyer, M. Schuler and D. R. Green, *Science*, 2004, **303**, 1010–1014.
283. R. Wadhwa, K. Taira and S. C. Kaul, *Cell Stress Chaperones*, 2002, **7**, 309–316.
284. D. M. Cyr and M. G. Douglas, *J. Biol. Chem.*, 1994, **269**, 9798–9804.
285. D. M. Cyr, T. Langer and M. G. Douglas, *Trends Biochem. Sci.*, 1994, **19**, 176–181.
286. K. Leonhard, A. Stiegler, W. Neupert and T. Langer, *Nature*, 1999, **398**, 348–351.
287. A. S. Savel'ev, L. A. Novikova, I. E. Kovaleva, V. N. Luzikov, W. Neupert and T. Langer, *J. Biol. Chem.*, 1998, **273**, 20596–20602.
288. L. E. Vickery and J. R. Cupp-Vickery, *Crit. Rev. Biochem. Mol. Biol.*, 2007, **42**, 95–111.
289. R. Lill, B. Hoffmann, S. Molik, A. J. Pierik, N. Rietzschel, O. Stehling, M. A. Uzarska, H. Webert, C. Wilbrecht and U. Muhlenhoff, *Biochim. Biophys. Acta*, 2012, **1823**, 1491–1508.
290. L. Walter and G. Hajnoczky, *J. Bioenerg. Biomembr.*, 2005, **37**, 191–206.
291. G. Szabadkai, K. Bianchi, P. Varnai, D. De Stefani, M. R. Wieckowski, D. Cavagna, A. I. Nagy, T. Balla and R. Rizzuto, *J. Cell Biol.*, 2006, **175**, 901–911.
292. J. Dudek, J. Benedix, S. Cappel, M. Greiner, C. Jalal, L. Muller and R. Zimmermann, *Cell. Mol. Life Sci.*, 2009, **66**, 1556–1569.
293. X. Chen, D. Easton, H. J. Oh, D. S. Lee-Yoon, X. Liu and J. Subjeck, *FEBS Lett.*, 1996, **380**, 68–72.
294. T. Dierks, J. Volkmer, G. Schlenstedt, C. Jung, U. Sandholzer, K. Zachmann, P. Schlotterhose, K. Neifer, B. Schmidt and R. Zimmermann, *EMBO J.*, 1996, **15**, 6931–6942.
295. A. Weitzmann, C. Baldes, J. Dudek and R. Zimmermann, *FEBS J.*, 2007, **274**, 5175–5187.
296. M. Dong, J. P. Bridges, K. Apsley, Y. Xu and T. E. Weaver, *Mol. Biol. Cell*, 2008, **19**, 2620–2630.
297. S. Oyadomari, C. Yun, E. A. Fisher, N. Kreglinger, G. Kreibich, M. Oyadomari, H. P. Harding, A. G. Goodman, H. Harant, J. L. Garrison, J. Taunton, M. G. Katze and D. Ron, *Cell*, 2006, **126**, 727–739.

298. K. Petrova, S. Oyadomari, L. M. Hendershot and D. Ron, *EMBO J.*, 2008, **27**, 2862–2872.
299. R. Ushioda, J. Hoseki, K. Araki, G. Jansen, D. Y. Thomas and K. Nagata, *Science*, 2008, **321**, 569–572.
300. A. Weitzmann, J. Volkmer and R. Zimmermann, *FEBS Lett.*, 2006, **580**, 5237–5240.
301. A. R. Osborne, T. A. Rapoport and B. van den Berg, *Annu. Rev. Cell Dev. Biol.*, 2005, **21**, 529–550.
302. T. A. Rapoport, *Nature*, 2007, **450**, 663–669.
303. B. D. Hamman, L. M. Hendershot and A. E. Johnson, *Cell*, 1998, **92**, 747–758.
304. B. D. Hamman, J. C. Chen, E. E. Johnson and A. E. Johnson, *Cell*, 1997, **89**, 535–544.
305. S. Liao, J. Lin, H. Do and A. E. Johnson, *Cell*, 1997, **90**, 31–41.
306. I. G. Haas and M. Wabl, *Nature*, 1983, **306**, 387–389.
307. C. Bies, R. Blum, J. Dudek, W. Nastainczyk, S. Oberhauser, M. Jung and R. Zimmermann, *Biol. Chem.*, 2004, **385**, 389–395.
308. K. Plath and T. A. Rapoport, *J. Cell Biol.*, 2000, **151**, 167–178.
309. J. L. Brodsky, *Cell*, 2012, **151**, 1163–1167.
310. C. J. Guerriero and J. L. Brodsky, *Physiol. Rev.*, 2012, **92**, 537–576.
311. S. Kawaguchi and D. T. Ng, *Cell*, 2007, **129**, 1230.
312. Y. Jin, W. Awad, K. Petrova and L. M. Hendershot, *EMBO J.*, 2008, **27**, 2873–2882.
313. B. Tirosh, M. H. Furman, D. Tortorella and H. L. Ploegh, *J. Biol. Chem.*, 2003, **278**, 6664–6672.
314. S. Olivari, T. Cali, K. E. Salo, P. Paganetti, L. W. Ruddock and M. Molinari, *Biochem. Biophys. Res. Commun.*, 2006, **349**, 1278–1284.
315. M. Hagiwara, K. Maegawa, M. Suzuki, R. Ushioda, K. Araki, Y. Matsumoto, J. Hoseki, K. Nagata and K. Inaba, *Mol. Cell*, 2011, **41**, 432–444.
316. C. K. Kassenbrock and R. B. Kelly, *EMBO J.*, 1989, **8**, 1461–1467.
317. Y. Fu, J. Li and A. S. Lee, *Cancer Res.*, 2007, **67**, 3734–3740.
318. S. Wang and R. J. Kaufman, *J. Cell Biol.*, 2012, **197**, 857–867.
319. M. Schroder and R. J. Kaufman, *Mutat. Res.*, 2005, **569**, 29–63.
320. P. Walter and D. Ron, *Science*, 2011, **334**, 1081–1086.
321. I. Tabas and D. Ron, *Nat. Cell Biol.*, 2011, **13**, 184–190.
322. D. Pincus, M. W. Chevalier, T. Aragon, E. van Anken, S. E. Vidal, H. El-Samad and P. Walter, *PLoS Biol.*, 2010, **8**, e1000415.
323. H. Yoshida, T. Matsui, A. Yamamoto, T. Okada and K. Mori, *Cell*, 2001, **107**, 881–891.
324. A. H. Lee, N. N. Iwakoshi and L. H. Glimcher, *Mol. Cell. Biol.*, 2003, **23**, 7448–7459.
325. D. T. Rutkowski, S. W. Kang, A. G. Goodman, J. L. Garrison, J. Taunton, M. G. Katze, R. J. Kaufman and R. S. Hegde, *Mol. Biol. Cell*, 2007, **18**, 3681–3691.

326. J. Hollien, J. H. Lin, H. Li, N. Stevens, P. Walter and J. S. Weissman, *J. Cell Biol.*, 2009, **186**, 323–331.
327. F. Urano, X. Wang, A. Bertolotti, Y. Zhang, P. Chung, H. P. Harding and D. Ron, *Science*, 2000, **287**, 664–666.
328. H. P. Harding, Y. Zhang, H. Zeng, I. Novoa, P. D. Lu, M. Calfon, N. Sadri, C. Yun, B. Popko, R. Paules, D. F. Stojdl, J. C. Bell, T. Hettmann, J. M. Leiden and D. Ron, *Mol. Cell*, 2003, **11**, 619–633.
329. H. Yamaguchi and H. G. Wang, *J. Biol. Chem.*, 2004, **279**, 45495–45502.
330. S. C. Cazanave, N. A. Elmi, Y. Akazawa, S. F. Bronk, J. L. Mott and G. J. Gores, *Am. J. Physiol. Gastrointest. Liver Physiol.*, 2010, **299**, G236–243.
331. H. P. Harding, I. Novoa, Y. Zhang, H. Zeng, R. Wek, M. Schapira and D. Ron, *Mol. Cell*, 2000, **6**, 1099–1108.
332. H. Yoshida, T. Okada, K. Haze, H. Yanagi, T. Yura, M. Negishi and K. Mori, *Mol. Cell. Biol.*, 2000, **20**, 6755–6767.
333. S. Gupta, A. Deepti, S. Deegan, F. Lisbona, C. Hetz and A. Samali, *PLoS Biol.*, 2010, **8**, e1000410.
334. A. J. Dorner, L. C. Wasley and R. J. Kaufman, *EMBO J.*, 1992, **11**, 1563–1571.
335. J. H. Otero, B. Lizak and L. M. Hendershot, *Semin. Cell Dev. Biol.*, 2010, **21**, 472–478.
336. J. E. Chambers, K. Petrova, G. Tomba, M. Vendruscolo and D. Ron, *J. Cell Biol.*, 2012, **198**, 371–385.
337. J. S. Long and K. M. Ryan, *Oncogene*, 2012, **31**, 5045–5060.
338. A. R. Goloudina, O. N. Demidov and C. Garrido, *Cancer Lett.*, 2012, **325**, 117–124.
339. R. F. Kelley, K. Totpal, S. H. Lindstrom, M. Mathieu, K. Billeci, L. DeForge, R. Pai, S. G. Hymowitz and A. Ashkenazi, *J. Biol. Chem.*, 2005, **280**, 2205–2212.
340. V. L. Gabai, K. Mabuchi, D. D. Mosser and M. Y. Sherman, *Mol. Cell. Biol.*, 2002, **22**, 3415–3424.
341. F. Guo, C. Sigua, P. Bali, P. George, W. Fiskus, A. Scuto, S. Annavarapu, A. Mouttaki, G. Sondarva, S. Wei, J. Wu, J. Djeu and K. Bhalla, *Blood*, 2005, **105**, 1246–1255.
342. Q. Pang, T. A. Christianson, W. Keeble, T. Koretsky and G. C. Bagby, *J. Biol. Chem.*, 2002, **277**, 49638–49643.
343. Q. Pang, W. Keeble, T. A. Christianson, G. R. Faulkner and G. C. Bagby, *EMBO J.*, 2001, **20**, 4478–4489.
344. H. Sakahira and S. Nagata, *J. Biol. Chem.*, 2002, **277**, 3364–3370.
345. M. Mayer, J. Reinstein and J. Buchner, *J. Mol. Biol.*, 2003, **330**, 137–144.
346. K. Cain, S. B. Bratton, C. Langlais, G. Walker, D. G. Brown, X. M. Sun and G. M. Cohen, *J. Biol. Chem.*, 2000, **275**, 6067–6070.
347. T. Gotoh, K. Terada, S. Oyadomari and M. Mori, *Cell Death Differ.*, 2004, **11**, 390–402.
348. A. M. Gorman, E. Szegezdi, D. J. Quigney and A. Samali, *Biochem. Biophys. Res. Commun.*, 2005, **327**, 801–810.

349. V. Veereshwarayya, P. Kumar, K. M. Rosen, R. Mestril and H. W. Querfurth, *J. Biol. Chem.*, 2006, **281**, 29468–29478.
350. E. Y. Komarova, E. A. Afanasyeva, M. M. Bulatova, M. E. Cheetham, B. A. Margulis and I. V. Guzhova, *Cell Stress Chaperones*, 2004, **9**, 265–275.
351. A. Saleh, S. M. Srinivasula, L. Balkir, P. D. Robbins and E. S. Alnemri, *Nat. Cell Biol.*, 2000, **2**, 476–483.
352. R. Steel, J. P. Doherty, K. Buzzard, N. Clemons, C. J. Hawkins and R. L. Anderson, *J. Biol. Chem.*, 2004, **279**, 51490–51499.
353. R. T. Uren, G. Dewson, C. Bonzon, T. Lithgow, D. D. Newmeyer and R. M. Kluck, *J. Biol. Chem.*, 2005, **280**, 2266–2274.
354. E. M. Creagh, R. J. Carmody and T. G. Cotter, *Exp. Cell Res.*, 2000, **257**, 58–66.
355. L. Ravagnan, S. Gurbuxani, S. A. Susin, C. Maisse, E. Daugas, N. Zamzami, T. Mak, M. Jaattela, J. M. Penninger, C. Garrido and G. Kroemer, *Nat. Cell Biol.*, 2001, **3**, 839–843.
356. K. Ruchalski, H. Mao, Z. Li, Z. Wang, S. Gillers, Y. Wang, D. D. Mosser, V. Gabai, J. H. Schwartz and S. C. Borkan, *J. Biol. Chem.*, 2006, **281**, 7873–7880.
357. M. P. Murphy, *Sci. Signal.*, 2012, **5**, 39.
358. M. Oren, *Cell Death Differ.*, 2003, **10**, 431–442.
359. J. I. Leu, P. Dumont, M. Hafey, M. E. Murphy and D. L. George, *Nat. Cell Biol.*, 2004, **6**, 443–450.
360. Q. Ke and M. Costa, *Mol. Pharmacol.*, 2006, **70**, 1469–1480.
361. D. Walerych, M. Napoli, L. Collavin and G. Del Sal, *Carcinogenesis*, 2012, **33**, 2007–2017.
362. C. Dai, L. Whitesell, A. B. Rogers and S. Lindquist, *Cell*, 2007, **130**, 1005–1018.
363. D. Walerych, M. B. Olszewski, M. Gutkowska, A. Helwak, M. Zylicz and A. Zylicz, *Oncogene*, 2009, **28**, 4284–4294.
364. M. Wiech, M. B. Olszewski, Z. Tracz-Gaszewska, B. Wawrzynow, M. Zylicz and A. Zylicz, *PLoS One*, 2012, **7**, e51426.
365. S. S. Gogate, N. Fujita, R. Skubutyte, I. M. Shapiro and M. V. Risbud, *J. Bone Miner. Res.*, 2012, **27**, 1106–1117.
366. G. L. Wang, B. H. Jiang, E. A. Rue and G. L. Semenza, *Proc. Natl Acad. Sci. USA*, 1995, **92**, 5510–5514.
367. W. G. Kaelin, Jr. and P. J. Ratcliffe, *Mol. Cell*, 2008, **30**, 393–402.
368. Z. Arany, L. E. Huang, R. Eckner, S. Bhattacharya, C. Jiang, M. A. Goldberg, H. F. Bunn and D. M. Livingston, *Proc. Natl Acad. Sci. USA*, 1996, **93**, 12969–12973.
369. B. H. Jiang, E. Rue, G. L. Wang, R. Roe and G. L. Semenza, *J. Biol. Chem.*, 1996, **271**, 17771–17778.
370. W. Luo, J. Zhong, R. Chang, H. Hu, A. Pandey and G. L. Semenza, *J. Biol. Chem.*, 2010, **285**, 3651–3663.
371. J. Zhou, T. Schmid, R. Frank and B. Brune, *J. Biol. Chem.*, 2004, **279**, 13506–13513.

372. W. Luo, J. Zhong, R. Chang, H. Hu, A. Pandey and G. L. Semenza, *J. Biol. Chem.*, 2010, **285**, 3651–3663.
373. G. L. Semenza, *Curr. Opin. Genet. Dev.*, 1998, **8**, 588–594.
374. R. R. Raval, K. W. Lau, M. G. Tran, H. M. Sowter, S. J. Mandriota, J. L. Li, C. W. Pugh, P. H. Maxwell, A. L. Harris and P. J. Ratcliffe, *Mol. Cell. Biol.*, 2005, **25**, 5675–5686.
375. B. Levine and J. Yuan, *J. Clin. Invest.*, 2005, **115**, 2679–2688.
376. A. M. Choi, S. W. Ryter and B. Levine, *N. Engl. J. Med.*, 2013, **368**, 651–662.
377. J. Nylandsted, M. Gyrd-Hansen, A. Danielewicz, N. Fehrenbacher, U. Lademann, M. Hoyer-Hansen, E. Weber, G. Multhoff, M. Rohde and M. Jaattela, *J. Exp. Med.*, 2004, **200**, 425–435.
378. M. Gyrd-Hansen, J. Nylandsted and M. Jaattela, *Cell Cycle*, 2004, **3**, 1484–1485.
379. T. Kirkegaard, A. G. Roth, N. H. Petersen, A. K. Mahalka, O. D. Olsen, I. Moilanen, A. Zylicz, J. Knudsen, K. Sandhoff, C. Arenz, P. K. Kinnunen, J. Nylandsted and M. Jaattela, *Nature*, 2010, **463**, 549–553.
380. M. Sherman, *Ann. NY Acad. Sci.*, 2010, **1197**, 152–157.
381. J. J. Jacobs and T. de Lange, *Cell Cycle*, 2005, **4**, 1364–1368.
382. I. B. Roninson, *Cancer Res.*, 2003, **63**, 2705–2715.
383. Y. Matsuda, *World J. Gastroenterol.*, 2008, **14**, 1734–1740.
384. M. Schwabe and M. Lubbert, *Curr. Pharm. Biotechnol.*, 2007, **8**, 382–387.
385. S. W. Lowe, E. Cepero and G. Evan, *Nature*, 2004, **432**, 307–315.
386. M. Narita and S. W. Lowe, *Cell Cycle*, 2004, **3**, 244–246.
387. C. Denoyelle, G. Abou-Rjaily, V. Bezrookove, M. Verhaegen, T. M. Johnson, D. R. Fullen, J. N. Pointer, S. B. Gruber, L. D. Su, M. A. Nikiforov, R. J. Kaufman, B. C. Bastian and M. S. Soengas, *Nat. Cell Biol.*, 2006, **8**, 1053–1063.
388. S. C. Kaul, S. Aida, T. Yaguchi, K. Kaur and R. Wadhwa, *J. Biol. Chem.*, 2005, **280**, 39373–39379.
389. S. C. Kaul, T. Yaguchi, K. Taira, R. R. Reddel and R. Wadhwa, *Exp. Cell Res.*, 2003, **286**, 96–101.
390. M. Rohde, M. Daugaard, M. H. Jensen, K. Helin, J. Nylandsted and M. Jaattela, *Gene. Dev.*, 2005, **19**, 570–582.
391. R. Wadhwa, S. Takano, M. Robert, A. Yoshida, H. Nomura, R. R. Reddel, Y. Mitsui and S. C. Kaul, *J. Biol. Chem.*, 1998, **273**, 29586–29591.
392. R. Qi, E. B. Sarbeng, Q. Liu, K. Q. Le, X. Xu, H. Xu, J. Yang, J. L. Wong, C. Vorvis, W. A. Hendricksoon, L. Zhou and Q. Liu, *Nat. Struct. Mol. Biol.*, 2013, **20**, 900–907.
393. A. Arakawa, N. Handa, M. Shirouzu and S. Yokoyama, *Protein Sci.*, 2011, **20**, 1367–1379.
394. M. Pellecchia, T. Szyperski, D. Wall, C. Georgopoulos and K. Wuthrich, *J. Mol. Biol.*, 1996, **260**, 236–250.

CHAPTER 4

Exploiting the Dependency of Cancer Cells on Molecular Chaperones

SWEE SHARP, JENNY HOWES AND PAUL WORKMAN*

Cancer Research UK Cancer Therapeutics Unit, Division of Cancer Therapeutics, The Institute of Cancer Research, London, SM2 5NG, UK
*Email: paul.workman@icr.ac.uk

4.1 Introduction: Stepping into the Limelight

The ongoing drama of the discovery and development of inhibitors of molecular chaperones is truly gripping – with an unpredictable script and heat-shock protein 90 (Hsp90) taking center stage as the main actor so far. Hsp90 has emerged into the limelight from unpromising beginnings where it was initially seen as an apparently rather dull housekeeping protein in which there was little or no interest within the pharmaceutical industry during the mid to late 1990 s, followed by a period of growing but generally uncommitted curiosity, to the situation today in which multiple pharma and biotech companies – as well as several academic drug discovery groups – are now actively involved. Thus over 500 patents on Hsp90 have been published (mostly since 2000); more than 20 small-molecule Hsp90 inhibitors have entered clinical trials in cancer patients (starting with 17-AAG/tanespimycin in 1999); clear objective evidence of clinical antitumor activity has been demonstrated, especially in certain subtypes of breast and lung cancer; and the leading Hsp90 drugs are now progressing towards regulatory studies.[1,2] Furthermore, in addition to cancer, there is also

---

RSC Drug Discovery Series No. 37
Inhibitors of Molecular Chaperones as Therapeutic Agents
Edited by Timothy Machajewski and Zhenhai Gao
© The Royal Society of Chemistry 2014
Published by the Royal Society of Chemistry, www.rsc.org

considerable enthusiasm for the application of Hsp90 inhibitors in other diseases as well as significant interest and activity in the therapeutic investigation of other chaperone and protein quality control targets – particularly Hsp70 and Hsp27 together with targets in the heat-shock pathway and further targets in the unfolded protein response in addition to the 26 S proteasome.[3–8]

An adequate therapeutic index is essential for any drug, so that beneficial effects can be achieved at well-tolerated doses. For anticancer drugs, this means that there needs to be a differential pharmacological activity towards tumor *versus* normal cells. Indeed it was a source of major concern when Hsp90 was first mooted as a cancer drug target that inhibiting its chaperone function would result in broad and unacceptably severe toxicity because of the large number of client proteins and biological pathways that would be affected. This concern was probably brought into sharper focus because at the turn of the new millennium the field of cancer drug discovery was moving away from broadly acting one-size-fits-all cytotoxics to an emphasis on highly targeted molecular therapeutics acting on, for example, a single oncogenic kinase and thereby achieving therapeutic selectivity by exploiting a unique oncogene addiction while minimizing side effects.[9,10] Hsp90 was clearly a riskier target to pursue at that time.

Over the last 15 years or so it has become clear as a result of an interconnecting combination of hypothesis-driven research on Hsp90 biology and observations made with pharmacologic Hsp90 inhibitory agents – both the original natural product inhibitors exemplified by geldanamycin and radicicol (Figure 4.1) and also the subsequent synthetic small-molecule inhibitors, employed as pathway-finding chemical tools and then as drugs in the clinic[11] – that Hsp90 inhibitors can indeed deliver genuine therapeutic selectivity for cancer *versus* normal cells. Although a precise universal mechanism has not been discovered – and indeed may not uniquely exist – there are a number of mechanisms that can contribute to the therapeutic index that can undoubtedly be achieved.[1,2,12–16] In particular, it is apparent that Hsp90 inhibitors benefit from delivering a combinatorial therapeutic exploitation of oncogene addiction and tumor stress pathway dependence. Moreover, although it is becoming obvious that particularly strong clinical responses are seen in settings where the oncogenic driver is an Hsp90 client protein that at the same time is also highly dependent upon Hsp90 and is rapidly depleted by Hsp90 inhibitors – best exemplified in the clinic to date with amplified and over-expressed ERBB2 (HER2) in breast cancer and mutant EGFR and translocated ELM4-ALK in non-small cell lung cancer (NSCLC) – it is nevertheless also apparent that the combinatorial effects of Hsp90 inhibitors on multiple clients (including mutant resistant alleles) and also on many different signaling pathways has the additional important advantage of overcoming or preventing resistance to targeted therapies. This is likely to become increasingly important with the recognition that drug resistance commonly arises from tumor heterogeneity, clonal evolution and genetic selection together with biochemical feedback loops – involving mechanisms that can be tackled with drug combinations, as part of which Hsp90 inhibitors represent an attractive and highly desirable

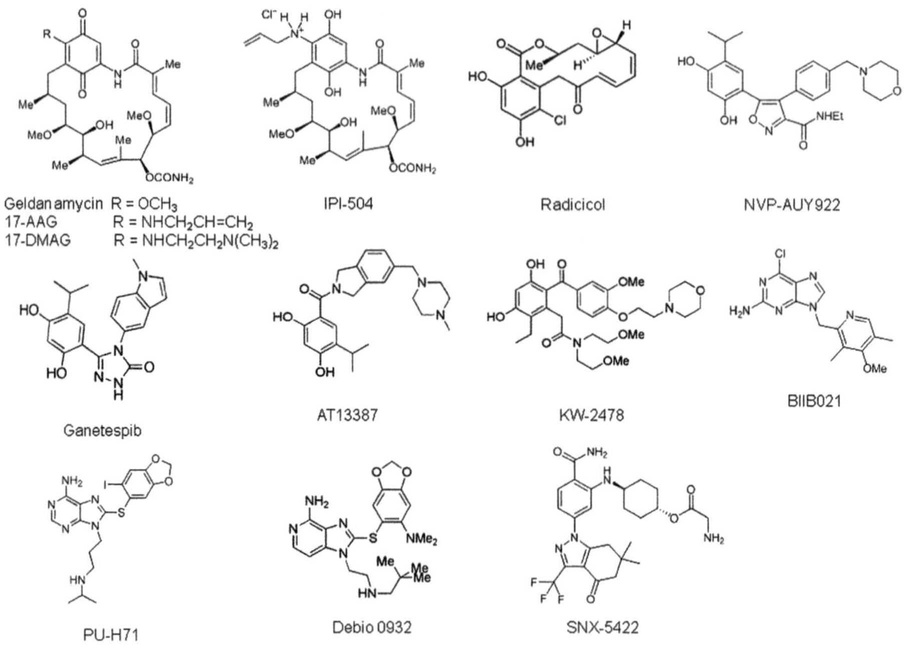

Figure 4.1 Chemical structures of Hsp90 inhibitors acting at the N-terminal nucleotide binding site.

feature.[17,18] Because of this there is a strong argument that Hsp90 inhibitors should be used upfront as part of a personalized combinatorial therapy approach to cancer treatment.

In the present chapter we discuss how it is possible to exploit the dependency of cancer cells on molecular chaperones to achieve selective anticancer effects. Reflecting the current state of play, the main focus will be on Hsp90, but at the end of the chapter we will also briefly discuss additional emerging targets such as Hsp70 and other Hsp90 co-chaperones. First we will review the brief essentials of Hsp90 biology and the evidence for its causal involvement in cancer, since these are critical to understand how therapeutic selectivity can be achieved with Hsp90 inhibitors and their potential to overcome or even prevent drug resistance. At the end of the chapter we will take a look at some interesting current developments and try to anticipate future directions.

4.2 Brief Essentials of Hsp90 Biology

Protein quality control pathways are critical for homeostasis in the crowded, protein-rich internal environment of the cell, and especially for ensuring tight regulation of the synthesis, folding, transport and degradation of proteins.[19,20] Failure to maintain the correct balance of these processes can result in cell death and is thought to cause or contribute to various diseases, including neurodegenerative conditions, aging and cancer.[21] Some of the key elements involved in protein quality control are depicted in Figure 4.2 including: 1) molecular chaperones that interact with nascent and incompletely folded or misfolded proteins to determine their fate; and 2) the ubiquitin-proteasome pathway for the degradation of misfolded proteins. These operate together to prevent protein aggregation. Also relevant (but not shown in Figure 4.2 for simplicity) are the unfolded protein response in the endoplasmic reticulum[22] and the autophagic system, which is involved in the removal of aggregates and other misfolded proteins.[23] Under normal cellular conditions, molecular chaperones ensure that correct protein folding is the expected fate for a newly synthesized protein. When cells are stressed (*e.g.* exposed to physiological pressures such as heat, cytotoxic drugs, UV irradiation, heavy metals, hypoxia or low pH), mature proteins undergo unfolding and the chaperone networks attempt to prevent protein degradation or aggregation. This is achieved by the

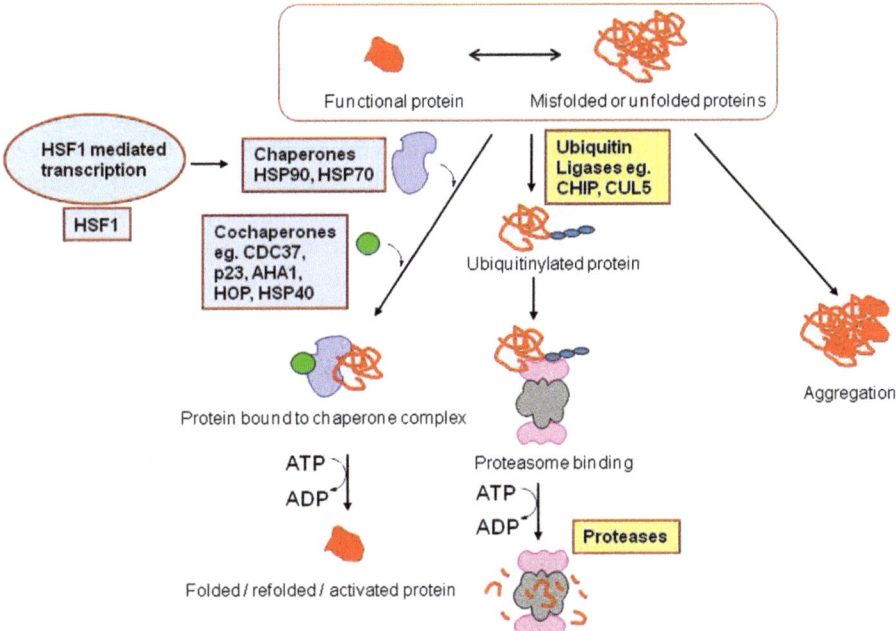

Figure 4.2 Cancer drug targets in protein homeostasis. The balance between protein folding (left) and degradation (center) is controlled by molecular chaperones (and regulatory co-chaperones), of which a key member is Hsp90. The expression of Hsp90 and other molecular chaperones, co-chaperones and components of the degradative machinery is regulated by heat-shock factor 1 (Hsf1), which controls the heat-shock response (extreme left). Protein triage decisions regulated by Hsp90 minimize protein aggregation (extreme right) unless the chaperone and degradative machinery are overwhelmed. Not shown for simplicity are the related unfolded protein responses in the endoplasmic reticulum or the autophagy pathway.

activation of the key transcription factor heat-shock factor 1 (Hsf1), resulting in a 2- to 3-fold increase in heat-shock proteins – featuring especially molecular chaperones, co-chaperones and anti-apoptotic proteins – in a concerted effort to shift the equilibrium towards protein folding, hence maintaining protein homeostasis and cell survival.[24] In situations where unfolding or aggregation is beyond recovery the degradation pathways are engaged. The heat-shock response and ubiquitin-proteasome pathway are clearly critical in protein homeostasis and chaperones like Hsp90 lie at the heart of protein triage decision-making (Figure 4.2).

But in addition to its major role in the triaging of proteins between the folding and degradation pathways, the molecular chaperone Hsp90 and its associated family of co-chaperones are also essential for the homeostatic biochemical regulation and correct functioning of many signal transduction and other proteins in the cell.[25,26] Hsp90 interacts with a plethora of cellular substrates or client proteins. To date, there are over 200 identified members of Hsp90's clientele (www.picard.ch). Many Hsp90 clients are signaling proteins, particularly transcription factors and protein kinases for which the function of the molecular chaperone in stability, activation and protein–protein interactions is well described.[1] In addition, Hsp90 is also reported to play an important role in the assembly of diverse multi-protein complexes, such as the small nucleolar ribonucleoprotein particle (snoRNP) containing small RNAs and conserved proteins needed for pre-RNA processing; the RNA polymerase II complex required for transcription; complexes containing phosphatidylinositol-3 kinase-related protein kinases (PIKKs) involved in nutrient sensing and the DNA damage response; complexes containing telomerase and telomere-capping components required to protect chromosome ends; the kinetochore complex that attaches the chromosomes to spindle microtubules during cell division; RNA-induced silencing complexes (RISC); and the 26S proteasome.[27] These and other examples illustrate the wide range of cellular functions that require Hsp90.

The mechanisms by which Hsp90 acts on its protein clientele to promote stabilization and activation have been covered in previous chapters in this volume, so only a brief recap is required here. Hsp90 has four isoforms, Hsp90α and Hsp90β located predominantly in the cytoplasm, GRP94 in the endoplasmic reticulum and the mitochondrial form TRAP1.[25,26] The same basic chaperone cycle mechanism is believed to apply to all four forms. Hsp90 operates as a homodimer. Each monomer comprises an ATP-binding N-terminal domain (where most inhibitors act), middle domain and C-terminal domain. Hsp90's chaperone function depends on a conformational cycle that is driven by binding and hydrolysis of ATP at the N-terminus, resulting in a sequential opening and closing of the chaperone dimer (Figure 4.3A, left-hand side). The altered chaperone conformation induced by ATP hydrolysis is believed to be transmitted to a required change in the structure of bound client protein.[28] The cycle is regulated by multiple co-chaperones, including Hsp40, Hsp70, HOP, CDC37, PP5 and P23 (Figure 4.3A). An additional co-chaperone AHA1 causes activation of ATPase activity.[29]

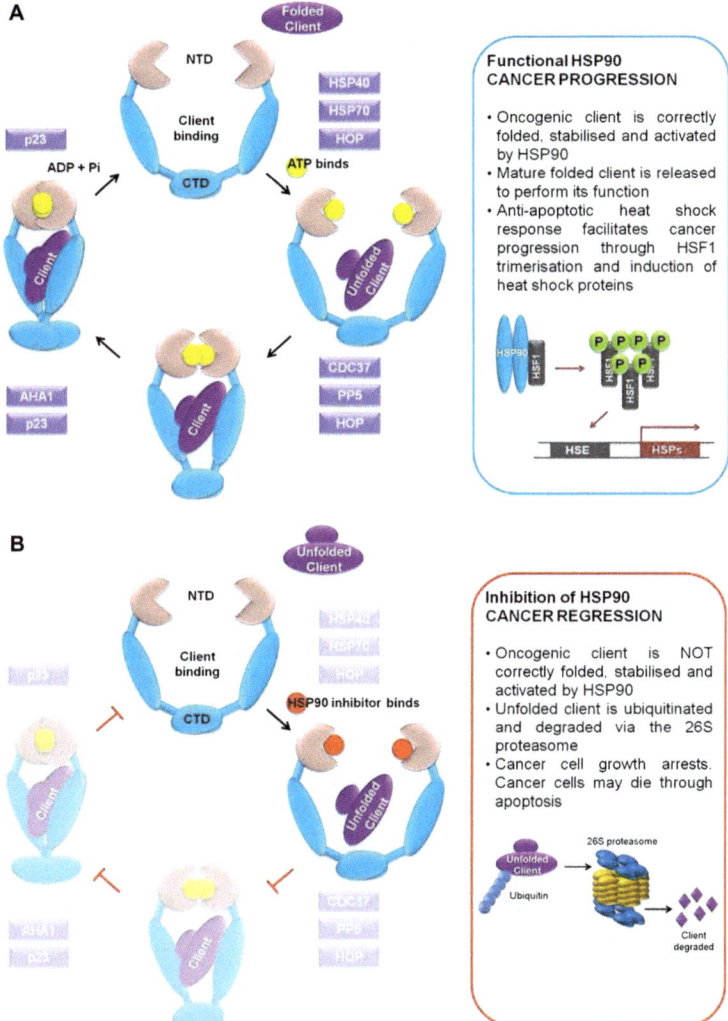

Figure 4.3 The Hsp90 chaperone cycle and its modulation by Hsp90 inhibitors. Panel A) Left-hand side: the chaperone cycle in the absence of inhibitors, showing the open and closing mechanism driven by ATP binding and hydrolysis and the role of regulatory co-chaperones, resulting in correct folding, stabilization and activation of the client protein. Right-hand side: upper text provides a brief summary of the role of Hsp90 in cancer progression and shown below this is the activation of Hsf1, following release from the inhibitory effects of Hsp90, to drive the heat-shock response. Panel B) Left-hand side: inhibition of the chaperone cycle by a small-molecule Hsp90 inhibitor that blocks ATP binding hydrolysis at the N-terminal site, resulting in failure to fold, stabilize and activate the client protein and causing ubiquitin-proteasome degradation of the client (right-hand side bottom). Text above this summarizes the effects on cancer cells, including cell cycle arrest and apoptosis.

Although the basic chaperone cycle is quite well understood, many questions remain that are very relevant both to the basic biology and also to future drug discovery. In particular, the critical and fundamental mechanism by which Hsp90 distinguishes client proteins from non-clients is poorly understood. Protein kinases as a class are recognized by the co-chaperone CDC37, which recruits the many client kinases to Hsp90.[28,30,31] Importantly, recent results suggest that the recognition motif in the kinase is the presence of unstable regions, including in particular flexible loops such as are located at the critical hinge region that connects amino and carboxy termini of the kinase.[31,32] This mechanism is consistent with several mutational studies that have implicated features around the ATP-binding cleft as being involved in kinase recruitment to the Hsp90-CDC37 chaperone system.

Figure 4.3A (right-hand side) shows another noteworthy point, namely that Hsf1 requires release from the inhibitory chaperoning effects of Hsp90 in order to trigger its activation and induce the heat-shock transcriptional response. This is believed to occur under heat shock and other stresses because the unfolded proteins compete for Hsp90. Release of Hsf1 also occurs following inhibition of Hsp90 by typical small-molecule inhibitors of the chaperone.[33]

The effect of Hsp90 inhibitors on the chaperone cycle is illustrated in Figure 4.3B. Typical inhibitors, including all the drugs currently undergoing clinical evaluation, bind at the N-terminal nucleotide site and hence block the binding and hydrolysis of ATP.[25] This inhibits the chaperone cycle and prevents the maturation of the client protein (Figure 4.3B, left-hand side), resulting in loss of at least some co-chaperones from the complex and recruitment of ubiquitin ligases such as CHIP, leading to degradation of the client through the ubiquitin-proteasome system (Figure 4.3B, right-hand side).[34]

4.3 Hsp90 in Cancer and the Basis for Therapeutic Selectivity

There is accumulating evidence – of various different types – that collectively provides strong support for an important role of Hsp90 and other related players in cancer progression and in addition suggests a clear basis for a dependency of cancer cells on the chaperone and for the therapeutic selectivity of Hsp90 inhibitors for cancer *versus* healthy cells.[1]

Firstly, the expression of Hsp90 and other heat-shock proteins such as Hsp70 and Hsp60 is often abnormally high in cancer cells and cancer tissues from patients, as compared to the corresponding healthy counterparts.[35–40] Clinical studies have frequently associated high levels of Hsp90 expression with a poor clinical outcome, including in breast and NSCLC, which of note are the cancers where the clearest objective evidence of therapeutic activity has been seen with Hsp90 inhibitors to date.[41–43] Such evidence is suggestive of a causal relationship.

Next to be considered in the pathogenic link between Hsp90 and cancer is that many of the identified Hsp90 client proteins (www.picard.ch) are oncogenic.[1,2] Frequently cited and well-studied examples include the androgen receptor (AR) and estrogen receptor (ER) and protein kinases such as BCR-ABL, EML4-ALK, ERBB2, CRAF, BRAF, CDK4 and AKT. Also of special significance regarding cancer selectivity is that Hsp90 is required to support the activated forms of oncoproteins that are mutated, translocated, amplified or over-expressed in cancer, and indeed the genetically altered forms are often more dependent on Hsp90 than the wild-type, as exemplified by BRAF.[44,45] Other hypersensitive mutant proteins include v-SRC, BCR-ABL, mutant EGFR and rearranged ALK. Not only are many of Hsp90's oncogenic clientele known to be key driver proteins in human malignancies, providing the basis for Hsp90 inhibitors to exploit oncogene addiction, but furthermore the various oncogenic Hsp90 clients are involved in all of the important hallmarks of cancer[12,46] (Figure 4.4). This involvement supports the notion that Hsp90 inhibitors would not only exert cancer selectivity through exploitation of oncogene addiction together with combinatorial effects on multiple signaling pathways, but in addition would deliver a powerful package of diverse phenotypic anticancer effects, for example blocking invasion, metastasis and angiogenesis as well as causing cell cycle arrest and apoptosis. This could offer even greater cancer selectivity.

An additional factor contributing to the therapeutic selectivity of Hsp90 inhibitors is the highly reproducible observation that Hsp90 inhibitors accumulate in cancer *versus* normal tissues.[47,48] One explanation is the higher level

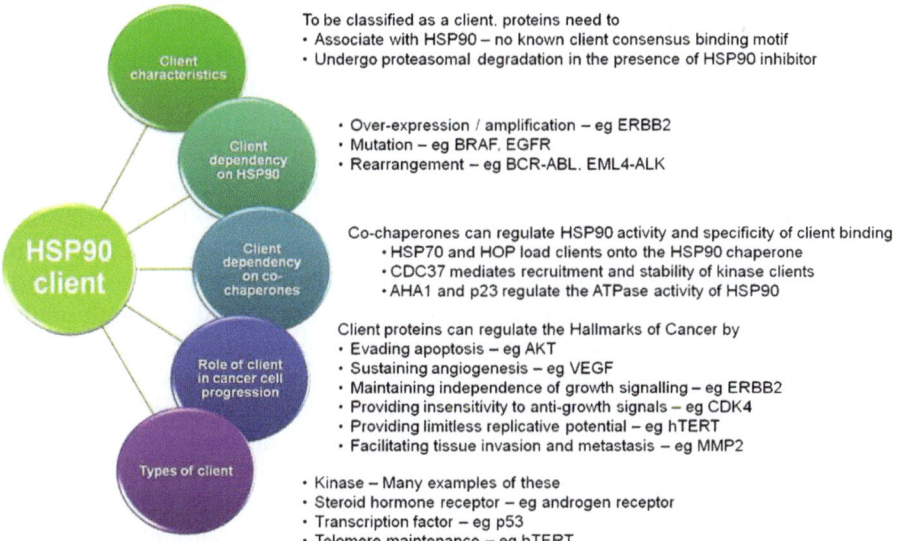

Figure 4.4 Key features of Hsp90 oncogenic client proteins. Illustrated are: 1) the defining characteristics of an Hsp90 client; 2) examples of oncogenic dependency; 3) roles of co-chaperones; 4) roles of client proteins in cancer progression and in supporting the hallmarks of cancer; and 5) types of client protein. Representative examples are shown.

of expression in cancer cells mentioned above while an additional possibility is that this therapeutically beneficial accumulation in malignant tissue relates to the intriguing observation that Hsp90 is present in malignant cells within an activated super-chaperone complex containing multiple co-chaperones that is more susceptible to pharmacological inhibition, whereas this super-complex is not seen in healthy cells.[49]

4.4 Exploiting the Hsp90 Dependency of Driver Oncoproteins by a Range of Hsp90 Inhibitors

The demonstration of the advantageous effects of Hsp90 inhibitors in exploiting Hsp90 dependency of key driver oncoproteins in cancer cells is inextricably linked to the pre-clinical and clinical evaluation of successive small-molecule inhibitors, and so this is how the findings will be presented in the present section. Thus this section also provides a useful summary of progress with a range of Hsp90 inhibitors alongside the key emerging findings made with these agents on the important role of oncoprotein dependence in the observed therapeutic activity in animal models and cancer patients. The chemical structures of the Hsp90 N-terminal Hsp90 inhibitors referred to in this section are shown in Figure 4.1.

In a now classic paper published in 1994, the benzoquinone ansamycin antibiotics geldanamycin and herbimycin were shown to revert v-Src-mediated oncogenic transformation.[50] Using a solid phase-immobilized geldanamycin derivative, the researchers were able to affinity precipitate Hsp90 as the molecular target with which geldanamycins interact and further demonstrated that they disrupted the Src-Hsp90 heteroprotein complex, leading to subsequent depletion of v-Src from the cell. These important results revealed the participation of Hsp90 in a multi-molecular chaperone complex formation that was required for the Src-mediated oncogenic transformation and suggested for the first time that Hsp90 could be a target for drug modulation.[50] Four years later, another macrocyclic natural product, radicicol, was shown to be a potent inhibitor of Hsp90 in vitro, displaying nanomolar affinity.[51,52] Both compounds bind to the N-terminal ATP/ADP-binding domain of Hsp90 and both inhibit the inherent ATPase activity of Hsp90, which is essential for its function,[53] resulting in depletion of Hsp90 oncogenic client proteins through the ubiquitin-proteasome pathway.[54,55] Crystal structure determinations of Hsp90 N-terminal domain complexed with geldanamycin or radicicol[52,53,56] provided unambiguous evidence for binding at the N-terminal ATP site and revealed the atomic detail of their remarkable nucleotide mimicry, which provided a rational basis for the design of potent second-generation small-molecule Hsp90 inhibitors, which all bind and are anchored through essentially the same set of water-mediated hydrogen bonding interactions, for example involving a critical active site aspartate and other key residues.

Although geldanamycin and radicicol were shown to be too unstable and toxic for clinical use, they proved to be invaluable chemical probes that have been employed to understand the function of Hsp90 and, equally important, to validate Hsp90 as a cancer drug target. In addition, these chemical tools have provided us with a molecular biomarker signature of Hsp90 inhibition comprising depletion of client proteins along with mechanism-based induction of heat-shock proteins, particularly Hsp72.[57]

The closely related geldanamycin analogue 17-AAG (17-allylamino-17-demethoxygeldanamycin, tanespimycin, KOS-953; Kosan, Bristol-Myers Squibb; Figure 4.1) was the first-in-class Hsp90 inhibitor to enter clinical trials. Pre-clinical studies showed that 17-AAG had significantly improved stability and pharmacokinetic properties compared to geldanamycin along with greater tolerability and therapeutic activity in human tumor xenograft models.[58]

Pharmacokinetic and pharmacodynamic results from the Phase I clinical trials of 17-AAG were extremely informative in showing that the drug achieved satisfactory plasma exposures at well-tolerated doses[59] and provided proof-of-mechanism for target inhibition in peripheral blood mononuclear cells and tumor biopsies, as demonstrated by what is now known to be the characteristic molecular signature, mentioned above, of depletion of client proteins (in this case CRAF and CDK4) together with Hsf1-dependent induction of Hsp72.[57,59] This signature was used as part of the Pharmacological Audit Trail (PhAT)[60,61] that was implemented in the Phase I studies of 17-AAG and future Hsp90 inhibitors.[59]

17-AAG has shown evidence of clinical activity in several human cancers. The most impressive results were seen in Phase II studies in patients with ERBB2-positive, trastuzumab-refractory breast cancer,[62,63] for which group a 24% response rate was recorded by objective response evaluation criteria in solid tumors (RECIST) using a weekly schedule of 450 mg/m^2, with an overall clinical benefit rate (complete response + partial response + stable disease) of 57% in the same patient cohort. The therapeutic activity seen in these patients is rationalized on the basis that ERBB2, the key oncoprotein driver in this subset, is an especially sensitive Hsp90 client as demonstrated in both corresponding human tumor xenografts and a related genetically engineered mouse model, which is highly sensitive with regressions seen.[64–66] The results were extremely important in providing the first objective clinical validation for Hsp90 inhibitors and in particular because this was seen in a disease setting where the key driver oncogene encodes an oncoprotein that is also an Hsp90 client that is highly dependent on Hsp90 and is very sensitive to depletion by Hsp90 inhibitors and in a cancer in which high Hsp90 expression is an adverse prognostic factor.

Of interest given that BRAF and CRAF are known clients of Hsp90,[44,45] in a hypothesis-testing, biomarker-led Phase I study of 17-AAG there were two cases of prolonged disease stabilization in patients with treatment-refractory metastatic malignant melanoma.[59] A follow-up study explored the possible relationship between BRAF and NRAS mutations and the clinical response to 17-AAG in six melanoma patients who received pharmacologically active doses of 17-AAG in this Phase I clinical trial.[67] Interestingly, one patient with disease stabilization lasting 49 months harbored a (G13D) NRAS mutation with wild-type BRAF in the tumor tissue. A second patient with stable disease for

15 months had a tumor with a (V600E) BRAF mutation and wild-type NRAS. These preliminary results suggested that BRAF and NRAS mutation status might be predictive of response and should be determined in prospective Phase II studies of Hsp90 inhibitors in melanoma. Although the Phase II results were disappointing with the limited 17-AAG dose and schedule used,[68,69] it seems important to revisit melanoma with improved second-generation inhibitors. Interestingly, the initial observation described above of unusual activity in melanoma, which was subsequently linked to patient genotype, is an early example of a general approach that is now attracting considerable interest.

Other examples in which 17-AAG and related agents were evaluated in settings where a particular driver oncoprotein and Hsp90 client could be targeted include acute myelogenous leukemia (AML) patients in which the Hsp90 client tyrosine kinase FLT3 is frequently mutated as the key oncogenic event.[70] Mutant FLT3 is highly dependent on Hsp90 and isogenic AML cells expressing mutant FLT3 have heightened sensitivity to herbimycin and geldanamycin analogues.[71,72] Extensive pre-clinical and preliminary clinical data have clearly suggested a benefit of Hsp90 inhibitors in AML.[73,74] Laboratory studies suggest the exciting potential of blocking the molecular evolution of myelodysplastic syndromes into AML through the effects of Hsp90 inhibitors on several client proteins.[75,76] Hsp90 inhibitors are being evaluated in chronic myelogenous leukemia (CML) and chronic lymphocytic leukemia (CLL), where the Hsp90 clients BCR-ABL and ZAP70, respectively, are particularly sensitive to Hsp90 inhibition.[77–79]

Also noteworthy has been the encouraging activity seen using 17-AAG combined with the approved 26S proteasome inhibitory drug bortezomib in multiple myeloma.[80,81] Several relevant Hsp90 client proteins are targeted for degradation in the proteasome following Hsp90 inhibition, but combination with the proteasome inhibitor results in accumulation of unfolded clients, leading to overload of the protein degradation machinery with resulting apoptosis.[82] Multiple myeloma is especially sensitive to this combination owing to the high level of immunoglobulin production inducing a type of synthetic lethal effect. Hence there is considerable interest in using Hsp90 inhibitors in this disease.

Despite the promising results with 17-AAG, especially in breast cancer, its development was terminated by the company.[83] This was likely due to formulation issues leading to suboptimal doses of 17-AAG together with limited patent life and commercial factors.[68,69,83] In addition, intrinsic and acquired resistance to 17-AAG can be caused by low expression or an inactivating polymorphism of the *NQO1* (NAD(P)H/quinone oxidoreductase I) gene, which is needed to reduce the benzoquinone to the more potent hydroquinone.[84–86] This reductive metabolism is likely also responsible for the hepatotoxicity seen with 17-AAG, which is an additional limitation for this class of inhibitors.

Another geldanamycin analogue 17-DMAG (17-demethoxy-17-(2-dimethylamino)-ethylamino geldanamycin, alvespimycin; Bristol-Myers

Squibb; Figure 4.1) was reported to be less toxic, more water soluble and orally bioavailable.[87–90] Clinical activity observed with 17-DMAG included a complete response in castration-refractory prostate cancer (CRPC), a partial response in melanoma and stable disease in chondrosarcoma, CRPC and renal cancer.[91] Activity in CRPC is likely due to the degradation of the androgen receptor, which is the critical oncogenic driver and an Hsp90 client protein.[92] One partial response and six cases of stable disease >6 months were observed in patients with refractory ERBB2-positive metastatic breast cancer receiving 17-DMAG and trastuzumab.[93] Development of 17-DMAG has been discontinued, probably for reasons similar to those for 17-AAG and recognition that second-generation inhibitors were coming through.

Finally in the geldanamycin class, the highly soluble hydroquinone hydrochloride IPI-504 (retaspimycin hydrochloride; Infinity Pharmaceuticals; Figure 4.1) showed encouraging activity in NSCLC patients with the oncogenic ALK gene rearrangements.[94] This may be attributed to the fact that translocated ALK is an Hsp90 client and the oncogenic forms can be depleted by clinically achievable drug doses.

The early natural product Hsp90 inhibitors exemplified by geldanamycin and radicicol bind to the N-terminal ATPase domain of the chaperone, where they fit into the unusually shaped Bergerat fold pocket that is characteristic of the small GHKL (Gyrase, Hsp90, Histidine, Kinase, MutL) family, leading to a high degree of biochemical selectivity.[95] This has stimulated the successful search for synthetic small-molecule inhibitors with improved pharmaceutical and drug-like properties. The discovery and development of synthetic Hsp90 inhibitors has been accelerated by a battery of discovery techniques, including high-throughput biochemical and fragment-based screening of compound libraries, structure-based design using protein X-ray crystallography, biophysical methods and virtual screening.

Two distinct small-molecule scaffolds were initially discovered that have subsequently entered clinical trials, namely the resorcinol and the purine chemotypes. Radicicol proved too reactive and toxic for clinical development. Binding in a similar manner, the resorcylic pyrazoles exemplified by the original hit CCT018159 were discovered by high-throughput screening at The Institute of Cancer Research (ICR) and the structure-activity relationships (SAR) of early leads were understood by crystallography.[96,97] Subsequent structure-based design by ICR and Vernalis Ltd generated much more potent resorcinylic pyrazole/isoxazole amide analogues VER49009/VER50589,[98] leading in turn to the fully optimized clinical drug NVP-AUY922/VER52296[99,100] (Figure 4.1), which is now being developed clinically by Novartis. The binding of NVP-AUY922 to the N-terminal nucleotide binding pocket of Hsp90α is shown in Figure 4.5, with the resorcinol hydroxyls pointing deep into the base of the pocket and anchoring the binding, as with radicicol. Pre-clinical studies showed impressive potency and selectivity, together with prolonged plasma and tumor exposure and impressive *in vivo* therapeutic effects on primary tumor growth, invasion and metastasis, consistent with the prediction of modulating multiple cancer hallmarks.[100]

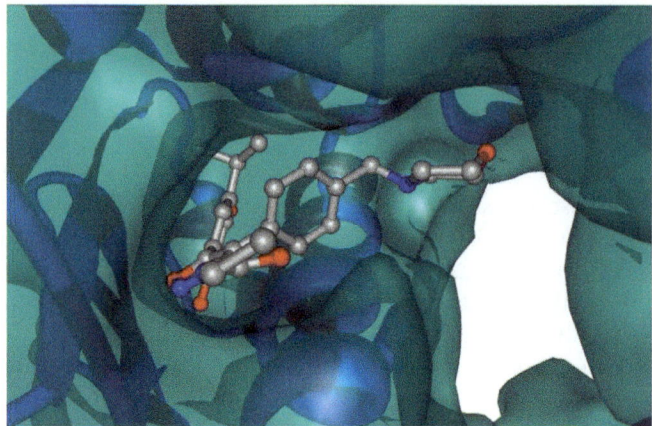

Figure 4.5 The binding of the Hsp90 inhibitor NVP-AUY922 to the N-terminal nucleotide binding pocket of Hsp90α. PDB code 2VCI.

Figure 4.6 shows an example of detailed molecular profiling of client proteins and heat-shock proteins following treatment of human melanoma and breast cancer cells with VER49009 and VER50589, closely related precursors of NVP-AUY922, results for which are illustrated here to exemplify molecular responses to Hsp90 inhibitors.[98] Four points are noteworthy and informative from these detailed results. First, as expected, combinatorial depletion of multiple client proteins is clearly seen. Second, there is a hierarchy of sensitivity among the client proteins – not all are depleted equally – with ERBB2 being by far the most sensitive and AKT and survivin the most resistant of those measured, while others like the RAF family and CDK4 are intermediate (with ERK2 acting as a non-client control). Thirdly, signal transduction is inhibited as shown by decreased phosphorylation of ERK and AKT. Fourthly, activation of the heat-shock response is exemplified by induction of Hsp72 and Hsp27. This pattern of molecular proteomic responses is typical of effects seen with Hsp90 inhibitors and smaller panels of such markers have been used extensively as a biomarker signature during discovery and development to ensure that observed effects are on target and that an appropriate level of inhibition is achieved.[57,59]

NVP-AUY922 is currently in Phase I/II single agent and combination trials as an intravenous agent in patients with a range of cancers including NSCLC, ERBB2-positive advanced breast cancer and gastric cancer, and also refractory gastrointestinal stromal tumors (GIST), which are frequently driven by the mutant KIT client protein (www.clinicaltrials.gov). A Phase I study showed acceptable tolerability, no significant hepatotoxicity (as designed by avoidance of the quinone) and a dose-dependent increase in plasma exposure, supporting once-a-week dosing.[101] Pharmacodynamic biomarkers confirmed depletion of client proteins and induction of Hsp72, and metabolic responses were shown by position emission tomography. Consistent with the 17-AAG experience, Phase II clinical data showed that NVP-AUY922 in combination with trastuzumab is safe

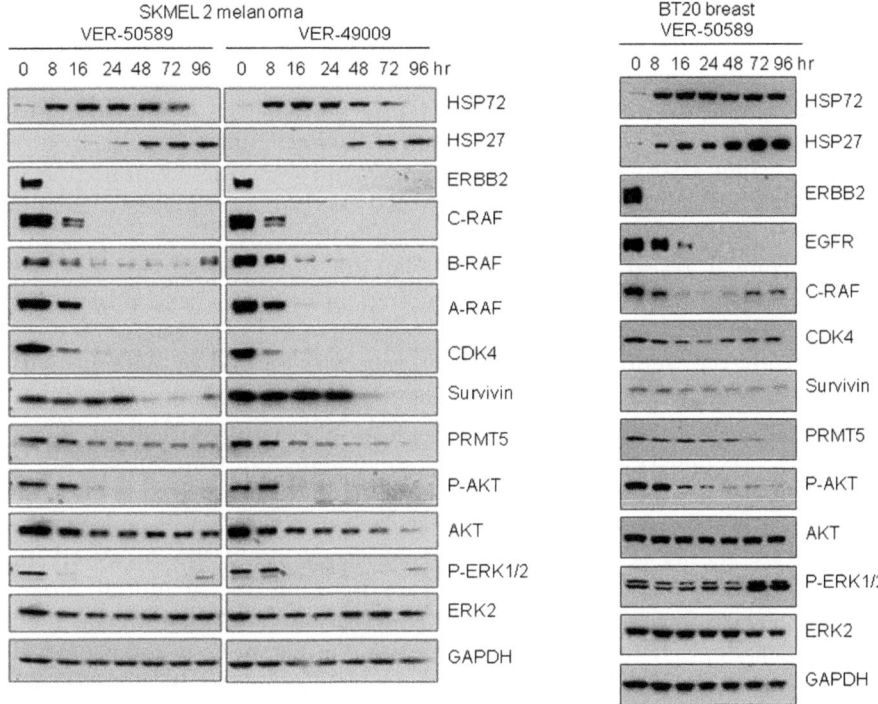

Figure 4.6 Combinatorial effects of Hsp90 inhibition in cancer cells. A hierarchy of differential depletion of multiple Hsp90 client proteins is apparent together with inhibition of signaling (phosphorylation) and activation of the heat-shock response in melanoma and breast cancer cell lines. Reproduced with permission from ref. 98.

and active (partial response was 23% and partial response plus stable disease was 74%) in patients with ERBB2-positive metastatic breast cancer who progressed on trastuzumab-based therapy.[102] This combination was well tolerated, with 5% of patients having diarrhea and eye disorders that were reversible with treatment interruption, dose reduction or discontinuation. The promising therapeutic activity in ERBB2-positive breast cancer is consistent with the pre-clinical data where NVP-AUY922 was shown to be active in trastuzumab-sensitive as well as both innate and acquired trastuzumab-resistant ERBB2-positive human breast cancer cells and human tumor xenograft models.[100,103,104]

NVP-AUY922 has also shown promising clinical activity as a single agent therapy in NSCLC, specifically EGFR mutant patients with acquired resistance to EGFR tyrosine kinase inhibitors, ALK-rearranged patients and EGFR/KRAS/ALK triple wild-type cases.[105] Further studies are ongoing in NSCLC and other malignancies.

The triazole-containing resorcinol STA9090 (ganetespib; Synta Pharmaceuticals; Figure 4.1) has been reported to overcome resistance to ALK inhibitors in ALK-rearranged NSCLC pre-clinical models. It showed combinatorial activity with ALK inhibitors and promising results in NSCLC cells driven by ALK amplification and ROS1 and RET kinase gene rearrangements.[106] Clinical activity was demonstrated with ganetespib in heavily pre-treated patients with advanced NSCLC, particularly in patients with tumors harboring ALK gene rearrangement.[107] Preliminary results from a randomized Phase IIb/III study of ganetespib and docetaxel combination in advanced NSCLC (the GALAXY Trial, NCT01348126) have suggested greater activity than single agent docetaxel in second-line therapy.[108]

In pre-clinical research, the resorcinylic dihydroxybenzamide AT13387 (Astex Therapeutics; Figure 4.1)[109] exhibited prolonged client protein depletion.[110] Among other studies, clinical trials of AT13387 alone or in combination with the androgen-suppressing drug abiraterone in patients with advanced prostate cancer are currently underway (www.clinicaltrials.gov).

A further resorcinol-related inhibitor KW-2478 (Kyowa Hakko Kirin; Figure 4.1)[111] is undergoing clinical trials, in combination with bortezomib in patients with relapsed and/or refractory multiple myeloma.

In addition to the resorcinols, the other early synthetic small-molecule inhibitors were purine scaffold compounds designed using the X-ray crystal structure of Hsp90 in research initially carried out by Memorial Sloan-Kettering Cancer Center and subsequently involving Conforma/Biogen Idec/Premiere Oncology. Understanding the precise binding mode of the prototype purine inhibitor PU3[112,113] led to the oral clinical candidates BIIB021 (CNF-2024), BIIB028 and PU-H71 (Figure 4.1). BIIB021 has undergone Phase I/II studies including patients with GIST and ER-positive metastatic breast cancer while PU-H71 (licensed to Samus Therapeutics) is being evaluated in clinical studies in advanced solid tumors, lymphoma and myeloproliferative disorders, which include the use of imaging biomarkers.[114] Activity of BIIB021 has been reported from the first completed Phase I study in CLL; this may be due to depletion of the Hsp90 client ZAP70.[115]

The purine-related, orally bioavailable Hsp90 inhibitor Debio 0932 (Debiopharm; formerly Curis CUDC-305; Figure 4.1) is currently undergoing clinical trial.[116] Interestingly, this agent not only showed extended tumor retention but also an ability to cross the blood-brain barrier effectively and reach therapeutic levels to demonstrate activity in an intracranial glioblastoma model.

The clinical development of the orally active 2-aminobenzamide compound SNX-5422 (Figure 4.1), which was discovered by Serenex using a novel chemical proteomics approach[117] and subsequently acquired by Pfizer, was discontinued due to ocular toxicity also seen with the resorcinols, but further clinical studies are now underway with Esanex. Other small-molecule Hsp90 inhibitors that have emerged include those from AstraZeneca, Exelixis, Daiichi Sankyo, Chroma Therapeutics and Merck.[114,118–120] This illustrates the high level of interest in Hsp90 as a drug target.

The Hsp90 inhibitors currently in the clinic all bind to the ATP-binding pocket in the N-terminal domain of Hsp90. However, other domains of Hsp90 can also be targeted. The coumarin antibiotics novobiocin, chlorobiocin and

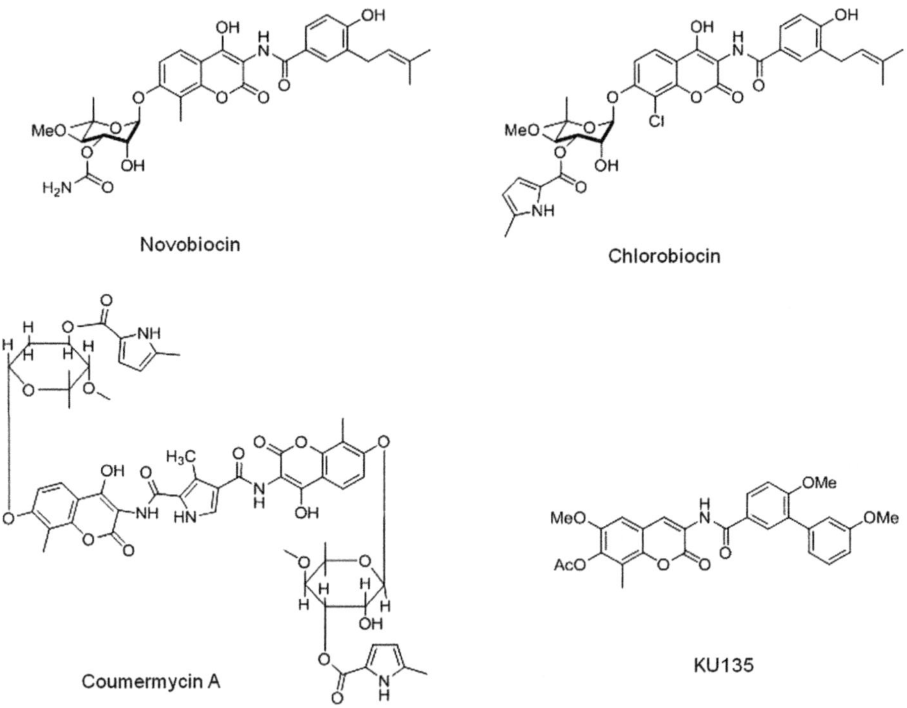

Figure 4.7 Chemical structures of C-terminal Hsp90 inhibitors.

coumermycin A (Figure 4.7) bind to the C-terminal domain of Hsp90 and are reported to exhibit certain unique effects on Hsp90 conformation, activity and interactions with co-chaperones and clients.[121–123] Interestingly, the novobiocin-derived C-terminal Hsp90 inhibitor KU135 (Figure 4.6) failed to up-regulate the expression of Hsp70, in contrast to the N-terminal Hsp90 inhibitors.[124–126] This could be advantageous since the heat-shock response is cytoprotective.[127] However, the C-terminal binders have yet to progress into clinical trials and the mechanism for the unusual behavior is not clear.

4.5 Exploiting the Non-oncogene Dependency of Cancer Cells on Stress Pathways

Recent developments with Hsp90 inhibitors have tended to focus on the progress made, especially in the clinic, on the exploitation of oncogene addiction, as discussed in the previous section. However, it is clear that Hsp90 inhibitors also provide an example of non-oncogene dependency.[128,129] As discussed previously,[2] a strength of Hsp90 inhibitors is the two-pronged exploitation cancer dependencies, involving both oncogene and non-oncogene vulnerabilities.

With regard to the non-oncogene dependency, the effects of Hsp90 inhibitors on proteostasis are most obvious in multiple myeloma, which is characterized by a potentially toxic overload of unfolded protein owing to massive immunoglobulin production.[80,130] 17-AAG and radicicol were shown to induce splicing of XBP1 in multiple myeloma cells, with the induction of CHOP and activation of ATF6, whereas bortezomib resulted in the induction of CHOP and activation of ATF6 with minimal effects on XBP1.[131] Expression levels of the molecular chaperones BiP and GRP94 were increased. At the cellular level, proliferation was inhibited and cell death induced with activation of JNK and caspase cleavage. Hsp90 inhibitors induce myeloma cell death at least in part *via* endoplasmic reticulum stress and the unfolded protein response death pathway. A non-oncogene dependency mechanism could well contribute to the mechanism of action of Hsp90 inhibitors, which may also occur in other tumor types, including induction of the unfolded protein response in cancers with secretory properties and this requires further study.

There is increasing evidence for the role of the transcription factor Hsf1 as a key player in cancer initiation and progression, which likely represents another important example of non-oncogene addiction[132] and is discussed later.

4.6 Lessons Learned and Future Directions

Side effects with single agent studies of Hsp90 inhibitors included fatigue, nausea, diarrhea and myalgias, with ocular effects also seen with potent agents, especially inhibitors in the resorcinol class.[90,133] However, in most instances the side effects have been manageable. Exposures to pharmacologically active drug concentrations have been achieved with multiple Hsp90 inhibitors and proof-of-concept for target modulation and downstream biochemical and cellular effects has been obtained using the classic biomarker signature of client protein depletion and the heat-shock response.[57,59] The latter is a more sensitive marker of Hsp90 target engagement but it is important that doses used clinically are shown to be sufficient to cause adequate depletion of biologically relevant clients. Despite the promise shown, no Hsp90 inhibitor so far received regulatory approval. So how do we best demonstrate the effectiveness of our current Hsp90 inhibitors in the clinic and how can we further increase their therapeutic effects?[1]

As discussed earlier in this chapter, there are already strong objective signs of clinical activity. Clear objective evidence of clinical benefit has been seen with several Hsp90 inhibitors in ERBB2-positive breast cancer and EGFR mutant and ALK rearranged NSCLC and there is an emerging view that these are settings in which initial approvals might best be sought. Other cancers where exploiting addiction to a specific Hsp90 client may prove beneficial include ERBB2-driven cancers other than breast, including gastric and esophageal malignancies; mutant KIT-driven GIST; AR-driven prostate cancer; mutant BRAF-driven malignancies such as melanoma and colorectal cancer; FLT3-driven AML; and CML resistant to BCR-ABL *via* kinase mutation. Interestingly, Hsp90 inhibitors have been reported to be effective against Kaposi sarcoma by enhancing the degradation of LANA, the viral co-receptor EphA2 and other client proteins.[134] Exploiting the unfolded protein response in multiple myeloma is also a promising approach because of the unique vulnerability in this cancer (see earlier).

The activity of the present Hsp90 inhibitors as single agents in the clinic may be enhanced by inclusion within molecular therapeutic combination therapies, including drugs targeted directly to the oncogenic client protein, as already exemplified with 17-AAG and trastuzumab in ERBB2-positive breast cancer. This concept has led to design of clinical trials of Hsp90 inhibitors in combination with inhibitors of the relevant oncogenic Hsp90 client proteins in order to enhance the therapeutic impact through a double hit on the oncogenic driver while potentially also blocking additional oncogenic drivers and feedback loops, bypass pathways and the emergence of resistance alleles of target kinases that remain dependent on Hsp90.[106]

Based on the promising clinical results now emerging from the long journey with Hsp90 inhibitors,[2] alternative therapeutic strategies are now being pursued to target the Hsp90 complex (Figure 4.8). These approaches, discussed below, could be useful in combination with Hsp90 inhibitors or as potential stand-alone treatments.

As we have noted earlier, Hsp90 inhibitors cause a mechanism-based on-target activation of the Hsf1-mediated heat-shock response, for example increasing transcription of genes encoding Hsp70 family members and Hsp27 and raising corresponding protein levels.[135] The extent of this response has been revealed by global microarray-based gene expression profiling at the mRNA level[136–138] and also by mass-spectrometry-based proteomic

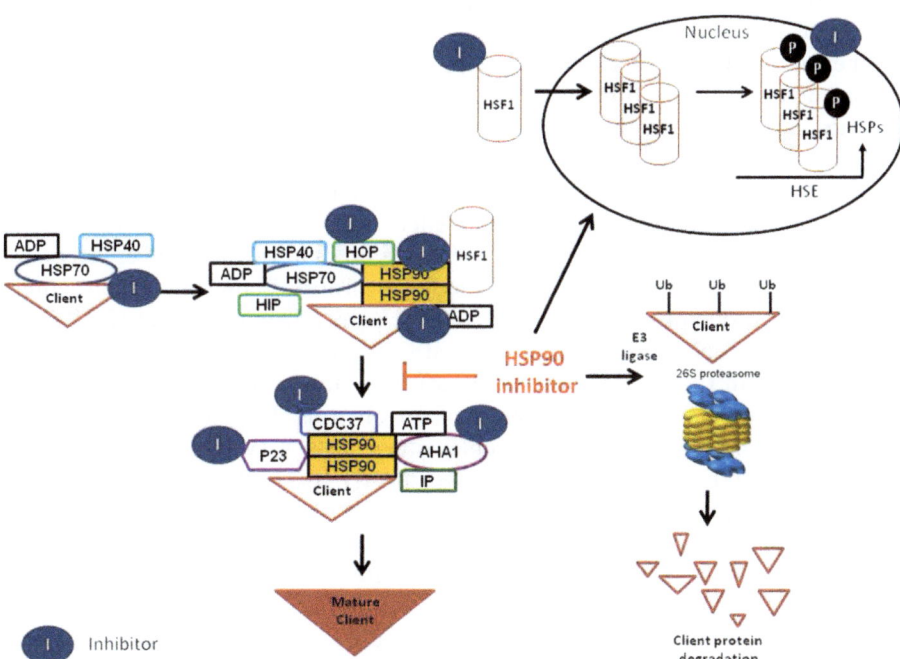

Figure 4.8 The potential opportunities for pharmacological intervention with various components of the Hsp90 chaperone super-complex and related activities.

analysis.[137,139] While this heat-shock response has provided useful pharmacodynamic biomarkers for Hsp90 drug discovery and clinical development,[57,59,68,91] at the same time the expression of heat-shock proteins and other induced cytoprotective factors undoubtedly constitutes an undesirable feedback system that protects cancer cells from the therapeutic effects of Hsp90 inhibition.

As an example, in addition to acting as a supportive co-chaperone for Hsp90, Hsp70 isoforms can contribute to tumor cell survival *via* their anti-apoptotic functions.[140,141] Hence inhibiting Hsp70 could potentially increase the capacity of Hsp90 inhibitors to induce cell death rather than cytostasis. In fact, dual siRNA knockdown of the constitutive Hsc70 and inducible Hsp72 isoforms resulted in a significant increase in tumor-specific sensitization to Hsp90 inhibition *in vitro*, and remarkably also caused Hsp90 inhibition and tumor-selective apoptosis in the absence of a pharmacological Hsp90 inhibitor.[127,142] Consistent with these knockdown studies, a small-molecule Hsc70/Hsp72 inhibitor potentiated Hsp90 inhibitor-induced apoptosis in colon cancer cells.[143,144] Hence there is considerable current interest in the discovery of Hsp70 inhibitors, albeit that this is a much more technically challenging target than Hsp90 because of the properties of the ATP binding/hydrolysis site in Hsp70.[5,145]

In addition to Hsp70, Hsp27 shows promise as a target since knockdown causes anticancer effects and sensitizes to Hsp90 inhibition.[146,147] As a non-ATP-binding low-molecular-weight chaperone, Hsp27 is an extremely challenging drug target but an antisense approach shows promise in early clinical studies for prostate and other cancers.[148]

Whereas inhibition of Hsp70 or Hsp27 would address only one, albeit an important one, of the protective heat-shock responses, there is now strong evidence that targeting the complete Hsf1-mediated transcriptional program could be valuable[149,150] (Figure 4.8). Knockdown of Hsf1 can cause a marked increase in the sensitivity of cancer cells to Hsp90 inhibition.[151] A recent study showed that a striking combinational effect was observed when Hsf1 was knocked down in combination with Hsp90 inhibitors in various cancer cell lines and tumor models in mice.[152] Pharmacologic proof-of-concept for targeting Hsf1 has been demonstrated by the use of chemical tools such as KNK437, which elicited enhanced Hsp90-induced apoptosis in multiple myeloma models.[145] As in the case of other transcription factors that are not regulated by direct ligand binding, the absence of an obvious druggable site on Hsf1 makes the discovery of small-molecule antagonists very challenging.[4,153] A range of compounds have been identified mostly by cell-based screening but these are not ideal.[4,151]

Interest in Hsf1 as a therapeutic target has been stimulated by recent reports that high Hsf1 levels are associated with poor prognosis in breast and other cancers and that Hsf1 drives a transcriptional program that is distinct from heat shock in order to support highly malignant human cancers.[154–156] Furthermore, an important earlier study showed that Hsf1 is a powerful multi-faceted modifier of carcinogenesis, with Hsf1 knockout blocking tumor initiation in mouse models driven by RAS mutation or by loss of p53, and silencing of Hsf1 antagonizes multiple hallmark features of cancer cells.[132,149,157]

Since the induction of the heat-shock response reduces the effectiveness of Hsp90 inhibitors the further exploration of C-terminal Hsp90 inhibitors that do not induce Hsp70 is also warranted as a means to circumvent the protective feedback loop.[126]

As with knockdown of Hsp70 and Hsf1, a number of pre-clinical studies have shown that silencing of Hsp90 co-chaperones, which are often over-expressed in cancer, such as AHA1, CDC37 and P23, can also increase cancer cell sensitivity to Hsp90 inhibitors.[158–160] Taken together, there is considerable support for increasing efforts to identify and validate inhibitors of Hsf1, Hsp70, Hsp27 and Hsp90 co-chaperones, either as single agents or in combination with Hsp90 inhibitors. However, careful attention needs to be paid to druggability.

4.7 Concluding Remarks

To return to our earlier theatrical metaphor for Hsp90 inhibitors, we may now be moving into the final act of the play. We hope that a happy ending will be achieved with the regulatory approval of an Hsp90 inhibitor. Even so there will be sequels, involving further refinements to the design of Hsp90 inhibitors and how they are used. At that point Hsp90 will step back from the limelight as the star performer and become an ensemble player, performing alongside other rising stars, which will likely be other molecular targets such as those discussed in the previous sections.

In terms of potential refinements to the design of Hsp90 inhibitors, one question still to be resolved is the preferred selectivity profile across the four isoforms. Current Hsp90 drugs should probably be considered as pan-Hsp90 isoform inhibitors, although in general they are more potent towards Hsp90α and Hsp90β, the two major Hsp90 isoforms that are commonly considered to be the most important targets for cancer. However, the importance of the other two isoforms (GRP94 and TRAP1) remains unclear. It is possible that future Hsp90 inhibitors that show different isoform selectivity patterns could deliver distinct biological and clinical effects, for example in terms of therapeutic activity and side effects. Further research is required to address this.

In addition to modulating additional chaperones and co-chaperones, as discussed in the previous section, research suggests that modifying the many post-translational modifications that regulate Hsp90 activity could also have therapeutic benefit.[161] A more detailed understanding of the phosphorylation of Hsp90 may provide additional strategy to enhance therapeutic benefit. For example, WEE1-induced phosphorylation of Hsp90 on the tyrosine 100 residue increases Hsp90's ability to chaperone several oncogenic kinases, including ERBB2, CRAF, CDK4 and WEE1 itself, and decreases its ability to bind to Hsp90 inhibitors.[1,162]

It is conceivable that the scheduling of Hsp90 inhibitors in the clinic may not be optimum at present, for example with respect to sparing the deletion of clients whose loss is not desired. Prolonged Hsp90 inhibition has been shown to cause increased mutation rates in germ cells, where Hsp90 is required to suppress transposon activity,[163,164] through the loss of activity of the Hsp90 client protein, arginine methyltransferase PRMT5.[137] Prolonged Hsp90 inhibition could also lead to deregulation of several tumor suppressor pathways *via* effects on proteins such as interferon regulatory factor 1 (IRF-1), LATS1 and LATS2.[165–167] In addition, a mutant form of retinoblastoma protein that retains ∼50% of the tumor suppressor activity of wild-type retinoblastoma protein is also regulated by Hsp90.[168] Careful monitoring will be required to evaluate the effects of long-term administration of Hsp90 inhibitors.

The cumulative effect of an Hsp90 inhibitor will depend on several factors and these will be influenced by the dose and duration of treatment, the differences in sensitivity of the hierarchy of client proteins in cancer cells and the dependence of the tumor on the expression of the clients in terms of both oncogene and non-oncogene dependency. Taken together, these results suggest that maximum clinical benefit that could be achieved from Hsp90 inhibitors may be determined (or limited) by the clinical trial designs.

Earlier in the chapter we have discussed how judicious use of Hsp90 inhibitors might help to overcome or even prevent resistance to targeted therapies, including kinase inhibitors and hormonal agents.[17] There is of course the possibility that cancer cells could develop resistance to Hsp90 inhibitors themselves. As mentioned earlier, pre-clinical studies have shown that loss of activity or expression of the oxidoreductase NQO1 can confer resistance to quinone-containing ansamycin Hsp90 inhibitors, notably 17-AAG, but cross-resistance to non-quinone drugs such as NVP-AUY922 is not seen.[86] In addition, this class of benzoquinone inhibitors are also substrates for P-glycoprotein-mediated drug efflux,[84,169] but again this is avoided by second-generation inhibitors. To date drug-resistant alleles of Hsp90 have not been identified in cancer and it has proved difficult to make tumor cell lines resistant to Hsp90 inhibitors such as NVP-AUY922 by continuous exposure using the conventional experimental protocol. The structure of Hsp90 suggests that it would be difficult to mutate residues in the nucleotide binding site that would decrease the binding of inhibitors without also blocking the binding of ATP, which would be lethal and so not tolerated. However, changes to the amino acid residues in the active site that reduce inhibitor binding have been identified in organisms that produce natural product inhibitors. A naturally occurring single amino acid change in the ATP-binding pocket of Hsp90 from the fungus *Humicola fuscoatra* has been reported to confer partial resistance to the radicicol it produces and other resorcinol-based synthetic Hsp90 inhibitors but not to geldanamycin.[170] Conversely, *Streptomyles hydroscopicus* has evolved an Hsp90 family protein with amino acid changes that result in partial resistance to the geldanamycin that the organism produces, but not to the resorcinols radicicol or NVP-AUY922.[171] Although the evolution of these partially resistant forms of Hsp90 has been seen in the producing microorganisms, equivalent findings have not been published for cancer cell models or patients. Of note is that if equivalent resistance mechanisms did occur, it should be possible to switch to an alternative Hsp90 chemotype to which cross-resistance is not seen.

One thing we can say for sure is that results from laboratory and clinical Hsp90 research will continue to surprise us. For example, a recent study[172] reported the finding that CDC37 antagonizes ATP binding to client protein kinases, suggesting a role for the Hsp90-CDC37 complex in regulating kinase activity. Even more surprising, the binding of CDC37 to client kinases is itself antagonized by clinically approved ATP-competitive inhibitors of those kinases, including vemurafenib, sorafenib, lapatinib and erlotinib. In cancer cells treated with these inhibitors this leads to deprivation of the target protein kinases, such as BRAF, ERBB2 and EGFR from access to Hsp90-CDC37 chaperone complex, resulting in their degradation *via* the ubiquitin-proteasome pathway. These surprising findings may have profound implications for protein kinase biology and cancer treatment. For example, they suggest that at least part of the efficacy of clinically used protein kinase inhibitors may result from chaperone deprivation and kinase degradation, analogous but not identical to the effects of Hsp90 inhibitors. It is possible that the impact of this effect could be optimized in the clinic by modifying the dose and schedule of kinase inhibitors. In addition, the results provide a further rationale for combining kinase and Hsp90 inhibitors in the clinic to maximize chaperone deprivation and client degradation in cancer tissue.

What other twists and turns might be possible in the convoluted plot of the Hsp90 molecule chaperone drama? No doubt many. Hopefully, we will see increasing evidence of the efficacy of Hsp90 in cancer patients, refinement of their use and progress with drugging next-generation chaperone targets for the treatment of cancer and potentially other diseases.

Conflicts of Interest

All the authors are employees of The Institute of Cancer Research, which has a commercial interest in Hsp90 inhibitors and operates a Reward to Inventors scheme. Paul Workman and Swee Sharp benefited from research funding (to PW) from Vernalis for the discovery of Hsp90 inhibitors and intellectual property from this program was licensed to Vernalis Ltd and Novartis. Paul Workman has been a consultant to Novartis, is a founder and scientific advisory board member of Chroma Therapeutics and is scientific advisory board member of Astex Pharmaceuticals and Nextech Invest.

Acknowledgements

We thank many colleagues and collaborators for helpful discussions and apologize to numerous authors whose work could not be cited because of length restrictions. We thank Dr Ian Collins for help with Figure 4.1 and Professor Keith Jones for Figure 4.5. We are grateful to Cancer Research UK for long-term program support for our work on Hsp90 and other chaperone targets. Paul Workman is a Cancer Research UK Life Fellow and Jenny Howes is funded by a Cancer Research UK PhD studentship. We acknowledge NHS support to our NIHR Biomedical Research Centre.

References

1. L. Neckers and P. Workman, *Clin. Cancer Res.*, 2012, **18**, 64–76.
2. J. Travers, S. Sharp and P. Workman, *Drug Discov. Today*, 2012, **17**, 242–252.
3. M. V. Powers and P. Workman, *FEBS Lett.*, 2007, **581**, 3758–3769.
4. E. de Billy, M. V. Powers, J. R. Smith and P. Workman, *Cell Cycle*, 2009, **8**, 3806–3808.
5. M. V. Powers, K. Jones, C. Barillari, I. Westwood, R. L. van Montfort and P. Workman, *Cell Cycle*, 2010, **9**, 1542–1550.
6. A. P. Arrigo, S. Simon, B. Gibert, C. Kretz-Remy, M. Nivon, A. Czekalla, D. Guillet, M. Moulin, C. Diaz-Latoud and P. Vicart, *FEBS Lett.*, 2007, **581**, 3665–3674.
7. A. Zoubeidi and M. Gleave, *Int. J. Biochem. Cell Biol.*, 2012, **44**, 1646–1656.
8. W. Xolalpa, P. Perez-Galan, M. S. Rodriguez and G. Roue, *Curr. Pharma. Des.*, 2013, **19**, 4053–4093.
9. I. B. Weinstein and A. Joe, *Cancer Res.*, 2008, **68**, 3077–3080; discussion 3080.
10. T. A. Yap and P. Workman, *Annu. Rev. Pharmacol. Toxicol.*, 2012, **52**, 549–573.
11. P. Workman and I. Collins, *Chem. Biol.*, 2010, **17**, 561–577.
12. A. Maloney and P. Workman, *Expert Opin. Biol. Ther.*, 2002, **2**, 3–24.
13. P. Workman, *Cancer Lett.*, 2004, **206**, 149–157.
14. P. Workman, F. Burrows, L. Neckers and N. Rosen, *Ann. NY Acad. Sci.*, 2007, **1113**, 202–216.
15. J. Trepel, M. Mollapour, G. Giaccone and L. Neckers, *Nat. Rev. Cancer*, 2010, **10**, 537–549.
16. Y. Miyata, H. Nakamoto and L. Neckers, *Curr. Pharm. Des.*, 2013, **19**, 347–365.
17. B. Al-Lazikani, U. Banerji and P. Workman, *Nat. Biotechnol.*, 2012, **30**, 679–692.
18. D. Gonzalez de Castro, P. A. Clarke, B. Al-Lazikani and P. Workman, *Clin. Pharmacol. Ther.*, 2013, **93**, 252–259.
19. F. U. Hartl, A. Bracher and M. Hayer-Hartl, *Nature*, 2011, **475**, 324–332.
20. C. M. Dobson, *Nature*, 2003, **426**, 884–890.
21. E. T. Powers, R. I. Morimoto, A. Dillin, J. W. Kelly and W. E. Balch, *Annu. Rev. Biochem.*, 2009, **78**, 959–991.
22. E. L. Davenport, G. J. Morgan and F. E. Davies, *Cell Cycle*, 2008, **7**, 865–869.
23. A. Lilienbaum, *Int. J. Biochem. Mol. Biol.*, 2013, **4**, 1–26.
24. R. Voellmy and F. Boellmann, *Adv. Exp. Med. Biol.*, 2007, **594**, 89–99.

25. L. H. Pearl, C. Prodromou and P. Workman, *Biochem. J.*, 2008, **410**, 439–453.
26. M. Taipale, D. F. Jarosz and S. Lindquist, *Nat. Rev. Mol. Cell Biol.*, 2010, **11**, 515–528.
27. T. Makhnevych and W. A. Houry, *Biochim. Biophys. Acta*, 2012, **1823**, 674–682.
28. C. K. Vaughan, U. Gohlke, F. Sobott, V. M. Good, M. M. Ali, C. Prodromou, C. V. Robinson, H. R. Saibil and L. H. Pearl, *Mol. Cell*, 2006, **23**, 697–707.
29. B. Panaretou, G. Siligardi, P. Meyer, A. Maloney, J. K. Sullivan, S. Singh, S. H. Millson, P. A. Clarke, S. Naaby-Hansen, R. Stein, R. Cramer, M. Mollapour, P. Workman, P. W. Piper, L. H. Pearl and C. Prodromou, *Mol. Cell*, 2002, **10**, 1307–1318.
30. Y. Kimura, S. L. Rutherford, Y. Miyata, I. Yahara, B. C. Freeman, L. Yue, R. I. Morimoto and S. Lindquist, *Gene. Dev.*, 1997, **11**, 1775–1785.
31. M. Taipale, I. Krykbaeva, M. Koeva, C. Kayatekin, K. D. Westover, G. I. Karras and S. Lindquist, *Cell*, 2012, **150**, 987–1001.
32. R. S. Samant and P. Workman, *Nature*, 2012, **490**, 351–352.
33. L. Whitesell, R. Bagatell and R. Falsey, *Curr. Cancer Drug Targets*, 2003, **3**, 349–358.
34. P. Connell, C. A. Ballinger, J. Jiang, Y. Wu, L. J. Thompson, J. Hohfeld and C. Patterson, *Nat. Cell Biol.*, 2001, **3**, 93–96.
35. Y. Yufu, J. Nishimura and H. Nawata, *Leuk. Res.*, 1992, **16**, 597–605.
36. R. Ralhan and J. Kaur, *Clin. Cancer Res.*, 1995, **1**, 1217–1222.
37. D. R. Ciocca, G. M. Clark, A. K. Tandon, S. A. Fuqua, W. J. Welch and W. L. McGuire, *J. Natl Cancer Inst.*, 1993, **85**, 570–574.
38. E. Kimura, R. E. Enns, J. E. Alcaraz, J. Arboleda, D. J. Slamon and S. B. Howell, *J. Clin. Oncol.*, 1993, **11**, 891–898.
39. M. Santarosa, D. Favaro, M. Quaia and E. Galligioni, *Eur. J. Cancer*, 1997, **33**, 873–877.
40. M. M. McCarthy, E. Pick, Y. Kluger, B. Gould-Rothberg, R. Lazova, R. L. Camp, D. L. Rimm and H. M. Kluger, *Ann. Oncol.*, 2008, **19**, 590–594.
41. E. Pick, Y. Kluger, J. M. Giltnane, C. Moeder, R. L. Camp, D. L. Rimm and H. M. Kluger, *Cancer Res.*, 2007, **67**, 2932–2937.
42. M. I. Gallegos Ruiz, K. Floor, P. Roepman, J. A. Rodriguez, G. A. Meijer, W. J. Mooi, E. Jassem, J. Niklinski, T. Muley, N. van Zandwijk, E. F. Smit, K. Beebe, L. Neckers, B. Ylstra and G. Giaccone, *PLoS One*, 2008, **3**, e0001722.
43. C. F. Li, W. W. Huang, J. M. Wu, S. C. Yu, T. H. Hu, Y. H. Uen, Y. F. Tian, C. N. Lin, D. Lu, F. M. Fang and H. Y. Huang, *Clin. Cancer Res.*, 2008, **14**, 7822–7831.
44. S. da Rocha Dias, F. Friedlos, Y. Light, C. Springer, P. Workman and R. Marais, *Cancer Res.*, 2005, **65**, 10686–10691.
45. O. M. Grbovic, A. D. Basso, A. Sawai, Q. Ye, P. Friedlander, D. Solit and N. Rosen, *Proc. Natl Acad. Sci. USA*, 2006, **103**, 57–62.

46. D. Hanahan and R. A. Weinberg, *Cell*, 2011, **144**, 646–674.
47. G. Chiosis and L. Neckers, *ACS Chem. Biol.*, 2006, **1**, 279–284.
48. P. Workman, *Mol. Cancer Ther.*, 2003, **2**, 131–138.
49. A. Kamal, L. Thao, J. Sensintaffar, L. Zhang, M. F. Boehm, L. C. Fritz and F. J. Burrows, *Nature*, 2003, **425**, 407–410.
50. L. Whitesell, E. G. Mimnaugh, B. De Costa, C. E. Myers and L. M. Neckers, *Proc. Natl Acad. Sci. USA*, 1994, **91**, 8324–8328.
51. S. V. Sharma, T. Agatsuma and H. Nakano, *Oncogene*, 1998, **16**, 2639–2645.
52. T. W. Schulte, S. Akinaga, S. Soga, W. Sullivan, B. Stensgard, D. Toft and L. M. Neckers, *Cell Stress Chaperones*, 1998, **3**, 100–108.
53. S. M. Roe, C. Prodromou, R. O'Brien, J. E. Ladbury, P. W. Piper and L. H. Pearl, *J. Med. Chem.*, 1999, **42**, 260–266.
54. E. G. Mimnaugh, C. Chavany and L. Neckers, *J. Biol. Chem.*, 1996, **271**, 22796–22801.
55. C. Schneider, L. Sepp-Lorenzino, E. Nimmesgern, O. Ouerfelli, S. Danishefsky, N. Rosen and F. U. Hartl, *Proc. Natl Acad. Sci. USA*, 1996, **93**, 14536–14541.
56. J. R. Porter, J. Ge, J. Lee, E. Normant and K. West, *Curr. Top. Med. Chem.*, 2009, **9**, 1386–1418.
57. U. Banerji, M. Walton, F. Raynaud, R. Grimshaw, L. Kelland, M. Valenti, I. Judson and P. Workman, *Clin. Cancer Res.*, 2005, **11**, 7023–7032.
58. T. W. Schulte and L. M. Neckers, *Cancer Chemother. Pharmacol.*, 1998, **42**, 273–279.
59. U. Banerji, A. O'Donnell, M. Scurr, S. Pacey, S. Stapleton, Y. Asad, L. Simmons, A. Maloney, F. Raynaud, M. Campbell, M. Walton, S. Lakhani, S. Kaye, P. Workman and I. Judson, *J. Clin. Oncol.*, 2005, **23**, 4152–4161.
60. P. Workman, *Curr. Pharm. Des.*, 2003, **9**, 891–902.
61. T. A. Yap, S. K. Sandhu, P. Workman and J. S. de Bono, *Nat. Rev. Cancer*, 2010, **10**, 514–523.
62. S. Modi, A. T. Stopeck, M. S. Gordon, D. Mendelson, D. B. Solit, R. Bagatell, W. Ma, J. Wheler, N. Rosen, L. Norton, G. F. Cropp, R. G. Johnson, A. L. Hannah and C. A. Hudis, *J. Clin. Oncol.*, 2007, **25**, 5410–5417.
63. S. Modi, A. Stopeck, H. Linden, D. Solit, S. Chandarlapaty, N. Rosen, G. D'Andrea, M. Dickler, M. E. Moynahan, S. Sugarman, W. Ma, S. Patil, L. Norton, A. L. Hannah and C. Hudis, *Clin. Cancer Res.*, 2011, **17**, 5132–5139.
64. P. N. Munster, D. C. Marchion, A. D. Basso and N. Rosen, *Cancer Res.*, 2002, **62**, 3132–3137.
65. A. D. Basso, D. B. Solit, P. N. Munster and N. Rosen, *Oncogene*, 2002, **21**, 1159–1166.
66. L. M. Rodrigues, Y. L. Chung, N. M. Al Saffar, S. Y. Sharp, L. E. Jackson, U. Banerji, M. Stubbs, M. O. Leach, J. R. Griffiths and P. Workman, *BMC Res. Notes*, 2012, **5**, 250.

67. U. Banerji, A. Affolter, I. Judson, R. Marais and P. Workman, *Mol. Cancer Ther.*, 2008, **7**, 737–739.
68. S. Pacey, M. Gore, D. Chao, U. Banerji, J. Larkin, S. Sarker, K. Owen, Y. Asad, F. Raynaud, M. Walton, I. Judson, P. Workman and T. Eisen, *Invest. New Drugs*, 2012, **30**, 341–349.
69. D. B. Solit, I. Osman, D. Polsky, K. S. Panageas, A. Daud, J. S. Goydos, J. Teitcher, J. D. Wolchok, F. J. Germino, S. E. Krown, D. Coit, N. Rosen and P. B. Chapman, *Clin. Cancer Res.*, 2008, **14**, 8302–8307.
70. E. Weisberg, R. Barrett, Q. Liu, R. Stone, N. Gray and J. D. Griffin, *Drug Resist. Updates*, 2009, **12**, 81–89.
71. Y. Minami, H. Kiyoi, Y. Yamamoto, K. Yamamoto, R. Ueda, H. Saito and T. Naoe, *Leukemia*, 2002, **16**, 1535–1540.
72. Q. Yao, R. Nishiuchi, Q. Li, A. R. Kumar, W. A. Hudson and J. H. Kersey, *Clin. Cancer Res.*, 2003, **9**, 4483–4493.
73. H. Reikvam, E. Ersvaer and O. Bruserud, *Curr. Cancer Drug Targets*, 2009, **9**, 761–776.
74. J. E. Lancet, I. Gojo, M. Burton, M. Quinn, S. M. Tighe, K. Kersey, Z. Zhong, M. X. Albitar, K. Bhalla, A. L. Hannah and M. R. Baer, *Leukemia*, 2010, **24**, 699–705.
75. P. Flandrin-Gresta, F. Solly, C. M. Aanei, J. Cornillon, E. Tavernier, N. Nadal, F. Morteux, D. Guyotat, E. Wattel and L. Campos, *Oncotarget*, 2012, **3**, 1158–1168.
76. A. S. Martins, F. E. Davies and P. Workman, *Oncotarget*, 2012, **3**, 1054–1056.
77. C. Peng, J. Brain, Y. Hu, A. Goodrich, L. Kong, D. Grayzel, R. Pak, M. Read and S. Li, *Blood*, 2007, **110**, 678–685.
78. T. O'Hare, C. A. Eide and M. W. Deininger, *Expert Opin. Investig. Drugs*, 2008, **17**, 865–878.
79. J. E. Castro, C. E. Prada, O. Loria, A. Kamal, L. Chen, F. J. Burrows and T. J. Kipps, *Blood*, 2005, **106**, 2506–2512.
80. P. G. Richardson, A. A. Chanan-Khan, S. Lonial, A. Y. Krishnan, M. P. Carroll, M. Alsina, M. Albitar, D. Berman, M. Messina and K. C. Anderson, *Br. J. Haematol.*, 2011, **153**, 729–740.
81. P. G. Richardson, C. S. Mitsiades, J. P. Laubach, S. Lonial, A. A. Chanan-Khan and K. C. Anderson, *Br. J. Haematol.*, 2011, **152**, 367–379.
82. E. G. Mimnaugh, W. Xu, M. Vos, X. Yuan and L. Neckers, *Mol. Cancer Res.*, 2006, **4**, 667–681.
83. C. L. Arteaga, *Clin. Cancer Res.*, 2011, **17**, 4919–4921.
84. L. R. Kelland, S. Y. Sharp, P. M. Rogers, T. G. Myers and P. Workman, *J. Natl Cancer Inst.*, 1999, **91**, 1940–1949.
85. W. Guo, P. Reigan, D. Siegel, J. Zirrolli, D. Gustafson and D. Ross, *Cancer Res.*, 2005, **65**, 10006–10015.
86. N. Gaspar, S. Y. Sharp, S. Pacey, C. Jones, M. Walton, G. Vassal, S. Eccles, A. Pearson and P. Workman, *Cancer Res.*, 2009, **69**, 1966–1975.

87. G. Kaur, D. Belotti, A. M. Burger, K. Fisher-Nielson, P. Borsotti, E. Riccardi, J. Thillainathan, M. Hollingshead, E. A. Sausville and R. Giavazzi, *Clin. Cancer Res.*, 2004, **10**, 4813–4821.
88. V. Smith, E. A. Sausville, R. F. Camalier, H. H. Fiebig and A. M. Burger, *Cancer Chemother. Pharmacol.*, 2005, **56**, 126–137.
89. M. J. Egorin, T. F. Lagattuta, D. R. Hamburger, J. M. Covey, K. D. White, S. M. Musser and J. L. Eiseman, *Cancer Chemother. Pharmacol.*, 2002, **49**, 7–19.
90. D. B. Solit, S. P. Ivy, C. Kopil, R. Sikorski, M. J. Morris, S. F. Slovin, W. K. Kelly, A. DeLaCruz, T. Curley, G. Heller, S. Larson, L. Schwartz, M. J. Egorin, N. Rosen and H. I. Scher, *Clin. Cancer Res.*, 2007, **13**, 1775–1782.
91. S. Pacey, R. H. Wilson, M. Walton, M. M. Eatock, A. Hardcastle, A. Zetterlund, H. T. Arkenau, J. Moreno-Farre, U. Banerji, B. Roels, H. Peachey, W. Aherne, J. S. de Bono, F. Raynaud, P. Workman and I. Judson, *Clin. Cancer Res.*, 2011, **17**, 1561–1570.
92. D. B. Solit, F. F. Zheng, M. Drobnjak, P. N. Munster, B. Higgins, D. Verbel, G. Heller, W. Tong, C. Cordon-Cardo, D. B. Agus, H. I. Scher and N. Rosen, *Clin. Cancer Res.*, 2002, **8**, 986–993.
93. K. Jhaveri, K. Miller, L. Rosen, B. Schneider, L. Chap, A. Hannah, Z. Zhong, W. Ma, C. Hudis and S. Modi, *Clin. Cancer Res.*, 2012, **18**, 5090–5098.
94. L. V. Sequist, S. Gettinger, N. N. Senzer, R. G. Martins, P. A. Janne, R. Lilenbaum, J. E. Gray, A. J. Iafrate, R. Katayama, N. Hafeez, J. Sweeney, J. R. Walker, C. Fritz, R. W. Ross, D. Grayzel, J. A. Engelman, D. R. Borger, G. Paez and R. Natale, *J. Clin. Oncol.*, 2010, **28**, 4953–4960.
95. P. Chene, *Nat. Rev. Drug Discov.*, 2002, **1**, 665–673.
96. K. M. Cheung, T. P. Matthews, K. James, M. G. Rowlands, K. J. Boxall, S. Y. Sharp, A. Maloney, S. M. Roe, C. Prodromou, L. H. Pearl, G. W. Aherne, E. McDonald and P. Workman, *Bioorg. Med. Chem. Lett.*, 2005, **15**, 3338–3343.
97. S. Y. Sharp, K. Boxall, M. Rowlands, C. Prodromou, S. M. Roe, A. Maloney, M. Powers, P. A. Clarke, G. Box, S. Sanderson, L. Patterson, T. P. Matthews, K. M. Cheung, K. Ball, A. Hayes, F. Raynaud, R. Marais, L. Pearl, S. Eccles, W. Aherne, E. McDonald and P. Workman, *Cancer Res.*, 2007, **67**, 2206–2216.
98. S. Y. Sharp, C. Prodromou, K. Boxall, M. V. Powers, J. L. Holmes, G. Box, T. P. Matthews, K. M. Cheung, A. Kalusa, K. James, A. Hayes, A. Hardcastle, B. Dymock, P. A. Brough, X. Barril, J. E. Cansfield, L. Wright, A. Surgenor, N. Foloppe, R. E. Hubbard, W. Aherne, L. Pearl, K. Jones, E. McDonald, F. Raynaud, S. Eccles, M. Drysdale and P. Workman, *Mol. Cancer Ther.*, 2007, **6**, 1198–1211.
99. P. A. Brough, W. Aherne, X. Barril, J. Borgognoni, K. Boxall, J. E. Cansfield, K. M. Cheung, I. Collins, N. G. Davies, M. J. Drysdale, B. Dymock, S. A. Eccles, H. Finch, A. Fink, A. Hayes, R. Howes,

R. E. Hubbard, K. James, A. M. Jordan, A. Lockie, V. Martins, A. Massey, T. P. Matthews, E. McDonald, C. J. Northfield, L. H. Pearl, C. Prodromou, S. Ray, F. I. Raynaud, S. D. Roughley, S. Y. Sharp, A. Surgenor, D. L. Walmsley, P. Webb, M. Wood, P. Workman and L. Wright, *J. Med. Chem.*, 2008, **51**, 196–218.
100. S. A. Eccles, A. Massey, F. I. Raynaud, S. Y. Sharp, G. Box, M. Valenti, L. Patterson, A. de Haven Brandon, S. Gowan, F. Boxall, W. Aherne, M. Rowlands, A. Hayes, V. Martins, F. Urban, K. Boxall, C. Prodromou, L. Pearl, K. James, T. P. Matthews, K. M. Cheung, A. Kalusa, K. Jones, E. McDonald, X. Barril, P. A. Brough, J. E. Cansfield, B. Dymock, M. J. Drysdale, H. Finch, R. Howes, R. E. Hubbard, A. Surgenor, P. Webb, M. Wood, L. Wright and P. Workman, *Cancer Res.*, 2008, **68**, 2850–2860.
101. C. Sessa, G. I. Shapiro, K. N. Bhalla, C. Britten, K. S. Jacks, M. Mita, V. Papadimitrakopoulou, T. Pluard, T. A. Samuel, M. Akimov, C. Quadt, C. Fernandez-Ibarra, H. Lu, S. Bailey, S. Chica and U. Banerji, *Clin. Cancer Res.*, in press.
102. D. R. A. Kong, S. Ahmed, J. T. Beck, R. López López, L. Biganzoli, A. Armstrong, M. Aglietta, E. Alba, M. Campone, M. Akimov, A. Matano, C. Lefebvre and S. C. Lee, *ASCO*, 2012, Abstract No. 530.
103. Z. A. Wainberg, A. Anghel, A. M. Rogers, A. J. Desai, O. Kalous, D. Conklin, R. Ayala, N. A. O'Brien, C. Quadt, M. Akimov, D. J. Slamon and R. S. Finn, *Mol. Cancer Ther.*, 2013, **12**, 509–519.
104. M. R. Jensen, J. Schoepfer, T. Radimerski, A. Massey, C. T. Guy, J. Brueggen, C. Quadt, A. Buckler, R. Cozens, M. J. Drysdale, C. Garcia-Echeverria and P. Chene, *Breast Cancer Res.*, 2008, **10**, R33.
105. T. M. E. B. Garon, F. Barlesi, L. Gandhi, L. V. Sequist, S.-W. Kim, H. J. M. Groen, B. Besse, E. Smit, D.-W. Kim, M. Akimov, E. Avsar, S. Bailey and E. Felip, *ASCO*, 2012, Abstract No. 7543.
106. J. Sang, J. Acquaviva, J. C. Friedland, D. L. Smith, M. Sequeira, C. Zhang, Q. Jiang, L. Xue, C. M. Lovly, J. P. Jimenez, A. T. Shaw, R. C. Doebele, S. He, R. C. Bates, D. R. Camidge, S. W. Morris, I. El-Hariry and D. A. Proia, *Cancer Discov.*, 2013, **3**, 430–443.
107. M. A. Socinski, J. Goldman, I. El-Hariry, M. Koczywas, V. Vukovic, L. Horn, E. Paschold, R. Salgia, H. West, L. V. Sequist, P. Bonomi, J. R. Brahmer, L. C. Chen, A. B. Sandler, C. P. Belani, T. R. Webb, H. Harper, M. Huberman, S. Ramalingam, K. K. Wong, F. Teofilovici, W. Guo and G. I. Shapiro, *Clin. Cancer Res.*, 2013.
108. B. Z. Suresh, S. Ramalingam, G. D. Goss, C. Manegold Sr., R. Rosell, V. Vukovic, I. El-Hariry, F. Teofilovici, S. Mulcahey, W. Guo and D. A. Fennell, *ESMO*, 2012, Abstract #1248P_PR.
109. A. J. Woodhead, H. Angove, M. G. Carr, G. Chessari, M. Congreve, J. E. Coyle, J. Cosme, B. Graham, P. J. Day, R. Downham, L. Fazal, R. Feltell, E. Figueroa, M. Frederickson, J. Lewis, R. McMenamin, C. W. Murray, M. A. O'Brien, L. Parra, S. Patel, T. Phillips, D. C. Rees, S. Rich, D. M. Smith, G. Trewartha, M. Vinkovic, B. Williams and A. J. Woolford, *J. Med. Chem.*, 2010, **53**, 5956–5969.

110. B. Graham, J. Curry, T. Smyth, L. Fazal, R. Feltell, I. Harada, J. Coyle, B. Williams, M. Reule, H. Angove, D. M. Cross, J. Lyons, N. G. Wallis and N. T. Thompson, *Cancer Sci.*, 2012, **103**, 522–527.
111. T. Nakashima, T. Ishii, H. Tagaya, T. Seike, H. Nakagawa, Y. Kanda, S. Akinaga, S. Soga and Y. Shiotsu, *Clin. Cancer Res.*, 2010, **16**, 2792–2802.
112. G. Chiosis, M. N. Timaul, B. Lucas, P. N. Munster, F. F. Zheng, L. Sepp-Lorenzino and N. Rosen, *Chem. Biol.*, 2001, **8**, 289–299.
113. L. Wright, X. Barril, B. Dymock, L. Sheridan, A. Surgenor, M. Beswick, M. Drysdale, A. Collier, A. Massey, N. Davies, A. Fink, C. Fromont, W. Aherne, K. Boxall, S. Sharp, P. Workman and R. E. Hubbard, *Chem. Biol.*, 2004, **11**, 775–785.
114. K. Jhaveri, T. Taldone, S. Modi and G. Chiosis, *Biochim. Biophys. Acta*, 2012, **1823**, 742–755.
115. K. Lundgren, H. Zhang, J. Brekken, N. Huser, R. E. Powell, N. Timple, D. J. Busch, L. Neely, J. L. Sensintaffar, Y. C. Yang, A. McKenzie, J. Friedman, R. Scannevin, A. Kamal, K. Hong, S. R. Kasibhatla, M. F. Boehm and F. J. Burrows, *Mol. Cancer Ther.*, 2009, **8**, 921–929.
116. R. Bao, C. J. Lai, H. Qu, D. Wang, L. Yin, B. Zifcak, R. Atoyan, J. Wang, M. Samson, J. Forrester, S. DellaRocca, G. X. Xu, X. Tao, H. X. Zhai, X. Cai and C. Qian, *Clin. Cancer Res.*, 2009, **15**, 4046–4057.
117. A. Rajan, R. J. Kelly, J. B. Trepel, Y. S. Kim, S. V. Alarcon, S. Kummar, M. Gutierrez, S. Crandon, W. M. Zein, L. Jain, B. Mannargudi, W. D. Figg, B. E. Houk, M. Shnaidman, N. Brega and G. Giaccone, *Clin. Cancer Res.*, 2011, **17**, 6831–6839.
118. Y. L. Janin, *Drug Discov. Today*, 2010, **15**, 342–353.
119. M. A. Biamonte, R. Van de Water, J. W. Arndt, R. H. Scannevin, D. Perret and W. C. Lee, *J. Med. Chem.*, 2010, **53**, 3–17.
120. H. J. Patel, S. Modi, G. Chiosis and T. Taldone, *Expert Opin. Drug Discov.*, 2011, **6**, 559–587.
121. B. G. Yun, W. Huang, N. Leach, S. D. Hartson and R. L. Matts, *Biochemistry*, 2004, **43**, 8217–8229.
122. A. Donnelly and B. S. Blagg, *Curr. Med. Chem.*, 2008, **15**, 2702–2717.
123. R. K. Allan, D. Mok, B. K. Ward and T. Ratajczak, *J. Biol. Chem.*, 2006, **281**, 7161–7171.
124. S. N. Shelton, M. E. Shawgo, S. B. Matthews, Y. Lu, A. C. Donnelly, K. Szabla, M. Tanol, G. A. Vielhauer, R. A. Rajewski, R. L. Matts, B. S. Blagg and J. D. Robertson, *Mol. Pharmacol.*, 2009, **76**, 1314–1322.
125. Y. Shiotsu, L. M. Neckers, I. Wortman, W. G. An, T. W. Schulte, S. Soga, C. Murakata, T. Tamaoki and S. Akinaga, *Blood*, 2000, **96**, 2284–2291.
126. A. K. Samadi, X. Zhang, R. Mukerji, A. C. Donnelly, B. S. Blagg and M. S. Cohen, *Cancer Lett.*, 2011, **312**, 158–167.
127. M. V. Powers, P. A. Clarke and P. Workman, *Cancer Cell*, 2008, **14**, 250–262.
128. N. L. Solimini, J. Luo and S. J. Elledge, *Cell*, 2007, **130**, 986–988.
129. J. Luo, N. L. Solimini and S. J. Elledge, *Cell*, 2009, **136**, 823–837.

130. P. G. Richardson, A. A. Chanan-Khan, M. Alsina, M. Albitar, D. Berman, M. Messina, C. S. Mitsiades and K. C. Anderson, *Br. J. Haematol.*, 2010, **150**, 438–445.
131. E. L. Davenport, H. E. Moore, A. S. Dunlop, S. Y. Sharp, P. Workman, G. J. Morgan and F. E. Davies, *Blood*, 2007, **110**, 2641–2649.
132. C. Dai, S. Dai and J. Cao, *J. Cell. Physiol.*, 2012, **227**, 2982–2987.
133. M. P. Goetz, D. Toft, J. Reid, M. Ames, B. Stensgard, S. Safgren, A. A. Adjei, J. Sloan, P. Atherton, V. Vasile, S. Salazaar, A. Adjei, G. Croghan and C. Erlichman, *J. Clin. Oncol.*, 2005, **23**, 1078–1087.
134. W. Chen, S. H. Sin, K. W. Wen, B. Damania and D. P. Dittmer, *PLoS Pathog.*, 2012, **8**, e1003048.
135. J. Zou, Y. Guo, T. Guettouche, D. F. Smith and R. Voellmy, *Cell*, 1998, **94**, 471–480.
136. P. A. Clarke, I. Hostein, U. Banerji, F. D. Stefano, A. Maloney, M. Walton, I. Judson and P. Workman, *Oncogene*, 2000, **19**, 4125–4133.
137. A. Maloney, P. A. Clarke, S. Naaby-Hansen, R. Stein, J. O. Koopman, A. Akpan, A. Yang, M. Zvelebil, R. Cramer, L. Stimson, W. Aherne, U. Banerji, I. Judson, S. Sharp, M. Powers, E. deBilly, J. Salmons, M. Walton, A. Burlingame, M. Waterfield and P. Workman, *Cancer Res.*, 2007, **67**, 3239–3253.
138. M. Zajac, G. Gomez, J. Benitez and B. Martinez-Delgado, BMC Med, *Genomics*, 2010, **3**, 44.
139. R. S. Samant, P. A. Clarke and P. Workman, *Cell Cycle*, 2012, **11**, 1301–1308.
140. D. D. Mosser and R. I. Morimoto, *Oncogene*, 2004, **23**, 2907–2918.
141. C. Garrido, M. Brunet, C. Didelot, Y. Zermati, E. Schmitt and G. Kroemer, *Cell Cycle*, 2006, **5**, 2592–2601.
142. M. V. Powers and P. Workman, *Endocr. Relat. Cancer*, 2006, **13**(1), S125–135.
143. A. J. Massey, D. S. Williamson, H. Browne, J. B. Murray, P. Dokurno, T. Shaw, A. T. Macias, Z. Daniels, S. Geoffroy, M. Dopson, P. Lavan, N. Matassova, G. L. Francis, C. J. Graham, R. Parsons, Y. Wang, A. Padfield, M. Comer, M. J. Drysdale and M. Wood, *Cancer Chemother. Pharmacol.*, 2010, **66**, 535–545.
144. M. V. Powers, P. A. Clarke and P. Workman, *Cell Cycle*, 2009, **8**, 518–526.
145. E. L. Davenport, A. Zeisig, L. I. Aronson, H. E. Moore, S. Hockley, D. Gonzalez, E. M. Smith, M. V. Powers, S. Y. Sharp, P. Workman, G. J. Morgan and F. E. Davies, *Leukemia*, 2010, **24**, 1804–1807.
146. P. Rocchi, P. Jugpal, A. So, S. Sinneman, S. Ettinger, L. Fazli, C. Nelson and M. Gleave, *BJU Int.*, 2006, **98**, 1082–1089.
147. A. K. McCollum, C. J. Teneyck, B. M. Sauer, D. O. Toft and C. Erlichman, *Cancer Res.*, 2006, **66**, 10967–10975.
148. V. Baylot, C. Andrieu, M. Katsogiannou, D. Taieb, S. Garcia, S. Giusiano, J. Acunzo, J. Iovanna, M. Gleave, C. Garrido and P. Rocchi, *Cell Death Dis.*, 2011, **2**, e221.
149. P. Workman and E. de Billy, *Nat. Med.*, 2007, **13**, 1415–1417.

150. N. Zaarur, V. L. Gabai, J. A. Porco, Jr., S. Calderwood and M. Y. Sherman, *Cancer Res.*, 2006, **66**, 1783–1791.
151. L. Whitesell and S. Lindquist, *Expert Opin. Ther. Targets*, 2009, **13**, 469–478.
152. Y. Chen, J. Chen, A. Loo, S. Jaeger, L. Bagdasarian, J. Yu, F. Chung, J. Korn, D. Ruddy, R. Guo, M. E. McLaughlin, F. Feng, P. Zhu, F. Stegmeier, R. Pagliarini, D. Porter and W. Zhou, *Oncotarget*, 2013, **4**, 816–829.
153. M. N. Patel, M. D. Halling-Brown, J. E. Tym, P. Workman and B. Al-Lazikani, *Nat. Rev. Drug Discov.*, 2013, **12**, 35–50.
154. S. Santagata, R. Hu, N. U. Lin, M. L. Mendillo, L. C. Collins, S. E. Hankinson, S. J. Schnitt, L. Whitesell, R. M. Tamimi, S. Lindquist and T. A. Ince, *Proc. Natl Acad. Sci. USA*, 2011, **108**, 18378–18383.
155. M. L. Mendillo, S. Santagata, M. Koeva, G. W. Bell, R. Hu, R. M. Tamimi, E. Fraenkel, T. A. Ince, L. Whitesell and S. Lindquist, *Cell*, 2012, **150**, 549–562.
156. E. de Billy, J. Travers and P. Workman, *Oncotarget*, 2012, **3**, 741–743.
157. C. Dai, L. Whitesell, A. B. Rogers and S. Lindquist, *Cell*, 2007, **130**, 1005–1018.
158. J. R. Smith, P. A. Clarke, E. de Billy and P. Workman, *Oncogene*, 2009, **28**, 157–169.
159. J. L. Holmes, S. Y. Sharp, S. Hobbs and P. Workman, *Cancer Res.*, 2008, **68**, 1188–1197.
160. F. Forafonov, O. A. Toogun, I. Grad, E. Suslova, B. C. Freeman and D. Picard, *Mol. Cell. Biol.*, 2008, **28**, 3446–3456.
161. M. Mollapour and L. Neckers, *Biochim. Biophys. Acta*, 2012, **1823**, 648–655.
162. M. Mollapour, S. Tsutsumi, A. C. Donnelly, K. Beebe, M. J. Tokita, M. J. Lee, S. Lee, G. Morra, D. Bourboulia, B. T. Scroggins, G. Colombo, B. S. Blagg, B. Panaretou, W. G. Stetler-Stevenson, J. B. Trepel, P. W. Piper, C. Prodromou, L. H. Pearl and L. Neckers, *Mol. Cell*, 2010, **37**, 333–343.
163. V. K. Gangaraju, H. Yin, M. M. Weiner, J. Wang, X. A. Huang and H. Lin, *Nat. Genet.*, 2011, **43**, 153–158.
164. V. Specchia, L. Piacentini, P. Tritto, L. Fanti, R. D'Alessandro, G. Palumbo, S. Pimpinelli and M. P. Bozzetti, *Nature*, 2010, **463**, 662–665.
165. A. Choo, P. Palladinetti, T. Passioura, S. Shen, R. Lock, G. Symonds and A. Dolnikov, *Curr. Gene Ther.*, 2006, **6**, 543–550.
166. M. A. John, W. Tao, X. Fei, R. Fukumoto, M. L. Carcangiu, D. G. Brownstein, A. F. Parlow, J. McGrath and T. Xu, *Nat. Genet.*, 1999, **21**, 182–186.
167. F. D. Camargo, S. Gokhale, J. B. Johnnidis, D. Fu, G. W. Bell, R. Jaenisch and T. R. Brummelkamp, *Curr. Biol.*, 2007, **17**, 2054–2060.

168. Y. Park, A. Kubo, T. Komiya, A. Coxon, K. Beebe, L. Neckers, P. S. Meltzer and F. J. Kaye, *Cell Cycle*, 2008, **7**, 2384–2391.
169. A. K. McCollum, C. J. TenEyck, B. Stensgard, B. W. Morlan, K. V. Ballman, R. B. Jenkins, D. O. Toft and C. Erlichman, *Cancer Res.*, 2008, **68**, 7419–7427.
170. C. Prodromou, J. M. Nuttall, S. H. Millson, S. M. Roe, T. S. Sim, D. Tan, P. Workman, L. H. Pearl and P. W. Piper, *ACS Chem. Biol.*, 2009, **4**, 289–297.
171. S. H. Millson, C. S. Chua, S. M. Roe, S. Polier, S. Solovieva, L. H. Pearl, T. S. Sim, C. Prodromou and P. W. Piper, *FASEB J.*, 2011, **25**, 3828–3837.
172. S. Polier, R. S. Samant, P. A. Clarke, P. Workman, C. Prodromou and L. H. Pearl, *Nat. Chem. Biol.*, 2013, **9**, 307–312.

CHAPTER 5

The Discovery of BIIB021 and BIIB028

KAREN LUNDGREN[a] AND MARCO A. BIAMONTE*[b]

[a] K Lundgren Pharmacology, 1804 Garnet Ave, Unit 246, San Diego, CA 92109, USA; [b] Drug Discovery for Tropical Diseases, 10835 Road to the Cure, Suite 230, San Diego, CA 92121, USA
*Email: marco.biamonte@ddtd.org

5.1 Background

In 1998–1999, Cancer Research UK and the National Cancer Institute initiated the first clinical trials with an Hsp90 inhibitor, namely 17-amino,17-desmethoxygeldanamycin (17-AAG). Soon thereafter, a few companies started competing in the Hsp90 space, with the intention of delivering the next generation of inhibitors and of solving some of the limitations associated with 17-AAG. The limitations included (i) the low oral bioavailability, which required an intravenous administration, (ii) the low solubility, which complicated the parenteral formulation, (iii) the hepatotoxicity due to the quinone ring and (iv) the cost of the fermentation process involved in the manufacture of 17-AAG.

Conforma Therapeutics, founded in 2000–2001 and acquired by Biogen-Idec in 2006, was one of the start-ups with an Hsp90 program. One of the Conforma assets was the Hsp90 inhibitor PU3 (Structure **5.1**, Scheme 5.1), in-licensed from the Memorial Sloane Kettering Cancer Center (MSKCC).[1,2] Although PU3 was only weakly active ($EC_{50} = 40$ μM), we viewed it as a

---

RSC Drug Discovery Series No. 37
Inhibitors of Molecular Chaperones as Therapeutic Agents
Edited by Timothy Machajewski and Zhenhai Gao
© The Royal Society of Chemistry 2014
Published by the Royal Society of Chemistry, www.rsc.org

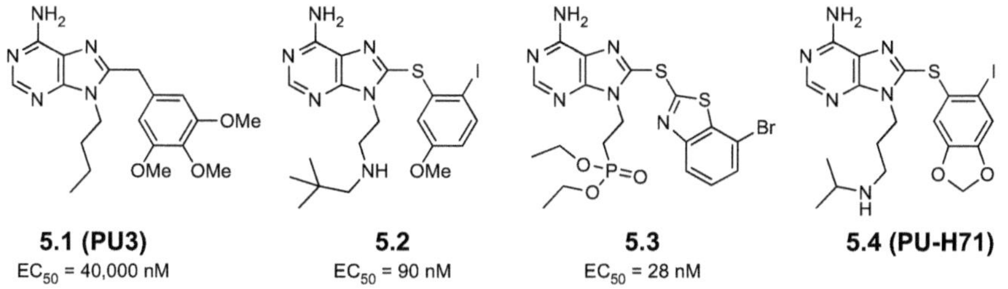

Scheme 5.1 PU3, the first synthetic Hsp90 inhibitor, and the early inhibitors **5.2–5.4**.

promising starting point to develop better and improved Hsp90 inhibitors. It had the advantage of being orally bioavailable and devoid of a quinone moiety, thus addressing two limitations of 17-AAG. Furthermore, it was the only fully synthetic Hsp90 inhibitor available at the time, offering a potentially lower cost of production.

This chapter describes the efforts that led from PU3 to the two clinical candidates BIIB021 (Phase II) and BIIB028 (Phase I). Unless otherwise noted, all the potency data (EC_{50} values) reported herein refer to the following cell-based assay:[3] MCF-7 breast-cancer cells were incubated for 16 hours with a given Hsp90 inhibitor, and the Hsp90 inhibitor led to the degradation of the Hsp90 client HER2, a receptor located on the surface of the cell and therefore easily monitored by flow cytometry using a phycoerythrin-labeled (*i.e.* fluorescent) antibody. We felt that this HER2 degradation assay was more meaningful than a binding assay, as it included all the cofactors and co-chaperones that modulate the activity of Hsp90 under physiological conditions. In addition, the degradation of HER2 is a clinically relevant end point, since HER2 drives 25% of all breast tumors and is the molecular target of the approved breast-cancer drug trastuzumab (Herceptin).

The early work on the optimization of PU3 has been published,[4,5] and is outside the scope of this chapter. Suffice it to say that the aromatic substituents were revised, the methylene linker was replaced with a sulfur atom using novel chemistry to construct the purine–sulfur bond[7,8] and an amino group was introduced in the side-chain to provide much-needed solubility. The outcomes were 8-sulfanyladenines **5.2** ($EC_{50} = 90$ nM, Scheme 5.1)[4] and **5.3** ($EC_{50} = 28$ nM, Scheme 5.1),[5] two molecules nearly 1000 times more potent than the starting point. The MSKCC reached independently a similar conclusion with compound **5.4** (PU-H71, Scheme 5.1).[6] Sulfanyladenine **5.2** was a milestone in our program, in that it was the first compound to consistently inhibit tumor growth in mice models upon oral administration. However, very large doses of **5.2** (2×100 mg/kg/day) were required to obtain *in vivo* efficacy. A complex metabolism with seven major metabolites, an impractically high number to monitor in clinical trials, was responsible at least in part for the low efficacy. Therefore, while attempting to further optimize the potency and metabolism of **5.2**, we also sought in parallel a new series.

5.2 The Discovery of BIIB021

5.2.1 BIIB021 Criteria

For our first clinical candidate, we sought an Hsp90 inhibitor that would be:

a) Orally bioavailable in rat, with $F > 20\%$.
b) Efficacious in mouse at doses below 100 mg/kg/day.
c) Easily synthesized.
d) Devoid of a quinone ring.

5.2.2 BIIB021 Design

In order to discover a new series distinct from that of purine inhibitors **5.1–5.4**, the options were limited. We did not have access to high-throughput screens, nor did we have an X-ray co-crystal structure to perform structure-based drug design. We therefore opted to identify a pharmacophore model that could guide us in a scaffold-hopping endeavor. By reviewing the extensive SAR performed around PU3, the following pharmacophore model emerged (Scheme 5.2). It appeared that:

1. The NH_2-C=N fragment (red) was indispensable, presumably to bind to the hinge region.
2. The phenyl ring (blue) needed to be located 5 Å away from the NH_2-C=N fragment.
3. A purine ring (black), rich in H-bond acceptors, was necessary.
4. A side-chain (R–CH_2–) provided additional potency.
5. A halogen substituent (X) also provided extra potency.

With a pharmacophore model at hand, the next challenge was to find a new series that would fit said model. One of the scientists (Dr. S. Kasibhatla) realized that the pharmacophore model could be leveraged by using the following three-step operation that we call a "disconnect-rotate-reconnect" technique (Scheme 5.3).[7]

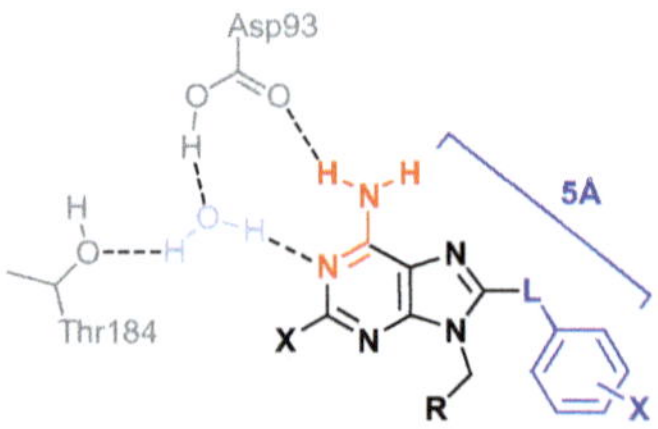

Scheme 5.2 Pharmacophore model.

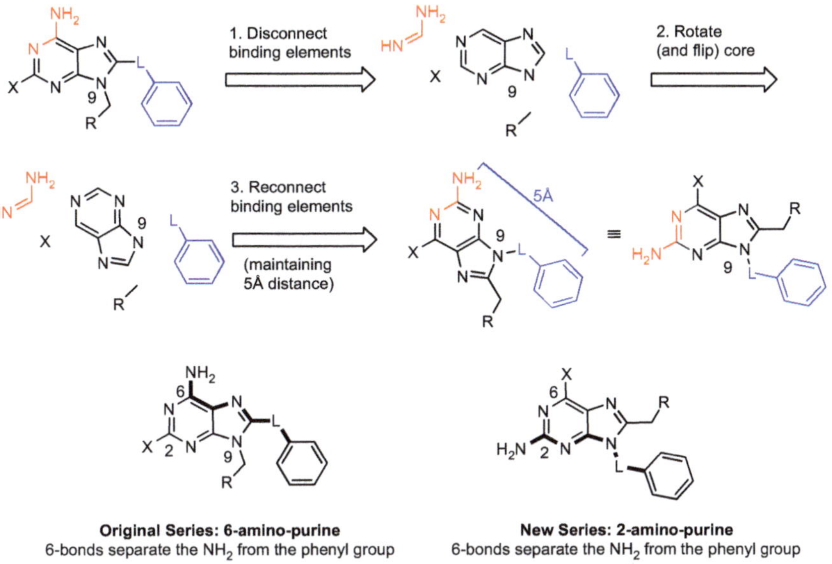

Scheme 5.3 Generation of a new series *via* a "Disconnect-Rotate-Reconnect" technique.

1) The critical binding elements were disconnected from the scaffold (purine ring).
2) The scaffold was rotated (and flipped).
3) The critical binding elements were reconnected onto the scaffold.

The resulting molecule was still a purine, and the distance between the NH_2 group and the aromatic ring was preserved. One simple way to verify that the distances were unchanged was to count the number of bonds that separate the NH_2 group from the phenyl group (six bonds in each case). Alternatively, old-fashioned plastic models showed that the two series could be overlapped, and proved to be a simple yet powerful technique.

The new series (2-aminopurines) was obviously reminiscent of the original one (6-aminopurines), as it featured the same scaffold and binding elements. However, the connectivity was different, and the binding elements "sprouted" from the scaffold at different positions. This introduced deviations in the SAR, which could be used to further improve the potency of the molecule. The most important difference was the SAR around the halogen atom. The latter only marginally improved the potency of the original series but was mandatory for potency in the new series. In addition, the side-chain R-CH_2- was best removed in the new series. As for the SAR around the phenyl ring, it appeared that the potency improved with multiple contiguous aromatic substituents (**5.5–5.7**, Scheme 5.4).

The new connectivity also simplified the chemistry. The aromatic group, being connected to the purine scaffold ring *via* an ArCH_2–N bond, as opposed to an ArCH_2–C or ArS–CH2 bond, was easily introduced by an alkylation step,

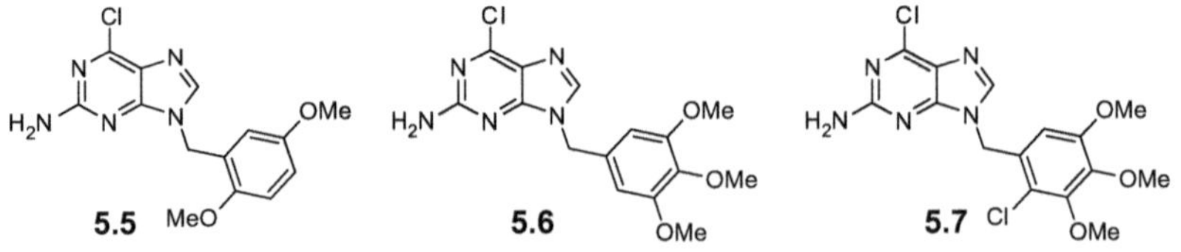

5.5 EC$_{50}$ = 3000 nM

5.6 EC$_{50}$ = 2000 nM

5.7 EC$_{50}$ = 50 nM

Scheme 5.4 Selected 2-aminopurines. The potency increases with the number of contiguous substituents.

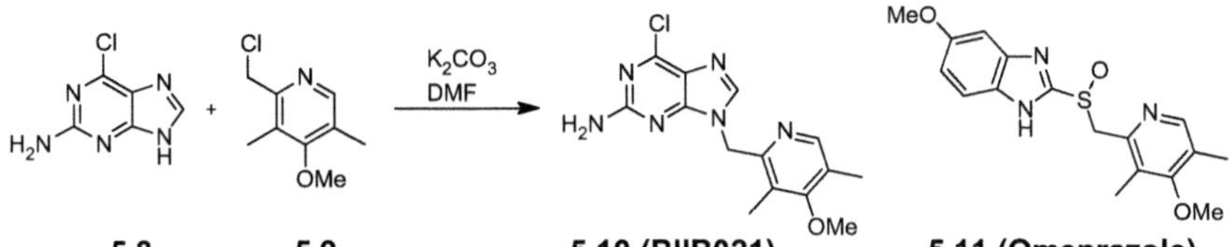

5.8 5.9 5.10 (BIIB021) EC$_{50}$ = 38 nM 5.11 (Omeprazole)

Scheme 5.5 Preparation of BIIB021.

thus allowing the exploration of a wider range of aromatic groups. Since contiguous aromatic substituents improved the potency, we attempted the alkylation of the commercial 6-chloropurine **5.8** (Scheme 5.5) with the commercially available pyridine **5.9**. The resulting derivative **5.10** proved to be potent ($EC_{50} = 38$ nM). Furthermore, the pyridine nitrogen improved the solubility of the series, and **5.10** was subsequently nominated as our clinical candidate BIIB021.[7]

An optimized process to synthesize BIIB021 on kg-scale has been reported.[8] Interestingly, chloromethylpyridine **5.9** is cheap and it enters the preparation of the blockbuster anti-acid omeprazole (**5.11**, Scheme 5.5), indicating that the 3,5-dimethyl-4-methoxypyridine motif is devoid of intrinsic toxicological liabilities.

5.2.3 BIIB021 Selectivity

Hsp90 exists as two isoforms: Hsp90α is an inducible form over-expressed in cancer cells, while Hsp90β is the constitutive form. Both Hsp90α and Hsp90β are found in the cytoplasm. There are also two paralogues: Grp94, localized in the endoplasmic reticulum lumen, and TRAP1, confined to mitochondria.

The binding assays were performed on Hsp90α as described previously.[3] BIIB021 was selective for Hsp90α ($IC_{50} = 5.1 \pm 2.0$ nM) and Hsp90β ($IC_{50} = 13.2 \pm 4.3$ nM), over Grp94 (233 ± 15 nM) and TRAP-1 ($IC_{50} = 109 \pm 37$ nM).

BIIB021 did not inhibit a panel of 11 kinases at 10 μM (Aurora-A, CHK2, IKKα, MAPK1, MAPK2, MEK1, PDK1, Plk3, PI3K, c-RAF and cSRC). This is likely due to the fact that the N-terminal ATP binding site of Hsp90 has a unique geometry known as the "Bergerat fold" (Figure 5.1a), found only in

ATPases belonging to the GHKL family (gyrase, Hsp90, histidine kinase, MutL), and not in kinases.[9]

5.2.4 BIIB021 Pharmacokinetics

Single-dose pharmacokinetic studies were performed in mice, rats and dogs (Table 5.1). BIIB021 was administered as its mono-mesylate salt, but the doses reported refer to the amount of BIIB021 as its *free base*. The AUC was dose proportional in mice from 37.5 mg/kg to 200 mg/kg and in rats from 10 mg/kg to 51 mg/kg. BIIB021 was orally available and was absorbed rapidly from solution in all species tested. Mean time to maximum plasma concentration (t_{max}) values were 0.08 h in the mouse, 0.1–0.5 h in the rat and 0.5–1.0 h in the dog. Bioavailability (F) ranged from 12–23% in the mouse, 46–82% in the rat and 13–49% in the dog.

The pharmacokinetics of BIIB021 after i.v. dosing was characterized by a mean total clearance of approximately 70% and 45% of the liver blood flow (LBF) in mice and rats, respectively, whereas in the dog it exceeded LBF (126%). The volume of distribution at steady state was moderate to high across species (70% to 580% of total body water). Mean pharmacokinetic parameters were similar after a single-dose, or after 28 consecutive days of dosing (data not shown).

In vitro and *in vivo* data indicated that the metabolic profile of BII021 was qualitatively similar in mice, rats, dogs and humans. CYP2C19 was determined

Table 5.1 Single-dose pharmacokinetic parameters after single-dose administration of BIIB021 mono-mesylate. BIIB021 mono-mesylate was pre-dissolved in 0.1 N HCl, except for the dog p.o. arms, where BIIB021 mono-mesylate was administered in capsules to mirror the clinical formulation. The doses reported in this table reflect the amount of BIIB021 in its free base form. Animals were fed, except for the oral dog experiments where they were fasted. Gender: m = male, f = female.

Species	Route	Dose (mg/kg)	Gender	C_{max} (µg/mL)	t_{max} (h)	$AUC_{(0-\infty)}$ (µg·h/mL)	V_{ss} (L/kg)	Cl_{tot} (L/h/kg)	$t_{\frac{1}{2}}$ (h)	F (%)
Mouse	i.v.	8.3	4 f	21.6	0.02	2.2	0.54	3.8	0.8	
Mouse	p.o.	37.5	4 f	3.1	0.083	1.2			0.5	12
		75	4 f	10.2	0.083	2.3			0.7	12
		150	4 f	10.0	0.083	4.8			0.8	18
		200	4 f	19.5	0.083	9.1			0.5	23
Rat	i.v.	6.5	3 f	7.6	0.02	4.3	3.9	1.5	2.9	
Rat	p.o.	10	4 m	1.4	0.5	3.0			1.9	46
		10	4 f	3.3	0.3	5.4			1.4	82
		51	6 f	9.6	0.1	25.8			3.9	77
Dog	i.v.	5	2 m/2 f	3.6	0.12	2.1	2.2	2.4	0.8	
Dog	p.o.	10	4 m	0.36	0.56	0.53			1.5	13
		50	1 m/1f	6.4	1.0	12.3			1.1	49
		150	1 m/1f	1.6	0.5	13.4			1.5	22

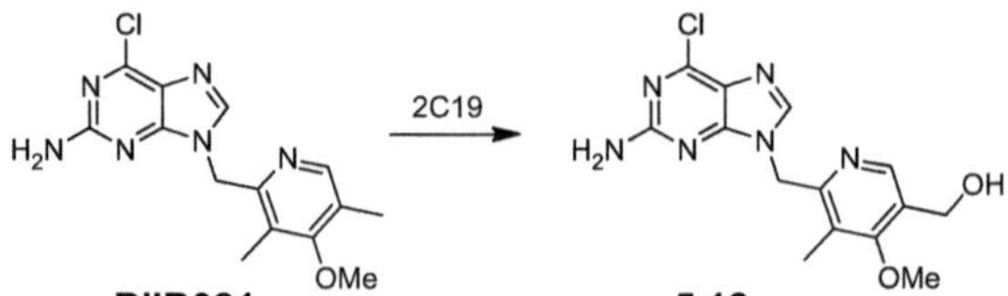

Scheme 5.6 Metabolism of BIIB021.

to be the primary enzyme responsible for the metabolism of BIIB021, while CYP3A4 played a negligible role. The primary metabolite in all species was the 5-hydroxymethylpyridine **5.12** (Scheme 5.6). The dog differed from the other species by producing the glucuronide of the 5-hydroxymethylpyridine **5.12** as the predominant metabolite.

5.2.5 BIIB021 Pharmacodynamics

Inhibition of Hsp90 prevents proper folding of client proteins, which results in loss of activity and ubiquitination, and ultimately proteasomal degradation of the client. Inhibition of Hsp90 has also been shown to induce a heat-shock response. To demonstrate that BIIB021 acts through the expected mechanism of action *in vivo*, the levels of Hsp90 client proteins and heat-shock proteins were analyzed after BIIB021 was administered orally to mice bearing BT474 human xenograft tumors. Again, BIIB021 was dosed as its mono-mesylate salt, but the doses reported refer to the amount of BIIB021 free-base. The tumors were excised at specific time points, tumor lysates were prepared and the proteins of interest were analyzed in Western blot experiments.

In a dose-response study, mice received single oral doses of 30, 60 or 120 mg/kg of BIIB021 and the tumors were collected after 6 hours. Treatment with BIIB021 resulted in decreased levels of Hsp90 clients HER2 and Raf-1 at a dose as low as 30 mg/kg. With increasing doses the degradation effects also increased, including effects on AKT, cdk4 and cdk6. The expression levels of the heat-shock proteins, Hsp70 and Hsp27, increased in tumors from treated animals in a dose-dependent manner (Figure 5.1, Panel A).

In a time course experiment, mice received a single oral dose of BIIB021 at 150 mg/kg and the tumors were collected at 6, 24, 48 and 72 hours after the dose (Figure 5.1, Panel B). The levels of Hsp90 client proteins, such as HER2, EGFR and AKT, were decreased in tumor samples as early as 6 hours after dosing. The levels of client protein expression gradually returned to baseline in 24 to 72 hours after dose administration. As expected, expression of Hsp70 increased in tumors from treated animals, and it remained elevated for the duration of the experiment. BIIB021 showed both hallmarks of a highly active Hsp90 inhibitor *in vivo*, namely Hsp90 client protein degradation and heat-shock protein induction. These effects were both dose- and time-dependent.

It is important to notice that in spite of the rapid clearance of BIIB021, the pharmacodynamic effects lasted for at least 24 hours. This illustrates an

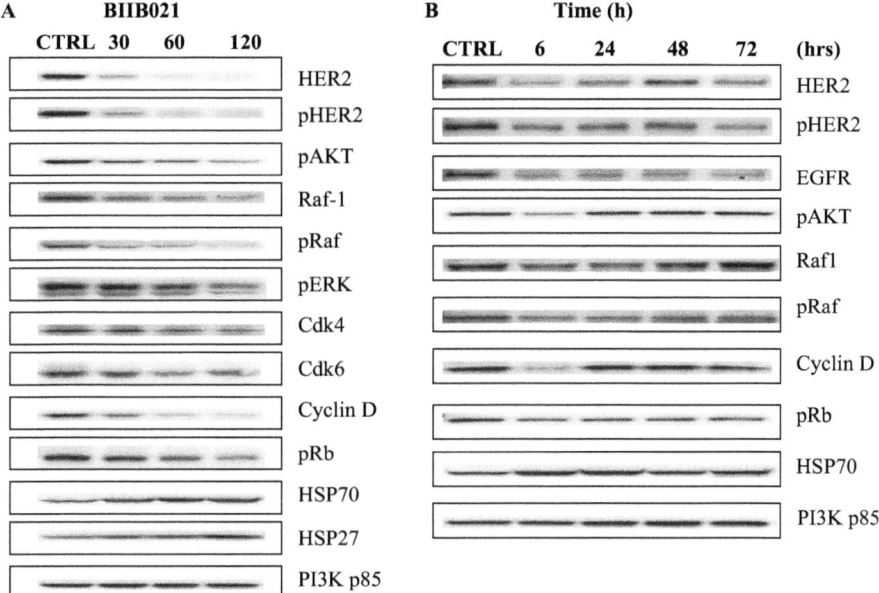

Figure 5.1 Pharmacodynamic effect of BIIB021 on expression and/or phosphorylation of selected proteins in BT474 tumors. (A) From mice harvested 6 hours after a single oral administration at 30, 60 or 120 mg/kg. (B) From mice 6 to 72 hours after a single oral administration at 150 mg/kg.

important point pertaining to all Hsp90 inhibitors. Though they may be short lived, they rapidly induce the degradation of the relevant clients and the cell requires 24–48 h to resynthesize them. A continuous inhibition of Hsp90 is therefore not necessary, and compounds with short half-lives can provide a long-lasting pharmacodynamic effect. This is why BIIB021 has been colloquially called a "hit-and-run" inhibitor, and there is evidence that the hit-and-run approach improves the tolerability of Hsp90 inhibitors (see Section 5.4).

5.2.6 BIIB021 *in Vivo* Efficacy

BIIB021 was tested for efficacy in human tumor xenografts grown in nude mice. The Tumor Growth Inhibition (TGI) was defined as the change in mean treated-tumor volume, divided by the change in mean control tumor volume, multiplied by 100 and subtracted from 100%.

A number of carcinomas were tested, and BIIB021 was active in most of them (see also Table 5.4). We report here the data for the N87 human stomach carcinoma model. BIIB021 was administered orally at doses of 31, 62.5 and 125 mg/kg once daily for five consecutive days per week for five weeks as its mesylate dissolved in 0.1 N HCl. The doses refer to the equivalent amount of BIIB021 free base. BIIB021 showed significant activity (Figure 5.2). On the final study day, 46% TGI ($p=0.02$) was observed at 31 mg/kg, 65% TGI

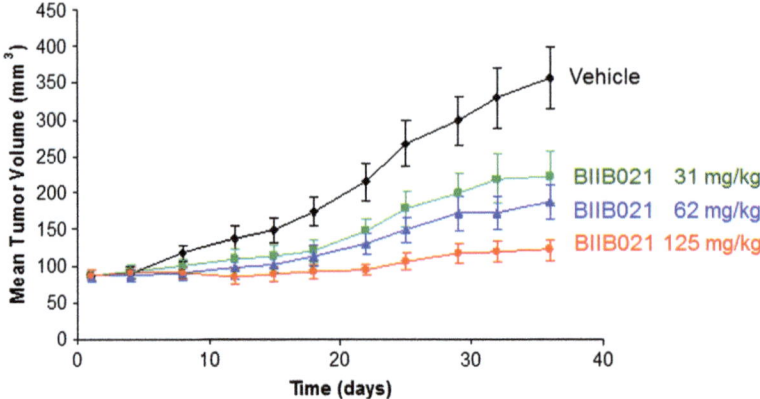

Figure 5.2 Inhibition of tumor growth in the N87 gastric tumor model. Athymic mice bearing established N87 gastric tumor xenografts were treated with vehicle control or BIIB021 monomesylate. The doses, based on the free-base, were 31, 62 or 125 mg/kg. Tumors were measured using Vernier calipers on the indicated days and tumor volumes are reported as mean volume ± SEM tumor.

($p = 0.002$) was observed at 62.5 mg/kg and 87% TGI at 125 mg/kg ($p = 0.0001$). The mean body weights of the control and treated groups did not differ significantly over the course of dosing. The maximum tolerated dose for *repeated* daily dosing was 125 mg/kg/day, above which weight loss and diarrhea were observed.

5.2.7 BIIB021 Summary

BIIB021 arose from a scaffold hopping exercise, where the binding elements were rotated around the purine core. This "disconnect-rotate-reconnect" technique proved to be a simple, but powerful, way of leveraging the known SAR. BIIB021 met our criteria of efficacy, oral bioavailability and low cost, and it became the first oral Hsp90 inhibitor to be tested in man. It was advanced to a Phase II trial in hormone receptor positive breast cancer (HR + BC) in combination with exemestane (Aromasin) at a dose of 450 mg per patient (once weekly) or 150 mg per patient (thrice weekly).

5.3 The Discovery of BIIB028

5.3.1 BIIB028 Criteria

For the second generation of Hsp90 inhibitors, we wanted to further improve:

(a) the potency and efficacy relative to BIIB021,
(b) the tolerability relative to BIIB021.

5.3.2 BIIB028 Design

5.3.2.1 Second-generation Oral Inhibitors: Improving the Potency

The goal was to identify an Hsp90 inhibitor with improved potency and tolerability over BIIB021. We first focused on improving the potency, which turned out to be a challenging endeavor, as all attempts to modify the substituents R_a–R_f (**5.13**, Scheme 5.7) typically led to a drastic decrease in potency. Similarly, replacement of the atoms X and Y (**5.13**, Scheme 5.7) by a nitrogen atom also led to a loss in potency.

Yet, it seemed logical to expect that at least one part of the molecule would face the solvent, and that modifications of the solvent-exposed area of the molecule would be the most tolerable. We felt that the solvent-exposed area may be used to our advantage, to improve the drug properties and possibly to gain potency by making new interactions at the rim of the binding pocket. Since all modifications around R_a–R_f, X and Y systematically decreased the potency of the inhibitors, we deduced that these positions were in close contact with the protein and not pointing towards a solvent-exposed area. The solvent-exposed area was therefore pointing in another direction and, by elimination, we surmised that the solvent was facing the direction of the N7 lone pair (Scheme 5.8). This initial hypothesis was later validated by an X-ray structure (Figure 5.3).

In order to substitute the molecule along the desired vector, it was necessary to replace N7 with a carbon atom. To this effect, we prepared pyrrolo[2,3-d]pyrimidines (= deazapurines) based on scaffold **5.21** (Scheme 5.9) where the iodo group would enable a range of cross-couplings. We devised an efficient

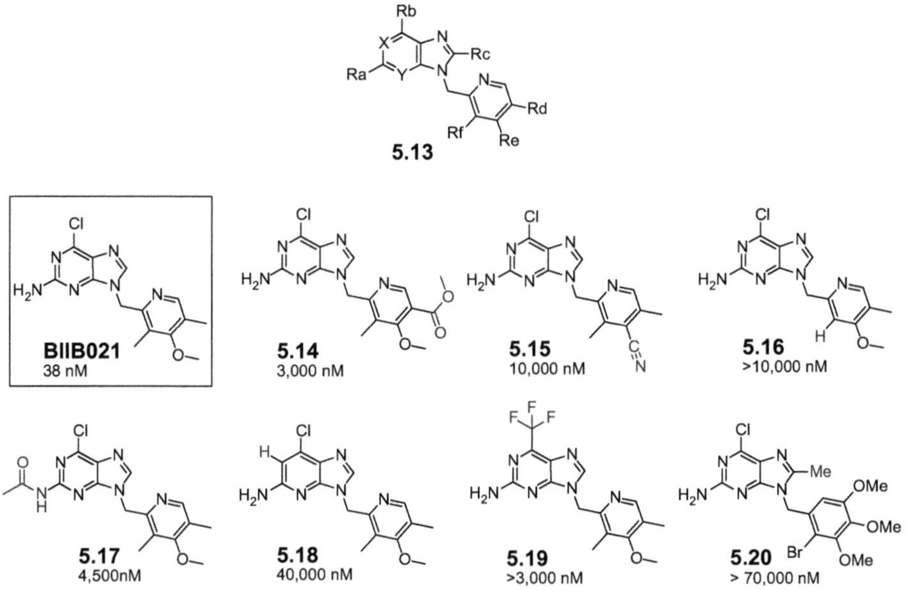

Scheme 5.7 Modifications at positions Ra–Rf, X and Y lead to a loss of potency.

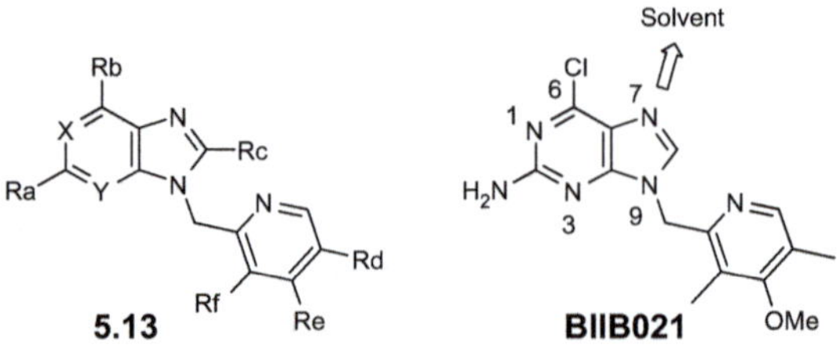

Scheme 5.8 Modifications of **5.13** in positions Ra–Rf led to a decrease in potency, suggesting that the solvent exposed area of the molecule was pointing in the same direction as the N7 lone pair of BIIB021.

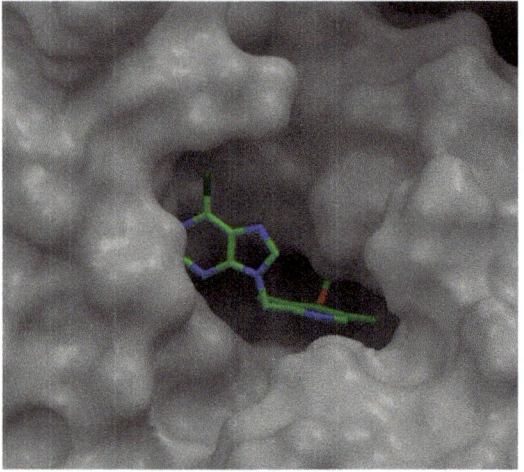

Figure 5.3 The X-ray structure of BIIB021 in complex with Hsp90 shows that N7 faces a solvent exposed area (pdb code: 3QDD).

synthesis of **5.21**, and used it to synthesize C5 modifications.[10] Remarkably, simple substituents such as H, Ph and i-Pr were tolerated (**5.23**–**5.24**), and only led to a marginal loss of potency (EC_{50} = 100–250 nM) as compared with BIIB021 (Scheme 5.9).[10]

Upon further experimentation, it appeared that alkyne substituents improved the potency, as in alkynes **5.25** (Scheme 5.10, EC_{50} = 26 nM) and **5.26** (EC_{50} = 17 nM). Of all the alkynes prepared, the most potent ones proved to be the homopropargylic alcohols **5.27** (EC_{50} = 9 nM) and **5.28**, also known as EC144 (EC_{50} = 14 nM).[10] A crystal structure of **5.28** in complex with Hsp90 revealed that the added potency of the alkynes was due to a favorable hydrophobic interaction between the C≡C fragment and the adjacent side-chains of Met-98 and Leu-107.[11] There were no other obvious additional interactions. For instance, the distance between the OH group of **5.28** and the Lys-58 amino group was too large (4.3 Å) for a hydrogen bond, although the solution structure may differ from the crystal structure, granting the side-chain of

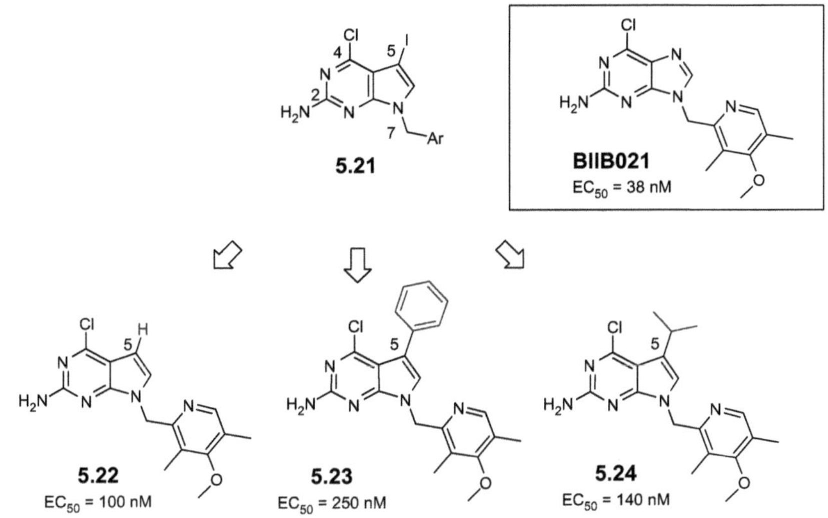

Scheme 5.9 Modifications at C5 retain potency.

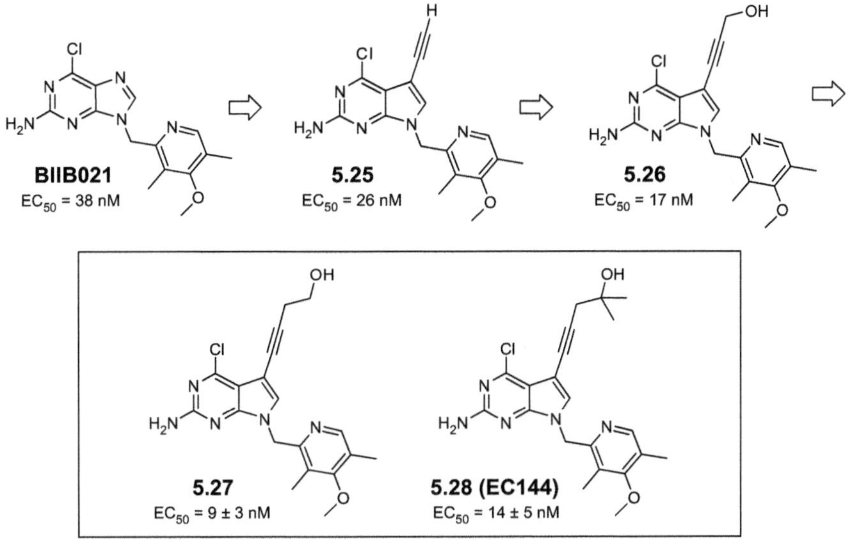

Scheme 5.10 Optimized 5-alkylnylpyrrolo[2,3-d]pyrimidines.

Lys-58 enough flexibility to form a hydrogen bond under physiological conditions (Figure 5.4).

5.3.2.2 Second-generation Oral Inhibitors: Selectivity

Primary alcohol **5.27** was selective for Hsp90α ($K_i = 0.50$ nM) *versus* Grp94 ($K_i = 11$ nM) and TRAP-1 ($K_i = 22$ nM), and did not inhibit a panel of 285 human kinases when tested at 10 μM.

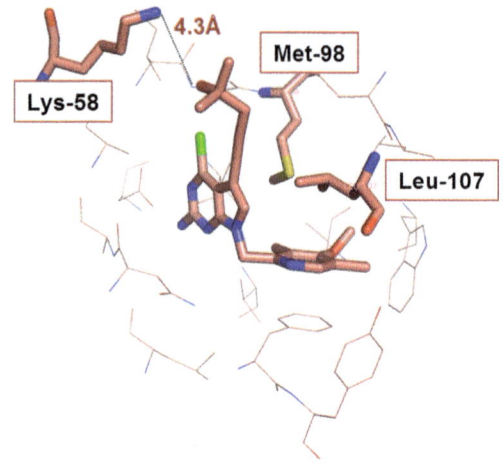

Figure 5.4 X-ray structure of tertiary alcohol **5.28** in complex with Hsp90α (pdb code: 3NMQ).

Table 5.2 Pharmacokinetic data for selected Hsp90 inhibitors in CD-1 mouse after a single oral or intravenous administration. The pro-drug SNX-5422 was administered and the PK parameters refer to its active metabolite SNX-2112.

Cmpd	ROA	Dose (mg/kg)	Cmax,Co (ng/mL)	Tmax (hr)	AUC (ng*hr/mL)	$t_{1/2}$ (hr)	Cl (mL/min/kg)	Vdss (L/kg)
BIIB021	p.o.	100	10 000	0.083	4800	0.8		
5.27	p.o.	25	424	0.25	327	5.0		
5.28	p.o.	25	1556	0.5	4419	1.5		
SNX5422	p.o.	25	1724	1.0	8517	2.9		
AUY922	i.v.	5	2180		2409	6.6	32.6	9.5

Similarly, tertiary alcohol **5.28** was selective for Hsp90α ($K_i = 0.20$ nM) versus Grp94 ($K_i = 255$ nM) and TRAP-1 ($K_i = 61$ nM), and did not inhibit a panel of 285 human kinases ($IC_{50} > 10$ μM).

5.3.2.3 Second-generation Oral Inhibitors: Pharmacokinetics

In mice, the maximum tolerated dose (MTD) of alkynes **5.27** and **5.28** for a single oral administration was approximately 30 mg/kg, while for daily dosing it was 10 mg/kg. Table 5.2 reports the mouse pharmacokinetic data for a single oral dose of **5.28**, compared with the data for other known Hsp90 inhibitors.

Incubation of **5.27** and **5.28** with rat liver microsomes (S9 fraction) showed in each case two major metabolites, resulting from oxidation at the propargylic position, and at the pyridine 5-Me position, respectively (Scheme 5.11). Oxidation at the propargylic position retained most of the activity (16–25 nM),

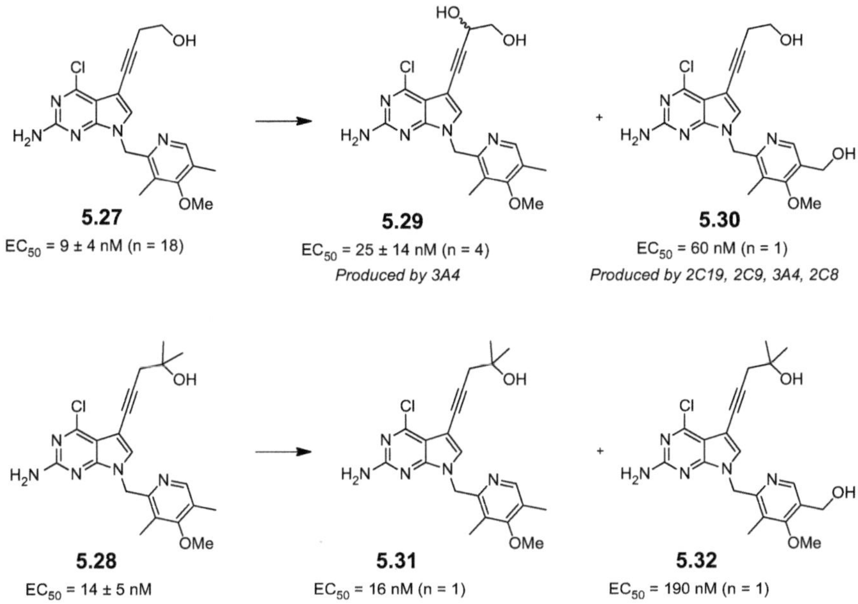

5.27
EC$_{50}$ = 9 ± 4 nM (n = 18)

5.29
EC$_{50}$ = 25 ± 14 nM (n = 4)
Produced by 3A4

5.30
EC$_{50}$ = 60 nM (n = 1)
Produced by 2C19, 2C9, 3A4, 2C8

5.28
EC$_{50}$ = 14 ± 5 nM

5.31
EC$_{50}$ = 16 nM (n = 1)

5.32
EC$_{50}$ = 190 nM (n = 1)

Scheme 5.11 Metabolism of alcohols **5.27** and **5.28**.

while oxidation of the pyridine 5-Me group led to a loss in potency (60–190 nM). The microsomes studies suggested that the main metabolite in mouse is the propargylic alcohol **5.29**, while in all other species (rat, dog, monkey, human) the major metabolite is the hydroxymethylpyridine **5.30**.

5.3.2.4 Second-generation Oral Inhibitors: Efficacy

In an N87 gastric tumor xenograft model, **5.28** caused tumor stasis when dosed five days per week at 5 mg/kg, and induced partial tumor regressions at 10 mg/kg (Figure 5.5). This is a significant increase in potency when compared to BIIB021, which was not quite as effective, even at 120 mg/kg, and which did not cause partial regressions. Similarly, **5.27** also proved to be active orally at 2 and 4 mg/kg (Figure 5.5).

5.3.2.5 Second-generation Oral Inhibitors: Summary

In summary, potency was gained relative to BIIB021 both *in vitro* and *in vivo* by taking advantage of the solvent-exposed area. Compounds **5.27** and **5.28** were efficacious in mouse at doses 20 times lower than for BIB021. Considering that the exposure of compound **5.27** is much lower than for **5.28**, the efficacy of **5.27** is presumably partly driven by its metabolite **5.29**. The MTD of **5.27** and **5.28** was also lower than for BIIB021, and consistent with an on-target mechanism of toxicity. For the first time, minor tumor regressions were noted in our program.

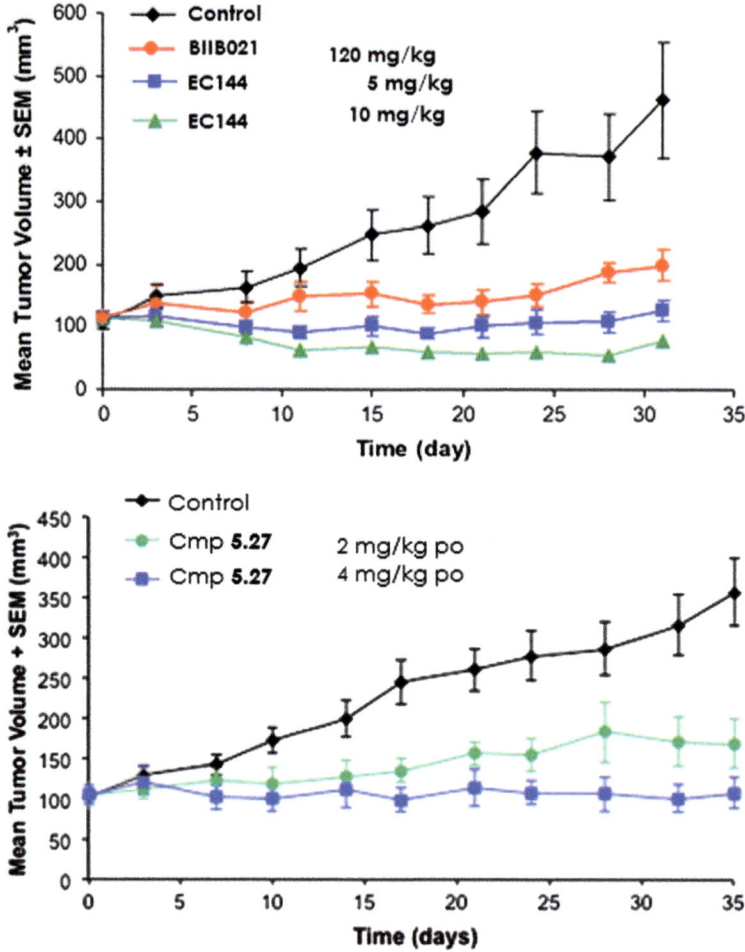

Figure 5.5 Inhibition of tumor growth in the N87 gastric tumor model. Athymic mice bearing established N87 gastric tumor xenografts were treated with vehicle control, BIIB021, alcohol **5.27** or alcohol **5.28** orally q.d.×5 for 4 weeks. Compounds were formulated in 0.1 N HCl. Tumors were measured using Vernier calipers on the indicated days and tumor volumes are reported as mean volume ± SEM tumor for groups of 8 mice. Panel A. Activity of gem-dimethyl alcohol **5.28**. Panel B. Activity of primary alcohol **5.27**.

5.3.2.6 Second-generation Inhibitors: Improving the Tolerability

Having succeeded in identifying compounds more potent and efficacious than BIIB021, we turned to improving the tolerability of the series. However, the toxicity increased proportionally with the efficacy for every Hsp90 inhibitor tested, irrespective of its chemical class, suggesting that the toxicity is consistently mediated by Hsp90 (on-target).

Since the dose-limiting toxicity of BIIB021 in rodents was gastrointestinal (diarrhea, bleeding), we anticipated that delivering the compound intravenously may be beneficial, as this would decrease the concentration of drug in the gastrointestinal tract. Intravenous delivery required water-soluble compounds, but **5.27** and **5.28** were too insoluble for this purpose ($\sim 5\ \mu g/mL$ at neutral pH). It was therefore imperative to increase the aqueous solubility of **5.27** and **5.28**.

Replacing the OH group of **5.27** and **5.28** with polar, charged functional groups such as ammoniums (R–OH → R–N$^+$HR^{1}R^2) or carboxylates (R–OH → R–CO$_2^-$) did not sufficiently improve the solubility. Similarly, glycine pro-drugs of **5.27** and **5.28** (R–OH → R–O–CO-CH$_2$-NH$_3^+$) failed at improving the solubility. Clearly, the single ionic charge provided by ammoniums or carboxylates group was insufficiently solubilizing for our purposes. We therefore turned to pro-drugs that would carry two charges at physiological pH and phosphate pro-drugs (R–OH → R–O–PO$_3^{--}$), by virtue of their two negative charges, became the natural candidates. After extensive experimentation, we were able to produce gram amounts of BIIB028 (**5.33**, Scheme 5.12), the phosphate pro-drug of primary alcohol **5.27**. The phosphorylation was best performed by reacting **5.27** with POCl$_3$ in PO(OEt)$_3$, the latter acting both as solvent and as catalyst. The isolation of BIIB028, however, was difficult because BIIB028 was too water soluble to use regular extraction techniques. After extensive experimentation, we discovered that by conducting the reaction under concentrated conditions, diluting the reaction mixture with water, and adjusting the final pH to 2–3, BIIB028 would crystallize out of solution in >98% purity. The conditions were further optimized by our process chemistry group.[12] This procedure could not be adapted to the phosphorylation of the unreactive tertiary alcohol **5.28**.

5.3.3 BIIB028 Pharmacokinetics

As desired, BIIB028 proved to be highly soluble (>1 mg/mL at pH = 7). It was also an ideal pro-drug: chemically stable towards hydrolysis, but enzymatically rapidly labile in serum. The pharmacokinetic parameters of BIIB028 were obtained in mouse, rat and dog. When administered to mice intravenously at 20 mg/kg, BIIB028 was converted into its active metabolite **5.27**, with a half-life of 5 minutes (Table 5.3). The latter had an AUC of 1917 ng*hr/mL and was eliminated with $t_{1/2} = 1.3$ h. Assuming a 100% conversion of BIIB028 to **5.27**,

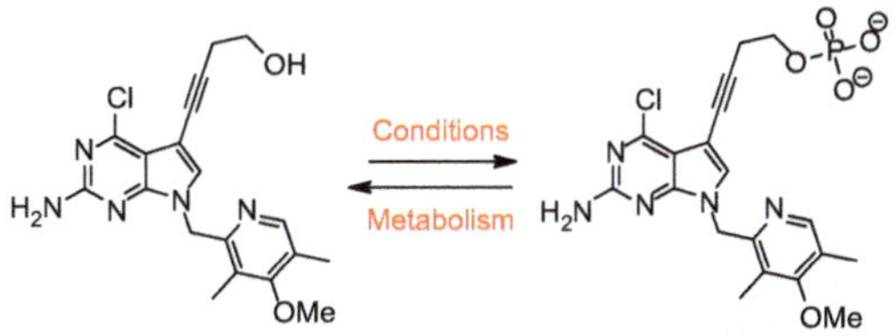

Scheme 5.12 Preparation of BIIB028. Conditions: 4 equiv. POCl$_3$, PO(OEt)$_3$, 0 °C, 45 min., then dilute with water, adjust pH to 3 with NaOH, and let stand at 0 °C to induce precipitation.

Table 5.3 Pharmacokinetic parameters of BIIB028 and its active metabolite **5.27** after single-dose administration of BIIB028 to mouse, rat and dog. The $t_{1/2}$, Cl and V values for the active metabolite **5.27** were calculated assuming a 100% conversion of BIIB028 to **5.27**.

Species	Gender	Dose (mg/kg)	Analyte	Cmax,Co (ng/mL)	t_{max} (hr)	AUC (ng*hr/mL)	$t_{1/2}$ (hr)	Cl (L/hr/kg)	Vz (L/kg)	Vss (L/kg)
Mouse	6M/0F	20	BIIB028	399	0.083	137	0.086	146	18.2	
Mouse	6M/0F	20	**5.27**	5080	0.083	1917	1.3	10.4	19.7	1.9
Rat	2M/2F	3	BIIB028	301	0.083	71	0.055	44		3.4
Rat	2M/2F	3	**5.27**	935	0.083	1238	2.5	5.3	16.7	
Dog	2M/2F	0.4	BIIB028	2825	0.02	238	0.050	1.8		0.13
Dog	2M/2F	0.4	**5.27**	382	0.02	299	7.0	14.1	14.1	

the clearance of the latter was 10.4 L/hr/kg and the volume of distribution $V_z = 19.7$ L/kg. Compared to the Liver Blood Flow (LBF) and the Total Body Water (TBW), this corresponded to a clearance $= 1.9 \times$ LBF and volume of distribution $= 27 \times$ TBW. In rat, a dose of 3 mg/kg of BIIB028 gave the active metabolite **5.27** rapidly, with a half-life of 3 minutes. The active metabolite **5.27** had an AUC $= 1238$ ng*hr/mL and was eliminated with $t_{1/2} = 2.5$ h. Its clearance was 5.3 L/hr/kg ($= 1.6 \times$ LBF) and the volume of distribution $V_z = 16.7$ L/kg ($= 25 \times$ TBW). In dog, a dose of 0.4 mg/kg of BIIB028 was rapidly converted ($t_{1/2} = 5$ min) to the active metabolite **5.27**. The active metabolite **5.27** had an AUC $= 299$ ng*hr/mL and was eliminated with $t_{1/2} = 7.0$ h. The clearance of **5.27** was 14.1 L/hr/kg ($= 7.6 \times$ LBF) and the volume of distribution $V_z = 14.1$ L/kg ($= 23 \times$ TBW).

5.3.4 BIIB028 Pharmacodynamics

For pharmacodynamic studies, mice bearing BT474 tumors were treated with a single dose of BIIB028 as a single IP injection. BIIB028 behaved similarly to BIIB021: the Hsp90 clients were degraded, Hsp70 and Hsp27 were up-regulated and PI3K, which is not an Hsp90 client, was unaffected. The data will be reported separately.

5.3.5 BIIB028 *in Vivo* Efficacy

For efficacy studies, BIIB028 was administered *via* intraperitoneal rather than intravenous injection. We had shown that the two routes of administration were bioequivalent (data not shown) and the IP route simplified the pharmacology work. The efficacy of BIIB028 was evaluated in the N87 tumor model, with a twice weekly administration to reflect a possible clinical schedule (a higher frequency of dosing would be too impractical in clinical settings). The doses were 7.5, 15, 30 and 60 mg/kg IP, the latter being the maximum tolerated dose for the twice weekly schedule. We were extremely pleased to note tumor regressions at the maximum tolerated dose (Figure 5.6), an effect that was

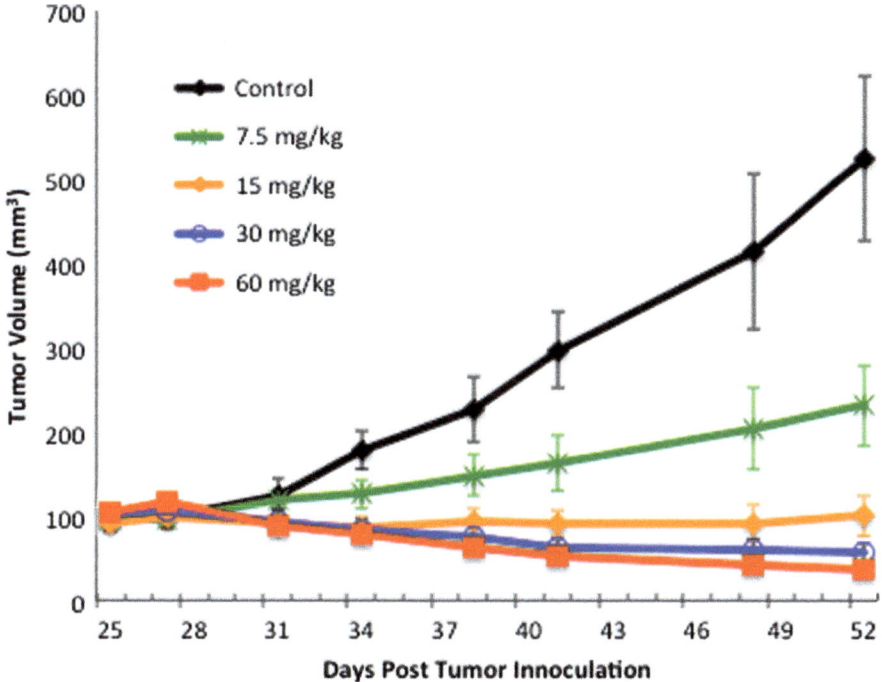

Figure 5.6 Inhibition of tumor growth in the N87 gastric tumor model. Athymic mice bearing established N87 gastric tumor xenografts were treated with vehicle control or BIIB028 at 10, 20, 30 or 60 mg/kg, i.p., q.d.×5 for 4 weeks. Tumors were measured using Vernier calipers on the indicated days and tumor volumes are reported as mean volume ± SEM tumor.

previously never seen as dramatically. In addition, the dose could be decreased by a factor of eight, namely to 7.5 mg/kg, while still significantly inhibiting tumor growth (TGI = 69%). Thus, we had succeeded in improving the tolerability of BIIB028 in a rodent model. Additional mouse xenograft models using cell-lines derived from breast, colon, pancreas, prostate, myeloma or NSCLC malignancies showed that BIIB028 demonstrates a superior TGI as compared with BIIB021 as a single-agent (Table 5.4).

5.3.6 BIIB028 Summary

BIIB028 represented an improvement over BIIB021 in terms of both potency and tolerability. The potency was increased by growing BIIB021 in the direction of the solvent-accessible area, which required modifying the scaffold from a purine to a deazapurine. Alkynes **5.27** and **5.28** proved to be 3–4 times more potent than BIIB021 *in vitro*, and 20–30-fold more efficacious in mouse. The tolerability was gained by delivering the drug intravenously. This required the use of a phosphate pro-drug to reach the required solubility. In mouse, BIIB028 induced tumor regressions at its maximum tolerated dose (120 mg/kg/

Table 5.4 Comparison of BIIB021 and BIIB028 activity in human tumor xenograft models. BIIB021 and BIIB028 were tested at MTD in a variety of human tumor xenografts representative of different tumor types. BIIB021 was administered p.o. and BIIB028 was given i.p.

Tumor line	Tumor type	BIIB021 dose (mg/kg)	BIIB01 schedule	TGI	p-Value	BIIB028 dose (mg/kg)	BIIB028 schedule	TGI	p-Value
N87	Stomach	125	q.d.×5	87	0.0001	60	BIW	133	0.0004
BT474	Breast	120	q.d.×5	94	<0.0001	30	BIW	105	<0.0001
MCF7	Breast	150	TIW	59	0.02	40	BIW	83	<0.0001
SKOV3	Ovarian	124	q.d.×5	63	0.01	40	BIW	91	<0.0001
CWR22 Rv1	Prostate	150	TIW	48	0.04	60	BIW	81	0.002
HT29	Colon	120	q.d.×5	56	0.01	50	BIW	85	<0.0001
A549	NSCLC	150	TIW	5	ns	60	BIW	99	0.0006
H1650	NSCLC	150	TIW	66	0.0001	60	BIW	101	<0.0001
Panc1	Pancreas	124	q.d.×5	51	0.001	60	BIW	78	0.0002
RPMI-8226	M. Myeloma	124	q.d.×5	30	ns	60	BIW	91	0.0001

week) and still retained a remarkable activity when dosed at 1/8 of the MTD. In contrast, BIIB021 stopped tumor growth, but did not induce regressions at its MTD (625 mg/kg/week), and lost efficacy below its MTD. Since BIIB028 met our criteria of improved potency and tolerability, it was advanced to Phase I clinical trials.

5.4 On the Tolerability of Hsp90 Inhibitors

In order to understand better the factors that improved the mouse tolerability of Hsp90 inhibitors, we selected a panel (Scheme 5.13) of two intravenous drugs (BIIB028, AUY-922) and two oral drugs (BIIB021, SNX-5422) and tested them in parallel in the N87 tumor model (Figure 5.7). With BIIB028, the tumors remained small at 60 mg/kg BIW, which is the MTD. Decreasing the dose to 30 or 15 mg/kg retained a substantial activity. With the intravenous drug AUY-922, the tumors remained small at 60 mg/kg BIW (its MTD), but decreasing the dose by a factor of 2 or 4 led to a striking loss of efficacy. Hence, i.v. delivery alone was not sufficient to warrant a high therapeutic index. The oral drugs BIIB02 and SNX-5422, even when dosed at their MTD (150 mg/kg TIW and 25 mg/kg TIW, respectively) were not as efficacious as BIIB028, suggesting that absolute potency still matters. Furthermore, the efficacy of BIIB021 diminished upon decreasing the dose by a factor of 2, contrary to BIIB028. SNX-5422 was dosed at its MTD (25 mg/kg TIW) and gave results comparable to AUY922.

There are not many properties that can account for the superior efficacy and tolerability of BIIB028 as compared to the other Hsp90 inhibitors. One differentiator is that the active metabolite of BIIB028 (primary alcohol **5.27**) is not a PGP substrate, as evidenced by Caco-2 studies ($B \to A/A \to B$ ratio = 1). In contrast, AUY-922 and SNX-5422 are strongly effluxed, with a $B \to A/A \to B$

The Discovery of BIIB021 and BIIB028

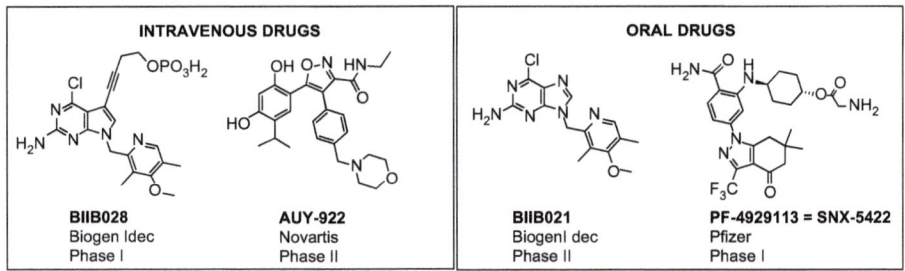

Scheme 5.13 Clinical Hsp90 inhibitors selected for head-to-head comparison.

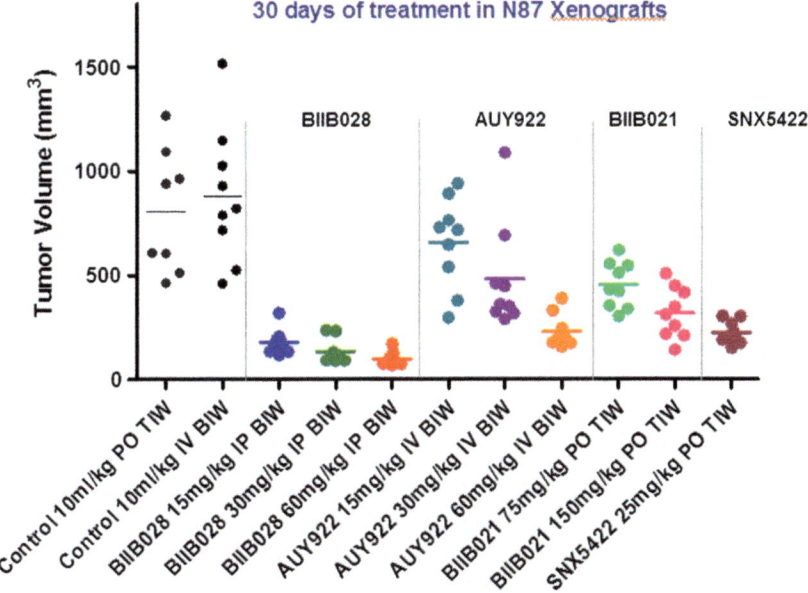

Figure 5.7 Comparison of four clinical Hsp90 inhibitors in N87 gastric tumor xenograft model. Tumors were grown to 100 mm³, and the mice were then treated with BII028, NVP-AUY922, BIIB021 or SNX-5422 three times (TIW) or two times (BIW) per week for 30 days. Individual tumor volumes are reported.

ratio of 15 and 35, respectively (Table 5.5). The second differentiator is that BIIB028 and BIIB021 are cleared fairly rapidly, with a $t_{1/2}$ of approximately 1 hour, while AUY-922 and SNX-5422 are cleared more slowly, with $t_{1/2} = 6.6$ and 2.9 h, respectively (Table 5.5).

Based on these observations, we tentatively suggest that what matters for the tolerability of Hsp90 inhibitors is the <u>combination</u> of (i) i.v. delivery, (ii) short $t_{1/2}$ and (iii) low efflux. The i.v. delivery decreases the local concentration of Hsp90 in the GI tract and therefore decreases the severity of gastrointestinal side effects. The short half-life is beneficial because, while Hsp90 inhibitors

Table 5.5 Main differentiators between BIIB028, BIIB021, AUY922, and SNX-5422. BIIB028 is characterized by a lack of efflux and a short half-life.

	HER-2 (nM)	CACO-2 Papp $A \rightarrow B$ ($\times 10^{-6}$ cm/s)	CACO-2 $B \rightarrow A / A \rightarrow B$	Mouse $t_{1/2}$ (h)
5.27 (metabolite of BIIB028)	9 ± 3	75	0.9	~1 (upon BIIB028 i.v.)
BIIB021	38 ± 13	20	1.5	0.8 (p.o.)
AUY-922	7 ± 1	1.9	15	6.6 (i.v.)
SNX-5422	19 ± 3	1.1	35	2.9 (p.o.)

induce the degradation of client proteins within 1 h, the cell requires 24–48 hours to resynthesize the client proteins. Thus, a brief exposure to the drug (1 h) results in a durable PD response (24–48 h) while minimizing possible side effects, whether off-target or on-target. Finally, a low efflux is preferable to a high efflux because the inhibitor resides for longer in the targeted organ (tumor) as opposed to the other compartments (blood, GI tract, tissues, *etc.*). Metabolites could also be invoked, but it is unclear what would be the property of the metabolite that the parent does not have.

Of note, in human, ocular toxicity has been observed with Hsp90 inhibitors of different chemotypes (SNX-5422 where it is dose-limiting, NVP-AUY922 where it is not dose-limiting), but has not been observed with BIIB021. This difference could also be ascribed to differences in the half-lives.

5.5 Conclusion

Starting from the PU3 series, a scaffold-hopping technique that we call "disconnect-rotate-reconnect" allowed us to create a novel purine series, from which BIIB021 emerged as the first oral Hsp90 inhibitor to reach the clinic. The potency was further improved by taking advantage of the solvent-exposed area, thus providing inhibitors **5.27** and **5.28** (EC144). The tolerability was improved by using a pro-drug that could be injected intravenously. The phosphate pro-drug BIIB028 proved to be ideal for its high solubility, chemical stability and enzymatic lability. The active metabolite of BIIB028, homopropargylic alcohol **5.27**, is also characterized by a lack of efflux from Caco-2 cells, and a relatively short half-life in mouse (1 h). Because of its remarkable activity in mouse models, we believe that BIIB028 has the potential to be a best-in-class.

References

1. G. Chiosis, M. N. Timaul, B. Lucas, P. N. Munster, F. F. Zheng, L. Sepp-Lorenzino and N. Rosen, *Chem. Biol.*, 2001, **8**, 289–299.
2. B. Lucas, N. Rosen and G. Chiosis, *J. Comb. Chem.*, 2001, **3**, 518–520.
3. K. Lundgren, H. Zhang, J. Brekken, N. Huser, R. E. Powell, N. Timple, D. J. Busch, L. Neely, J. L. Sensintaffar, Y. C. Yang, A. McKenzie,

J. Friedman, R. Scannevin, A. Kamal, K. Hong, S. R. Kasibhatla, M. F. Boehm and F. J. Burrows, *Mol. Cancer Ther.*, 2009, **8**, 921–929.
4. M. A. Biamonte, J. Shi, K. Hong, D. C. Hurst, L. Zhang, J. Fan, D. J. Busch, P. L. Karjian, A. A. Maldonado, J. L. Sensintaffar, Y. C. Yang, A. Kamal, R. E. Lough, K. Lundgren, F. J. Burrows, G. A. Timony, M. F. Boehm and S. R. Kasibhatla, *J. Med. Chem.*, 2006, **49**, 817–828.
5. L. Zhang, J. Fan, K. Vu, K. Hong, J. Y. Le Brazidec, J. Shi, M. Biamonte, D. J. Busch, R. E. Lough, R. Grecko, Y. Ran, J. L. Sensintaffar, A. Kamal, K. Lundgren, F. J. Burrows, R. Mansfield, G. A. Timony, E. H. Ulm, S. R. Kasibhatla and M. F. Boehm, *J. Med. Chem.*, 2006, **49**, 5352–5362.
6. R. M. Immormino, Y. Kang, G. Chiosis and D. T. Gewirth, *J. Med. Chem.*, 2006, **49**, 4953–4960.
7. S. R. Kasibhatla, K. Hong, M. A. Biamonte, D. J. Busch, P. L. Karjian, J. L. Sensintaffar, A. Kamal, R. E. Lough, J. Brekken, K. Lundgren, R. Grecko, G. A. Timony, Y. Ran, R. Mansfield, L. C. Fritz, E. Ulm, F. J. Burrows and M. F. Boehm, *J. Med. Chem.*, 2007, **50**, 2767–2778.
8. X. Shi, H. Chang, W. F. Kiesman and M. Grohmann, *13th International Conference on The Scale-up of Chemical Processes*, 2012, July 9–12, Lake Maggiore, Italy.
9. R. Dutta and M. Inouye, *Trends Biochem. Sci.*, 2000, **25**, 24–28.
10. J. Shi, R. Van de Water, K. Hong, R. B. Lamer, K. W. Weichert, C. M. Sandoval, S. R. Kasibhatla, M. F. Boehm, J. Chao, K. Lundgren, N. Timple, R. Lough, G. Ibanez, C. Boykin, F. J. Burrows, M. R. Kehry, T. J. Yun, E. K. Harning, C. Ambrose, J. Thompson, S. A. Bixler, A. Dunah, P. Snodgrass-Belt, J. Arndt, I. J. Enyedy, P. Li, V. S. Hong, A. McKenzie and M. A. Biamonte, *J. Med. Chem.*, 2012, **55**, 7786–7795.
11. T. J. Yun, E. K. Harning, K. Giza, D. Rabah, P. Li, J. W. Arndt, D. Luchetti, M. A. Biamonte, J. Shi, K. Lundgren, A. Manning and M. R. Kehry, *J. Immunol.*, 2011, **186**, 563–575.
12. D. G. Walker, M. J. Humora, W. Kiesman, A. Joshi, T. Clifford and S. Chopade, *244th National ACS Meeting*, 2012, Aug. 19, Poster 209.

CHAPTER 6

Discovery and Development of Ganetespib

WEIWEN YING

Synta Pharmaceuticals, Corp., 45 Hartwell Avenue, Lexington, MA 02421, USA
Email: wying@syntapharma.com

6.1 Ganetespib

6.1.1 Chemical Description

Ganetespib (formerly known as STA-9090) is a novel resorcinolic triazolone compound. Its IUPAC name is 5-[2,4-dihydroxy-5-(1-methylethyl)phenyl]-2,4-dihydro-4-(1-methyl-1H-indol-5-yl)-3H-1,2,4-triazole-3-one. Ganetespib is structurally distinct from the geldanamycin class of Hsp90 inhibitors, such as 17-AAG and 17-DMAG, and shares some structural similarity with the natural product Hsp90 inhibitor radicicol. The indole moiety of the molecule is important for the correct conformation of ganetespib that enhances its binding affinity. The chemical structure is shown in Figure 6.1A. With a molecular weight of 364.4, ganetespib is considerably smaller than the geldanamycins, and most of the second-generation Hsp90 inhibitors. Ganetespib is relatively hydrophobic. Ganetespib is a stable and non-hygroscopic white powder and can be stored at room temperature.

6.1.2 Screening Process

Hsp90 was considered a niche oncology target in the 1990s due to its association with many critical biological functions and the safety concern of targeting

RSC Drug Discovery Series No. 37
Inhibitors of Molecular Chaperones as Therapeutic Agents
Edited by Timothy Machajewski and Zhenhai Gao
© The Royal Society of Chemistry 2014
Published by the Royal Society of Chemistry, www.rsc.org

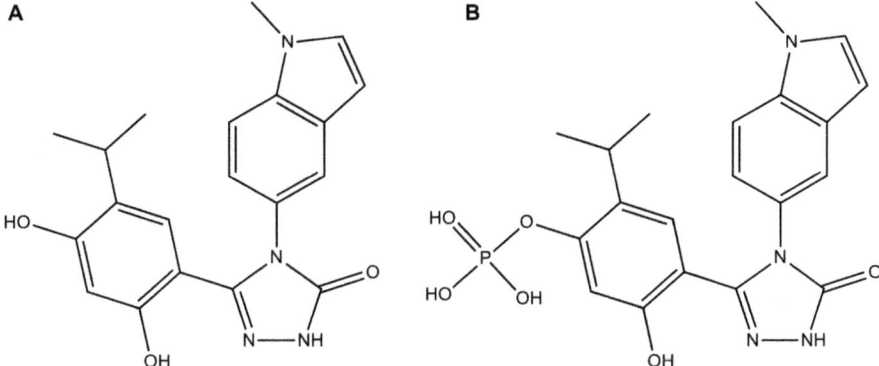

Figure 6.1 Chemical structures of ganetespib and its phosphate pro-drug STA-1474. A. Ganetespib; B. STA-1474.

such an essential protein. Early proof-of-concept evidence was provided by work on geldanamycin and its derivatives, including 17-AAG and 17-DMAG.[1] However the geldanamycins proved to be a difficult class of compounds to develop due to an adverse safety profile and other developmental challenges involving manufacturing and formulation. Targeting Hsp90 for oncology application presents an ideal opportunity for early-stage biotech companies.

Hsp90 contains a unique ATP binding pocket, termed the Bergerat fold, which is found in members of the ATPase/Kinase GHKL (Gyrase, Hsp90, Histidine, Kinase, MutL) superfamily.[2] The co-crystal structures of both geldanamycin and radicicol bound to the ATP pocket in the Hsp90 N-terminal were resolved in the late 1990s and thus became available for *in silico* screening.[3,4]

The initial screening efforts at Synta began with the pharmacophore generation using the SBF (Structure-Based Focus) module in the Cerius2 package from Accelrys.[5] A collection of pharmacophores was generated using both co-crystal structures of geldanamycin and radicicol with the Hsp90 N-terminal. Pharmacophore scoring was performed using the Catalyst module[5] from Accelrys against the Synta 3D compound library. Initial hits included quercetrin, which contains a phendiol moiety. An Hsp90 client protein degradation assay confirmed that quercetrin was a weak Hsp90 inhibitor. Subsequent screening on an expanded collection of phendiol-containing compounds led us to the scaffold of resorcinol-triazole molecules.

After a few years of lead optimization with inputs from chemistry, biology, DMPK, discovery toxicology and pre-formulation, we identified ganetespib, which contains an indole branch, as our lead candidate. Ganetespib entered clinical development in late 2007.

6.1.3 Co-crystal Structure with Hsp90 N-terminal

Ganetespib binds to the ATP pocket at the N-terminus of Hsp90.[6] A number of conformations including an "open" or "closed" conformation (in reference to the position of the ATP binding pocket lid) can be identified in Hsp90 N-terminal co-crystal structures. We have obtained the co-crystal structure for ganetespib bound to the "closed" conformation of the Hsp90 N-terminus; however, we anticipate that ganetespib can also access the ATP pocket in the "open" conformation due to its small size. This lack of restriction for binding to the Hsp90 ATP pocket may be one of the reasons that ganetespib demonstrates higher *in vitro* potency compared to the geldanamycin analogs.

The X-ray co-crystal structure of ganetespib bound to Hsp90 confirmed important hydrogen bonding interactions. Hydrogen bonds involving the resorcinol hydroxyl group with Asp93 and the carbonyl group of triazolone with Lys58 are also seen in the geldanamycin family. In ganetespib, the 2-hydroxyl of resorcinol is within hydrogen bond distance to both oxygen atoms of the carboxylic group in Asp93, resulting in a stronger interaction. The N_2 of triazolone forms a water-bridged hydrogen bond with Asp93 to provide additional hydrogen bonding. Water bridge hydrogen bonds between 4-hydroxyl of resorcinol and Leu48 and Ser52 were found to be critical for binding efficiency in the lead optimization efforts. The carbonyl group of triazolone in ganetespib is of particular structural importance. In addition to the direct hydrogen bond with Lys58, it forms a hydrogen bond with Gly97, a distinguishing feature from the geldanamycin analogs. Further, it also interacts with Thr184 and Asp102 through water bridge hydrogen bonding.

6.2 Pre-clinical Pharmacology

6.2.1 *In Vitro* and *In Vivo* Single Agent Activity

The *in vitro* cytotoxic activity of ganetespib was determined using a panel of over 50 transformed cell lines, derived from both hematological and solid tumors.[6] Ganetespib was potently cytotoxic in the majority of lines examined, typically with IC_{50} values in the low nanomolar range. Ganetespib demonstrated a 20-fold greater potency than 17-AAG with median IC_{50} values of 14 nM *vs.* 280 nM, respectively. This difference in sensitivity was even more pronounced in the subset of hematological malignancies, which showed a 47.5-fold difference (median IC_{50} values of 10 nM *vs.* 475 nM). Notably, ganetespib retained potency against cell lines expressing mutated kinases that confer resistance to kinase inhibitors currently in clinical practice.

Hsp90 occupancy by ganetespib was examined in the non-small-cell lung cancer (NSCLC) cell line NCI-H1975 using a quantitative occupancy assay. This method involved titration of increasing concentrations of ganetespib to cells in culture prior to lysis and incubation with saturating concentrations of deuterated ganetespib (ganetespib-D3) to unoccupied binding sites. The ratio of Hsp90-bound ganetespib/ganetespib-D3 was measured by LC-MS/MS. Occupancy values were produced over a 0.5–512 nM range of concentrations, which showed that ganetespib binding to Hsp90 saturated between 64 and 128 nM. Kinetic analysis showed that the association of ganetespib with Hsp90 was relatively fast under saturating conditions, reaching equilibrium within 5 minutes of ganetespib exposure.

Cell cycle analyses revealed that ganetespib induced marked accumulation in the G2/M phase within 24 hours in NCI-H1975 cells, with a concomitant loss of S phase. The viable cell population remained blocked for at least 72 hours; however, over this period, the percentage of apoptotic cells increased. Subsequently cells were exposed to increasing concentrations of ganetespib for 6 to 72 hours. Apoptosis was measured using activated caspase 3/7 levels and compared to cell viability. No effects were seen 6 hours after treatment. However, the marked loss of viability following exposure to ganetespib was observed 24–48 hours post-treatment and this correlated with the increased apoptotic induction. These results suggest that ganetespib-induced cytotoxicity is mediated by an irreversible commitment to apoptosis, which is likely subsequent to growth arrest and effects on the cell cycle.

The exposure time of ganetespib required to induce cytotoxic responses *in vitro* has been investigated. Cells were exposed to ganetespib for various periods of time (5 minutes to 24 hours), washed to remove the drug and then re-cultured in medium. Exposure to ganetespib for only 1 hour resulted in cytotoxicity IC_{50} value of 510 nM for NCI-H1975 NSCLC cells. Even a 5-minute exposure to ganetespib in NCI-H1975 cells still resulted in an IC_{50} value <1 µM, a concentration that is achievable *in vivo*. These findings indicated that cell viability was quickly affected by ganetespib treatment, and suggest that even brief drug exposure may be sufficient to affect tumor growth.

This *in vitro* activity profile translated into broad *in vivo* efficacy as demonstrated by the following examples. MV4-11 acute myeloid leukemia (AML) cells express the Hsp90 client protein FLT3, an oncogenic driver and the most common genetic alteration associated with AML. This cell line was highly sensitive to ganetespib exposure *in vitro* (IC_{50} value of 4 nM). Ganetespib was further administered intravenously to MV4-11 tumor-bearing mice once weekly at doses of 100 mg/kg and 125 mg/kg. These two treatment regimens resulted in significant tumor regression (85% and 94%, respectively). Moreover, tumors were undetectable in 37.5% of ganetespib-treated animals at the end of the three-week dosing period. Amplification of the c-MET receptor tyrosine kinase occurs in approximately 20% of gastric carcinomas. MET is a known Hsp90 client. In xenograft models of human c-MET amplified MKN45 gastric carcinoma, ganetespib treatment was highly efficacious, with a 50 mg/kg dose three times per week resulting in 92% inhibition of tumor growth.

6.2.2 Anti-angiogenic Properties of Ganetespib

Targeting angiogenic pathways represents an important strategy for combating cancer since angiogenesis is required for cancer progression from an *in situ* lesion to invasive and metastatic disease. Hypoxia inducible factor 1 alpha (HIF-1α) is an Hsp90 client that plays an important role in angiogenesis through controlling the expression of multiple angiogenic factors including vascular endothelial growth factor (VEGF), stromal cell-derived factor 1 (SDF-1) and placental growth factor (PGF). Hence, inhibition of HIF-1α activity dramatically inhibits tumor vascularization.

We employed the NCI-H1975 NSCLC xenograft model to study the anti-angiogenic properties of ganetespib. Female SCID mice were injected subcutaneously with tumor cells in Matrigel.[6] Animals were treated with vehicle or a single bolus injection of ganetespib (125 mg/kg). Tumors were excised 24 hours after dosing, processed and subject to quantitative image analysis for HIF-1α expression, markers of proliferation (bromodeoxyuridine), apoptosis (TUNEL), vasculature (CD31), hypoxia (pimonidazole) and perfusion (DiOC7[3]). The data showed that a single dose of ganetespib greatly reduced the expression of HIF-1α in NSCLC tumor xenografts, as far as 150 μm from the nearest blood vessel, suggesting that ganetespib can penetrate deep into the hypoxic regions of tumor tissue. Ganetespib also had a marked effect on the tumor vasculature, reducing CD31 (endothelial cell marker) expression by 39%, coordinate with a reduction in cell proliferation and an increase in both apoptosis and hypoxia.

6.2.3 Ganetespib in Non-small-cell Lung Cancer

NSCLC is the most common type of lung cancer and the leading cause of cancer mortality worldwide. Intensive research into the underlying biology of this disease has been critical in the development of novel molecularly targeted therapeutics. Many oncogenic drivers in NSCLC are established Hsp90 clients that are dependent on the chaperone for correct folding and conformational stability.

Ganetespib has demonstrated efficacy in EGFR-dependent NSCLC xenograft mouse models.[7] Interestingly, in NCI-H1975 xenografts, levels of mutant EGFR(L858R/T790M) display a complete recovery five days following exposure to a single-dose of ganetespib, suggesting that once-weekly administration of ganetespib as single agent may not be adequate to effectively suppress mutant EGFR(L858R/T790M) signaling. Indeed, re-expression of mutant EGFR is associated with a return of tumor cell proliferation and reversal of apoptosis. Therefore, a sustained reduction in client protein expression may be required for efficient cell death in oncoprotein-driven NSCLC. In agreement with this premise, ganetespib was more efficacious using a daily dosing regimen five times per week in the same model, which induced tumor regressions instead of tumor growth inhibition. Depletion of the proteins c-MET and CDK4 by ganetespib in NCI-H1975 xenografts showed similar kinetics to EGFR, including a return of expression despite persistent drug concentration in tumor. A number of contributing factors may account for these observations, including the re-expression of client proteins as reducing Hsp90 inhibitory activity over time, re-synthesis of Hsp90, altered intracellular compartmentalization of ganetespib or increased assembly of available Hsp90 into an active co-chaperone/substrate bound complex. In addition, induction of other stress proteins such as Hsp70 and Hsp27 may contribute to client re-expression.

Some clients exhibit higher sensitivity to decreases in Hsp90 activity with more rapid and complete depletion than others. A prime example is

HER2. *In vitro*, HER2 became depleted within 6 hours and protein levels did not fully recover for a six-day period after ganetespib exposure. This efficient and prolonged depletion of HER2 following ganetespib treatment was confirmed in xenografts using transformed Ba/F3 cells exhibiting IL-3 independence *via* mutant HER2 expression. *In vivo*, single agent ganetespib caused robust depletion of mutant HER2 following initial dosing, which translated to tumor growth inhibition after two weeks of treatment and ultimately to tumor regressions after four weeks of drug exposure. These results suggest a potential efficacy of ganetespib against NSCLCs driven by mutant HER2.

NSCLC tumors driven by oncogenic KRAS respond poorly to current therapies. A panel of lung cancer cell lines harboring a diverse spectrum of KRAS mutations was tested with ganetespib. Ganetespib was potently cytotoxic in all lines, with concomitant destabilization of KRAS signaling effectors.[8] KRAS mutations result in the activation of the RAF/MAPK/ERK and PI3K/AKT signaling cascades. Combinations of low-dose ganetespib with MEK or PI3K/mTOR inhibitors resulted in superior cytotoxic activity than single agents alone. *In vivo*, the antitumor efficacy of ganetespib was potentiated by the PI3K/mTOR inhibitor BEZ235 in A549 NSCLC xenografts. Further, ganetespib suppressed activating feedback signaling loops that occurred in response to MEK and PI3K/mTOR inhibition. Moreover, ganetespib sensitized mutant KRAS NSCLC cells to standard of care chemotherapeutics including antimitotic agents, topoisomerase inhibitors and alkylating agents.

Crizotinib was approved by the Food and Drug Administration (FDA) for the treatment of advanced, ALK-positive NSCLC in 2011. Although most patients with ALK-positive NSCLC gain substantial clinical benefit from crizotinib, the benefit is relatively short-lived because of the development of acquired resistance. Clinical data have revealed that Hsp90 inhibitors, including ganetespib, can produce encouraging objective responses in NSCLC patients whose tumors contain ALK rearrangements and who have progressed on previous treatments. Cell viability and signaling cascades were examined in ALK-driven NSCLC cell lines treated with ganetespib, crizotinib or the combination of ganetespib and crizotinib.[9] Ganetespib was 50-fold more potent than crizotinib in killing H3122 NSCLC (EML4-ALK expressing) cells. When ganetespib and crizotinib were combined together at sublethal doses, they displayed strong synergistic anticancer activity. Ganetespib showed similar potency in additional cell lines driven by constitutively active ALK or ROS1 kinase fusions, due to the abrogation of their oncogenic kinase activity. Ganetespib effectively inhibits the activity of ALK and ROS1 kinases, known to be associated with several tumor types, resulting in potent single-agent activity irrespective of the mutational status of ALK. Mechanistically, ganetespib induces the degradation of EML4-ALK protein by targeting the chaperone dependency of ALK, rather than the kinase directly. Ganetespib demonstrates superior ALK+ tumor growth suppression compared to crizotinib and is equally potent in ALK inhibitor-sensitive and -resistant cancer cells, regardless of ALK mutation status. The complementary mechanisms of action and promising pre-clinical results therefore support the potential use of ganetespib and crizotinib in combination and such a strategy is currently being explored at an investigator-sponsored clinical trial.

Overall, ganetespib displays significant pre-clinical potency with potential for activity in several NSCLC subsets defined by their addiction to individual oncoproteins. Optimization of dosing schedules and integration with tyrosine kinase inhibitor-based therapy could be key to the success of the broad application of Hsp90 inhibitors in NSCLC patients.

6.2.4 Ganetespib in Hematological Malignancies

Inhibition of Hsp90 by ganetespib results in the destabilization of a broad range of oncogenic kinases that are often over-expressed or mutated in hematological cancers.[10] For example nucleophosmin-anaplastic lymphoma kinase (NPM-ALK), expressed by the anaplastic large cell lymphoma (ALCL) cell line Karpas 299, is degraded rapidly in the presence of ganetespib *in vitro*, resulting in loss of cellular viability. Similar results were seen in other NPM-ALK-driven ALCL cell lines including SU-DHL-1 and SR-786, with IC_{50} values less than 20 nM. The stability of other kinases commonly expressed in hematological malignancies, such as BCR-ABL, FLT3 and c-KIT, were also shown to be highly sensitive to ganetespib, with treatment resulting in potent cell death of cell lines addicted to these kinases.

In vivo, ganetespib was highly effective in a subcutaneous xenograft model of diffuse large B-cell lymphoma (SU-DHL-4 cells) with resulting %T/C values of 26, 4, –90 and –93 when dosed at 25, 50, 75 and 100 mg/kg twice per week, respectively. Importantly, dosing at 75 and 100 mg/kg ganetespib two times per week over a three-week cycle resulted in 25% and 50% of the animals in each group being tumor free by the end of the study, respectively.

The aberrant over-expression of Wilms tumor 1 (WT1) in myeloid leukemia plays an important role in blast cell survival and resistance to chemotherapy. High expression of WT1 is also associated with relapse and shortened disease-free survival in patients. In the human leukemic cell line K562, WT1 was found to co-localize with Hsp90 in the nucleus and endogenous WT1 could be coimmunoprecipitated using an anti-Hsp90 antibody.[11] Hsp90 inhibition by ganetespib reduces the expression of WT1 protein in a dose-dependent manner in myeloid leukemia cell lines. Ganetespib down-regulates WT1 with consequent reduced expression of WT1 target proteins c-MYC and BCL-2. In addition, the down-regulation of WT1 by ganetespib was observed in primary myeloid leukemia blasts isolated from AML patients. Hsp90 inhibition caused ubiquitination and subsequent proteasome-dependent degradation of the WT1 protein. Furthermore, silencing of WT1 with shRNA potentiated apoptosis by chemotherapeutic agents and further sensitized leukemia cells to ganetespib. In a WT1-expressing leukemia xenograft model, single-dose ganetespib could significantly down-regulate expression of WT1 protein and inhibit tumor growth. While there is currently no established therapy that durably inhibits WT1 oncogenic functions, targeting WT1 expression by ganetespib may offer new strategies to limit the survival-promoting effects of WT1 in myeloid leukemias.

Ganetespib treatment results in sustained depletion of JAK2, including the constitutively active JAK2^{V617F} mutant, with subsequent loss of STAT activity and reduced STAT-target gene expression.[12] The concomitant impact of ganetespib on both cell growth and cell division signaling translates to potent antitumor efficacy in mouse models of xenografts of JAK/STAT-driven leukemia.

6.2.5 Combination Studies with Ganetespib

Combination-based therapies have proven to be a successful treatment option for many cancer patients. Due to its broad effects on multiple oncogenic pathways and favorable safety profile, strategies for combining common therapeutics with ganetespib have been investigated.

Combining ganetespib with the standard of care taxane agents paclitaxel and docetaxel has shown potential for improved therapeutic benefit in NSCLC. Median effect analysis suggested that combinations of paclitaxel or docetaxel with ganetespib in NCI-H1975 cells were highly synergistic. Mice bearing NCI-H1975 xenografts were treated with ganetespib and these taxanes, both as single agents and in combination.[13] Weekly administration of suboptimal doses of ganetespib (50 mg/kg) and paclitaxel (7.5 mg/kg) reduced tumor growth by 45% and 62%, respectively. Concurrent treatment with both drugs at the same dose level resulted in a significant enhancement of antitumor activity, blocking tumor growth by 93%. Importantly, combination treatment was well tolerated. In a separate study, the combinatorial effect of ganetespib with docetaxel was examined. Mice were treated with ganetespib at 100 mg/kg, which inhibited tumor growth by 85%, and a similar level of inhibition (87%) was seen when docetaxel was administered at 5 mg/kg. Combined treatment with ganetespib and docetaxel resulted in a significantly improved antitumor response, resulting in 24% tumor regression as compared to the growth inhibition seen in the single agent cohorts. Similarly impressive results were obtained using another NSCLC xenograft model HCC827. When dosed as a single agent at 75 mg/kg, ganetespib potently inhibited HCC827 tumor growth (T/C 26%). Docetaxel (4 mg/kg) also inhibited tumor growth in this model (T/C 46%). While the two agents were administered concurrently, tumor growth was completely abrogated. Taken together, these data indicate that the combination of ganetespib with taxanes results in a superior therapeutic response compared to the single agent activity of any of the compounds alone.

Radiation is an important standard therapy for locally unresectable cancers. However, radio-resistance and repair of sublethal radiation damage can limit its efficacy. Inhibition of Hsp90 by ganetespib potently destabilizes proteins required for both DNA repair and cell cycle checkpoints, but without altering the integrity of DNA.[14,15] Monotherapy treatment with either ganetespib or 2 Gray (Gy) ionizing irradiation resulted in moderate reductions in tumor growth rates in a cervical cancer xenograft model. In contrast, combining ganetespib with 2 Gy irradiation resulted in substantial tumor regression. Increasing the dose of radiation in the combination arm to 4 Gy further enhanced tumor regression, resulting in a 50% reduction in tumor volume. In a separate study using colorectal cancer xenografts, ganetespib potentiated the effects of ionizing radiation resulting in concordant decreases in colony formation, decreased cellular migration and an increase in necrosis.

Small-cell lung cancer (SCLC) is initially highly sensitive to chemotherapy; however, responses in patients with metastatic disease are typically of short duration and resistance inevitably occurs. Ganetespib inhibits SCLC cell growth *via* induction of persistent G2/M arrest and caspase 3-dependent cell death.[16] In a combination study, ganetespib was found to synergize with doxorubicin, which is commonly used in SCLC chemotherapy. The ganetespib plus doxorubicin combination effect was profound. In comparison to either agent alone, combinatorial treatment resulted in significantly more tumor regression using human H82 SCLC mouse xenografts.

6.2.6 Broad Activity of Ganetespib in Different Tumor Types

The broad biological activity of Hsp90 results from its ability to chaperone a vast number of client proteins that play a central pathogenic role in human cancer. It is established that Hsp90 has over 200 clients and substrates. Not surprisingly, ganetespib has been studied extensively in treating a diverse collection of tumor types.

Metastatic pheochromocytoma represents a major clinical challenge in the field of neuroendocrine oncology; however, recent molecular characterization of this disease has highlighted potential new treatment strategies involving the targeting of Hsp90. Hsp90 is over-expressed in malignant pheochromocytoma. Ganetespib has demonstrated potent inhibition of proliferation and migration in pheochromocytoma cell lines through the degradation of key Hsp90 clients.[17] Studies have shown that the down-regulation of HIF-1α protein by ganetespib, together with ganetespib-induced cytotoxicity, are consistent with hypoxia as representing a critical adaptive response promoting survival in pheochromocytoma. Moreover, ganetespib exhibits dose-dependent cytotoxicity in primary pheochromocytoma cells. Using mouse model of metastatic pheochromocytoma, ganetespib demonstrated efficacy in reducing metastatic burden and increasing survival.

Hsp90 inhibitors have been proposed as an important alternative approach for modulating the canonical RAS/MAPK pathway. In this regard ganetespib has been studied in a number of melanoma cell lines that are highly dependent on the RAF/MEK/ERK signaling axis for growth and survival.[18] Ganetespib down-regulated the expression of multiple signal transduction proteins, resulting in pronounced decrease in phosphorylation of AKT and ERK1/2 in a panel of cutaneous melanoma cell lines, including those harboring BRAF and NRAS mutations. Ganetespib has exhibited potent cytotoxic activity in all melanoma cell lines tested to date, with IC_{50} values between 37.5 and 84 nM. Significantly, ganetespib is active in BRAF mutated melanoma cells that have acquired resistance to selective BRAF inhibitors.

A key feature of prostate cancer is that tumors remain critically dependent on androgen. Hsp90 is essential for the stability and function of androgen receptor (AR), which has been causally implicated in the pathogenesis of prostate cancer. Ganetespib potently decreases cellular viability in prostate cancer cell lines, irrespective of their androgen sensitivity or receptor status.[19] In prostate cancer cells ganetespib exerted concomitant effects on mitogenic and survival pathways, as well as direct modulation of cell cycle regulators, to induce growth arrest and apoptosis. *In vivo*, ganetespib displayed potent antitumor efficacy in both AR-negative and -positive xenografts, including those derived from the 22Rv1 prostate cancer cell line that co-expresses full-length and variant receptors.

Most gastrointestinal stromal tumors (GIST) express mutant KIT or PDGFRA oncoproteins, which are established targets of small-molecule tyrosine kinase inhibitors. Clinical resistance to imatinib or sunitinib in GIST has been associated with the acquisition of heterogeneous secondary mutations in the KIT/PDGFRA ATP binding pocket or activation loop, which in turn maintains the constitutively activated state of these kinases. Kinase mutants were biochemically profiled for imatinib and ganetespib sensitivity in Ba/F3 cells transformed by mutant KIT constructs and in GIST cell lines.[20] As many as eight different secondary KIT imatinib-resistance mutations were detected in individual patients whose GISTs progressed after imatinib therapy and all mutations remained sensitive to ganetespib. Ganetespib was as effective against the primary plus secondary mutations in combination, when compared to primary imatinib-sensitive mutations alone. Importantly, the potency of ganetespib was undiminished against activating KIT mutations including secondary kinase-domain resistance mutations, as typically occur in imatinib-resistant GIST clones. In addition, ganetespib potently inhibited cell growth in the 17-AAG-resistant GIST882B cell line. Based on these results, ganetespib entered Phase I clinical evaluation in GIST patients. One GIST patient with a PDGFR^{D842V} mutation has experienced a reduction in tumor size and disease stabilization.

6.3 Pre-clinical Pharmacokinetics and Toxicology

6.3.1 Tissue Distribution in Tumor-bearing Mice

The pharmacokinetic parameters of ganetespib were evaluated *in vivo* using mice bearing NCI-H1975 NSCLC xenografts. An important finding relates to the accumulation of ganetespib within tumors. Ganetespib selectively accumulates in tumors relative to normal tissues. Ganetespib was administered as a single dose intravenously at 125 mg/kg and its elimination kinetics were determined in tumor, liver, lung and plasma over a 6-day time period utilizing HPLC/MS-MS.[21] Ganetespib rapidly distributed from the bloodstream into tissues with a short plasma half-life of 3 hours. The half-life in normal liver and lung was 5–6 hours. In contrast, the half-life of ganetespib in tumors was 58.3 hours, 10- to 19-fold longer than that in normal tissues or plasma, respectively. This observation is similar to the other resorcinol-containing Hsp90 inhibitors such as NVP-AUY922 and AT13387.[22] Remarkably, at six days after dosing, the tumor concentration of ganetespib remained 215-fold higher than the median IC_{50} of 6.5 nM, which has been determined as necessary for anti-proliferative cytotoxicity against a broad NSCLC cell line panel.

6.3.2 Lack of Liver and Cardiac Toxicities

Liver toxicity represents a major impediment to the clinical development of many Hsp90 inhibitors, particularly those of the first-generation geldanamycin class. Structurally, ganetespib lacks the hydroquinone moiety in geldanamycin, which is believed to be the main cause of hepatotoxicity in this class of compounds. The hepatotoxicity profile of ganetespib was evaluated in rats based on changes in the liver enzymes aspartate aminotransferase (AST) and alanine aminotransferase (ALT).[6] Animals were treated with repeated administration of ganetespib from 25 to 75 mg/kg/d for five days. No changes in the levels of either enzyme were observed in the ganetespib-treated animals, even at the highest dose of 75 mg/kg, which is higher than the efficacious dose range for this compound. In contrast, a dose-dependent and marked elevation of ALT and AST was seen with 17-DMAG-treated rats, at doses 12.5 times lower than ganetespib. In accordance with the lack of enzymatic induction, there were no discernible morphologic changes in the hepatocytes of animals treated with ganetespib.

The cardiovascular effects of escalating doses of ganetespib on electrophysiological and mechanical properties were evaluated in isolated New Zealand white rabbit hearts using the Langendorff assay.[6] Ganetespib exerted no significant physiological effects other than a minimal reduction in AV conduction (lengthening PQ interval) and a minor reduction in heart rate (increased RR interval) over the concentrations 10^{-8}–10^{-5} M. There was no change in the QTc(F) intervals at concentrations of ganetespib between 10^{-8} and 10^{-5} M when compared to baseline or vehicle. Similarly, there was no change in the QRS duration after exposure to concentrations of ganetespib ranging from 10^{-8} to 10^{-6} M, when compared to baseline or vehicle; however, an increase in the duration of the QRS was noted after exposure to the 10^{-5} M concentration. At 10^{-4} M, the highest concentration tested, ganetespib lengthened PQ interval and QRS duration; however, this concentration was approximately 3000-fold higher than the unbound C_{max} in the 125 mg/kg dose in the NCI-H1975 tumor-bearing mouse pharmacokinetic studies. Other cardiac electrophysiological parameters and mechanical properties, including left ventricular developed pressure, were not significantly altered following exposure to ganetespib, while expected physiological changes with the positive control quinidine were observed.

6.3.3 Absence of Ocular Toxicity

Ocular toxicity has emerged as one of the major safety concerns in developing Hsp90 inhibitors. In human clinical trials, patients treated with a selected number of Hsp90 inhibitors have reported visual disorders including blurred vision, flashes, delayed light/dark accommodation and photophobia. These adverse effects involving injury to the retina may be attributable to photoreceptor degeneration and cell death, as previously reported in dogs following repeated doses of SNX-5422[23]. On the contrary, ganetespib has demonstrated promising clinical activity without ocular toxicity.

A rodent model to determine risks of ocular toxicity induced by targeted Hsp90 inhibition has been developed.[24] This experimental model includes evaluating microscopic retinal pathology, heat-shock protein modulation and profiles of retinal drug exposure. Two different chemical classes of Hsp90 inhibitors have been selected as the study articles: 17-DMAG and 17-AAG representing the prototypical geldanamycin class and ganetespib and NVP-AUY922 representing second-generation resorcinolic compounds.

Both 17-DMAG and NVP-AUY922 induced strong yet restricted retinal Hsp70 expression and promoted photoreceptor cell death. In contrast, neither 17-AAG nor ganetespib elicited photoreceptor injury. When the relationship between drug distribution and photoreceptor degeneration was examined, 17-DMAG and NVP-AUY922 showed significant retinal accumulation, with high retina/plasma (R/P) ratios and slow elimination rates, such that over 50% of each drug present at 30 minutes post-injection was retained in the retina 6 hours post-dose. For 17-AAG and ganetespib, retinal elimination was rapid (94% and 71% of drug eliminated from the retina at 6 hours, respectively) which resulted in lower R/P ratios.

Thus it appears that the drug retina/plasma exposure ratio and elimination rate profiles play crucial roles in ocular toxicity and can be used as indicators of Hsp90 inhibitor-induced damage in rats. Hsp90 plays an important role in the retina and prolonged Hsp90 inhibition can lead to vision disorders such as those that have been seen in the clinical setting. The ocular toxicity potential of each of the Hsp90 inhibitors defined in the rodent model was entirely consistent with their observed clinical profile in humans.

6.4 STA-1474, a Phosphate Pro-drug of Ganetespib

Phosphate pro-drugs of ganetespib have been developed in order to improve water solubility and other physicochemical properties. There are a number of regions on ganetespib to which a phosphate group may be attached. In STA-1474, the phosphate is linked to the 4-OH position of the resorcinol moiety. The chemical structure is shown in Figure 6.1B. STA-1474 has been selected over other ganetespib phosphate pro-drugs for further evaluation. STA-1474 is highly soluble and stable in physiological buffers. It readily converts to ganetespib in microsomes and in plasma. *In vivo*, STA-1474 also readily metabolizes to ganetespib in both mice and dogs.

Osteosarcoma (OSA) is the most common malignant bone tumor in dogs and children, and exhibits a similar clinical presentation and molecular biology in both species. Canine and human OSA cell lines and normal canine osteoblasts were treated with STA-1474 and evaluated for effects on proliferation, apoptosis and stability of established Hsp90 client proteins.[25] Treatment with STA-1474 resulted in loss of cell viability, inhibition of cell proliferation and induction of apoptosis. STA-1474 exhibited selectivity for OSA cells *versus* normal canine osteoblasts. This selectivity of STA-1474 for neoplastic cells has important implications for patient well-being, particularly in reducing drug-related toxicity. STA-1474 down-regulated the expression of phosphorylated (p)-MET/MET, p-AKT/AKT and p-STAT3. Mice bearing canine OSA xenografts were treated with STA-1474, where it was found to activate caspase-3, down-regulate p-MET/MET and p-AKT/AKT levels and consequently induce tumor regression.

A Phase I clinical study was conducted to extend these observations and to investigate the safety and efficacy of STA-1474 in dogs with spontaneous tumors.[26] In this trial, dogs with spontaneous tumors of a variety of cancer types received STA-1474 under one of three different dosing schemes. Pharmacokinetics, toxicities, biomarker changes and tumor responses were assessed. Toxicities were primarily gastrointestinal in nature consisting of diarrhea, vomiting, inappetence and lethargy. Up-regulation of Hsp70 protein expression was noted in both tumor specimens and PBMCs within 7 hours of drug administration. Measurable objective responses were observed in dogs with malignant mast cell disease (n = 3), osteosarcoma (n = 1), melanoma (n = 1) and thyroid carcinoma (n = 1), for an overall response rate of 24% (6/25). Stable disease (>10 weeks) was seen in three dogs, for a resultant overall biological activity of 36% (9/25).

This study provides direct evidence that STA-1474 and its parent compound ganetespib have biological and therapeutic activity in a relevant large animal model of spontaneous cancer.

6.5 Clinical Update

Ganetespib first entered clinical trials in late 2007. Many clinical evaluations are being conducted using ganetespib as a monotherapy, in both solid tumor and hematological malignancies. Notably, a Phase II trial evaluating the combination of ganetespib with docetaxel for treating second-line NSCLC patients has been conducted in a global, randomized and multi-center setting. In addition, ganetespib is being studied in many investigator sponsored trials, the majority of which are proof-of-concept studies across a variety of tumor types.

6.5.1 Ganetespib in Combination with Docetaxel in Treatment of NSCLC Patients

At the 2012 Congress of the European Society for Medical Oncology (ESMO), investigators reported results from the second interim efficacy analysis of the Phase IIb portion of the GALAXY trial.[27] There were 172 adenocarcinoma patients in the clinical database at the time of the September 10, 2012 cutoff for this analysis.

An increase in overall survival was observed in adenocarcinoma patients treated with ganetespib plus docetaxel. A median overall survival of 7.4 months was observed in the docetaxel control arm, while median overall survival had not been reached in the ganetespib arm. Results for docetaxel were consistent with results from prior second-line NSCLC therapy trials.

Objective response rate and progression-free survival in adenocarcinoma patients were also improved: from 8% to 16%, and from 2.8 months to 4.2 months, in the control arm vs. ganetespib arm, respectively.

Results in several GALAXY patient subpopulations, defined by pre-specified clinical and biomarker characteristics, showed a substantially improved survival difference between the control arm and ganetespib arm, as compared with the difference in the all-comer (intent-to-treat) adenocarcinoma patient population. These findings have been incorporated into the design of the Phase III portion of the GALAXY program, with the objective of enriching for patients likely to derive the greatest benefit from ganetespib treatment.

A favorable safety profile was observed with the ganetespib plus docetaxel combination in adenocarcinoma patients. Transient, mild-to-moderate diarrhea was the most common adverse event, consistent with observations from other clinical trials evaluating ganetespib.

6.5.2 Efficacy in NSCLC Patients Harboring ALK Rearrangement

Results from a multi-center Phase II study of ganetespib monotherapy in NSCLC patients have been reported.[28]

Ninety-nine patients with a median of two prior systemic therapies were enrolled. There are three cohorts in the study: cohort A, mutant EGFR; cohort B, mutant KRAS; and cohort C, no known EGFR or KRAS mutations. Patients were treated with 200 mg/m^2 ganetespib by intravenous infusion once weekly for three weeks followed by one week of rest. Ninety-eight patients were assigned to cohort A (n = 15), B (n = 17) or C (n = 66), with progression-free survival (PFS) rates at 16 weeks of 13.3%, 5.9% and 19.7%, respectively. There are four patients (4%) who achieved partial response (PR). All four patients had disease that harbored ALK gene rearrangement in crizotinib-naïve patients enrolled to cohort C. In addition to the antitumor activities observed in ALK+ patients, ganetespib showed modest clinical activity in patients with mutant KRAS. In cohort B, 47% of patients had tumor shrinkage. In cohort A, stable disease (SD) was achieved in 40% of patients.

Ganetespib demonstrated a manageable safety profile. Eight patients (8.1%) experienced treatment-related serious adverse events (AE); two of these resulted in death. The most common AEs were diarrhea, fatigue, nausea and anorexia.

6.5.3 Clinical Benefits in Patients of Various Cancers

Ganetespib has demonstrated encouraging signs of clinical activity in breast cancer patients, including confirmed partial responses in both a triple negative breast cancer patient and an HER2-positive breast cancer patient.[29]

In Phase I ganetespib monotherapy trials testing alternate dosing regimens in solid tumors, preliminary signs of clinical activity have been observed, including a durable PR in a patient with rectal carcinoma, a durable PR in a patient with melanoma and tumor shrinkage with prolonged SD in several patients including melanoma, NSCLC, renal cell carcinoma and GIST, who were treated for 16 to 48 weeks.[30]

Phase I clinical trials of ganetespib in hematological malignances have also been conducted. Clinical responses include one patient with CML and one patient with AML who had best responses of hematological improvement lasting two and three months, respectively. In addition, four patients with myelofibrosis had stable disease as best response; one of these responses lasted for seven months and the patient went on to receive an allogenic stem cell transplant.[31,32]

6.6 Conclusion

Ganetespib is a novel, injectable, synthetic small molecule. Several unique chemical structural features confer selective advantages for ganetespib over other Hsp90 inhibitors. These features include smaller size, higher lipophilicity, rigid and proper conformation and key hydrogen bonds that enhance the binding affinity between ganetespib and the chaperone protein.

Ganetespib binds to Hsp90 N-terminal ATP pocket. Functional Hsp90 inhibition in this manner causes its client proteins to adopt aberrant conformations, which are then targeted for ubiquitination and degradation by the proteasome. Ganetespib has been shown to induce the degradation of a wide variety of important signaling proteins, including ALK, BCR-ABL, BRAF, CRAF, CDK4, HIF-1α, KIT, EGFR and HER2. *In vitro*, ganetespib is a potent inducer of cell death in many cancer cell lines, and *in vivo* inhibits the growth of human tumor cell lines in mouse xenograft models.

Ganetespib selectively accumulates in tumors and induces long-lasting client protein degradation and subsequent tumor growth inhibition in a lung cancer xenograft model. In tumor bearing mice, ganetespib has a half-life of 58 hours in tumor *versus* 4–6 hours in other tissues (lung, liver and plasma). The results from safety and pharmacology studies show acceptable selectivity as well as acceptable cardiac effects; in particular there has been no evidence of qualitative or quantitative ECG changes. Moreover, ganetespib lacks the liver toxicity commonly associated with geldanamycin analogs. In terms of ocular toxicity, and unlike 17-DMAG and NVP-AUY922, ganetespib was not associated with retinal dysfunction in the rat model, was rapidly eliminated from retinal tissue and did not cause photoreceptor cell apoptosis.

Ganetespib is being studied in ten company-sponsored clinical trials as of March 2013. In addition, ganetespib is being studied in 15 investigator-sponsored trials, the majority of which are proof-of-concept studies across a variety of tumor types.

As of 20 September, 2012, 630 patients have received at least one dose of intravenous (IV) ganetespib as part of one of 25 studies. A total of 148 have been treated in company-sponsored combination therapy studies (ganetespib with docetaxel).

Clinical benefits have been observed in trials using ganetespib as a monotherapy. Significantly, in a Phase IIb global, randomized, multi-center study of ganetespib in combination with docetaxel, good tolerability for the combination as well as meaningful improvements in overall survival in adenocarcinoma patients have been observed.

Acknowledgment

The author thanks Richard Bates who provided editorial assistance during preparation of this manuscript.

References

1. For a comprehensive review on Hsp90 inhibitor development, see: L. Neckers and P. Workman, *Clin. Cancer Res.*, 2012, **18**, 64.
2. A. Bergerat, B. de Massy, D. Gadelle, P. C. Varoutas, A. Nicolas and P. Forterre, *Nature*, 1997, **386**, 414.
3. C. E. Stebbins, A. A. Russo, C. Schneider, N. Rosen, F. U. Hartl and N. P. Pavletich, *Cell*, 1997, **89**(2), 239.
4. S. M. Roe, C. Prodromou, R. O'Brien, J. E. Ladbury, P. W. Piper and L. H. Pearl, *J. Med. Chem.*, 1999, **42**, 260.
5. http://accelrys.com/resource-center/case-studies/archive/studies/catshape_two_ligands.html. Last accessed 29 April, 2013.
6. W. Ying, Z. Du, L. Sun, K. P. Foley, D. A. Proia, R. K. Blackman, D. Zhou, T. Inoue, N. Tatsuta, J. Sang, S. Ye, J. Acquaviva, L. Shin-Ogawa, Y. Wada, J. Barsoum and K. Koya, *Mol. Cancer Ther.*, 2012, **1**, 475.
7. T. Shimamura, S. A. Perera, K. P. Foley, J. Sang, S. J. Rodig, T. Inoue, L. Chen, D. Li, J. Carretero, Y. C. Li, P. Sinha, C. D. Carey, C. L. Borgman, J. P. Jimenez, M. Meyerson, W. Ying, J. Barsoum, K. K. Wong and G. I. Shapiro, *Clin. Cancer Res.*, 2012, **18**, 4973.
8. J. Acquaviva, D. L. Smith, J. Sang, J. C. Friedland, S. He, M. Sequeira, C. Zhang, Y. Wada and D. A. Proia, *Mol. Cancer Ther.*, 2012, **11**, 2633.
9. J. Sang, J. Acquaviva, J. C. Friedland, D. L. Smith, M. Sequeira, C. Zhang, Q. Jiang, L. Xue, C. M. Lovly, J. P. Jimenez, A. T. Shaw, R. C. Doebele, S. He, R. C. Bates, D. R. Camidge, S. W. Morris, I. El-Hariry and D. A. Proia, *Cancer Discov.*, 2013, **3**, 430.
10. W. Ying, D. Proia, S. He, J. Sang, K. Foley, Z. Du, R. Blackman, Y. Wada, L. Sun and K. Koya, *52nd American Society of Hematology Annual Meeting (Orlando, FL)*, 2010, Abstract #2899.
11. H. Bansal, S. Bansal, M. Rao, K. P. Foley, J. Sang, D. A. Proia, R. K. Blackman, W. Ying, J. Barsoum, M. R. Baer, K. Kelly, R. Swords, G. E. Tomlinson, M. Battiwalla, F. J. Giles, K. P. Lee and S. Padmanabhan, *Blood*, 2010, **116**, 4591.

12. D. A. Proia, K. P. Foley, T. Korbut, J. Sang, D. Smith, R. C. Bates, Y. Liu, A. F. Rosenberg, D. Zhou, K. Koya, J. Barsoum and R. K. Blackman, *PLoS One*, 2011, **6**, e18552.
13. D. A. Proia, J. Sang, S. He, D. L. Smith, M. Sequeira, C. Zhang, Y. Liu, S. Ye, D. Zhou, R. K. Blackman, K. P. Foley, K. Koya and Y. Wada, *Invest. New Drugs*, 2012, **30**, 2201.
14. J. Sang, D. Smith, T. Korbut, C. Zhang, D. A. Proia and Y. Wada, *102nd American Association for Cancer Research Annual Meeting (Orlando, FL)*, 2011, Abstract #2677.
15. H. Mahaseth, G. P. Nagaraju, R. Diaz, L. D. Taliaferro-Smith, J. C. Landry, N. F. Saba and B. F. El-Rayes, *103rd American Association for Cancer Research Annual Meeting (Chicago, IL)*, 2012, Abstract #2326.
16. D. H. Lee, C. H. Lai, Y. Wang and G. Giaccone, *14th World Conference on Lung Cancer (Amsterdam, the Netherlands)*, 2011, Abstract #P2.028.
17. A. Giubellino, C. Sourbier, M. J. Lee, B. Scroggins, P. Bullova, M. Landau, W. Ying, L. Neckers, J. B. Trepel and K. Pacak, *PLoS One*, 2013, **8**, e56083.
18. X. Wu, M. E. Marmarelis and F. S. Hodi, *PLoS One*, 2013, **8**, e56134.
19. S. He, C. Zhang, A. A. Shafi, M. Sequeira, J. Acquaviva, J. C. Friedland, J. Sang, D. L. Smith, N. L. Weigel, Y. Wada and D. A. Proia, *Int. J. Oncol.*, 2013, **42**, 35.
20. J. A. Fletcher, M. Debiec-Rychter, S. Swank, J. Morgan, J. Morgan, B. Liegl, T. A. Rege, M. C. Heinrich, C. L. Corless, R. Donsky, C. D. M. Fletcher, C. P. Raut, A. Marino-Enriquez, R. K. Blackman, D. Proia, J. Sang, D. Smith, K. P. Foley, W. Ou, Y. Wang and G. D. Demetri, *International Conference on Molecular Targets and Cancer Therapeutics (Boston, MA)*, 2009, Abstract #B184.
21. K. P. Foley, R. K. Blackman, T. Korbut, M. Nagai, D. A. Proia, J. Sang, D. Smith, W. Ying, C. Zhang, H. Zhang and K. Koya, *101st American Association for Cancer Research Annual Meeting (Washington, DC)*, 2010, Abstract #2638.
22. J. Lyons, J. Curry, T. Smyth, I. Harada, L. Fazal, M. Reule, B. Graham and N. Thompson, *AACR-NCI-EORTC International Conference on Molecular Targets and Cancer Therapeutics (Boston, MA)*, 2009, Abstract #A217.
23. J. R. Infante, G. J. Weiss, S. Jones, R. Tibes, J. C. Bendell, N. Brega, V. Torti, D. D. Von Hoff, H. A. Burris III and R. K. Ramanathan, *International Conference on Molecular Targets and Cancer Therapeutics (Berlin, Germany)*, 2010, Abstract #375.
24. D. Zhou, F. Teofilovici, Y. Liu, J. Ye, W. Ying, L. Shin-Ogawa, T. Inoue, W. Lee, A. Adjiri-Awere, L. Kolodzieyski, N. Tatsuta, Y. Wada and A. J. Sonderfan, *2012 American Society of Cancer Oncology Annual Meeting (Chicago, IL)*, 2012, Chicago, IL Abstract #3086.
25. J. K. McCleese, M. D. Bear, S. L. Fossey, R. M. Mihalek, K. P. Foley, W. Ying, J. Barsoum and C. A. London, *Int. J. Cancer*, 2009, **125**, 2792.

26. C. A. London, M. D. Bear, J. McCleese, K. P. Foley, R. Paalangara, T. Inoue, W. Ying and J. Barsoum, *PLoS One*, 2011, **6**, e27018.
27. S. S. Ramalingam, B. Zaric, G. Goss, C. Manegold Sr., R. Rosell, V. Vukovic, I. El-Hariry, F. Teofilovici, S. Mulcahey, W. Guo and D. A. Fennell, *European Society for Medical Oncology (ESMO) 2012 Congress (Vienna, Austria)*, 2012, Abstract #1248P_PR.
28. M. A. Socinski, J. Goldman, I. El-Hariry, M. Koczywas, V. Vukovic, L. Horn, E. Paschold, R. Salgia, H. West, L. V. Sequist, P. Bonomi, J. R. Brahmer, L. C. Chen, A. B. Sandler, C. P. Belani, T. R. Webb, H. Harper, M. Huberman, S. Ramalingham, K. K. Wong, F. Teofilovici, W. Guo and G. I. Shapiro, *Clin. Cancer Res.*, 2013, Epub ahead of print.
29. K. Jhaveri, S. Chandarlapaty, D. Lake, T. Gilewski, M. Robson, S. Goldfarb, P. Drullinsky, S. Sugarman, C. Wasserheit-Leiblich, J. Fasano, M. E. Moynahan, G. D'Andrea, K. Lim, L. Reddington, S. Haque, S. Patil, L. Bauman, V. Vukovic, I. El-Hariry, C. Hudis and S. Modi, *34th Annual San Antonio Breast Cancer Symposium (San Antonio, TX)*, 2011, Abstract #P1-17-08.
30. J. W. Goldman, R. N. Raju, G. A. Gordon, I. El-Hariry, F. Teofilivici, V. M. Vukovic, R. Bradley, M. D. Karol, Y. Chen, W. Guo, T. Inoue and L. S. Rosen, *BMC Cancer*, 2013, **13**, 152.
31. J. E. Lancet, B. D. Smith, R. Bradley, R. S. Komrokji, F. Teofilovici and D. A. Rizzieri, *52nd American Society of Hematology Annual Meeting (Orlando, FL)*, 2010, Abstract #3294.
32. S. Padmanabhan, K. R. Kelly, M. Heaney, S. Hodges, S. Chanel, M. Frattini, R. Swords, R. Bradley, F. Teofilovici and D. J. DeAngelo, *52nd American Society of Hematology Annual Meeting (Orlando, FL)*, 2010, Abstract #2898.

CHAPTER 7

Discovery of the Serenex Hsp90 Inhibitor, SNX5422

TIMOTHY HAYSTEAD* AND PHILIP HUGHES

Department of Pharmacology and Cancer Biology, Duke University Medical Center, Durham, NC 27710, USA
*Email: timothy.haystead@dm.duke.edu

7.1 Introduction

SNX 5422 is a synthetic, orally bioavailable inhibitor of Hsp90 discovered by the biotechnology company Serenex Inc. (Durham NC, USA) using the chemoproteomic platform proteome mining.[1–4] The molecule is derived from an indoline scaffold and is therefore structurally distinct from all other known Hsp90 inhibitors reported in the literature.[4,5] The molecule also inhibits with nM potency the related heat-shock proteins GRP94 and TRAP-1, although it has yet to be established to what extent these actions contribute to its biological effects *in vivo*.[6] SNX5422 is a pro-drug with oral bioavailability and is hydrolyzed to the active parent compound SNX2112 upon uptake. In cell-based and initial animal studies, and subsequent full pre-clinical work up, SNX5422 exhibited all the hallmarks of a *bona fide* Hsp90 inhibitor (*i.e.* degradation of the expected portfolio of Hsp90-dependent clients, induction of Hsp70, tumor growth arrest). In general, the compound was well tolerated in animals and showed potent efficacy in a variety of xenograph models of human cancer, both alone and in combination with other existing anticancer agents. The compound was therefore advanced to Phase I safety studies. At the time of writing SNX5422 had completed two Phase I clinical trials conducted by the National Cancer Institutes (Bethesda, MD, USA) and Pfizer (who acquired

7.2 Proteome Mining

Proteome mining enables hundreds of enzymes and receptors representing multiple gene families that are normally expressed in cells to be screened *en masse* against drug-like molecules to identify novel chemical starting points for drug-discovery programs. Simultaneously, potential off-target liabilities can also be identified at the earliest stages of the drug-discovery process. It was within this context that the initial scaffold that led to the clinical candidate SNX5422 was discovered.[2,3] The principles of proteome mining are based on the assumption that all drug-like molecules selectively compete with a natural cellular ligand for a binding site on a protein target. In a proteome mine, natural ligands are tethered on a solid phase surface (usually Sepharose) at high density and in an orientation that sterically favors reversible interaction with their protein targets. The affinity of the ligand for its protein targets must be in the exchangeable range (10–50 μM). This range is the normal affinity of most physiological substrates for their protein targets and ensures efficient recovery of the proteins from complex mixtures while being optimal for affinity-based drug screens, since compound libraries can be screened at soluble concentrations (<1 mM). Below this range (<1 μM) affinities are too great for useful screening once a molecule is immobilized. Above 50 μM proteins are recovered too weakly.

Once immobilized the tethered ligand is used to capture a targeted proteome from complex protein mixtures (tissue or cell extract) and bound proteins (targeted proteome) evaluated for specificity by high-throughput protein sequencing using mass spectrometry (MS). The captured proteome can then be screened against libraries of small molecules and if any molecule within the library shows some affinity for any proteins within the captured proteome (includes therapeutic target and potential off-target liabilities), these will be competitively released from the tethered ligand (Figure 7.1). Identification of the liberated proteins by high-throughput mass spectrometry (MS) enables immediate evaluation of the relative pharmacological importance of each of the liberated proteins. Of interest are therapeutic targets. In instances where one has eluted an attractive therapeutic target with several other proteins, literature and searches of the

OMIM (Online Mendelian Inheritance in Man) and SNP (Single nucleotide polymorphism) databases may be used to evaluate whether inhibition of a particular protein is likely to have liabilities *in vivo*. Some off-targets may have no pharmacological consequences within a particular therapeutic dosing window. Indeed, almost all drugs in current use have dose-dependent liabilities, although, within their respective therapeutic windows, this does not preclude their usefulness. An important aspect of proteome mining, therefore, is that the same assay can be used to determine the relative potencies of all the off-targets relative to the therapeutic target. This information is also an invaluable driver for subsequent quantitative structure activity relationship (QSAR) studies should

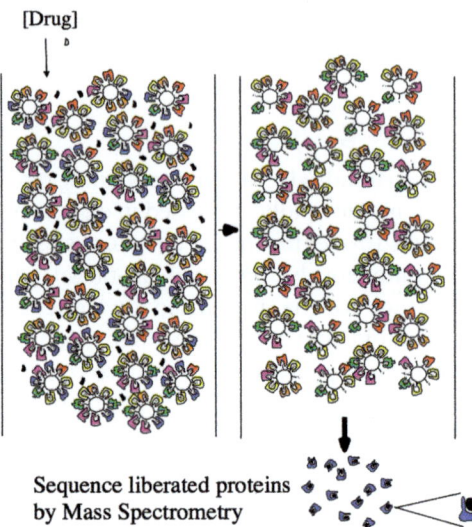

Figure 7.1 The principles of proteome mining. An affinity resin is constructed comprising an immobilized natural ligand (*e.g.* ATP) tethered in an orientation that favors interactions with its nature substrates. The resin is charged with a targeted proteome from a cell or tissue extract and washed to remove non-specifically associated proteins. The captured proteome is then challenged with small molecules to defined selectively eluted proteins. Eluted proteins are identified by mass spectrometry.

one wish to remove the identified off-target liabilities through the iterative process prior to animal studies. In instances where a particular off-target cannot be readily separated through iterative chemistry without loss of efficacy, knowledge of the off-target profile can enable development of biomarker assays (or metabolite profiles) to monitor potential toxicity during therapy.

For the discovery of the SNX 5422 scaffold immobilized ATP, linked through its gamma phosphate, was utilized.[8] Purine-containing nucleotides have multi-functional roles in many aspects of cellular metabolism. They form the precursors of RNA and DNA, function as regulators of many allosteric enzymes, mediate both extracellular and intracellular signals, participate in dehydrogenase and carboxylase reactions in the form of NAD+, NADP+ and Co enzyme A and are substrates (in the form of ATP and GTP) for stress-induced proteins (heat-shock proteins) and motor proteins (myosin), protein kinases and non-protein kinases.[3,9] Consistent with the ubiquitous roles of purine-utilizing enzymes primary sequence motif searches of the current databases reveals that over 2000 proteins can bind purines. We refer to this as the "purinome".[3] Importantly, the purinome contains both established drug targets as well as cutting-edge drug-discovery targets. Established targets include dihydrofolate reductase (methotrexate) and HMG CoA reductase (statins). Cutting-edge targets include protein kinases and stress-induced proteins such as Hsp90. Protein kinases are generally recognized as the frontier of drug discovery and many have likened this enzyme class to GPCRs in terms of the richness of potential therapeutic targets.[10,11] In addition to containing therapeutic targets, searches of the OMIM database (http://www.nslij-genetics.org/search_omim.html) reveal that the human purinome contains over 1500 enzymes associated with inherited metabolic disorders. Many classical examples include glucose 6 phosphate dehydrogenase deficiency (Favism),[12] branched-chain α-keto acid dehydrogenase deficiency (Maple Syrup Urine Disease),[13] hepatic phosphorylase kinase deficiency (Glycogen Storage Disease VIII),[14] 3-α-methylcrotonyl-CoA carboxylase 1 deficiency (MCC1 deficiency)[15] and pyruvate carboxylase deficiency (Leigh syndrome).[16] In Graves et al. we first demonstrated that when ATP was immobilized through its γ-phosphate it could be used to reversibly capture a large number of purine utilizing enzymes expressed in human blood red cells and Plasmodium falciparum.[2]

Correct orientation of the molecule was found to be critical for the selective capture of the targeted proteome. Immobilization of ATP at its ribose or secondary N6 amine sterically disfavored recovery of any of the proteins that bound γ-phosphate linked ATP as determined by MS. Subsequently, over the years we have expanded our analysis of the types of proteins that can be captured with γ-phosphate linked ATP media using various animal and human tissues and identified several hundreds of distinct enzymes and proteins that bind these media (Table 7.1). Small G proteins are not recovered; however, this is expected, since this class of proteins has low affinity for ATP relative to GTP/GDP and binds these purines in a different orientation from proteins captured on γ-linked ATP media.[17] The crystal structures of some of the identified proteins shown in Table 7.1 have been solved and explain why such a diverse array of purine-utilizing gene family members can be selectively recovered on γ-phosphate immobilized ATP. In every case, from dehydrogenase to kinase to heat-shock protein, the purine moieties of each respective nucleotide are found to be buried within the nucleotide-binding pockets of these proteins with the α, β or γ phosphates exposed on the solvent-accessible surface of the protein.[9,18,19]

Table 7.1 Numbers of proteins from various species identified to bind gamma phosphate linked ATP.

Gene class	Total number	Nucleotide
Kinases	518	ATP
CTK	34	ATP
Carboxylases	26	CoA
Dehydrogenases	456	NAD/NADP
Helicases	217	ATP
Lipases	78	ATP
ATPases	453	ATP
Motor proteins	22	ATP
Non-conventional purine utilizing enzymes	357	ATP, ADP, AMP, GTP
Purinergic receptors	6	Adenosine
	11	ATP
Deaminases	85	ATP
Sulfotransferase	40	ATP
Synthetase	213	ATP

7.2.1 Discovery of SNX5422 by Proteome Mining

Screening by proteome mining of porcine lung and liver with an initial library of 5000 compounds, specifically chosen to target purine-binding proteins, revealed a number of ligand–protein pairs including 463 compounds and 77 distinct proteins.[4] A follow-up library of 3000 additional compounds was purchased based on chemical similarity to the 463 original hits. Interestingly, proteome mining of this library gave a similar hit rate to the original library.

From the numerous compound–protein interactions identified in the original screen, 88 compounds across 20 distinct scaffolds eluted the chaperone protein Hsp90. The silver-stained gels and structures for the strongest Hsp90 eluting compounds are shown in Figure 7.2.

None of the compounds shown in Figure 7.2 showed any cellular activity in standard cancer cell proliferation assays. Of the six best hits, two (**2** and **6**) have adenine substructures. All of the compounds elute other proteins to some degree. Most of the other proteins eluted by **2** were kinases. The other major protein eluted by compound **5** is an alcohol dehydrogenase. Compound **6** also elutes synapsin. Synapsin regulates neurotransmitter release and has been shown to have an ATP binding domain.[20] This very undesirable off-target binding precluded any further consideration of **6**. Based on potency and a lack of strong off-target binding, structure activity relationship (SAR) campaigns were initiated on compounds **1–3**.

Analogs of compound **1** gave a very flat SAR, showing little improvement in potency, no cell activity and no client effects. Analogs of compound **2**

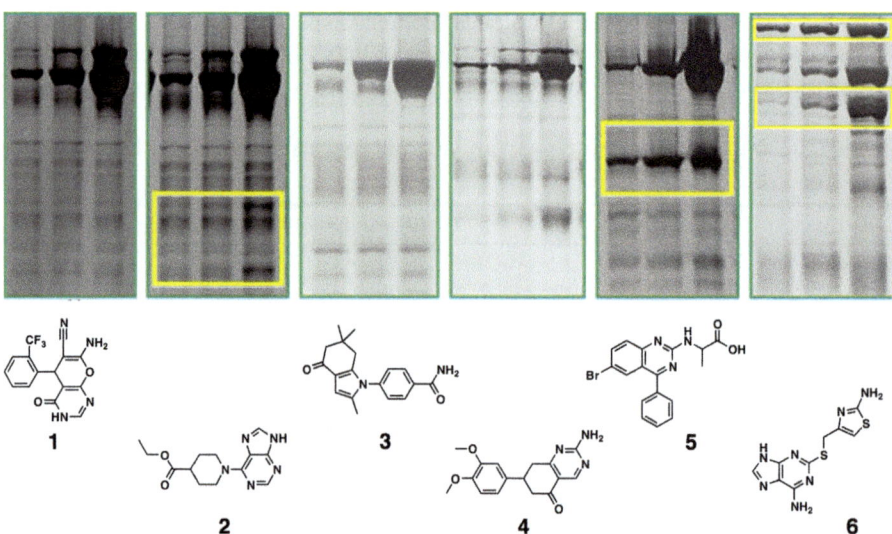

Figure 7.2 Hsp90 hits from a proteomic survey. The silver-stained gels of Hsp90 elutions at various concentrations (left to right, 20, 50, 500 µM). The yellow boxes indicate off-target hits.

showed enhanced potency, but continued to have selectivity issues and elicited no cell activity or client effects. However, analogs of compound **3** showed enhanced affinity for Hsp90, while maintaining good selectivity. Therefore, the medicinal chemistry focus shifted entirely to analogs of **3** because of its novelty, low molecular weight and synthetic tractability despite no cellular activity.

7.3 Lead Optimization Studies Leading to SNX2112

To help guide future optimization strategies for analogs, computational models were developed.[5] At the time, X-ray crystal structures were available for an engineered N-terminal ATP binding domain with a variety of bound ligands. The structures fall into two groups based on the secondary structure of the protein at residues from 100 to 120 as shown in Figure 7.3. With geldanamycin (GM) residues 100 through 120 form two helical domains separated by an irregular domain.[21–23] However, with PU3, residues 100 through 120 form a single helix, which gives rise to a deeper pocket by the movement of Leu 107 out of the pocket.[24]

Compound **3** was docked into both structures with the assumption that the benzamide would bind similarly to the adenine of ATP forming hydrogen bonds with Asp 93. Although both models were considered, the model with the PU3 protein structure was preferred due to better fit in the pocket with the dimethyl group of compound **3** mimicking the displaced Leu 107. The model suggested that *ortho-* substitution to the benzamide was reasonable with the possibility of orienting the carbonyl of the amide. Substitution on the 2 and 3 positions of the pyrole seemed feasible. Also, replacing the indolone with an indazolone seemed reasonable and offered some flexibility in the synthesis of peripherally substituted analogs.

A general retro-synthetic analysis to deliver the desired analogs is shown in Scheme 7.1. Analogs were constructed either by route A, coupling the

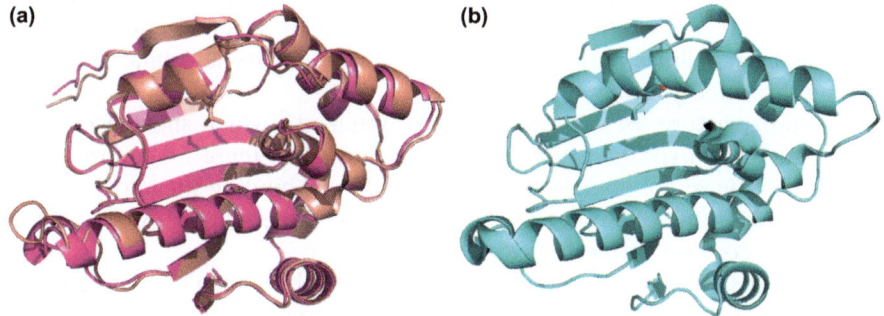

Figure 7.3 Cartoon of Hsp90 N-terminal domains when bound to a) GM and ADP and b) PU3. Note the change in the alpha helix and the movement of Leu 107 going from a) to b).

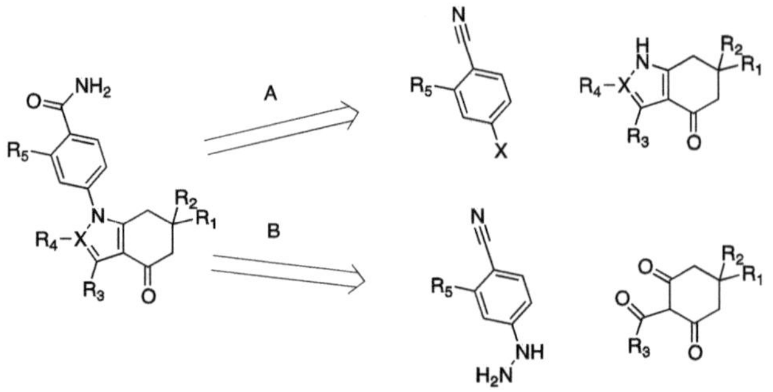

Scheme 7.1 Generalized approach to the synthesis of Hsp90 inhibitor analogs of compound 3.

cyano-activated aryl halide with the appropriate indolone or indazolone, or by route B, coupling an aryl hydrazine with an appropriate acyl dimedone, forming an indazolone in the process. Coupling reactions were usually realized *via* nucleophilic aromatic substitution of aryl fluorides or by palladium mediated Buchwald–Hartwig couplings to aryl bromides. Insertion of R_5 was accomplished either before or after ring coupling, but always before nitrile hydrolysis, and was also generally accomplished by nucleophilic aromatic substitution or by palladium-mediated couplings. The desired indolones and indazolones were prepared by modifications of literature methods.[5]

Initial synthetic studies confirmed the requirement for the carboxamide as N-substituted amides, the carboxylic acid and esters, the intermediate nitrile and other functionalities replacing the carboxamide all lead to loss of binding. Guided by the modeling studies described above, one early structure activity relationship (SAR) investigation focused on substitution of the lead compound 3 in the position *ortho* to the benzamide. As shown in Table 7.2, substitution with oxygen, sulfur and carbon derivatives did not lead to useful moderation of activity. However, nitrogen substitution gave rise to enhanced activity and with a series of aniline derivatives, activity in a cellular HER2 degradation activity was first seen. It is not entirely clear whether the nitrogen substitution afforded some enhanced cellular permeability or if the cell activity is primarily due to reaching threshold potency.

A crystal structure for compound 9 was obtained,[4] which showed that the original modeling hypothesis was correct. As shown in Figure 7.4, compound 9 fits into the N-terminal domain of Hsp90 at the ATP binding site and gives rise to the extended alpha helix seen with PU3.[24] The dimethyl group on 9 nicely occupies the position formerly held by Leu 107.

Further SAR studies explored the role of methyl substitution on compound 9. As shown in Table 7.3, removal of R_2 gave a compound of lower potency in binding and cellular assays. The same was true when R_3 and R_4 were removed. Moving the methyl from R_2 to R_1 led to a small change in Hsp90 binding but gave a substantial change in the cellular activity.

Table 7.2 Activities of *ortho* substitution on lead compound **3**.

Compound	R_5	HER2 degradation[a] IC_{50} (μM)	Hsp90 affinity[b] K_d (μM)
17-AAG		0.003	0.087
3	H	>100	3.73
4	BuO	>10	>50
5	Pentyl	>10	1.96
6	PhS	>10	1.06
7	BuNH	>10	0.39
8	PhNH	6.4	0.37
9	2,3,5-(MeO)$_3$PhNh	0.49	0.22

[a]HER2 imaging assay using AU565 cells.[4]
[b]8-point non-enzymatic ATPase assay.[4]

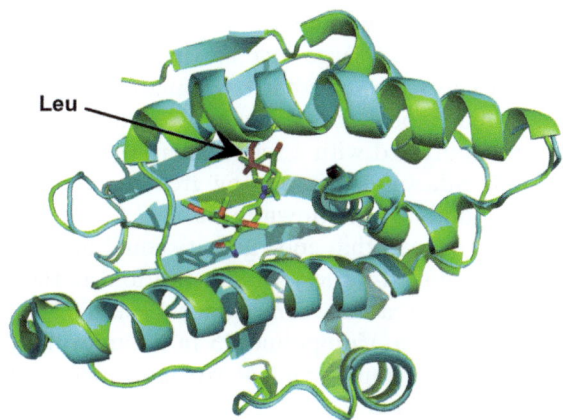

Figure 7.4 Overlap of Hsp90 bound to PU3 (1YUG) and compound 9 (3MNR). Only the ligand from 3MNR is shown for clarity. The Leu 107 from the ADP bound structure (1BYQ) is shown to illustrate how compound **9** displaces it on binding.

7.3.1 Development of SNX5422, an Orally Bioavailable Hsp90 Inhibitor

Compound **12** represented the best compound to date but it showed no oral availability or activity in a mouse xenograph tumor model. There were also

Table 7.3 Analysis of peripheral methyl groups.

Compound	R_1	R_2	R_3, R_4	HER2 degradation[a] IC$_{50}$ (μM)	Hsp90 affinity[b] K_d (μM)
17-AAG				0.003	0.087
9	H	Me	Me	0.49	0.22
10	H	H	Me	1.83	0.59
11	Me	H	H	1.61	0.94
12	Me	H	Me	0.046	0.39

[a]HER2 imaging assay using AU565 cells.[4]
[b]8-point non-enzymatic ATPase assay.[4]

concerns related to the possible metabolic liabilities of the electron-rich aniline and pyrole. A major advance came with the replacement of the trimethoxy-aniline with a *trans*-4-hydroxycyclohexylamino group to give compound **13**. This new compound showed a large increase in binding affinity and HER2 activity. Compound **13** also showed some bioavailability (10%) and activity in a mouse xenograft tumor model on i.p. dosing. Remaining concerns regarding the pyrole in **34** were addressed with a series where the indolone was replaced with less-electron-rich indazolones, imparting greater stability, polarity (clogP values) and, perhaps, solubility as shown in Table 7.4.

The data in Table 7.4 suggest that good activity could be maintained with the indazolone and that a limited variety of substitution at the C-3 position is tolerated. For a variety of reasons, compounds **13** and **15** (SNX-2112) were chosen for further evaluation. Compound **15** in its amorphous form showed good bioavailability (40%, mouse) and *in vivo* antitumor activity in mouse xenograft models following oral dosing.

A parallel effort was made to further explore substitution of the position *ortho* to the benzamide functionality. Although a large number of compounds were made and examined, none showed a better balance of activities than **15** and were, therefore, not pursued further.

Compound **15** was compared with 17-AAG, in a number of cell proliferation assays as well as some client impact assays.[25] The results, shown in Table 7.5, indicate potency against a broad range of cancers as well as a tighter potency range, when compared to 17-AAG. It was also compared in additional assays for client effects and Hsp70 induction with the results shown in Table 7.6.

Early evaluation of **15** was performed with an amorphous form of the compound. Unfortunately, it was later discovered that compound **15** has a very

Table 7.4 Replacement of the indolone with various indazolones.

Compound	R_1	X	HER2 degradation[a] IC_{50} (nM)	Hsp90 affinity[b] K_d (nM)
17-AAG			3	87
13	Me	CH	7	27
14	Me	N	30	16
15	CF_3	N	10	16
16	Et	N	9	20
17	CHF_2	N	16	98
18	c-PrMe	N	46	41
19	i-Pr	N	203	177
20	Me	CMe	64	86

[a]HER2 imaging assay using AU565 cells.[4]
[b]8-point non-enzymatic ATPase assay.[4]

Table 7.5 Comparison of **15** (SNX-2112) with 17-AAG in proliferation assays.

Cell line	17-AAG (IC_{50} nM)	15 (IC_{50} nM)
A375 (melanoma)	1287	23
LNCAP (prostate)	82	3
MCF-7 (breast)	0.3	53
HT-29 (colon)	0.1	3
SW620 (colon)	328	3
SK-MEL-5 (melanoma)	134	6
PC-3 (prostate)	9	17
MDA-MB-231 (breast)	194	6
NCI-H460 (NSCLC)	255	18
HCT-15 (colon)	1487	52
K-562 (erythroleukemia)	126	6

Table 7.6 Comparison of **15** (SNX-2112) with 17-AAG in cellular process assays.

Client assay	17-AAG (IC_{50} nM)	15 (IC_{50} nM)
Hsp70 induction (A375 cells)	4	2
p-S6 degradation (A375 cells)	32	1
HER2 degradation (AU565 cells)	3	11
p-ERK degradation (AU565 cells)	8	41

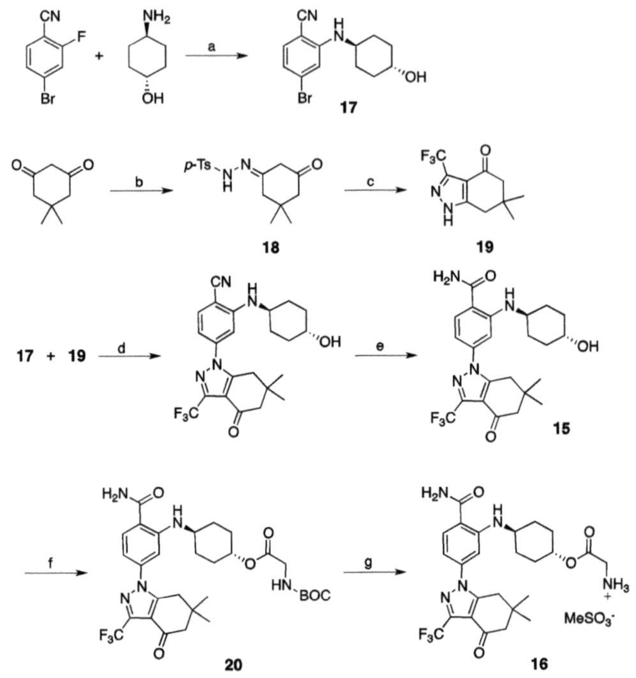

Scheme 7.2 Process synthesis of **16** (SNX-5422). a) Hunig's base, DMSO, 125 °C, 85%; b) *p*-TsNH$_2$NH$_3$, *p*-TsOH (cat.), THF, 73%; c) i) trifluoroacetic anhydride, THF, Et$_3$N, ii) MeOH, NaOH, 60%; d) Cu(I)I, DMEDA, dioxane 98 °C, 73%; e) EtOH, DMSO, H$_2$O$_2$, NaOH, 93%; f) BOC-Gly-OH, EDCI, DMAP, CH$_2$Cl$_2$, 87%; g) MeSO$_3$H, AcOH, MTBE, 80%.

stable crystalline form with greatly diminished aqueous solubility and no oral bioavailability. To improve solubility, two pro-drugs, β-alanine and glycine esters of the cyclohexanol hydroxyl, were evaluated. Both had improved aqueous solubility and oral availability. However, the β-alanine derivative was only partially hydrolyzed in plasma, whereas the glycine ester **16** (SNX-5422) showed rapid and complete hydrolysis on both oral and i.v. administration. For this reason, compound **16** was chosen for further pre-clinical evaluation.

7.3.2 Synthesis of SNX5422

A variety of routes can be used for the synthesis of compound **16**, most devised with the goal of allowing flexibility for analog synthesis. The final route, developed with considerations for scale-up, was recently reported and is shown in Scheme 7.2.[26]

7.4 Pre-clinical Studies with SNX5422

When added to various tumor cell lines the parent compound SNX2112 bears all the hallmarks of a classical Hsp90 inhibitor such as geldanamycin.[27,28] In

HER2+ breast cancer cells one can readily measure HER2 degradation within 6 hours and induction of Hsp70 expression. This phenomenon could also be observed along with growth arrest in xenographs following IP administration of SNX2112 to animals bearing HER2 positive human breast tumors.[4] Our desire was to develop a synthetic Hsp90 that could be readily formulated for oral administration. Unfortunately, due to crystallization issues, it was not possible to do this with SNX2112, hence the development of SNX5422 as described. In cell and animal studies SNX5422 was to prove to be equally effective as the parent compound SNX2112. In animals and humans, SNX5422 behaves as a pro-drug. When administered PO to mice or rats it is rapidly absorbed in the stomach and appears in the serum as SNX2112 with no evidence of the glycine ester. This is also true in humans.

When SNX5422 was administered p.o. to female rats (30 mg/kg), the active parent SNX2112 peaks within the serum (C_{max} 2427 ng/ml) at 120 min. (T_{max}) and is subsequently cleared within 12 hours ($t_{1/2}$ 6 hr, clearance 1560 ml/hr/kg). The parent exhibits an apparent large volume of distribution of 13,600 ml/kg.[4] Similar pharmacokinetic properties were also observed in mice and both species exhibited maximum tolerated doses in the 30–40 mg/kg range upon every-other-day dosing. In multiple mouse xenograph models SNX5422 when dosed orally displayed significant antitumor activity, discriminating it from other classes of Hsp90 inhibitors such as the geldanamycins. Generally, in most human xenograph models, at therapeutic doses SNX5422 produced tumor growth arrest. Upon drug withdrawal, the tumors begin to grow again. This is a typical response to most Hsp90 inhibitors and is also observed in humans.[29] However, most advocates of Hsp90 would agree that the real therapeutic application for clinically relevant Hsp90 inhibitors will be in combination with other cutting-edge therapies such as tyrosine kinase inhibitors,[28,29] although there is evidence in animal models that established chemotherapeutics (*e.g.* cisplatin) or radiation treatment can be highly synergistic in combination with Hsp90 inhibitor such as SNX5422.

Figure 7.5 illustrates some of the highlights of combination studies in which SNX5422 was dosed alone and in combination with the antibody-based drug trastuzumab (TZB) targeting in a HER2+ BT-474 human breast cancer xenograph model.[4] At subtherapeutic doses (10 mg/kg), SNX5422 and TZB (3 mg/kg) exhibit modest tumor growth arrest over a 20-day dose period (every-other-day dosing). Upon withdrawal tumor growth rates are restored to vehicle rates. However, in combination, significant synergism is observed, which persists for over 160 days following drug withdrawal.

7.5 Clinical Studies with SNX5422

At the time of writing there were two completed Phase I clinical trials conducted by the National Cancer Institutes (NCI) and Pfizer (Inc.). Of these studies only the NCI study has been reported completely with peer review.[7] The outcomes of the Pfizer study are, however, discussed within this article in limited detail. In the NCI study 33 patients with various malignancies were dose

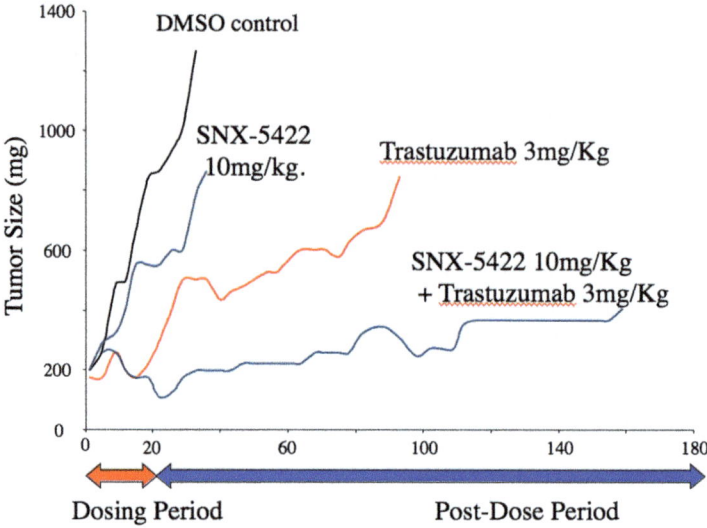

Figure 7.5 SNX-5422 is highly synergistic against BT474 breast xenograft in combination with Herceptin (trastuzumab). These data were previously reported in a different format in Fadden et al. 2010.[4]

escalated by oral administration twice weekly. The purpose of the Phase I study was to determine dose-limiting toxicities, the MTD and general pharmacokinetic and safety profile of SNX5422 (also referred to as PF-04929113). Dosing studies were continued up to 177 mg/m^2 and terminated. Of the 33 patients in the study 15 patients showed long-lasting stable disease and 17 were unresponsive and showed progressive disease. In general on an every-other-day dosing schedule SNX5422 was well tolerated and no adverse outcomes (grade 3 or 4) were reported in this study. Pharmacokinetic studies following oral administration showed the parent compound SNX2112 (also referred to as PF-04928473) attained maximal concentration in serum at 1–3 hr with a $t_{1/2}$ that ranged from 8–15 hours. Overall, SNX2112 showed linear kinetics in humans with elevated dosing and serum levels that correlated well with Hsp70 induction as measured by western blot in PBMCs (peripheral blood mononuclear cells). The study generally concluded that in responsive patients the pharmacological actions of SNX5422 were consistent with its actions as an Hsp90 inhibitor and outcomes were similar to most other clinically relevant Hsp90 inhibitors.

A common issue with many clinically active Hsp90 inhibitors is a reversible ocular toxicity. In the NCI study, SNX5422-related ocular toxicity was not seen. Although the specific details of the Pfizer Phase I study are not in the public domain, the authors of the NCI study do report that a reversible ocular effect was observed in 4 patients (of 44 in the study), manifest as reduced night vision.[7] These effects in humans were, however, observed only after an aggressive every-day dosing regimen. Dosing limiting toxicities for the Pfizer study were reported at 80 mg/m^2 following the daily dosing regimen. Clearly these two clinical studies indicate that the dosing regimen for SNX5422 needs to be carefully monitored and restricted to every-other-day dosing. Although the molecular mechanisms of the ocular toxicities are not known, proteomic studies in our laboratory using affinity resins based on an immobilized version of SNX2112 suggest enzymes involved in the metabolism of retinoic acid (vitamin A) may be clients for Hsp90.[6]

Acknowledgements

Discovery of SNX5422 was a team effort that was coordinated by Dr Steven Hall at Serenex Inc. (2000–2008). The following personnel from Serenex contributed to the discovery and development of SNX5422: Patrick Fadden, Kenneth H. Huang, James M. Veal, Paul M. Steed, Amy F. Barabasz, Briana Foley, Mei Hu, Jeffrey M. Partridge, John Rice, Anisa Scott, Laura G. Dubois, Tiffany A. Freed, Melanie A. Rehder Silinski, Thomas E. Barta, Philip F. Hughes, Andy Ommen, Wei Ma, Emilie D. Smith, Angela Woodward Spangenberg, Jeron Eaves, Gunnar J. Hanson, Lindsay Hinkley, Matthew Jenks, Meredith Lewis, James Otto, Gijsbertus J. Pronk, Katleen Verleysen, Timothy A. Haystead and Steven E. Hall. Timothy Haystead was the scientific founder of Serenex and developed the proteome mining platform. In 2012 Steven Hall founded Esanex Inc. to continue the clinical development of SNX5422.

References

1. J. S. Duncan, T. A. Haystead and D. W. Litchfield, *Meth. Mol. Biol.*, 2012, **795**, 119–134.
2. P. R. Graves, J. J. Kwiek, P. Fadden, R. Ray, K. Hardeman, A. M. Coley, M. Foley and T. A. Haystead, *Mol. Pharmacol.*, 2002, **62**, 1364–1372.
3. T. A. Haystead, *Curr. Top. Med. Chem.*, 2006, **6**, 1117–1127.
4. P. Fadden, K. H. Huang, J. M. Veal, P. M. Steed, A. F. Barabasz, B. Foley, M. Hu, J. M. Partridge, J. Rice, A. Scott, L. G. Dubois, T. A. Freed, M. A. Silinski, T. E. Barta, P. F. Hughes, A. Ommen, W. Ma, E. D. Smith, A. W. Spangenberg, J. Eaves, G. J. Hanson, L. Hinkley, M. Jenks, M. Lewis, J. Otto, G. J. Pronk, K. Verleysen, T. A. Haystead and S. E. Hall, *Chem. Biol.*, 2010, **17**, 686–694.
5. K. H. Huang, J. M. Veal, R. P. Fadden, J. W. Rice, J. Eaves, J. P. Strachan, A. F. Barabasz, B. E. Foley, T. E. Barta, W. Ma, M. A. Silinski, M. Hu, J. M. Partridge, A. Scott, L. G. DuBois, T. Freed, P. M. Steed, A. J. Ommen, E. D. Smith, P. F. Hughes, A. R. Woodward, G. J. Hanson, W. S. McCall, C. J. Markworth, L. Hinkley, M. Jenks, L. Geng, M. Lewis, J. Otto, B. Pronk, K. Verleysen and S. E. Hall, *J. Med. Chem.*, 2009, **52**, 4288–4305.
6. P. F. Hughes, J. J. Barrott, D. A. Carlson, D. R. Loiselle, B. L. Speer, K. Bodoor, L. A. Rund and T. A. Haystead, *Bioorg. Med. Chem.*, 2012, **20**, 3298–3305.
7. A. Rajan, R. J. Kelly, J. B. Trepel, Y. S. Kim, S. V. Alarcon, S. Kummar, M. Gutierrez, S. Crandon, W. M. Zein, L. Jain, B. Mannargudi,

W. D. Figg, B. E. Houk, M. Shnaidman, N. Brega and G. Giaccone, *Clin. Cancer Res.*, 2011, **17**, 6831–6839.
8. C. M. Haystead, P. Gregory, T. W. Sturgill and T. A. Haystead, *Eur. J. Biochem./FEBS*, 1993, **214**, 459–467.
9. W. Eventoff and M. G. Rossmann, *CRC Crit. Rev. Biochem.*, 1975, **3**, 111–140.
10. G. Manning, D. B. Whyte, R. Martinez, T. Hunter and S. Sudarsanam, *Science*, 2002, **298**, 1912–1934.
11. P. Cohen, *Nat. Rev. Drug Discov.*, 2002, **1**, 309–315.
12. A. S. Alving, P. E. Carson, C. L. Flanagan and C. E. Ickes, *Science*, 1956, **124**, 484–485.
13. J. H. Menkes, P. L. Hurst and J. M. Craig, *Pediatrics*, 1954, **14**, 462–467.
14. F. Huijing, *FEBS Lett.*, 1970, **10**, 328–332.
15. M. R. Baumgartner, M. F. Dantas, T. Suormala, S. Almashanu, C. Giunta, D. Friebel, B. Gebhardt, B. Fowler, G. F. Hoffmann, E. R. Baumgartner and D. Valle, *Am. J. Hum. Genet.*, 2004, **75**, 790–800.
16. S. Monnot, V. Serre, B. Chadefaux-Vekemans, J. Aupetit, S. Romano, P. De Lonlay, J. M. Rival, A. Munnich, J. Steffann and J. P. Bonnefont, *Hum. Mutat.*, 2009, **30**, 734–740.
17. D. E. Coleman and S. R. Sprang, *Meth. Enzymol.*, 1999, **308**, 70–92.
18. D. R. Knighton, N. H. Xuong, S. S. Taylor and J. M. Sowadski, *J. Mol. Biol.*, 1991, **220**, 217–220.
19. D. R. Knighton, J. H. Zheng, L. F. Ten Eyck, V. A. Ashford, N. H. Xuong, S. S. Taylor and J. M. Sowadski, *Science*, 1991, **253**, 407–414.
20. L. Esser, C. R. Wang, M. Hosaka, C. S. Smagula, T. C. Sudhof and J. Deisenhofer, *EMBO J.*, 1998, **17**, 977–984.
21. C. E. Stebbins, A. A. Russo, C. Schneider, N. Rosen, F. U. Hartl and N. P. Pavletich, *Cell*, 1997, **89**, 239–250.
22. J. P. Grenert, W. P. Sullivan, P. Fadden, T. A. Haystead, J. Clark, E. Mimnaugh, H. Krutzsch, H. J. Ochel, T. W. Schulte, E. Sausville, L. M. Neckers and D. O. Toft, *J. Biol. Chem.*, 1997, **272**, 23843–23850.
23. W. M. Obermann, H. Sondermann, A. A. Russo, N. P. Pavletich and F. U. Hartl, *J. Cell Biol.*, 1998, **143**, 901–910.
24. L. Wright, X. Barril, B. Dymock, L. Sheridan, A. Surgenor, M. Beswick, M. Drysdale, A. Collier, A. Massey, N. Davies, A. Fink, C. Fromont, W. Aherne, K. Boxall, S. Sharp, P. Workman and R. E. Hubbard, *Chem. Biol.*, 2004, **11**, 775–785.
25. T. W. Schulte and L. M. Neckers, *Canc. Chemother. Pharmaco.*, 1998, **42**, 273–279.
26. S. Duan, S. Venkatraman, X. Hong, K. Huang, L. Ulysse, B. I. Mobele, A. Smith, L. Lawless, A. Locke and R. Garigipati, *Org. Process Res. Dev.*, 2012, **16**, 1787–1793.
27. L. Neckers and P. Workman, *Clin. Cancer Res.*, 2012, **18**, 64–76.
28. P. Workman, F. Burrows, L. Neckers and N. Rosen, *Ann. NY Acad. Sci.*, 2007, **1113**, 202–216.
29. J. J. Barrott and T. A. Haystead, *FEBS J.*, 2013, **280**, 1381–1396.

CHAPTER 8
Discovery of NVP-AUY922

PAUL A. BROUGH,[a] JOSEPH SCHOEPFER,[b] ANDREW MASSEY[a] AND MICHAEL RUGAARD JENSEN*[b]

[a] Vernalis Ltd, Granta Park, Great Abington, Cambridge, CB21 6GB, UK;
[b] Novartis Institutes for BioMedical Research, CH-4057, Basel, Switzerland
*Email: michael_rugaard.jensen@novartis.com

8.1 Historical Context

In 2002 an Hsp90 inhibitor program was initiated at Ribotargets, Cambridge, UK. In 2003 Ribotargets and the Institute of Cancer Research (ICR), Sutton, UK, entered into a collaborative program to identify and develop Hsp90 inhibitors, and at this time all project IP and expertise was combined and future work prosecuted in collaboration across all scientific disciplines. Ribotargets had particular expertise with *in silico* screening, computational chemistry and structure-based drug design (SBDD); the ICR had been working on Hsp90 since the late 1990 s and had already published seminal papers in the area of Hsp90 structural biology and characterization. By late 2004 Ribotargets had become Vernalis *via* company mergers and had entered into a second collaboration with Novartis Institutes for Biomedical Research, Basel, Switzerland. At this point the joint Vernalis/ICR program had already identified several potent Hsp90 inhibitors from several different chemical classes, including the resorcinol inhibitor NVP-AUY922 (also known as VER-52296) see Figure 8.1. Subsequent work focused on preclinical development of NVP-AUY922 and the optimization of existing distinct chemical series and the identification of novel compounds as part of a plan to deliver second-generation inhibitors.

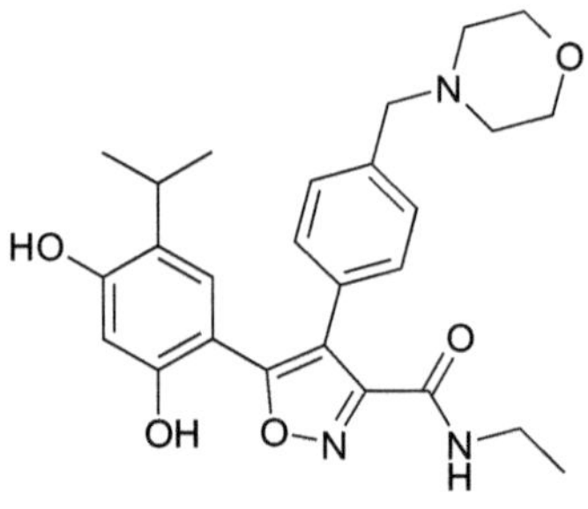

Figure 8.1 Chemical structure of NVP-AUY922 (VER-52296).

8.2 Hit Identification

At the inception of the Hsp90 project there were several compounds known in the literature that bound to the N-terminal ATP site of Hsp90, such as the natural products geldanamycin[1–3] and radicicol[1,4] and substrate mimetics such as PU3[5] (Figure 8.2). Whilst these compounds provided valuable tools for research and start points for early drug development programs, it was felt that new a hit ID campaign was required. Finding novel chemical compounds for starting points in any drug discovery program is a major challenge. High-throughput screening (HTS) has been a mainstay of Hit Identification for many years and has delivered many compounds to the clinic. However, there are significant issues with this process that have been widely discussed, and though it is likely these screening campaigns will always have a place in the drug-discovery arena there is a need for alternative techniques that can complement or replace HTS in specific instances. *In silico* screening and fragment-based screening have emerged over the last 10 to 15 years as alternative tools for Hit ID. Fragment screening has become established as an effective way to sample chemical diversity with a limited number of low-molecular-weight compounds (typically 110 to 250 daltons). The approach has been well reviewed elsewhere[6–10] and several biophysical methods are used to identify and characterize the fragment hits. The approach applied at Vernalis[10] included the design of a bespoke fragment library,[11] and subsequent identification of binding fragments using three distinct 1D NMR experiments.[12–15] To select only those fragments that bind to the N-terminal ATP binding site, a ligand-competition step using the known ligand PU3[5] (Figure 8.2) was included in the screening protocol. Competitive hits were further characterized and their binding mode established by Hsp90α ligand-protein X-ray crystallography. This structural information then allows for a structure-guided medicinal chemistry program to evolve the fragments to more potent drug-like molecules.

Figure 8.3 shows some of the 1D NMR derived hits from fragment screens against Hsp90 that were also shown to bind at the N-terminal ATP site by X-ray crystallography. Often in this process the low-molecular-weight compounds may be characterized by other biophysical methods but often the hits are of such low affinity that it may be difficult to establish affinity in the

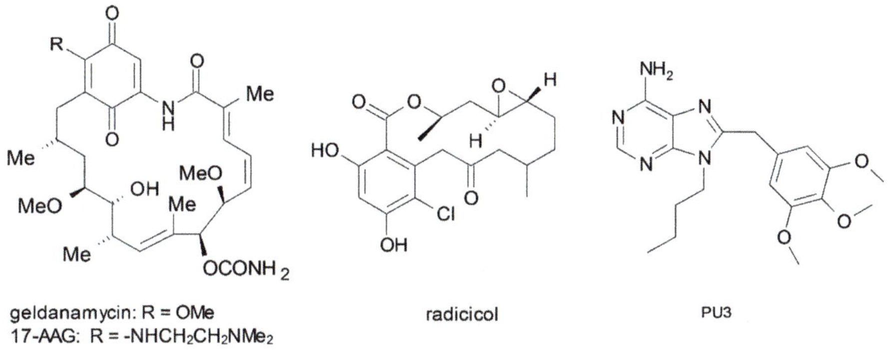

Discovery of NVP-AUY922 215

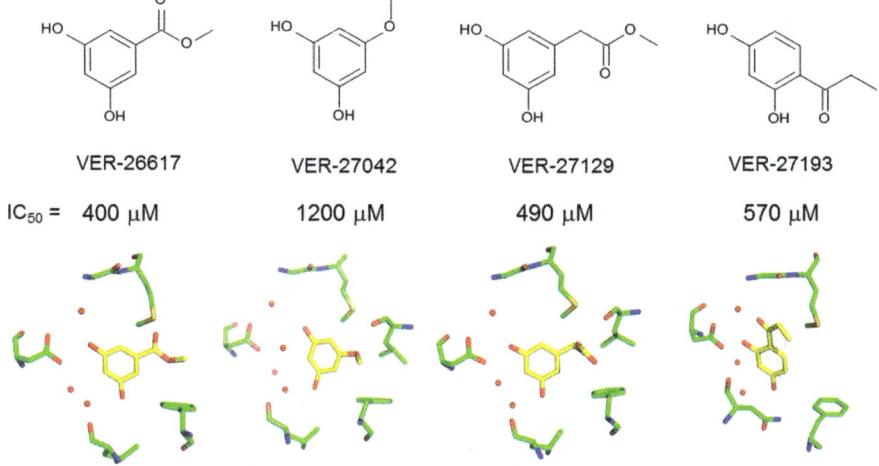

Figure 8.2 Hsp90 inhibitors that bind at the N-terminal ATP site of Hsp90.

Figure 8.3 Selected resorcinol fragments hits, affinity (IC_{50}) in ATPase assay and binding mode at ATP site determined by ligand-protein crystallography in Nt-Hsp90α (some residues removed for clarity).

traditional configuration of biological assays. The affinity becomes "on scale" sometimes only when the first round of synthetic chemistry evolves the fragments to larger compounds. The power of the technique resides in the ability of fragments to identify affinity "hot spots" that when combined with structural information, demonstrate potential for rapid potency enhancement. In the case of the resorcinol hits, however, we were able to generate affinity data from an ATP turnover assay. The compounds are of such weak affinity that they may not have been considered valuable hits (or not even identified at all) if they were the output of a traditional HTS screen. The key feature to acknowledge, however, is that though the affinity is weak, it is generated from a small number of atoms that can potentially contact the protein (directly or *via* water-mediated interactions) and thus may represent more efficient binding than other HTS hits that may be more potent but of much higher mass. This is the basis of the concept of "ligand efficiency" (LE), which enables ligands to be compared not on their potency alone, but also on the amount of free energy of binding contributed per heavy atom.[16–18]

These early data showed promise, as the fragment start points displayed a consistent binding mode (which recapitulated a key binding feature of radicicol), high ligand efficiency and, very importantly, had suitable "chemical handles" that could exploit vectors revealed by the X-ray structures. It is interesting to note that Astex independently used a fragment approach[19] to identify a series of resorcinol amides that delivered potent inhibitors to the clinic. Fragments from other chemical classes have been evolved to novel Hsp90 inhibitors at Vernalis[20,21] and elsewhere,[22] demonstrating the utility of this approach.

Another hit ID approach exploited in parallel at Ribotargets/Vernalis was that of *in silico* or "virtual" screening. In this method a prerequisite is structural information that enables a molecular docking technique to be used. Here we used the proprietary docking software rDOCK and a previously described docking protocol[22,23] to identify potential Hsp90 ligands. It was known that the shape and size of the ATP binding site can be altered upon binding of certain ligands as a consequence of a flexible loop between residues 104 and 111.[24] Therefore the structure of both the open and helical forms were used in the docking and in addition three highly conserved water molecules in the region of Asp93 were also treated as part of the receptor. A virtual library of 700,000 compounds was selected from a library of 3.5 million commercially available compounds.[25] The molecular docking gave an output of docking scores for each virtual ligand and this information and further filtering were used to identify 719 commercially available compounds, which were obtained and screened in an ATP turnover assay. We thus found a number of hits with weak to moderate affinity that in addition soaked into the ATP binding site of Hsp90 for X-ray crystallography, providing information for SBDD. Figure 8.4 shows a selection of validated hits with affinity in the 1 to 30 µM range in an ATPase turnover assay that was used during the hit identification phase.

A key screening hit was the resorcinol pyrazole compound highlighted in Figure 8.4, not only because it displayed the same binding mode with respect to resorcinol as found for our fragment hits but also because it had been

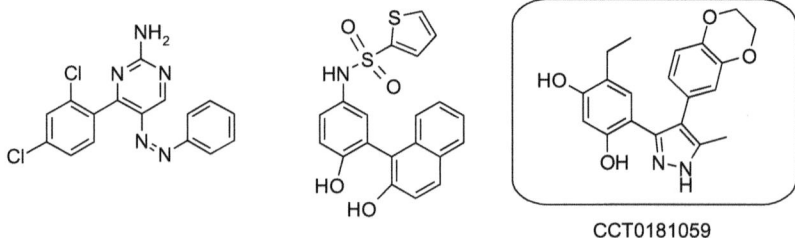

Figure 8.4 Selected validated hit compounds from *in silico* screening.

discovered by the ICR (CCT018159) as part of a medium-throughput screen of 65,000 compounds against Hsp90.[26] The ICR had subsequently initiated a lead optimization exercise in this area and shown that CCT0181159 inhibited the growth of HCT116 human colon tumor cells *in vitro*, up-regulated Hsp70 (a molecular signature of Hsp90 inhibition) and depleted Raf-1 levels (a known client protein of Hsp90). It was thus felt that CCT018159 represented a good start point for further work as part of a structure-based collaborative medicinal chemistry program.

After the identification of NVP-AUY922, another group published the independent identification of resorcinol-pyrazole scaffold inhibitors (G3130 and G3129, Figure 8.5), which were identified from a high-throughput TR-FRET-based assay against a library of 1 million compounds.[27]

A crucial part of NVP-AUY922 pharmacophore is the resorcinol (benzene-1,3-diol) moiety. Interestingly, this moiety is also constitutive of radicicol, a macrocyclic antifungal antibiotic (Figures 8.2 and 8.6). Radicicol was first described as a potent and specific inhibitor of p60v-src,[28] and it was only later shown[29] that this effect is due to the now well-established Hsp90-mediated destabilization of specific client proteins. Radicicol itself is a potent Hsp90

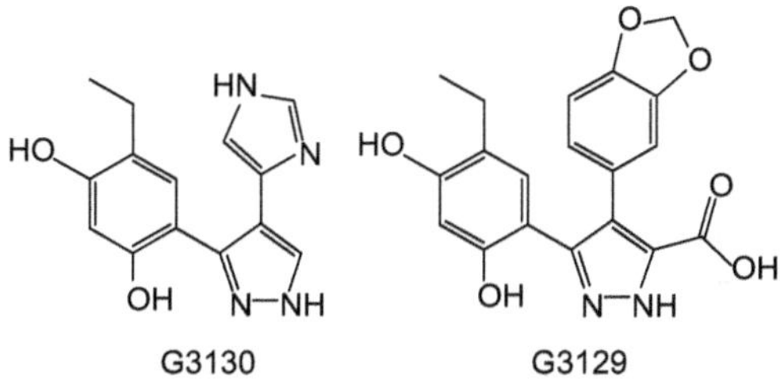

Figure 8.5 Structure of resorcinol-pyrazole inhibitors found by GNF.

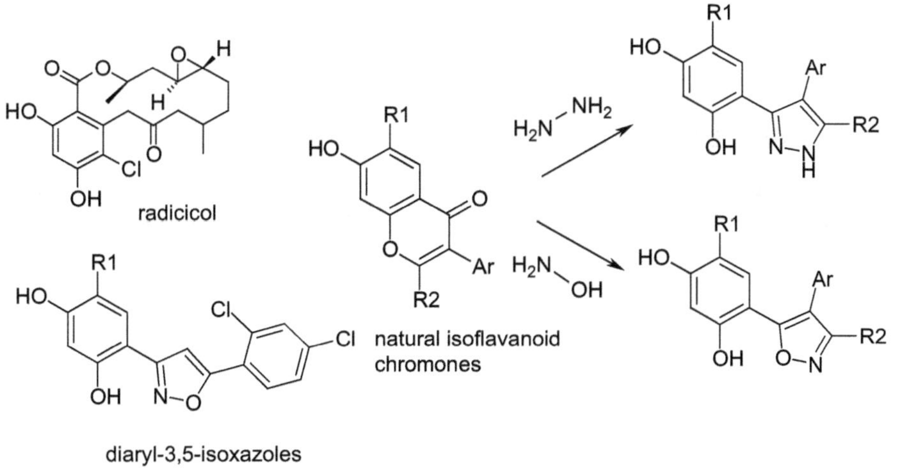

Figure 8.6 Historical resorcinol containing Hsp90 inhibitors.

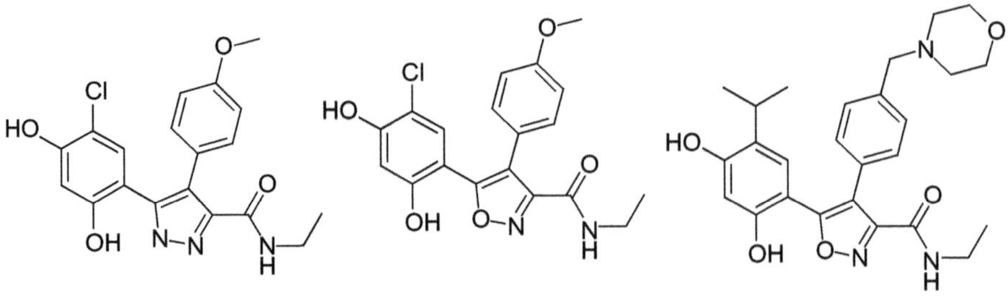

Figure 8.7 Structure-based design of resorcinol class of Hsp90 inhibitors.

inhibitor, with a K_d of 19 nM.[30] It also inhibits cell proliferation at double digit nM concentration; however, it did not exhibit *in vivo* antitumor activity.[31]

Also noteworthy is the bacteriostatic activity of aryl-3,5-isoxazole-resorcinol derivatives, which was reported several years ago.[32] Although these early compounds share part of the NVP-AUY922 pharmacophore, currently available data would not allow one to conclude that this anti-proliferative activity is due to Hsp90 inhibition.

Previously, the preparation of the diaryl-4,5-isoxazoles was achieved by reacting isoflavones with hydoxylamine (Figure 8.6).[33,34] No biological activity was reported for the compounds described in these publications. Reaction of the same flavones with hydrazine, however, afforded 3,4-diphenyl-1H-pyrazole compounds,[35] including CCT018159 (Figure 8.4), which was discovered by HTS and served as a basis for the structure-based design of the resorcinol class of Hsp90 inhibitors, including VER-49009[36] and VER-50589, which ultimately led to the identification of NVP-AUY922 (Figure 8.7).

The isoxazole has been replaced by different triazole,[37,38] pyrrole[39] and thiadiazole[40,41] ring systems and led to potent Hsp90 inhibitors. The structures of some key compounds from these classes of inhibitor are shown in Figure 8.8.

8.3 Lead Optimization

At the initiation of lead optimization phase of the program there were, however, several issues to address with the lead molecule CCT018159:

- Novelty: the compound was a commercially available screening compound
- Potency: low micromolar affinity in functional cell free assays (ATP turnover) and cell-based assays that looked at growth inhibition
- Selectivity against other targets (*e.g.* IC_{50} *vs.* GSK3β, $IC_{50} = 4\,\mu M$)
- Poor pharmacokinetic profile
- Low aqueous solubility
- P450 inhibition (CYP 3A4 inhibition, $IC_{50} < 1\,\mu M$)

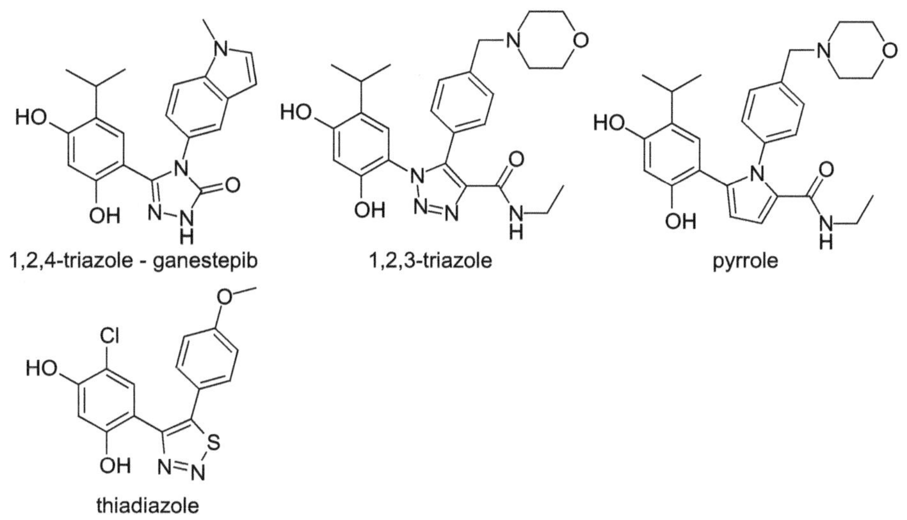

Discovery of NVP-AUY922

Figure 8.8 Structures of selected resorcinol-based Hsp90 inhibitors.

A known issue at this point in the project was the resilience of the ATPase assay, especially for more potent compounds. The identification of CCT018159 provided an opportunity to generate a chemical probe that could be used in a fluorescence polarization (FP) assay as a measure of competitive binding and enable better discrimination of affinity that was vital for SAR generation.

8.4 *In Vitro* Assays for Medicinal Chemistry Optimization

The *in vitro* biological testing cascade is critical to support medicinal chemistry efforts to improve target potency and cellular activity whilst maintaining drug-like properties and monitoring potential off-target liabilities. The initial biological screening cascade used in the early stages of the Hsp90 program is outlined in Figure 8.9A. Progression of the project and an increased understanding of the relationship between the *in vitro* and cellular assays allowed the biological screening cascade to be streamlined (Figure 8.9B) to reduce data delivery times.

In mammalian cells, Hsp90 exists as a dimer in a large, dynamic multi-protein complex with numerous co-chaperones. Purified, full-length human Hsp90 expressed in *E. coli* has little intrinsic ATPase activity. It has subsequently been shown that addition of purified co-chaperones, such as Aha I, can dramatically increase the ATP turnover rate of purified Hsp90. In the initial screen that led to the identification of CCT18159, yeast Hsp90 was utilized. This has much greater intrinsic ATPase activity than human Hsp90 and shares reasonable homology, especially within the active site, with the human enzyme. The assay quantifies the generation of orthophosphate from

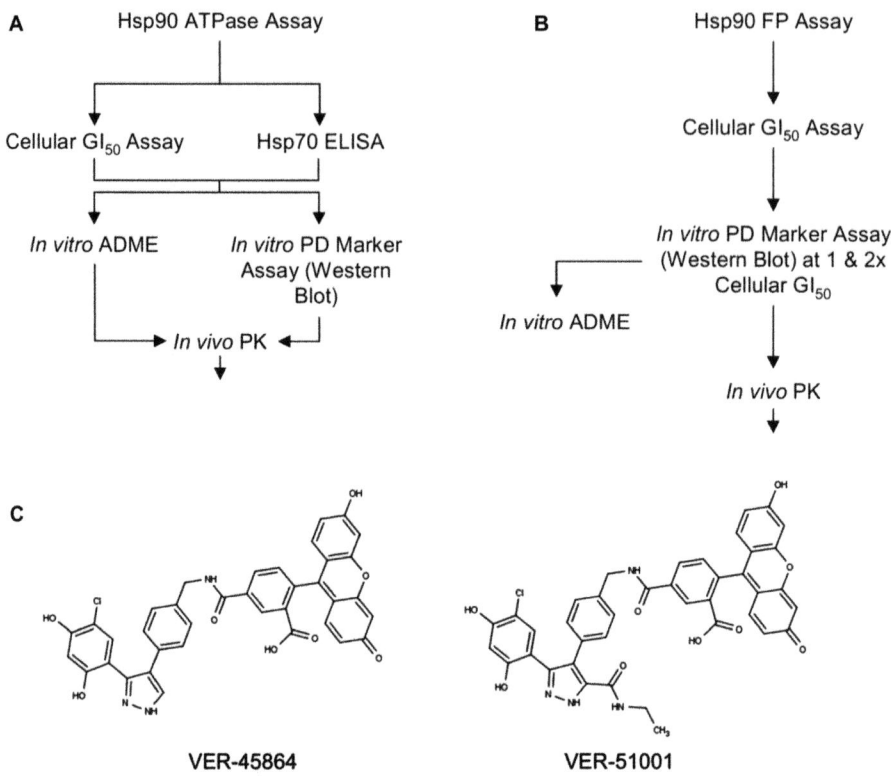

Figure 8.9 *In vitro* assay screening cascade. Outline of initial (A) and streamlined (B) *in vitro* screening cascades used to support the medicinal chemistry program. (C) Structure of the fluorescent polarization assay probes.

ATP through the formation of a green complex formed between malachite green, molybdate and free orthophosphate. Despite having greater enzymatic activity than the human enzyme, extensive optimization still resulted in an assay requiring a yeast Hsp90 concentration of 0.42 µM and overnight incubation at 37 °C. The assay was extremely sensitive to orthophosphate contamination but did routinely deliver Z's > 0.7. However, even with the extensive optimization this assay could only reliably determine the potency of compounds with an $IC_{50} > 0.2$ µM.

The lower limit of detection of the yeast Hsp90 ATPase assay was rapidly exceeded and a new *in vitro* assay was developed to support the medicinal chemistry efforts. Using the crystal structure of the initial hit (CCT18159) bound to the N-terminal ATP binding site of human Hsp90α, we developed a fluorescently labeled probe.[42] This probe allowed the subsequent development of a fluorescence polarization (FP) assay to monitor compound binding affinity to the ATP-binding site of human Hsp90. The initial probe (VER-45864), based on CCT18159 (Figure 8.9C), bound to human full-length Hsp90β with a K_d of 80 nM and a Z' > 0.9. Importantly, the incubation time of the assay was reduced from 18 hours for the ATPase assay to 60 minutes for the FP assay. The lower limit of this initial FP assay ($IC_{50} \sim 40$ nM) was soon exceeded by the resorcinol-based series of Hsp90 inhibitors. A new FP probe (VER-51001) was designed based on the potent 5-amide substituted compound VER-49009. This bound Hsp90 with a much higher affinity ($K_d = 6$ nM) than VER-45864. This increased affinity enabled a lower concentration of Hsp90β to be used in the assay allowing the determination of IC_{50}s down to around 10 nM. Selectivity was determined by cross-screening against the closely related Hsp90 family members Grp94 and TRAP-1 using an FP assay with VER-51001 again used as the fluorescent probe, another GHKL ATPase Topoisomerase-II, an unrelated, structurally distinct ATPase Hsp70 and a panel of protein kinases.

Surface plasmon resonance (SPR) is a powerful biophysical technique that allows the determination of target affinity (K_d) and the measurement of kinetics in an open system. Using this technique, the off rate (*i.e.* how tightly a compound binds to the target of interest) can be determined and may be a useful predictor of duration of activity in an *in vivo* setting. We have subsequently used this technique to retrospectively rationalize differences between our pyrzaole and isoxazole classes of Hsp90 inhibitors. Additionally, SPR has the potential to be an extremely useful tool for investigating selectivity between Hsp90α/β, Grp94 and TRAP-1 as it measures the direct interaction between compound and target.

8.5 *In Vitro* Biomarker Discovery/Development

The pharmacodynamic biomarker signature of Hsp90 inhibition comprises the concentration and time-dependent induction of Hsp70 coupled with the depletion of client proteins. Inhibition of Hsp90 results in the release of transcription factor heat-shock factor 1 (Hsf1) and the subsequent induction of Hsp70. An ELISA assay using cell lysates from treated mutant KRas HCT116 colon carcinoma cells coupled with a proliferation assay in HCT116 cells was used to drive the early medicinal chemistry program. The mechanism of action was subsequently confirmed using western blotting to monitor changes in client proteins and their subsequent downstream signaling pathways. Client proteins were selected according to the likely oncogenic drivers in a given cell line (for example, Raf-1 in KRas colon cancer, HER2 in HER2+ breast cancer, B-Raf in mutant B-Raf melanoma, c-Kit in GIST, EGFR in mutant EGFR driven lung cancer *etc.*). As the medicinal chemistry program progressed and an increasing number of compounds were screened in the cell-based assays, a close correlation between the various cell-based assays was established for active Hsp90 inhibitors that allowed the cellular screening cascade to be streamlined (Figure 8.9B). We determined that compounds that did not inhibit HCT116 cell proliferation were cell inactive. Subsequently, those that did not induce Hsp70 and down-regulate Raf-1 in HCT116 cells when treated at concentrations equivalent to 1- or 2-fold the cellular GI_{50} were inhibiting cellular proliferation *via* an off-target mechanism.

The classical method for following the cellular activity of Hsp90 inhibitors, as described above, is to follow the proteasome-dependent degradation of Hsp90 client proteins and the subsequent loss of pathway signaling. This, however, does not directly demonstrate that the catalytic activity of Hsp90 has been blocked. Effects on the Hsp90 catalytic cycle can be measured and followed more closely by evaluating the disruption of the complex between Hsp90 and p23. The binding of p23 to Hsp90 requires Hsp90 to be bound to ATP as p23 specifically recognizes Hsp90 complexed with ATP and not Hsp90 alone. An immunoprecipitation assay using an anti-Hsp90α antibody to precipitate Hsp90 complexes followed by western blotting to determine the presence of p23 in those complexes was developed. Hsp90 inhibitors induced Hsp90α and p23 dissociation in a concentration- and time-dependent fashion in the HER2-positive breast carcinoma cell line BT474. Dissociation between Hsp90α and p23 occurred rapidly with complete dissociation observed within 15 minutes. The dissociation of p23 from Hsp90α correlated closely with client protein (HER2) degradation and loss of HER2 signaling in this cell model.

We rapidly established the binding mode of CCT018159, which revealed key interactions to Hsp90 including water-mediated interactions (Figure 8.10). Closer inspection revealed a number of features that could be modified for potency and/or property enhancement without disrupting the critical binding interactions to the Asp93 residue and network waters.

Close inspection of the crystal structure indicated that the 5-methyl pyrazole substituent was located in a solvent channel but that it was close enough to the

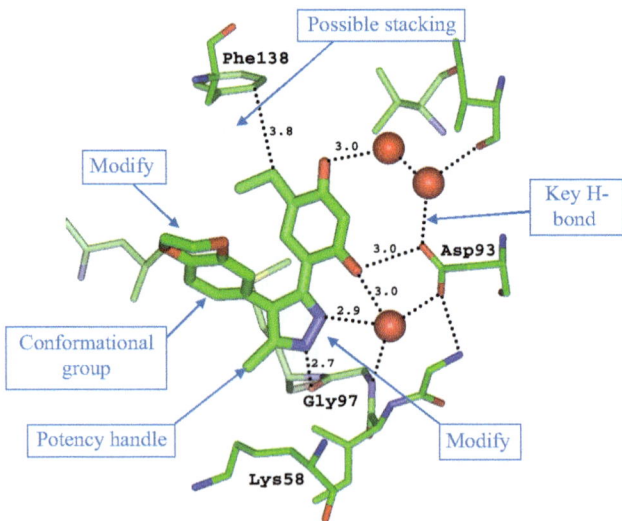

Figure 8.10 Binding mode of CCT0181059 bound to ATP site of Nt-Hsp90α [PDB code: 2BT0]. Structure shows key water-mediated and direct hydrogen bond interactions to Asp93 and possible sites for chemical modification. (Adapted with permission from *J. Med. Chem.*, 2005, **48**, 4212–4215. © 2005 American Chemical Society.)

protein for several potential interactions with an appropriate substituent. One of these points of potential interaction is the carbonyl oxygen of Gly97, which is located less than 4 Å away from methyl carbon. The carbonyl oxygen is 2.7 Å from the pyrazole N1 but out of plane of the peptide bond so allowing an electron pair to be available to act as an H-bond acceptor. A small solvent exposed hydrophobic patch is also in the vicinity of the 5-Me group. We selected amides as a moiety that could meet the H-bond donating requirement. Table 8.1 shows data for a small set of amide analogues of CCT0181059 (3, Table 8.1) along with corresponding data for geldanamycin and radicicol.[36]

The FP IC$_{50}$ data displayed in Table 8.1 showed significant enhancement of activity for the new 5-amide compounds compared to the 5-methyl parent compounds. This enhancement of affinity was coupled with an improvement in the compound ability to inhibit cancer cell growth. In HCT116 cells, one compound **11** (VER-49009) displayed a GI$_{50}$ value close to that for 17-AAG (0.26 µM for **11**, vs. 0.16 µM for 17-AAG). The enhancement in binding potency may be rationalized by considering the X-ray crystal structure of **11** bound to Hsp90 (Figure 8.11). Compound **11** retains the same key interaction

Table 8.1 Fluorescence polarization (FP), ATPase and growth inhibition data for resorcinol-pyrazole Hsp90 inhibitors.

Cpd No	R^1	R^2	R^3	R^4	FP IC$_{50}$ (μM)[b]	ATPase IC$_{50}$ (μM)[a,b]	GI$_{50}$ (μM)[c]
17-AAG	–	–		–	1.27	13.4	0.16
Radicicol	–	–		–	0.15	0.20	0.13
1	Et	–O(CH$_2$)$_2$O–		Me	0.28	5.70	5.8
2	H	–O(CH$_2$)$_2$O–		Me	4.73	23.0	34.8
3	Cl	–O(CH$_2$)$_2$O–		Me	0.21	0.65	14.9
4	H	OMe	H	Me	4.40	36.9	33.3
5	Br	OMe	H	Me	0.062	1.90	6.1
6	Cl	OMe	H	Me	0.150	1.7	6.7
7	Cl	OMe	H	COOH	0.51	2.3	>80
8	Cl	OMe	H	CONH$_2$	0.053	0.42	1.1
9	Cl	OMe	H	CONHMe	0.039	0.11	0.68
10	Br	OMe	H	CONHEt	0.070	0.390	0.938
11	Cl	OMe	H	CONHEt	0.025	0.14	0.26
12	Cl	OMe	H	CONH^{i}Pr	0.12	1.5	1.9

[a] ATPase activity determined with full-length yeast Hsp90 protein.
[b] GI$_{50}$, FP and ATPase IC$_{50}$ values determined are averages of at least n = 2.
[c] GI$_{50}$ data for HCT116 human colon cell line.

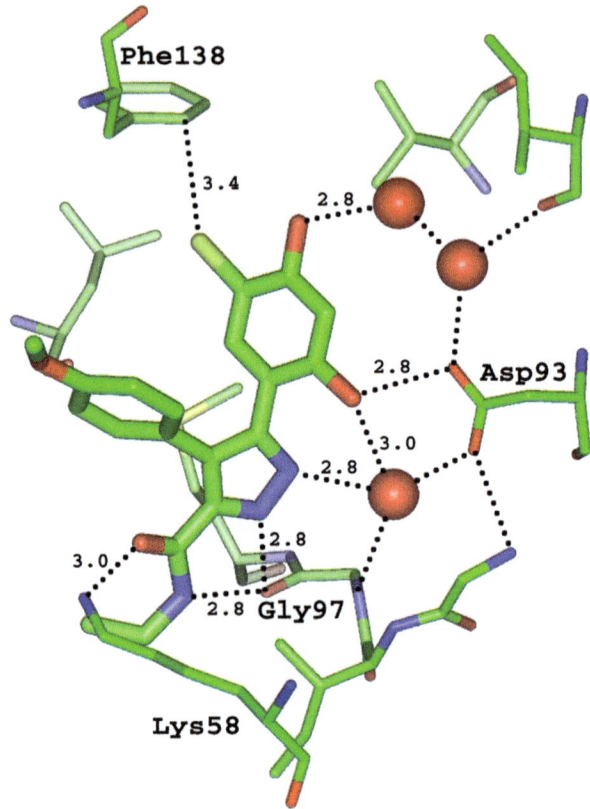

Figure 8.11 Binding mode of VER-49009 (11) bound to ATP site of Nt-Hsp90α [PDB code: 2BSM]. The structure shows key water-mediated and direct hydrogen bond interactions to Asp93 and hydrogen bonds from the ligand to residues Gly97 and Lys58.
(Reprinted with permission from *J. Med. Chem.*, 2005, **48**, 4212–4215. © 2005 American Chemical Society.)

as the water-mediated H-bonds to Asp93 that are seen with CCT0181059; however, as predicted by modeling studies, the amide compound **11** makes an H-bond with Gly97. In addition, there is a second H-bond to Lys58 from the amide; however, this residue side-chain is solvent exposed and unconstrained so it is probable that solvation and entropic penalties may counteract any benefit of this interaction to affinity enhancement. It is most probable that the H-bond from amide to Gly97 explains most of the affinity enhancement and indeed this is supported by structural data from analogous amides that did not make the Lys58 interaction. Compound **11** (VER-49009) displayed the molecular signature of Hsp90 inhibition, down-regulating the client protein Raf-1 and up-regulating the co-chaperone Hsp70.[36]

It was clear from the crystal structures of CCT0181059 and VER-49009 that the resorcinol hydroxyl groups and the nitrogen of the atoms of the pyrazole ring were deeply embedded in the binding pocket and formed a tight network of hydrogen bonds. It was deemed likely that only small modifications could be made to this core. However, the 4-aryl group was orientated towards solvent in a fairly open part of the binding site. This meant there was a possible opportunity to moderate physiochemical properties of the inhibitor series, despite minimal opportunity for large gains in potency. Both the *para* and *meta* positions of the 4-aryl ring appeared to be suitable positions for the attachment of solubility modifying groups and a number of analogues were made that addressed SAR at both positions.[43] Table 8.2 shows a representative selection of data from this exercise. Whilst several compounds (**13e**, **13f**) of the *para*-substituted series displayed affinity and anti-proliferative effect similar to VER-49009, most compounds were less potent. All of the *meta*-substituted examples were lower affinity than VER-49009, suggesting these changes were less well tolerated. Though the new set of compounds did not demonstrate potency enhancement over VER-49009, there was an improvement in compound solubility and metabolic stability as suggested by microsomal incubation assays in several species. Both these features were clearly desirable improvements towards the goal of achieving *in vivo* activity for the Hsp90 inhibitor program.

In parallel studies, several subseries were examined that investigated replacements of the 4-aryl moiety[44,45] (Figure 8.12). Whilst these series delivered potent compounds and structural insights, they did not match the profile of the 4-aryl compounds being developed concurrently, and more focus was placed on the more promising series.

From the structural information available for the pyrazole-resorcinol inhibitors, it was clear that the polar interaction in the deep pocket were well matched for protein and ligand and that little steric change would be tolerated in this region. However, the nitrogen of the pyrazole adjacent to resorcinol (N-1) appeared to be acting as a hydrogen bond acceptor and it was felt that an oxygen atom might make a suitable replacement. Pyrazoles may exist in tautomeric forms and replacing N-1 with an oxygen atom would lock the N-2 form and be preferable if the H-bond acceptance was the more important of the two H-bonding interactions made by the pyrazole or isoxazole with the protein. The first analogue made in this series was the isoxazole version of VER-49009, compound **15** (VER-50589; Figure 8.13).[43] Affinity data from FP assay showed that the pyrazole to isoxazole change was tolerated with essentially equipotent activity (though subsequent studies by ITC showed that the isoxazole had a *ca.* 17-fold lower K_d as measured by isothermal titration calorimetry.[46] Not surprisingly X-ray crystallography confirmed identical binding mode for the two compounds. A further advantage to the isoxazole VER-50589 was revealed when the activity in growth inhibition assays was compared for the two analogues.[43] Table 8.3 shows GI_{50} values in growth inhibition assays across a range of human cancer cell lines.

Calculated and measured physiochemical properties for the two compounds are, not surprisingly, very similar and it seemed unlikely that significantly improved growth inhibition for VER-50589 could be solely down to differences in

Table 8.2 Binding (FP assay) and cell growth inhibition (SRB assay) data for amino-methyl functionalized diaryl pyrazoles.[a]

Cmpd number	NR1R2	FP IC$_{50}$r[a] (μM)	GI$_{50}$[a] (HCT116) (μM)
13a	N-ethyl, N-propyl	0.146	0.315
13b	3-(dimethylcarbamoyl)piperidine amide	0.115	3.47
13c	methylsulfonylethylamino	0.035	2.162
13d	thiomorpholine dioxide	0.027	0.826
13e	morpholine	0.037	0.290
13f	prolinamide	0.057	0.275
13g	bis(2-hydroxyethyl)amino	0.142	7.80
14a	morpholine	0.222	2.69
14b	2-(piperidin-1-yl)ethylamino	1.00	25.14
14c	2-(trifluoromethyl)benzylamino	0.231	3.11
14d	4-(hydroxymethyl)piperidine	2.63	>80

Table 8.2 (*Continued*)

Cmpd number	NR1R2	FP IC$_{50}$r^a (μM)	GI$_{50}$^{a} (HCT116) (μM)
14e	—NH—CH$_2$—O—CH$_3$	0.728	12.93
14f	N-methylpiperazine acetyl	1.29	32.7
14g	—NH—CH$_2$CH$_2$-morpholine	0.431	12.54
14h	—N(CH$_3$)—CH$_2$CH$_2$OH	0.914	9.24
14i	(S)-2-(hydroxymethyl)pyrrolidinyl	1.66	11.37

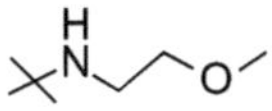

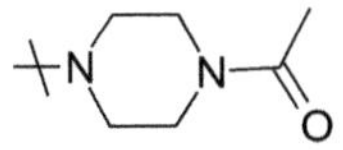

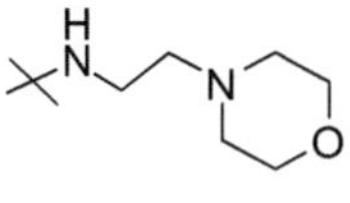

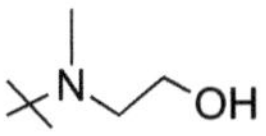

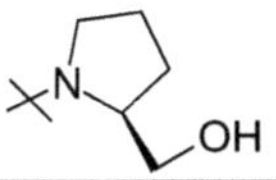

^{a}Data are the means of at least two independent determinations.

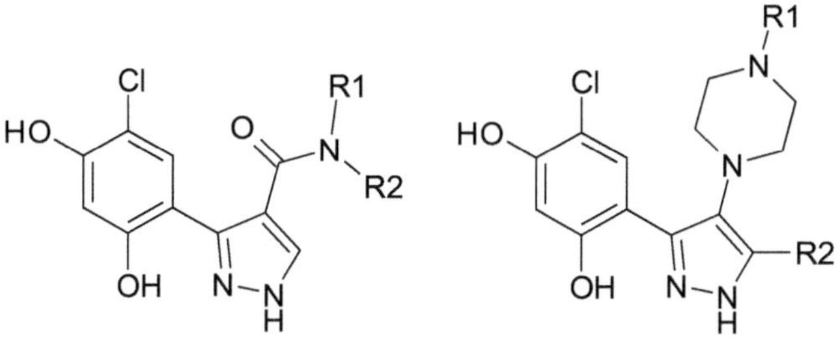

Figure 8.12 Non-aryl 4-substituted pyrazole-resorcinols.

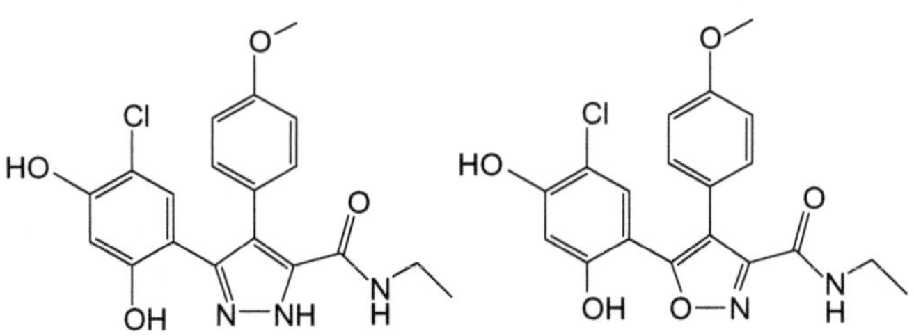

Figure 8.13 Resorcinol-pyrazole Hsp90 inhibitor 11 (VER-49009) and resorcinol-isoxazole Hsp90 inhibitor 15 (VER-50589).

cell uptake and local drug concentrations. Subsequent kinetic studies (Figure 8.14) showed that the two compounds had different binding kinetics; the isoxazole displayed longer residence time (slower off-rate) bound to the

Table 8.3 Binding (FP assay) and cell growth inhibition (SRB assay) data in various human cancer cell lines for compounds **11** and **15**.[a]

Cmpd Number	Human Hsp90β FP IC_{50} (μM)	HCT116 (colon) GI_{50} (μM)	DU145 (prostate) GI_{50} (μM)	PC3M (prostate) GI_{50} (μM)	SKMel28 (melanoma) GI_{50} (μM)	BT20 (breast) GI_{50} (μM)	U87MG (glioma) GI_{50} (μM)
11	0.025	0.36	1.2	2.2	0.48	0.55	1.2
15	0.028	0.12	0.23	0.22	0.045	0.059	0.028

[a]Values are reported as the means of at least two independent determinations.

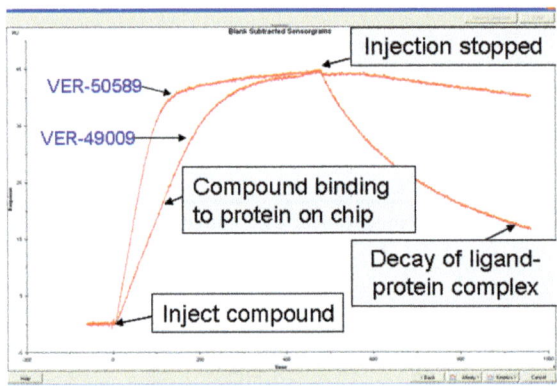

Figure 8.14 Overlay of sensograms from surface plasmon resonance experiments for VER-49009 and VER-50589 binding to immobilized Hsp90 (His tagged).

Hsp90 protein *in vitro*, which may provide an additional rationale for the higher potency observed in cellular assays. The next logical step was to expand the isoxazole series to incorporate the solubilizing functionally as had been previously explored in the pyrazole series (Table 8.2). Table 8.4 shows representative data from this exercise and by comparison with compounds from Table 8.2 shows the improved *in vitro* cell growth inhibition for isoxazoles translated to other members of the series (*e.g.* **13a** *vs.* **16f** and **13e** *vs.* **16d**) with similar affinity data. The observed preference for solubilizing group attachment at the *para-* over the *meta*-position of the 4-aryl ring that was seen for the pyrazole series held for the analogous isoxazoles as well, though the *meta* isoxazoles appeared more potent than the corresponding pyrazoles.

Previously it had been seen in a series closely related to VER-49009 that removal of a chloro or ethyl at the 5′ position of the resorcinol ring (to 5′-H) led to a 20-fold decrease in affinity. The potency loss can be rationalized by loss of interaction with the edge of a lipophilic pocket. This is the region of the protein that is flexible and able to switch to an open confirmation (the "PU3 loop"). We decided to explore this region more comprehensively in the resorcinol-isoxazole series and were able to establish that compounds with large lipophilic substituents incorporated at the 5′-position retained binding affinity similar to smaller alkyl substituents such as chloro, ethyl and isopropyl (though this

Table 8.4 Binding (FP assay) and cell growth inhibition (SRB assay) data for amino-methyl functionalized diaryl isoxazoles.[a]

Cmpd Number	NR1R2	FP IC$_{50}$[a] (μM)[a]	GI$_{50}$[a] (HCT116) (μM)[a]
16a	thiomorpholine dioxide	0.021	0.47
16b	NH-CH2CH2-SO2Me	0.014	1.98
16c	N-methylpiperazine	0.039	0.125
16d	morpholine	0.021	0.083
16f	pyrrolidine	0.064	0.056
16g	piperidine	0.019	0.061
16i	4-hydroxypiperidine	0.028	1.288
16j	pyrrolidine	0.039	0.118
16k	NMe2	0.018	0.069
17a	morpholine	0.127	0.634
17b	N-methylpiperazine	0.343	2.78
17c	N-acetylpiperazine	0.239	6.76

[a]Values are reported as the means of at least two independent determinations.

equipotency translated to a loss of ligand efficiency as previously discussed). Table 8.5 shows that a wide range of substituents were well tolerated with respect to affinity and displayed excellent anti-proliferative effects in HCT116

Table 8.5 Effects of modifications to the 5' position of the resorcinol on binding (FP assay) and cell growth inhibition (HCT116 cells; SRB assay).[a]

Cmpd number	X	Amine moiety	FP IC$_{50}$[a] (μM)	GI$_{50}$[a] (HCT116) (μM)
18a	–CH$_2$CH$_2$Ph	Morpholino	0.032	0.029
18b	Et	Morpholino	0.029	0.025
20b	Et	NEt$_2$	0.013	0.041
18c	–CH$_2$CH$_2$(4-FPh)	Morpholino	0.020	0.039
19c	–CH$_2$CH$_2$(4-FPh)	Piperidinyl	0.026	0.029
18d	–CH$_2$CH$_2$(3-FPh)	Morpholino	0.026	0.058
18e	Ph	Morpholino	0.022	0.152
19e	Ph	Piperidinyl	0.014	0.086
18f	2-FPh	Morpholino	0.039	0.087
19g	2-MePh	Piperidinyl	0.014	0.114
19h	4-FPh	Piperidinyl	0.036	0.035
18i	iso-Butyl	Morpholino	0.035	0.133
18j	tert-Butyl	Morpholino	0.008	0.070
19j	tert-Butyl	Piperidinyl	0.011	0.065
18k	iso-Propyl	Morpholino	0.021	0.016
20k	iso-Propyl	NEt$_2$	0.006	0.006
19k	iso-Propyl	Piperidinyl	0.006	0.040
21e	Ph	NHEt	0.074	0.091

[a]Values are reported as the mean of at least two independent determinations.

cells. The 5'-substituent was varied in parallel with a number of para-4-aryl solubilizing moieties previously shown to be favorable in cell-based assays.

At this stage of the project a large number of potent compounds had been identified in the isoxazole–pyrazole series; the next goal was to select compounds for *in vivo* efficacy testing. It had already been established by pharmacokinetic studies on resorcinol pyrazoles[46] that the resorcinol moiety was rapidly glucuronidated *in vivo* and that the traditional *in vitro* microsomal stability assays were not translational to an *in vivo* context, correlating poorly with plasma clearance. It was decided that the best way forward to was to make the selection based on *in vivo* data from tumor-bearing animals that would allow for the PK for the target tissue to be measured and contrasted with plasma PK. Cassette dosing[47] PK studies offered an attractive method for

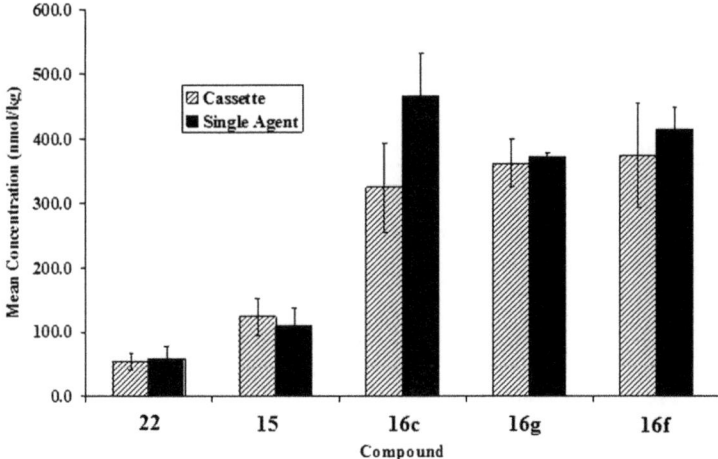

Figure 8.15 Cassette-dosing method validation, showing compound tumor concentration in mice bearing HCT116 subcutaneous human colon xenografts dosed at 4 mg per kg i.p. Compounds were dosed as single agents or cassette dosed with mixtures of five compounds.
(Adapted with permission from *J. Med. Chem.*, 2008, **51**, 196–218. © 2008 American Chemical Society.)

acquiring these data, especially considering we wished to assay a relatively large number of compounds. Prior to initiating this study though, it was important to validate the method by comparing data for five project compounds dosed singly or in combination (cassette). This study was conducted in HCT116 tumor-bearing mice by dosing i.p. and measuring tumor concentrations at a 6-hour time point. The data from the validation study are shown in Figure 8.15, and show the tumor concentrations to be similar in each experiment suggesting compound–compound interactions were not significant and thus the cassette-dosing method to be viable for further studies.

Table 8.6 shows data from three cassette PK experiments where potent resorcinol isoxazoles were dosed 4 mg per kg i.p. and tumor and plasma concentration measured at 6 hours.[43] The ratio of tumor to plasma concentrations is listed along with the maximum fold over the specific compound GI_{50} in HCT116 cells *in vitro*. Compound **15** was used as a PK standard and though it showed some variation in plasma levels it gave consistent results for tumors.

Data in Table 8.6 showed there was a big variation in tumor to plasma ratio (0.3 to 20) and that generally aliphatic 5′-resorcinols had higher tumor exposure than aromatic substituents. Chloro-substituted compounds also achieved high tumor concentrations though the isopropyl-substituted compound **18k** clearly demonstrated the highest ratio of tumor concentration to cellular GI_{50} and half-life of 9.5 hours in tumor. The X-ray crystal structure of **18k** (Figure 8.16) showed the expected binding mode when compared with other resorcinol compounds; the iso-propyl group at resorcinol 5′ position

Table 8.6 Pharmacokinetic data obtained by cassette dosing with isoxazole compounds (HCT116 tumors).

Cassette	Cmpd	AUClast tumor (hr*nmol/l)	AUClast plasma (hr*nmol/l)	Tumor:plasma ratio	Max fold above GI_{50}
Cassette 1	15	1332.8	1222.8	1.1	6.5
	16c	2522.9	1542.3	1.6	4.7
	22	597.5	1191.8	0.5	1.5
	16f	2930.4	925.8	3.2	12.0
	16g	2403.5	989.0	2.4	20.0
Cassette 2	15	1437.6	814.9	1.8	6.0
	18a	655.9	2547.2	0.3	4.2
	18b	1750.9	618.8	2.8	24.0
	19c	668.2	1327.4	0.5	5.6
	19m	685.5	1630.4	0.4	1.9
Cassette 3	15	1733.0	659.5	2.6	7.0
	21e	1248.0	3708.1	0.3	2.8
	18k	1720.0	754.2	2.3	35.0
	18j	751.1	792.2	0.9	2.9
	19c	994.8	966.7	1.0	2.8

makes a hydrophobic interaction with Lue107 and additionally the morpholine moiety makes interactions with Thr109 and Gly135.

Compound **18k** was potent in a range of human cancer cell line (selection Table 8.7) and subsequently was shown to have average GI_{50} of 9 ($\pm$9) nM in a panel of 45 cell lines with no clear relationship to mutational or client protein expression status.

On the basis of the available data compound **18k** was selected for efficacy studies in mouse models of cancer (see below). In summary, the medicinal chemistry program used a number of hit identification strategies and a lead optimization process that was driven by X-ray crystallographic structural information of compounds bound to Hsp90. The program was able to deliver a pre-clinical candidate rapidly on this basis. The compound evolution is summarized in Figure 8.17.

8.6 *In Vivo* Characterization

The oral bioavailability of NVP-AUY922 is very low in rodents despite a high intrinsic absorption and a low efflux ratio determined in an *in vitro* Caco-2 cell-based absorption assay. This is likely due to a high first pass metabolism and rapid clearance of the compound. Liver microsomes incubated with NVP-AUY922 showed comparable metabolic reactions in the rat and human, but a faster formation of metabolites in rats compared to human. Since oral bioavailability is low the clinical route of administration is i.v. However, the free base of NVP-AUY922 has very low solubility in water at physiological pH, which made it challenging to formulate for intravenous administration in rodents. To improve the solubility, the mesylate salt of NVP-AUY922 was

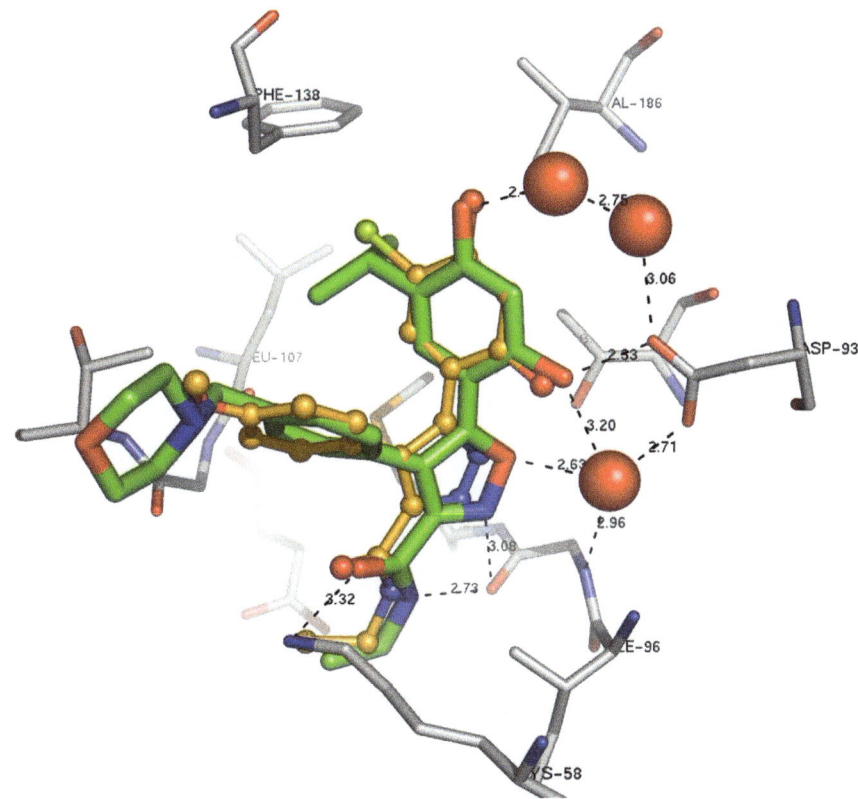

Figure 8.16 Overlay of X-ray structure of **18k** and **11** bound to the ATP binding site of human Hsp90α. Red spheres represent structurally conserved water molecules [PDB Code for **18k**: 2VCI].
(Reprinted with permission from *J. Med. Chem.*, 2008, **51**, 196–218. © 2008 American Chemical Society.)

Table 8.7 Binding (FP assay) and cell growth inhibition (SRB assay) in various human cancer cell lines for compound **18k**.[a]

Cmpd number	Human Hsp90β FP IC$_{50}$ (μM)	HCT116 (colon) GI$_{50}$ (μM)	DU145 (prostate) GI$_{50}$ (μM)	PC3M (prostate) GI$_{50}$ (μM)	SKMel28 (melanoma) GI$_{50}$ (μM)	SF268 (glioma) GI$_{50}$ (μM)	U87MG (glioma) GI$_{50}$ (μM)
18k	0.021	0.016	0.005	0.006	0.005	0.006	0.008

[a] Values are reported as the means of at least two independent determinations.

selected for further development since it has a relatively high solubility in water (>20 mg/ml). The mesylate salt of NVP-AUY922 has been used in all current clinical trials.

To expand on the initial pharmacokinetics cassette-dosing studies (Table 8.6), more extensive pharmacokinetics analyses were performed and

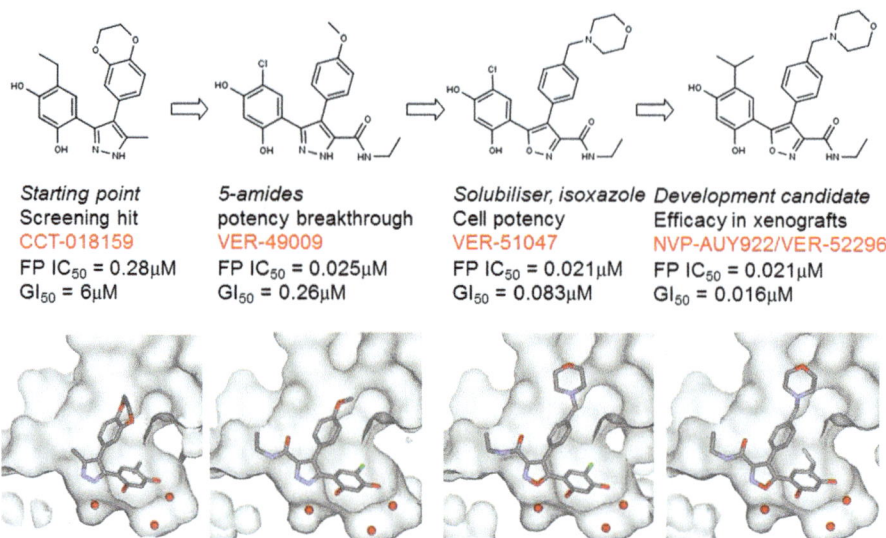

Figure 8.17 Summary of compound optimization from CCT-018159 to NVP-AUY922.

combined with evaluation of the corresponding pharmacodynamic readouts. During the preclinical development phase at Novartis the majority of the antitumor studies were performed in the human ductal breast carcinoma xenograft mouse model BT-474.[48] BT-474 xenografts are driven by ErbB2 signaling, depend on estrogen for proliferation and are highly sensitive to Hsp90 inhibitors *in vivo* and *in vitro*. Interestingly, in the BT-474 model we observed that NVP-AUY922 was cleared slowly from the xenografts compared to organs such as liver, heart, lung and muscle (Figure 8.18). The pharmacokinetic profile of NVP-AUY922 administered i.v. at 30 mg/kg is characterized by a biphasic decline in plasma, with a first and last concentration of 14.08 µmol/l at 0.083 h and 0.010 µmol/l at 24 h, respectively. The apparent terminal elimination half-life for NVP-AUY922 in plasma is 10 hours using the last three time points. The dose normalized plasma exposure following intravenous dosing with 20 mg/kg resulted in a similar value as with 30 mg/kg, indicating dose linearity within this range. The systemic exposure in tissues as presented by the AUC was 5-fold (muscle, heart), 7-fold (liver) and 10-fold (lung) greater than in plasma and the terminal elimination half-life from normal tissues was in the range of 5.5–8 hours. However, the tumor exposure was approximately 47-fold higher than that in plasma with a concentration of 2.22 nmol/g at 48 hours post-dose. This level corresponded approximately to 13% of the amount of C_{max} at 0.25 h (16.3 nmol/g) and resulted in a protracted apparent terminal half-life of 25–30 hours. In summary, the PK data indicated a rapid distribution of NVP-AUY922 from plasma into tissues, corroborated by a large volume of distribution. There was a preferential uptake into tumor tissue observed with a long-lasting exposure and a protracted terminal

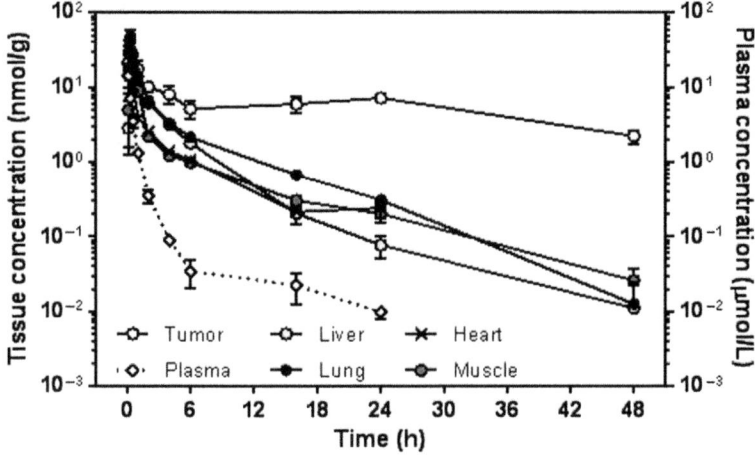

Figure 8.18 Pharmacokinetic profile of NVP-AUY922 after administration of a single dose. Female athymic mice bearing subcutaneous xenotransplants of the human ductal breast carcinoma BT-474 received a single dose (30 mg/kg; n = 4). The highest concentration (C_{max}), terminal half-life ($t_{1/2}$ elimination λ_z) and exposure as determined by the area under the curve (AUC) were determined for each organ, plasma and BT-474 xenograft (tumor).
(Reprinted with permission from *Breast Canc. Res.*, 2008, **10**, R33. © BMC.)

elimination. Similar findings were described using a WM266.4 melanoma xenograft model.[49] In the ErbB2 positive BT-474 model, we observed decreased ErbB2 levels in the tumor for about two days and increased Hsp70 levels for at least three days. This information was helpful in determining optimum dosing schedule for AUY922.

During the drug-discovery selection process for the optimal development candidate, NVP-AUY922 was administered intraperitonealy (i.p.) once per day to colon (HCT116), breast (BT-474), ovarian (A2780), glioblastoma (U87MG), prostate (PC3) and melanoma (WM266.4) tumor xenograft-bearing mice.[20,49] Apart from demonstrating the antitumor potential of the compound, these studies demonstrated that NVP-AUY922 was well tolerated as no significant body weight changes were observed when administered i.p. daily at dose levels up to 50 mg/kg.[20,49] Statistically significant antitumor effects, but not tumor regression, were observed at tolerated dose levels.

We wanted to elaborate further on the initial findings by investigating whether the effect observed on the PD markers correlated with an antitumor effect at well-tolerated doses and regimen in the BT-474 tumor model.[48] During establishment of the BT-474 model we noticed that untreated, tumor and estrogen pellet bearing mice generally either lost or gained minimum body weight during the experiment. The body weight loss is likely to make the animals more sensitive to the potential additional toxicity caused by the treatment. As such, this xenograft model is ideally suited to optimize the dosing regimen, which gives a good efficacy/tolerability balance. In order to simulate more closely the administration route to be used in clinical trials the administration schedule was optimized using intravenous administration of NVP-AUY922. In these studies, schedules ranging from daily to weekly schedules were evaluated. During these studies, we discovered that extending the recovery period in between dosing tended to improve tolerability without significantly affecting antitumor effect.[48] This suggested that achieving a high C_{max} and allowing a recovery period in between dosing was key to achieving a therapeutic window. Overall, we demonstrated that NVP-AUY922 has good pharmacokinetic properties, is efficacious and is well tolerated when administered as a single agent once per week in mouse xenograft models.

8.7 Clinical Development

NVP-AUY922 entered Phase I clinical trials in 2007 where it is being administered by intravenous infusion in a once-weekly schedule. Data from Phase I trials have not yet provided any objective responses according to RECIST criteria.

The Phase I data revealed important information about the compound, which built confidence in moving it into global Phase II trials. In a study where pharmacokinetics and pharmacodynamic relationship was evaluated in 36 patients, a dose-dependent induction of Hsp70 was observed in patient-derived PBMCs and the up-regulation observed at $40\,mg/m^2$ exceeded that observed in BT474 tumor-bearing mice.[48,50] Doses from 2 to $80\,mg/m^2$ have been evaluated in a number of Phase I trials and the recommended Phase II dose was set at $70\,mg/m^2$.[50–59] Weekly administration of $70\,mg/m^2$ is generally well tolerated. Some of the most frequently reported adverse events are diarrhea, night blindness, nausea, fatigue rash, decreased appetite and vomiting. The visual symptoms, including night blindness and blurred vision, which has been reported at doses from $22\,mg/m^2$, are generally reversible.[59]

NVP-AUY922 achieved clinical Proof of Concept (PoC) in April 2011.

References

1. P. Delmotte and J. Delmotte-Plaque, *Nature*, 1953, **171**, 344.
2. J. G. Supko, R. L. Hickman, M. R. Grever and L. Malspeis, *Canc. Chemother. Pharmacol.*, 1995, **36**, 305–315.

3. C. DeBoer, P. A. Meulman, R. J. Wnuk and D. H. Peterson, *The Journal of Antibiotics*, 1970, **23**, 442–447.
4. S. Soga, Y. Shiotsu, S. Akinaga and S. V. Sharma, *Current Cancer Drug Targets*, 2003, **3**, 359–369.
5. G. Chiosis, M. N. Timaul, B. Lucas, P. N. Munster, F. F. Zheng, L. Sepp-Lorenzino and N. Rosen, *Chem. Biol.*, 2001, **8**, 289–299.
6. H. Jhoti and A. R. Leach, ed., *Structure-Based Drug Discovery*, Springer, Dordrecht, The Netherlands, 2007.
7. M. Congreve, G. Chessari, D. Tisi and A. J. Woodhead, *J. Med. Chem.*, 2008, **51**, 3661–3680.
8. D. A. Erlanson, *Top. Curr. Chem.*, 2012, **317**, 1–32.
9. P. J. Hajduk and J. Greer, *Nat. Rev. Drug Discov.*, 2007, **6**, 211–219.
10. R. E. Hubbard, B. Davis, I. Chen and M. J. Drysdale, *Curr. Top. Med. Chem.*, 2007, **7**, 1568–1581.
11. N. Baurin, F. Aboul-Ela, X. Barril, B. Davis, M. Drysdale, B. Dymock, H. Finch, C. Fromont, C. Richardson, H. Simmonite and R. E. Hubbard, *J. Chem. Inform. Comput. Sci.*, 2004, **44**, 2157–2166.
12. C. Dalvit, P. Pevarello, M. Tato, M. Veronesi, A. Vulpetti and M. Sundstrom, *J. Biomol. NMR*, 2000, **18**, 65–68.
13. P. J. Hajduk, E. T. Olejniczak and S. W. Fesik, *J. Am. Chem. Soc.*, 1997, **119**, 12257–12261.
14. C. A. Lepre, J. M. Moore and J. W. Peng, *Chem. Rev.*, 2004, **104**, 3641–3676.
15. M. Mayer and B. Meyer, *Angew. Chem. Int. Ed.*, 1999, **38**, 1784–1788.
16. P. R. Andrews, D. J. Craik and J. L. Martin, *Journal of Medicinal Chemistry*, 1984, **27**, 1648–1657.
17. I. D. Kuntz, K. Chen, K. A. Sharp and P. A. Kollman, *Proc. Natl Acad. Sci. USA*, 1999, **96**, 9997–10002.
18. S. Schultes, C. de Graaf, E. E. J. Haaksma, I. J. P. de Esch, R. Leurs and O. Krämer, *Drug Discov. Today Tech.*, 2010, **7**, e157–e162.
19. C. W. Murray, M. G. Carr, O. Callaghan, G. Chessari, M. Congreve, S. Cowan, J. E. Coyle, R. Downham, E. Figueroa, M. Frederickson, B. Graham, R. McMenamin, M. A. O'Brien, S. Patel, T. R. Phillips, G. Williams, A. J. Woodhead and A. J. Woolford, *J. Med. Chem.*, 2010, **53**, 5942–5955.
20. P. A. Brough, X. Barril, J. Borgognoni, P. Chene, N. G. Davies, B. Davis, M. J. Drysdale, B. Dymock, S. A. Eccles, C. Garcia-Echeverria, C. Fromont, A. Hayes, R. E. Hubbard, A. M. Jordan, M. R. Jensen, A. Massey, A. Merrett, A. Padfield, R. Parsons, T. Radimerski, F. I. Raynaud, A. Robertson, S. D. Roughley, J. Schoepfer, H. Simmonite, S. Y. Sharp, A. Surgenor, M. Valenti, S. Walls, P. Webb, M. Wood, P. Workman and L. Wright, *J. Med. Chem.*, 2009, **52**, 4794–4809.
21. N. G. Davies, H. Browne, B. Davis, M. J. Drysdale, N. Foloppe, S. Geoffrey, B. Gibbons, T. Hart, R. Hubbard, M. R. Jensen, H. Mansell, A. Massey, N. Matassova, J. D. Moore, J. Murray, R. Pratt, S. Ray, A. Robertson, S. D. Roughley, J. Schoepfer, K. Scriven, H. Simmonite,

S. Stokes, A. Surgenor, P. Webb, M. Wood, L. Wright and P. Brough, *Bioorg. Med. Chem.*, 2012, **20** , 6770–6789.
22. S. D. Roughley and R. E. Hubbard, *J. Med. Chem.*, 2011, **54**, 3989–4005.
23. X. Barril, P. Brough, M. Drysdale, R. E. Hubbard, A. Massey, A. Surgenor and L. Wright, *Bioorg. Med. Chem. Lett.*, 2005, **15**, 5187–5191.
24. L. Wright, X. Barril, B. Dymock, L. Sheridan, A. Surgenor, M. Beswick, M. Drysdale, A. Collier, A. Massey, N. Davies, A. Fink, C. Fromont, W. Aherne, K. Boxall, S. Sharp, P. Workman and R. E. Hubbard, *Chem. Biol.*, 2004, **11**, 775–785.
25. N. Baurin, R. Baker, C. Richardson, I. Chen, N. Foloppe, A. Potter, A. Jordan, S. Roughley, M. Parratt, P. Greaney, D. Morley and R. E. Hubbard, *J. Chem. Inform. Comput. Sci.*, 2004, **44**, 643–651.
26. K. M. Cheung, T. P. Matthews, K. James, M. G. Rowlands, K. J. Boxall, S. Y. Sharp, A. Maloney, S. M. Roe, C. Prodromou, L. H. Pearl, G. W. Aherne, E. McDonald and P. Workman, *Bioorg. Med. Chem. Lett.*, 2005, **15**, 3338–3343.
27. A. Kreusch, S. Han, A. Brinker, V. Zhou, H. S. Choi, Y. He, S. A. Lesley, J. Caldwell and X. J. Gu, *Bioorg. Med. Chem. Lett.*, 2005, **15**, 1475–1478.
28. H. J. Kwon, M. Yoshida, Y. Fukui, S. Horinouchi and T. Beppu, *Cancer Res.*, 1992, **52**, 6926–6930.
29. T. W. Schulte, M. V. Blagosklonny, C. Ingui and L. Neckers, *J. Biol. Chem.*, 1995, **270**, 24585–24588.
30. S. M. Roe, C. Prodromou, R. O'Brien, J. E. Ladbury, P. W. Piper and L. H. Pearl, *J. Med. Chem.*, 1999, **42**, 260–266.
31. S. Soga, L. M. Neckers, T. W. Schulte, Y. Shiotsu, K. Akasaka, H. Narumi, T. Agatsuma, Y. Ikuina, C. Murakata, T. Tamaoki and S. Akinaga, *Cancer Res.*, 1999, **59**, 2931–2938.
32. K. S. R. K. Rao and N. V. S. Rao, *Indian J. Chem. B Org.*, 1968, **6**, 66–68.
33. M. S. R. Murthy and E. V. Rao, *Indian J. Chem. B Org.*, 1985, **24B**, 667–669.
34. S. P. Bondarenko, M. S. Frasinyuk and V. P. Khilya, *Chem. Nat. Comp.*, 2007, **43**, 402–407.
35. V. P. Khilya, A. Aitmawbetov, M. Ismailov and L. G. Grishko, *Khim. Prir. Soedin.*, 1994, 629–633.
36. B. W. Dymock, X. Barril, P. A. Brough, J. E. Cansfield, A. Massey, E. McDonald, R. E. Hubbard, A. Surgenor, S. D. Roughley, P. Webb, P. Workman, L. Wright and M. J. Drysdale, *J. Med. Chem.*, 2005, **48**, 4212–4215.
37. G. Giannini, W. Cabri, L. Vesci, M. L. Cervoni, C. Pisano, M. Taddei and S. Ferrini, 2012, WO 2012084602.
38. W. Ying, Z. Du, L. Sun, K. P. Foley, D. A. Proia, R. K. Blackman, D. Zhou, T. Inoue, N. Tatsuta, J. Sang, S. Ye, J. Acquaviva, L. S. Ogawa, Y. Wada, J. Barsoum and K. Koya, *Mol. Canc. Therapeut.*, 2012, **11**, 475–484.
39. D. U. Chimmanamada and W. Ying, 2009, WO2009148599.

40. I. Cikotiene, E. Kazlauskas, J. Matuliene, V. Michailoviene, J. Torresan, J. Jachno and D. Matulis, *Bioorg. Med. Chem. Lett.*, 2009, **19**, 1089–1092.
41. D. Matulis, I. Cikotiene, E. Kazlauskas and J. Matuliene, 2009, WO2009134110.
42. R. Howes, X. Barril, B. W. Dymock, K. Grant, C. J. Northfield, A. G. Robertson, A. Surgenor, J. Wayne, L. Wright, K. James, T. Matthews, K. M. Cheung, E. McDonald, P. Workman and M. J. Drysdale, *Anal. Biochem.*, 2006, **350**, 202–213.
43. P. A. Brough, W. Aherne, X. Barril, J. Borgognoni, K. Boxall, J. E. Cansfield, K. M. Cheung, I. Collins, N. G. Davies, M. J. Drysdale, B. Dymock, S. A. Eccles, H. Finch, A. Fink, A. Hayes, R. Howes, R. E. Hubbard, K. James, A. M. Jordan, A. Lockie, V. Martins, A. Massey, T. P. Matthews, E. McDonald, C. J. Northfield, L. H. Pearl, C. Prodromou, S. Ray, F. I. Raynaud, S. D. Roughley, S. Y. Sharp, A. Surgenor, D. L. Walmsley, P. Webb, M. Wood, P. Workman and L. Wright, *J. Med. Chem.*, 2008, **51**, 196–218.
44. X. Barril, M. C. Beswick, A. Collier, M. J. Drysdale, B. W. Dymock, A. Fink, K. Grant, R. Howes, A. M. Jordan, A. Massey, A. Surgenor, J. Wayne, P. Workman and L. Wright, *Bioorg. Med. Chem. Lett.*, 2006, **16**, 2543–2548.
45. P. A. Brough, X. Barril, M. Beswick, B. W. Dymock, M. J. Drysdale, L. Wright, K. Grant, A. Massey, A. Surgenor and P. Workman, *Bioorg. Med. Chem. Lett.*, 2005, **15**, 5197–5201.
46. S. Y. Sharp, C. Prodromou, K. Boxall, M. V. Powers, J. L. Holmes, G. Box, T. P. Matthews, K. M. Cheung, A. Kalusa, K. James, A. Hayes, A. Hardcastle, B. Dymock, P. A. Brough, X. Barril, J. E. Cansfield, L. Wright, A. Surgenor, N. Foloppe, R. E. Hubbard, W. Aherne, L. Pearl, K. Jones, E. McDonald, F. Raynaud, S. Eccles, M. Drysdale and P. Workman, *Mol. Canc. Therapeut.*, 2007, **6**, 1198–1211.
47. N. F. Smith, F. I. Raynaud and P. Workman, *Mol. Canc. Therapeut.*, 2007, **6**, 428–440.
48. M. R. Jensen, J. Schoepfer, T. Radimerski, A. Massey, C. T. Guy, J. Brueggen, C. Quadt, A. Buckler, R. Cozens, M. J. Drysdale, C. Garcia-Echeverria and P. Chene, *Breast Canc. Res. BCR*, 2008, **10**, R33.
49. S. A. Eccles, A. Massey, F. I. Raynaud, S. Y. Sharp, G. Box, M. Valenti, L. Patterson, A. de Haven Brandon, S. Gowan, F. Boxall, W. Aherne, M. Rowlands, A. Hayes, V. Martins, F. Urban, K. Boxall, C. Prodromou, L. Pearl, K. James, T. P. Matthews, K. M. Cheung, A. Kalusa, K. Jones, E. McDonald, X. Barril, P. A. Brough, J. E. Cansfield, B. Dymock, M. J. Drysdale, H. Finch, R. Howes, R. E. Hubbard, A. Surgenor, P. Webb, M. Wood, L. Wright and P. Workman, *Cancer Res.*, 2008, **68**, 2850–2860.
50. S. Ide, M. Motwani, M. R. Jensen, J. Wang, N. Huseinovic, P. Stiegler, X. Wang and C. Quadt, *ASCO Meeting Abstracts*, 2009, **27**, 3533.
51. J. C. Bendell, L. L. Hart, S. Pant, J. R. Infante, S. F. Jones, A. Mohyuddin, P. Murphy, J. Patton, W. C. Penley, D. S. Thompson and H. A. Burris, *ASCO Meeting Abstracts*, 2013, **31**, 475.

52. E. B. Garon, T. Moran, F. Barlesi, L. Gandhi, L. V. Sequist, S.-W. Kim, H. J. M. Groen, B. Besse, E. F. Smit, D.-W. Kim, M. Akimov, E. Avsar, S. Bailey and E. Felip, *ASCO Meeting Abstracts*, 2012, **30**, 7543.
53. M. L. Johnson, H. A. Yu, E. M. Hart, R. Worden, A. Rademaker, R. Gupta, C. Miller, J. D. Patel, M. G. Kris, V. A. Miller and G. J. Riely, *ASCO Meeting Abstracts*, 2012, **30**, 3083.
54. A. Kong, D. Rea, S. Ahmed, J. T. Beck, R. Lopez Lopez, L. Biganzoli, A. Armstrong, M. Aglietta, E. Alba, M. Campone, M. Akimov, A. Matano, C. Lefebvre and S.-C. Lee, *ASCO Meeting Abstracts*, 2012, **30**, 530.
55. Z. Qadir, J. Crown, M. R. Jensen, M. Clynes, D. Slamon and N. O'Donovan, *ASCO Meeting Abstracts*, 2009, **27**, e14573.
56. T. A. Samuel, C. Sessa, C. Britten, K. S. Milligan, M. M. Mita, U. Banerji, T. J. Pluard, P. Stiegler, C. Quadt and G. Shapiro, *ASCO Meeting Abstracts*, 2010, **28**, 2528.
57. C. P. Schroder, J. V. Pedersen, S. Chua, C. Swanton, M. Akimov, S. Ide, C. Fernandez-Ibarra, A. Dzik-Jurasz, E. De Vries, S. B. Gaykema and U. Banerji, *ASCO Meeting Abstracts*, 2011, **29**, e11024.
58. C. Sessa, S. K. Sharma, C. D. Britten, N. J. Vogelzang, K. N. Bhalla, M. M. Mita, T. J. Pluard, P. Stiegler, C. Quadt and G. I. Shapiro, *ASCO Meeting Abstracts*, 2009, **27**, 3532.
59. Y. Onozawa, T. Doi, K. Yamazaki, J. Watanabe, N. Fuse, T. Yoshino, M. Akimov, M. Robson, N. Boku and A. Ohtsu, *Mol. Canc. Therapeut.*, 2011, **10**, Abstract nr B98.

CHAPTER 9

Discovery and Selection of NVP-HSP990 as a Clinical Candidate

TIMOTHY D. MACHAJEWSKI, DANIEL MENEZES AND ZHENHAI GAO*

Novartis Institutes for BioMedical Research, 4560 Horton Street Emeryville, CA 94608, USA
*Email: zhenhai.gao@novartis.com

9.1 Introduction to the Target as a Cancer Therapeutic

Hsp90 chaperones, which possess a conserved ATP-binding site at their N-terminal domain, belong to a small ATPase subfamily known as the DNA Gyrase, Hsp90, Histidine Kinase and MutL (GHKL) subfamily.[1] The Hsp90 family of chaperones is comprised of four known members: Hsp90α and Hsp90β both located in the cytosol, GRP94 in the endoplasmic reticulum and TRAP1 in the mitochondria.[2] While the cellular functions of Grp94 and TRAP1 and their relevance to human diseases have just begun to be appreciated, the role of cytosolic Hsp90α and Hsp90β (referred to as "Hsp90" for the rest of chapter unless otherwise indicated) under physiological and pathological conditions has been more extensively studied. Hsp90 is responsible for folding/refolding, maintaining active conformation and ensuring correct translocation of its cellular client proteins. A recent elegant study by Taipale et al.[3] not only expands the repertoire of Hsp90-interacting proteins, but also provides mechanistic insights into the fundamental principles governing Hsp90 client recognition. The list of Hsp90-associated proteins has grown substantially from ~80 in 2003 to roughly over 600 in 2012 (the list can be found at http://www.picard.ch/download, which is regularly updated by Dr. Didier Picard's laboratory). The diverse array of Hsp90 client proteins encompasses kinases, transcription factors, GPCRs, enzymes and ion channels *etc*. As such, Hsp90 is involved in regulating a multitude of cellular processes such as growth, differentiation, apoptosis, autophagy, protein homeostasis, oncogenic transformation and chromatin remodeling through epigenetic mechanisms.[4]

The intrinsic ATPase activity is essential for Hsp90 to carry out its chaperoning (folding) functions and to complete the repeated cycles of loading and releasing client proteins. Small-molecule inhibitors that bind at the N-terminal ATP-binding site of Hsp90 disrupt the dynamic ATP-driven chaperone cycles and alter Hsp90 client interaction, leading to either degradation or aggregation of client proteins. Importantly, a subset of Hsp90 client proteins are *bona fide* oncogenes and inhibition of Hsp90 uniquely causes simultaneous destabilization of multiple oncoproteins critical for tumor cell growth and survival.[5] The complex genetic heterogeneity, genomic and epigenetic plasticity of tumor cells are now believed to allow development of resistance to otherwise effective targeted anticancer therapies. In this regard, Hsp90 inhibitors may offer an unparalleled advantage of combinatorial attack on multi-step oncogenesis with a single agent treatment, and may possibly make it harder for tumor cells to evolve drug resistance. This attractive concept in conjunction with proof-of-concept clinical activity exhibited by the first-in-class Hsp90 inhibitor 17AAG has fueled the fervor of both pharmaceutical and academic labs to develop Hsp90 inhibitors superior to 17AAG in the past decade.[2,6–8] The effort has led to the entry of an impressive 17 novel chemical entities into clinical evaluation.[2,9–19]

In this chapter, we will describe the discovery and preclinical development of one of the most potent oral Hsp90 inhibitors, NVP-HSP990, which entered human cancer trials in 2009. Because the pharmacological properties of NVP-HSP990 have been published,[20] we believe it will be more valuable to the scientific community to focus our discussion on integrated approaches and processes we have taken to identify and select NVP-HSP990 as our clinical candidate. Given the complexity and unpredictability of drug development, and inherent difficulty for rational drug discovery, we hope that as we gather and learn from more practical aspects of each individual drug program, we might be able to streamline the future drug-discovery process to more efficiently and effectively develop better small-molecule therapeutics targeting Hsp90 as well as other anticancer targets.

9.2 Biochemical and Cellular Assay Development

The initial attempts to identify novel Hsp90 inhibitors were hampered by the lack of sensitive and robust biochemical assays suitable for high-throughput screening (HTS) of a large compound library (over million collections). This is in part due to the extremely slow turnover rate of intrinsic ATPase activity of

Hsp90 and the low affinity of readily available tool compounds geldanamycin (GA) and 17-AAG for Hsp90 at the time. By selecting appropriate technology combined with careful optimization, we were able to develop the following three critical assays:[21]

1) Primary homogenous time-resolved fluorescence (HTRF) HTS assay using biotin-GA.
2) Secondary TRF (time-resolved fluorescence) binding assay using biotin-radicicol as the label. This assay is based on dissociation-enhanced lanthanide fluorescence immunoassay (DELFIA) technology.
3) Cell-based target modulation assays to monitor Hsp90-p23 dissociation, client protein degradation and Hsp70 induction using ELISA or TRF in-cell western technology.

The key to the success of the primary HTRF HTS assay resulted from the serendipitous discovery that DTT included as a matter of course in the assay buffer significantly boosted the assay window and signal strength. We determined subsequently by mass spectrometer that DTT converted GA to dihydro-GA, a reduced form that now is known to have much higher binding potency (about 25-fold) for Hsp90. Omission of DTT from the assay buffer led to a near complete collapse of the signal-to-noise window of the assay. The robustness, high sensitivity and homogenous nature of the HTRF assay make it an ideal choice for HTS. However, we realized later that some of the compounds identified by this assay turned out to be false positive. This is possibly due to non-specific binding of these compounds to the hydrophobic biotin-GA, which prevented biotin-GA from interacting with Hsp90 (hence the false positive results). To distinguish real hits from false positives, we implemented in our testing funnel a secondary heterogeneous TRF assay binding using a different label, biotin-radicicol. The TRF assay significantly helped eliminate false positive compounds. All compounds that were positive in the TRF binding assay also were active in cell-based target modulation assays.

In summary, we successfully established a series of biochemical and cellular assays that allowed us to implement an HTS campaign, rapidly identify and confirm HTS hits, generate structure-activity relationships (SAR) and evaluate cellular potency of Hsp90 inhibitors. These assays were instrumental to the ultimate success of our Hsp90 inhibitor program.

9.3 Library Screening

In 2003, when this project was initiated, there were several small-molecule Hsp90 inhibitors reported (Figure 9.1): geldanamycin and derivatives such as 17-AAG (R = CH_2CHCH_2NH) **1**, the natural product radiciol **2**, a series of recorcinol based inhibitors **3** and a series of purine based inhibitors **4**.[22]

Our search for a small-molecule inhibitor of Hsp90 began with a high-throughput screen of a collection of approximately 600 000 compounds consisting of purchased libraries and in-house synthesized compounds. Several hit

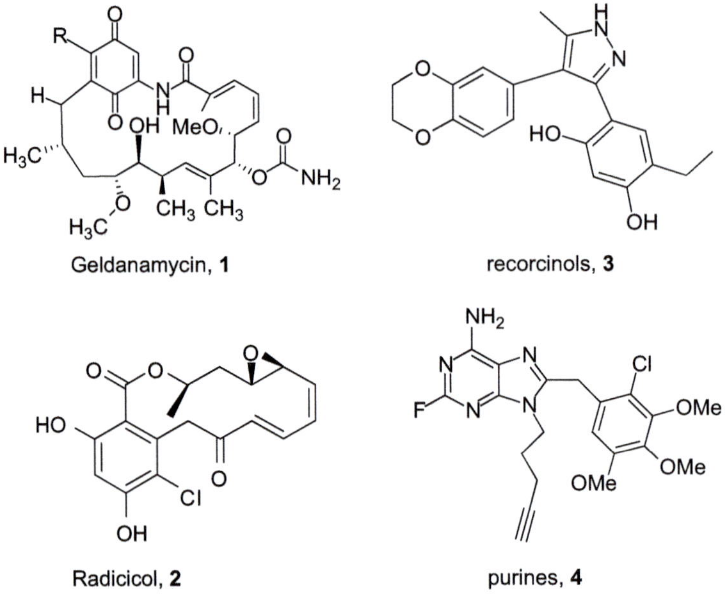

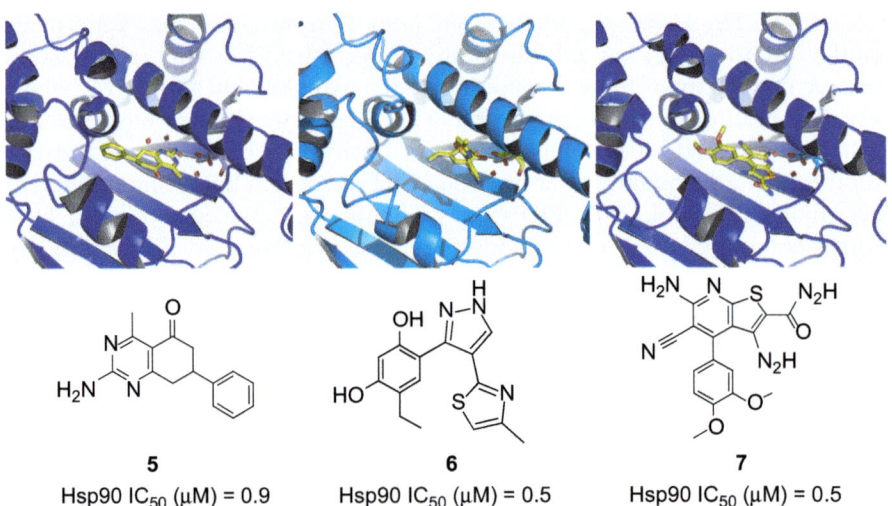

Figure 9.1 Hsp90 inhibitors *circa* 2003.

Figure 9.2 Hits from screening with co-crystal structure.

series were identified and confirmed as inhibitors using the assays described above. Three representative examples are shown in Figure 9.2. Each had a measured IC_{50} of ~1 µM or less. These compounds were subsequently co-crystallized with N-terminal Hsp90, confirming the mechanism of inhibition of binding to the ATP binding site.

Table 9.1 Key properties of confirmed hit series.

Compound	5	6	7
IC_{50}	$IC_{50} = 0.9\,\mu M$	$IC_{50} = 0.5\,\mu M$	$IC_{50} = 0.5\,\mu M$
LE	0.45	0.41	0.34
MW	253	301	369
cLogP	2.2	2.4	1.8
Solubility	79 μM		3.3 μM

9.3.1 Selection of Lead Series

Of the three hit series, we focused our efforts on the development of the series represented by **5** as a lead. Table 9.1 summarizes the key properties considered for evaluation of the series. This series of compounds had many features that made it an attractive starting point for hit to lead optimization studies: a molecular weight less than 300 (and a corresponding high ligand efficiency, LE,[23] of 0.42), good measured permeability with an indication of cell-based activity on the target ($EC_{50} \sim 20\,\mu M$ for induction of Hsp70 and $EC_{50} \sim 35\,\mu M$ for depletion of the Hsp90 client protein Raf-1) and good physicochemical properties (cLogP 2.2), which translated into an aqueous solubility of 79 μM.

The other two series were less attractive for several reasons. The series represented by **6** contained a recorcinol moiety and preliminary SAR (data not shown) showed that this functionality was required for target potency. A key requirement for the program was an orally bioavailable inhibitor and the phenolic functionality was anticipated and observed to result in significant Phase II metabolism and poor bioavailability. In addition, these and similar compounds were previously reported and pursued by several other organizations. The pyridothiophene series, represented by **7**, suffered from higher molecular weight and poor solubility. Despite our decision to not pursue these hit chemotypes, we leveraged the structural knowledge gained from co-crystals and SAR of related compounds in the library.

9.4 Structure and Modeling

9.4.1 Structure Analysis – Comparison with Known Inhibitors

With the availability of several co-crystal structures of our screened hits as well as those of the natural product inhibitors, geldanamycin and radicicol, we pursued a rational structure-based approach to optimization of inhibitor potency. A comparison of the binding interactions of the available co-crystal structures is shown in Figure 9.3.

Our analysis included all of the hydrogen bond interactions, either direct to protein or water mediated, as well as hydrophobic van der Waals interactions with the protein. This initial analysis highlighted the importance of the interaction with Asp93: all of the known inhibitors as well as the natural substrate ADP make this interaction. In fact, the importance of the water molecule

Compound	Comment	Asp93	H2O-1 to Asp93	H2O-2 to Asp93	Ser52	Leu48	Lys58	Thr184	Asn106	Phe138	Asn51	Lys112	Gly137	Val136	Phe138	Ile96
5		x	x		w	w			w		w				x		x
6	displaced H2O-3, H2O-4	x	x	x				x								x	x
Radicicol	displaced H2O-4	x	x	-	w	x	x									x	x
Geldenamycin		x	x	-	w	x				x		x	w		x	x
ADP		x	x	-	w			w		p	p,w	p	p			x	x
7	docking model	x	x		-		x									x	x

Black: Hydrogen bond interactions
Green: Hydrogen bond interactions to phosphate binding residues
Blue: hydrophobic interactions

x: direct interaction
w: water-mediated HB
p: phosphate-group

Columns highlighted in yellow are target areas expected to increase probability of success

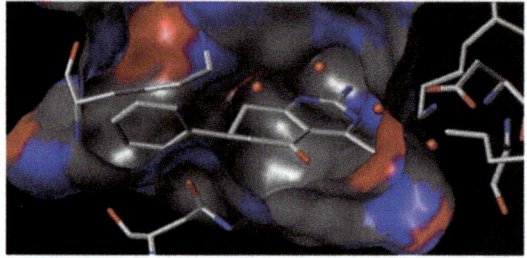

Figure 9.3 Analysis of protein–inhibitor interactions.

labeled water-1 was also highlighted, as it is present in all structures. Several compounds displaced one conserved water molecule and one (the recorcinol, **6**) displaced two conserved water molecules, suggesting that an improvement in binding energy could be gained by displacement of one or more of these water molecules. There were several other water-mediated interactions with Leu48 in most of the structures. Of note, too, was the hydrogen bond interaction that radicicol and geldanamycin made with Lys58. Observation of the hydrophobic interactions revealed that while many of the inhibitors made van der Waals contact with Phe138, our hit series **5** did not, suggesting an additional path to improved binding energy. The pursuit of these strategies toward improved potency is described below.

9.4.2 Design of Inhibitors to Displace Structural Waters

The initial SAR studies focused on the importance of the aminopyrimidine portion of the molecule and its interaction with the structural waters. As shown in Table 9.2, replacement of the 4-methyl group with either a smaller group (H, **11**) or a larger group (ethyl, **12**) resulted in a 10-fold loss of potency. Surprisingly, the much larger *p*-methoxybenzyl group, **13**, while somewhat less potent, was approximately as potent as the significantly smaller hydrogen. While an X-ray co-structure of this compound was not generated, we speculated that a change in binding mode occurred (Figure 9.4).

As expected from the X-ray structure, replacement of the 2-amine with a hydroxyl group, **8**, or methylation of the 2-amine, **9**, resulted in either a complete loss of activity or 10-fold loss compared to the parent compound,

Table 9.2 SAR for aminopyrimidine replacement.

Cmpd	Structure	Hsp90 IC$_{50}$ (μM)	LE	Cmpd	Structure	Hsp90 IC$_{50}$ (μM)	LE
5		0.89	0.45	11		8.52	0.39
8		9.74	0.37	12		10.23	0.35
9		>25	<0.26	13		4.46	0.28
10		>25	<0.26				

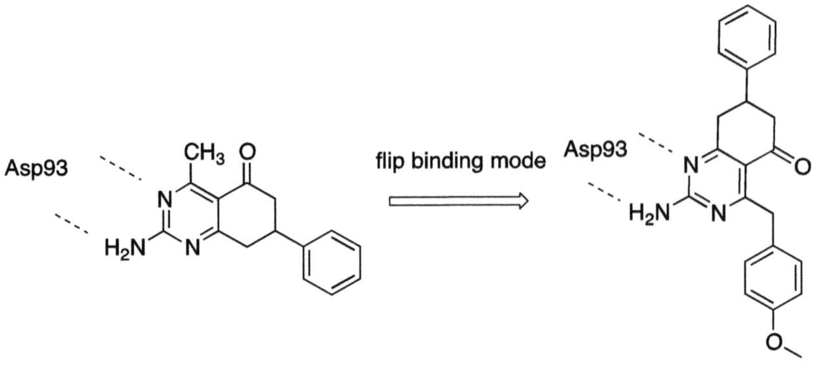

Figure 9.4 Proposed flipped binding mode.

consistent with the expectation that the amino pyrimidine made optimal contacts with Asp93 and the structural waters. Projection of a nitrile, known to be capable of mimicking the interactions of water, into the region occupied by "water-4" was shown by molecular modeling to provide a near perfect overlap, but these compounds were also inactive (compound **10**).

9.4.3 Attempted Optimization through Lys58

As shown in Figure 9.3, both geldanamycin and radicicol make interactions with Lys58. Computational studies suggested that appropriate trajectories for building interactions with this lysine residue existed from either the methyl group on the aminopyrimidine portion of the molecule or through replacement of the carbonyl group. Multiple compounds were prepared in an attempt to capture this interaction, but none achieved more affinity for the target.

9.4.4 Reorganization of the "Flexible Loop" Region

We next turned our attention to the SAR of the pendent aryl group (Table 9.3). Despite the apparent opportunities to make specific interactions with the protein in this region, we observed decreases in potency. A further analysis of the protein structures showed this region to be quite variable from structure to structure, as shown in Figure 9.5. As a result, the entropic cost in energy to rigidify a small portion of this region for a specific interaction with an inhibitor exceeds any potential gain in binding affinity.

A critical observation was made in examining the region of the protein surrounding Phe138. In addition to the hydrophobic Phe, this region is comprised of several Val and Leu residues. As previously observed, the potent inhibitors geldanamycin, radicicol and the recorcinol inhibitors make van der Waals contact with this region of the protein. Attempting to target this region to make similar van der Waals contacts, we prepared the biphenyl substituted compound **20** (Table 9.4).

Table 9.3 SAR table for C-ring modification.

Cmpd	R	Hsp90 IC$_{50}$ (μM)
5	phenyl	0.89
14	isopropyl	>25
15	4-chlorophenyl	0.91
16	4-methoxyphenyl	5.4
17	3-bromophenyl	2.3
18	2-imidazolyl	>10
19	3-pyridyl	>25

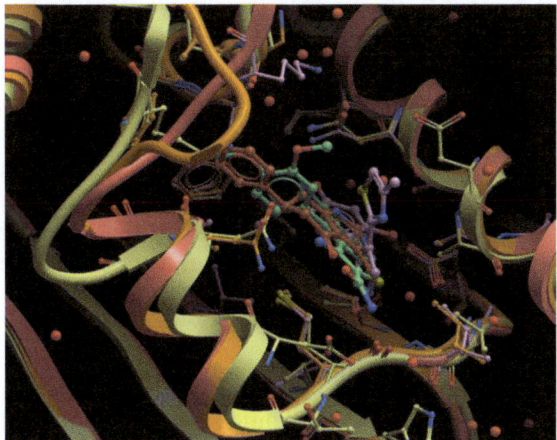

Figure 9.5 Variability across structures in "flexible loop" region of protein.

Table 9.4 SAR table for D-ring addition.

Cmpd	R	Hsp90 IC$_{50}$ (μM)	LE
5	H	0.89	0.45
20	phenyl	0.94	0.34
21	4-pyridyl	3.88	0.30
22	3-pyridyl	0.84	0.34
23	2-pyridyl	0.21	0.37
24	2-pyrazinyl	0.11	0.39

Despite no increase in potency, the X-ray co-structure revealed an important and distinct change in the crystal structure that both explained the similar potency to the parent compound and provided a new direction to improved potency.

As shown in Figure 9.6, the flexible loop rearranges to become a slightly distorted alpha-helix. Closer inspection shows that the conformational change is driven largely by the movement of Leu107, which in the disordered structure sits in the hydrophobic pocket defined by Phe138 and Trp162. In contrast, upon binding the biphenyl inhibitor, the phenyl group replaces Leu107 in this hydrophobic region, allowing for the reorganization into the more stable alpha-helix observed. No increase in binding affinity is observed because the increased van der Waals contact energy is compensated by the loss of entropy due to protein ordering. The new structure, however, provided several additional vectors for optimization that were pursued to success.

9.4.5 Optimization of Binding Affinity

Additional structural analysis and modeling suggested that optimal interaction of the aromatic ring with Phe138 requires a specific angle of the biphenyl-type group. In order to achieve that preferred conformation, the *ortho* CH was

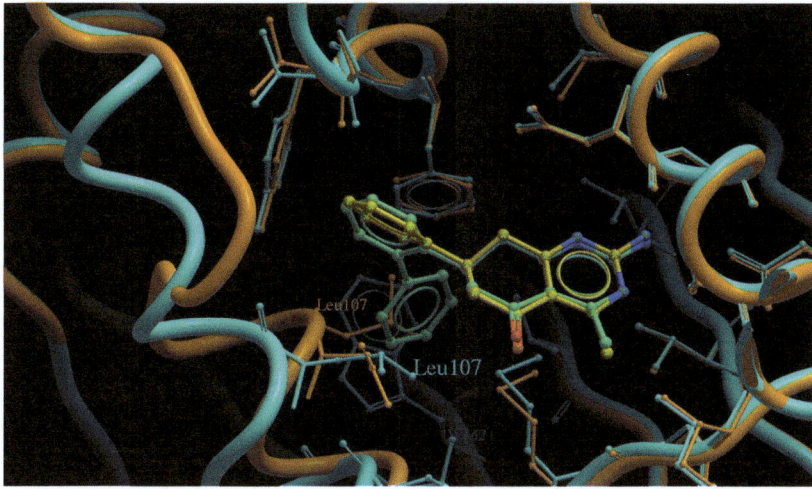

Figure 9.6 Hsp90:compound (**34**) co-structure superimposed with Hsp90:compound (**1**) co-structure. Compound **1** in the open conformation. Upon addition of the D-ring, the protein rearranges to "helical open conformation" with accompanying displacement of Leu107 and an increase in volume of hydrophobic void.

Table 9.5 SAR of D-ring *meta* substitution.

Cmpd	X	R_1	R_2	Hsp90 IC_{50} (μM)
25	N	F	OCH_3	0.015
26	N	F	CH_2CH_3	0.280
27	N	F	CF_3	0.390
28	N	H	OCH_3	0.070
29	N	H	OCH_2CH_3	1.4
30	CH	H	OCH_3	9.6

replaced with a nitrogen to form the pyridyl group. In addition, additional protein contact opportunities exist for small substitutions in the *meta* position.

We found the methoxy group to provide both the optimal size and geometry required for protein interaction. As shown in Table 9.5, a *meta* ethyl group was significantly less potent because the ground state conformation places the group out of the plane of the ring. For the same reason, replacing the ring nitrogen with a CH also reduces the binding affinity. An ethoxy group was too

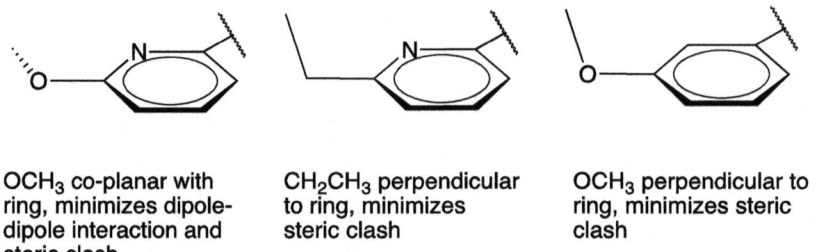

| OCH₃ co-planar with ring, minimizes dipole-dipole interaction and steric clash | CH₂CH₃ perpendicular to ring, minimizes steric clash | OCH₃ perpendicular to ring, minimizes steric clash |

Figure 9.7 Comparison of ground state conformations.

large and results in a significant decrease in potency. The optimal interaction is observed with the methoxy group adjacent to a ring nitrogen. A dipole-dipole interaction and minimization of steric clash places the methyl group in the bioactive conformation (Figure 9.7).

9.5 Lead Optimization

9.5.1 Testing Cascade – Selection of GTL16 Cell Line for *in Vitro* and *in Vivo* Optimization

Our *in vivo* selection of the GTL16 tumor cell line was driven by a desire to have a consistent measure of activity from the bench through the clinic. The GTL16 line is a gastric tumor cell line where proliferation is driven through the overexpression and amplification of the Hsp90 client receptor tyrosine kinase c-Met.[24] We initially envisioned a preliminary target indication of c-Met driven tumors and hence *ex vivo* activity in this cell line mediated through inhibition of Hsp90 might directly translate into efficacy in an *in vivo* PD model, efficacy in a mouse xenograft model and ultimately activity in the clinic.

Once we had optimized the biochemical potency to a consistently acceptable range (<100 nM), we started to optimize the lead into a potential drug candidate by tuning *in vivo* activity. During the course of the biochemical potency optimization, an increase in molecular weight was accompanied by a concomitant increase in lipophilicity. Based on the co-crystal structure of the inhibitor bound to Hsp90, there were several regions that offered an opportunity to improve upon the physicochemical properties while maintaining the desired level of binding affinity: we chose to focus on the C- and D-ring regions.

In an iterative fashion following compound design and synthesis, novel inhibitors were screened in our Hsp90 biochemical binding assay utilizing the biotinylated radicicol (described earlier). Compounds that met the *in vitro* biochemical potency cutoff of <0.1 μM were advanced into a cell-based target modulation assay in the GTL-16 cell line, measuring depletion of the client protein c-Met and up-regulation of Hsp70. Inhibitors that exhibited a cellular potency IC_{50} filter of <0.1 μM were then moved into a GTL16 proliferation assay. Selectivity was periodically assessed against Hsp90 isoforms and a broad kinase panel, along with a safety pharmacology panel (including hERG) (Figure 9.8).

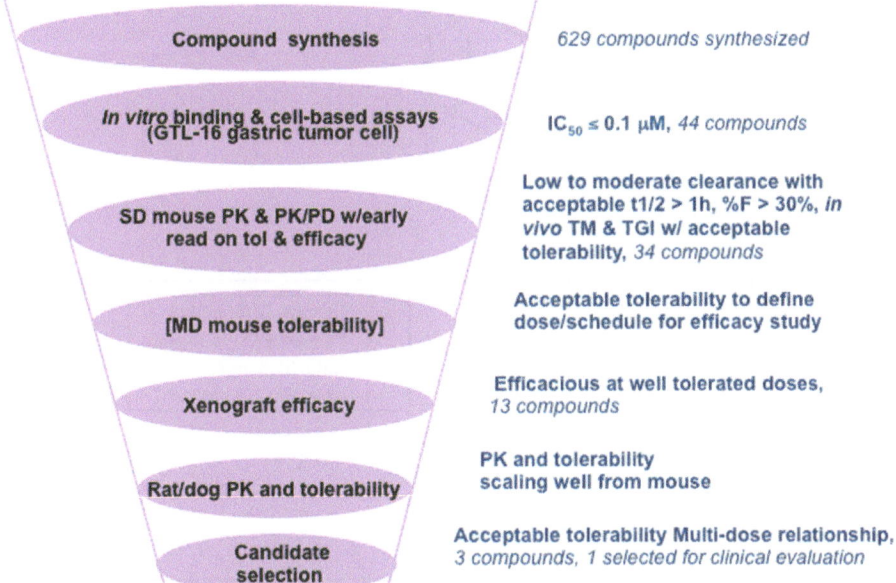

Figure 9.8 Lead optimization testing paradigm.

At this point, inhibitors exhibiting both biochemical and cellular potencies were tested against *in vitro* ADME assays, including a CYP panel, microsomal and hepatocyte stability and solubility. Inhibitors meeting flow chart criteria were moved into a series of *in vivo* assessments, initially a single-dose mouse PK. Inhibitors with bioavailability >30% and low to moderate CL were assessed in a single dose rising tolerability study in mice to enable dose selection for a subsequent single dose PK-PD study in the GTL16 model, tested at the maximal tolerated dose. The extent and duration of PD modulation (c-Met and Hsp70 in GTL16 tumors) were used to guide dose and schedule selection, based on the underlying hypothesis that continuous suppression of the target was required during the entire dosing interval to confer efficacy (Figure 9.8).[20]

As previously mentioned, the D-ring region offered opportunity to adjust the *in vivo* properties while maintaining binding affinity for the target. For example, Hsp90 inhibitors from the 6-membered D-ring series typically exhibited long *in vivo* terminal half-lives. In contrast, replacement of the 6-membered D-rings by 5-membered rings, such as a thioazole, altered the PK properties in which half-lives were generally shorter, requiring more frequent dosing in order to maintain the same extent and duration of Hsp90 inhibition *in vivo*. Both compounds had similar biochemical target potency. This allowed for a broad assessment of a series of compounds with a range of pharmacokinetic properties. The models for PK-PD and efficacy were refined with larger sets of compounds.[20] NVP-HSP990 (Figure 9.9) was selected for further profiling

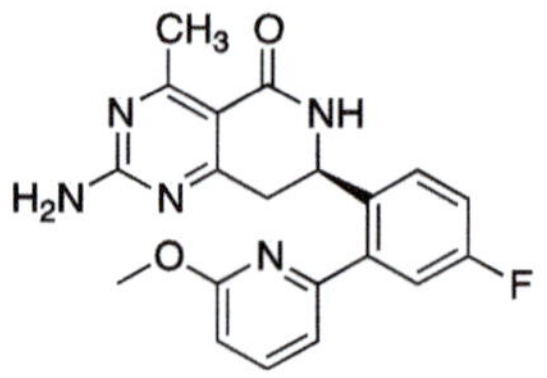

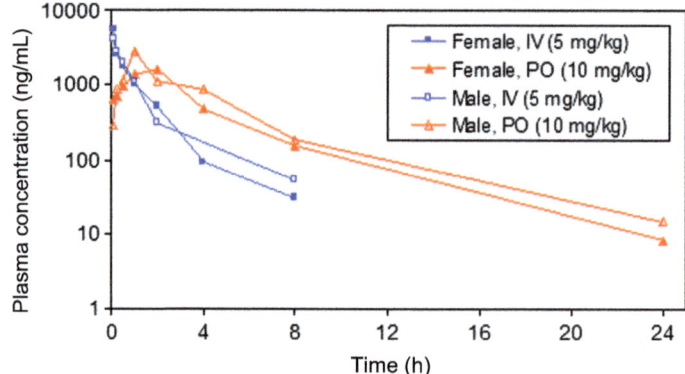

Figure 9.9 NVP-HSP990.

Figure 9.10 Pharmacokinetics profile of NVP-HSP990 in male and female CD1 mice.

Table 9.6 Pharmacokinetic parameters of NVP-HSP990.

Route (dose)	Intravenous (5 mg/kg)		Oral (10 mg/kg)	
Gender	Female	Male	Female	Male
$t_{1/2}$ (h)	2.5	4.0	3.4	3.9
T_{max} (h)	–	–	1.1	1.0
C_0 (ng/mL)	6648	5129	–	–
C_{max} (ng/mL)	–	3867	1280	2370
$AUC_{(0,last)}$ (ng*hr/ml)	4333	3900	6566	8833
Vss (l/kg)	2.0	2.8	–	–
CL (ml/min/kg)	19	21	–	–
%F	–	–	76	115

The plasma pharmacokinetics of NVP-HSP990 were evaluated in either female or male CD-1 mice after a 5 mg/kg intravenous (in 15 or 20% Captisol®) dose or 10 mg/kg oral (in PEG400) dose. Plasma bioanalytics were conducted by quantitative LC-MS, and PK data were analyzed using standard non-compartmental methods (WinNonLin).

based on its PK characteristics (intermediate half-life, AUC), which resulted in greater schedule flexibility and therapeutic index.

9.5.2 Relationship of PK-PD and Efficacy for NVP-HSP990

A summary of the single-dose plasma pharmacokinetics of NVP-HSP990 was characterized in CD1 mice (Figure 9.10, Table 9.6).

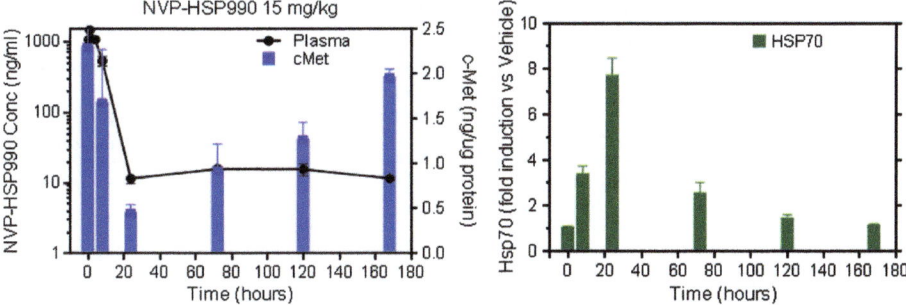

Figure 9.11 PK-PD relationship of NVP-HSP990 in GTL16 tumor model.

Following i.v. administration, NVP-HSP990 displayed moderately low clearance (19–21 ml/min/kg) with a plasma half-life ($t_{1/2}$) ranging from 2.5–4 h. Following oral dosing, NVP-HSP990 was absorbed rapidly with peak plasma concentrations observed at 1 h (T_{max}). NVP-HSP990 exhibited high (>76%) oral bioavailability in mice. The volume of distribution was also high (2.0–2.8 l/kg), indicating extensive tissue distribution. In addition, using an equilibrium dialysis mouse plasma protein binding assay, 74% of NVP-HSP990 was protein bound.

To elucidate the PK-PD relationship of NVP-HSP990, a single oral treatment of 15 mg/kg NVP-HSP990 against GTL-16 human gastric carcinoma xenografts caused degradation of the Hsp90 client protein c-Met and induction of Hsp70 lasting up to 120 hours (Figure 9.11). Additional Hsp90 clients like ERK and AKT pathways downstream of c-Met were also inhibited.[20]

9.5.3 PK/PD Optimization

The PK-PD relationship for NVP-HSP990 in the GTL16 model provided the framework for guiding the PD-efficacy profile, which allowed for dose/schedule optimization of q3d and once-a-week regimens.

9.5.4 *In Vivo* Efficacy of NVP-HSPP990 in Multiple Cancer Xenografts

Given the broad *in vitro* antitumor activity of NVP-HSP990, and that each cancer model is driven by a unique set of oncogenic Hsp90 client proteins, we profiled NVP-HSP990 in multiple cancer models. Table 9.7 shows the selection of tumor models driven by mutationally active or over-expressed client proteins. The antitumor activity of NVP-HSP990 was generally broad, ranging from tumor inhibitor to regressions in multiple models, irrespective of cancer cell type or genetic background. We established that NVP-HSP990 tumor responses *in vivo* were associated with Hsp90 inhibition and client protein inhibition.[20] The pre-clinical activity of NVP-HSP990 indicates efficacy in a

Table 9.7 NVP-HSP990 efficacy in subcutaneous mouse tumor xenograft models selected for expression of key client proteins.

Cancer type	Genotype/client expression	Cell line (mouse strain)	Dose, schedule tested	Efficacy (%T/C or regression)
Gastric	c-Met over-expression	GTL-16 (Nude)	5 mg/kg, q3d	−25%
			15 mg/kg, q7d	22%
	ErbB2+	N87 (Nude)	5 mg/kg, q7d	51%
Lung	Mut EGFR (L858R+T790M), c-Met over-expression	H1975 (Nude)	5 mg/kg, q3d	10%
			15 mg/kg, q6d	−33%
	Wt EGFR, c-Met, mut Ras	A549 (Nude)	15 mg/kg, q7d	13%
Ovarian	Mut PTEN (in house)	A2780 (Nude)	15 mg/kg, q7d	16%
Breast	ErbB2+/ER+	BT474 (SCID)	10 mg/kg, q7d	6%
Melanoma	Mut B-raf	A375 (Nude)	15 mg/kg, q7d	−49%
Leukemia	Mut Flt3-ITD	MV4;11 (Nude)	5 mg/kg, q3d	3%
			15 mg/kg, q6d	−5%

Human tumor xenograft models indicated in the able were implanted subcutaneously (s.c.) with 50% Matrigel™ in nude (Charles River Laboratories, Wilmington, MA) or SCID mice (Harlan, Livermore, CA). Mice were randomized into cohorts (10 mice/group for efficacy) when tumors reached 200–500 mm³. NVP-HSP990 was administered *via* oral gavage in a vehicle of 100% polyethylene glycol (PEG400) at indicated doses. Tumor volumes were obtained from caliper measurements converted using the formula: tumor volume $=\frac{1}{2}$ (length × [width]²). Efficacy, represented as a relative tumor inhibition, was calculated as %T/C = 100 × dT/dC, where, dT or dC = difference of mean tumor volume of drug treatment (T) or vehicle (C) on the final day of the study and the randomization volume.

variety of cancers with well-defined oncogenic clients, supporting clinical evaluation in a range of cancer lineages.

9.6 Summary of Discovery of HSP990

This chapter provides an overview of strategy for the discovery of HSP990 from hit identification to pre-clinical evaluation *in vivo*. One element to highlight in our approach was the application of structure-based drug design to rapidly optimize biochemical potency, and opportunities for fine-tuning the *in vitro* and *in vivo* properties. Our approach used PK-PD and efficacy relationships to guide the dose and schedule optimization. Given the relevance of Hsp90 oncogenic client proteins in cancer, we demonstrate a broad activity of HSP990 in multiple lineages. NVP-HSP990 is currently in early clinical development at Novartis.

References

1. R. Dutta and M. Inouye, *Trends Biochem. Sci.*, 2000, **25**, 24–28.
2. Z. Gao, C. Garcia-Echeverria and M. R. Jensen, *Curr. Opin. Drug Discov. Dev.*, 2010, **13**, 193–202.
3. M. Taipale, D. F. Jarosz and S. Lindquist, *Nat. Rev. Mol. Cell Biol.*, 2010, **11**, 515–528.
4. L. Neckers, *J. Biosci.*, 2007, **32**, 517–530.
5. P. Workman, F. Burrows, L. Neckers and N. Rosen, *Ann. NY Acad. Sci.*, 2007, **1113**, 202–216.
6. E. A. Ronnen, G. V. Kondagunta, N. Ishill, S. M. Sweeney, J. K. Deluca, L. Schwartz, J. Bacik and R. J. Motzer, *Invest. New Drugs*, 2006, **24**, 543–546.
7. S. Pacey, M. Gore, D. Chao, U. Banerji, J. Larkin, S. Sarker, K. Owen, Y. Asad, F. Raynaud, M. Walton, I. Judson, P. Workman and T. Eisen, *Invest. New Drugs*, 2010, **30**, 341–349.
8. D. B. Solit, I. Osman, D. Polsky, K. S. Panageas, A. Daud, J. S. Goydos, J. Teitcher, J. D. Wolchok, F. J. Germino, S. E. Krown, D. Coit, N. Rosen and P. B. Chapman, *Clin. Cancer Res.*, 2008, **14**, 8302–8307.
9. S. A. Eccles, A. Massey, F. I. Raynaud, S. Y. Sharp, G. Box, M. Valenti, L. Patterson, A. de Haven Brandon, S. Gowan, F. Boxall, W. Aherne, M. Rowlands, A. Hayes, V. Martins, F. Urban, K. Boxall, C. Prodromou, L. Pearl, K. James, T. P. Matthews, K. M. Cheung, A. Kalusa, K. Jones, E. McDonald, X. Barril, P. A. Brough, J. E. Cansfield, B. Dymock, M. J. Drysdale, H. Finch, R. Howes, R. E. Hubbard, A. Surgenor, P. Webb, M. Wood, L. Wright and P. Workman, *Cancer Res.*, 2008, **68**, 2850–2860.
10. H. Zhang, L. Neely, K. Lundgren, Y. C. Yang, R. Lough, N. Timple and F. Burrows, *Int. J. Cancer*, 2010, **126**, 1226–1234.
11. J. R. Porter, J. Ge, J. Lee, E. Normant and K. West, *Curr. Top. Med. Chem.*, 2009, **9**, 1386–1418.
12. T. Nakashima, T. Ishii, H. Tagaya, T. Seike, H. Nakagawa, Y. Kanda, S. Akinaga, S. Soga and Y. Shiotsu, *Clin. Cancer Res.*, 2010, **16**, 2792–2802.
13. L. V. Sequist, S. Gettinger, N. N. Senzer, R. G. Martins, P. A. Janne, R. Lilenbaum, J. E. Gray, A. J. Iafrate, R. Katayama, N. Hafeez, J. Sweeney, J. R. Walker, C. Fritz, R. W. Ross, D. Grayzel, J. A. Engelman, D. R. Borger, G. Paez and R. Natale, *J. Clin. Oncol.*, 2010, **28**, 4953–4960.
14. M. R. Jensen, J. Schoepfer, T. Radimerski, A. Massey, C. T. Guy, J. Brueggen, C. Quadt, A. Buckler, R. Cozens, M. J. Drysdale, C. Garcia-Echeverria and P. Chene, *Breast Canc. Res.*, 2008, **10**, R33.
15. Y. Wang, J. B. Trepel, L. M. Neckers and G. Giaccone, *Curr. Opin. Invest. Drugs*, 2010, **11**, 1466–1476.
16. A. J. Woodhead, H. Angove, M. G. Carr, G. Chessari, M. Congreve, J. E. Coyle, J. Cosme, B. Graham, P. J. Day, R. Downham, L. Fazal, R. Feltell, E. Figueroa, M. Frederickson, J. Lewis, R. McMenamin,

C. W. Murray, M. A. O'Brien, L. Parra, S. Patel, T. Phillips, D. C. Rees, S. Rich, D. M. Smith, G. Trewartha, M. Vinkovic, B. Williams and A. J. Woolford, *J. Med. Chem.*, 2010, **53**, 5956–5969.
17. J. R. Porter, C. C. Fritz and K. M. Depew, *Curr. Opin. Chem. Biol.*, 2010, **14**, 412–420.
18. S. K. Lyman, S. C. Crawley, R. Gong, J. I. Adamkewicz, G. McGrath, J. Y. Chew, J. Choi, C. R. Holst, L. H. Goon, S. A. Detmer, J. Vaclavikova, M. E. Gerritsen and R. A. Blake, *PLoS One*, 2011, **6**, e17692.
19. J. Trepel, M. Mollapour, G. Giaccone and L. Neckers, *Nat. Rev. Cancer*, 2010, **10**, 537–549.
20. D. L. Menezes, P. Taverna, M. R. Jensen, T. Abrams, D. Stuart, G. K. Yu, D. Duhl, T. Machajewski, W. R. Sellers, N. K. Pryer and Z. Gao, *Mol. Canc. Therapeut.*, 2012, **11**, 730–739.
21. Z. Gao and S. Fong, in *A Practical Guide to Assay Development and High-throughput Screening in Drug Discovery*, ed. T. Chen, A. W. Czarnik and Y. Bing, CRC Press, Boca Raton, FL, 2009, pp. 83–98.
22. T. D. Machajewski, X. Lin, A. B. Jefferson and Z. Gao, *Annu. Rep. Med. Chem.*, 2005, **40**, 263–276.
23. C. Abadzapatero and J. Metz, *Drug Discov. Today*, 2005, **10**, 464–469.
24. J. G. Christensen, R. Schreck, J. Burrows, P. Kuruganti, E. Chan, P. Le, J. Chen, X. Wang, L. Ruslim, R. Blake, K. E. Lipson, J. Ramphal, S. Do, J. J. Cui, J. M. Cherrington and D. B. Mendel, *Cancer Res.*, 2003, **63**, 7345–7355.

CHAPTER 10

Inhibitors of the Hsp90 C-terminus

HUIPING ZHAO AND BRIAN S. J. BLAGG*

Department of Medicinal Chemistry, 1251 Wescoe Hall Drive, Malott 4070, The University of Kansas, Lawrence, Kansas 66045-7563, USA
*Email: bblagg@ku.edu

10.1 Introduction

10.1.1 Hsp90-mediated Protein Folding Machinery

Heat-shock proteins (Hsps), originally named as a consequence of their ability to up-regulate upon exposure to elevated temperatures,[1,2] are a family of housekeeping molecular chaperones that maintain protein homeostasis within the crowded cellular macromolecular environment.[3] Under intrinsic or extrinsic cellular stress, such as heat, abnormal pH, nutrient deprivation and malignancy, these protective Hsps are over-expressed to ensure cell survival.[4] The 90 kDa heat-shock protein (Hsp90) folding machinery plays a pivotal role in this chaperone system. In fact, Hsp90 is ubiquitous and highly conserved in diversified organisms, ranging from prokaryotes to eukaryotes, and is the most abundant molecular chaperone in the cell, accounting for approximately ~1–2% of total protein in normal cells,[5] and is increased to ~4–6% in disease states such as cancer.[6] In addition, Hsp90 is widely expressed in many cellular compartments and four Hsp90 isoforms have been discovered, which include: Hsp90α (inducible and major form) and Hsp90β (constitutive and minor form), which are predominantly found in the cytosol, the 94 kDa glucose-regulated protein (Grp94), which resides in endoplasmic reticulum, and the Hsp75/tumor necrosis factor receptor protein 1 (TRAP-1), which is localized to the mitochondrial matrix.[4] The Hsp90 protein folding machinery is capable of folding newly synthesized polypeptides and refolding denatured proteins into their biologically active three-dimensional conformations, preventing/dissociating protein aggregates, and directing damaged proteins for degradation through the ubiquitin-proteasome pathway.[7,8]

Although the exact mechanism of Hsp90-mediated protein folding is still not fully understood, 20 years of research has substantially increased our knowledge of this chaperone system.[9–14] It has been acknowledged that the Hsp90 protein folding machinery is complex and involves a variety of co-chaperones, immunophilins and partner proteins for the conformational maturation of nascent polypeptides. It is postulated that Hsp40 and Hsp70 recognize and bind to the exposed hydrophobic amino acids of newly synthesized or misfolded client proteins (**I**) to stabilize and prevent aggregation (Figure 10.1).[15] With the assistance of Hop [Hsp70 and Hsp90 organizing protein containing tetratricopeptide repeats (TPRs) recognized by both Hsp70 and Hsp90],[16] the client protein is delivered from Hsp40/Hsp70 to Hsp90 for stabilization (**IV**), with simultaneous release of Hsp40/Hsp70 and Hop. Upon the association of co-chaperones and partner proteins in concert with immunophilins (FKBP51, FKBP-52 or Cyp-40), a heteroprotein complex is formed (**V**).[17] Following ATP binding to Hsp90 N-terminus, the N-terminal domains of the Hsp90 homodimer clamp around the bound client protein in an ATP-dependent manner (**VI**).[18] Upon the recruitment of co-chaperone p23, the client protein is folded at the expense of ATP hydrolysis (**VII**) in an undefined process that is mediated by p23.[19] The release of a mature three-dimensional and biologically active client

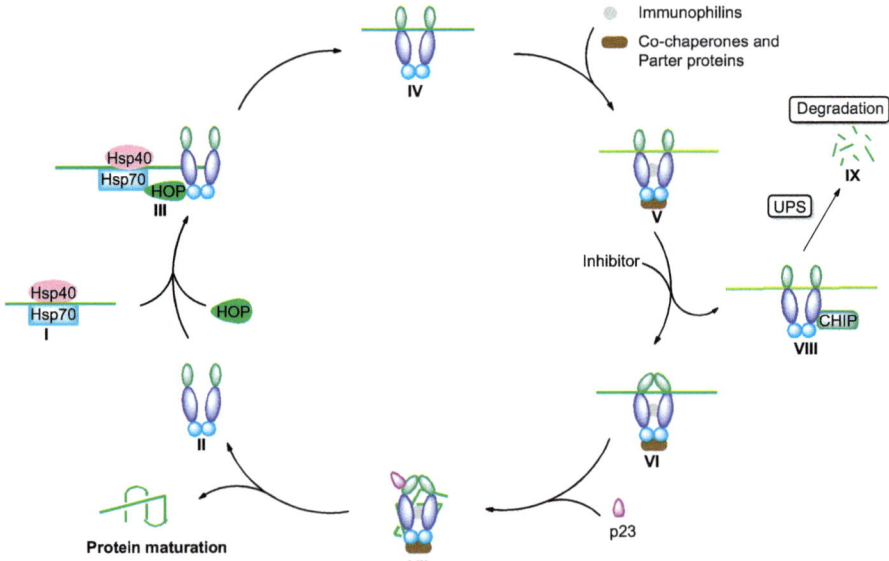

Figure 10.1   The Hsp90-mediated protein folding machinery.

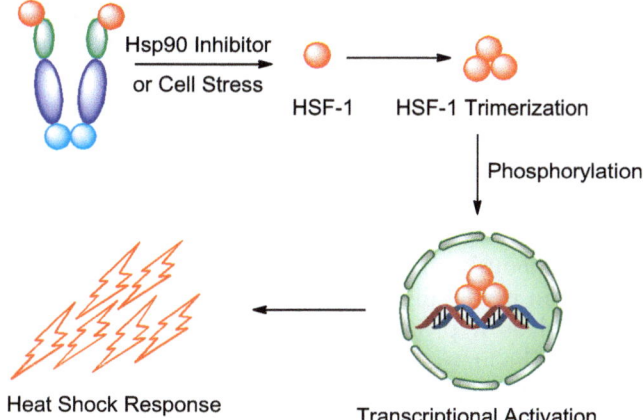

Figure 10.2 The heat-shock response induced by pharmacological inhibition of Hsp90.

protein regenerates the Hsp90 homodimer to complete the cycle. However, if the client protein is damaged or unable to undergo folding, CHIP (carboxy terminus of Hsp70-interacting protein), an important quality controller of this chaperone system, is recruited and binds Hsp90 (**VIII**) to direct client protein degradation (**IX**).[20–23]

In contrast, pharmacological inhibition of Hsp90 prevents the N-terminus from wrapping around the bound client protein, resulting in an unstable protein complex that is recognized by ubiquitin lygase and ultimately leads to the degradation of client proteins by proteasome (**IX**, Figure 10.1),[4,24,25] a process detrimental to cell survival. Hsp90 also tightly regulates the activity of HSF-1, a master regulator of the heat-shock response.[26] Hsp90 inhibition simultaneously dissociates HSF-1 (Figure 10.2), which is then phosphorylated, trimerized and translocated to the nucleus to bind the heat-shock elements and activate the transcription of a set of pro-survival heat-shock proteins that include Hsp27, Hsp40, Hsp70 and Hsp90. Consequently, these Hsps can expand the cellular buffering capacity and enhance cell survival.[26–28] These two signature effects resulting from pharmacological inhibition of Hsp90 have potential applications to the treatment of a wide array of disease states, including cancer and neurodegenerative diseases, respectively.[29–31]

10.1.2 Hsp90 Modulation for the Treatment of Cancer and Neurodegenerative Diseases

Mutational analyses have demonstrated that the activity of structurally diversified transcription factors, signal transduction kinases and steroid hormone receptors are highly dependent upon Hsp90.[32,33] To date, extensive research has identified more than 200 protein substrates that have either limited or extensive dependency on Hsp90 for their biological activity.[34] Cancer cells are particularly sensitive to small molecules that inhibit Hsp90 function, wherein the pro-survival Hsp90 protein folding machinery is recruited to fold mutated proteins in an effort to facilitate cell survival, growth and uncontrolled proliferation. These diverse oncogenic proteins are distributed across all six hallmarks of cancer and are highly dependent upon the Hsp90 protein folding machinery for their stability and cellular activities.[24,25,35] In fact, many of these oncogenic proteins, such as Raf, HER2 and Akt, are independently pursued as anticancer targets.[36] However, inhibition of the Hsp90 protein folding machinery can simultaneously disrupt all of these Hsp90-dependent oncogenic substrates, which can eventually lead to cancer cell death. As a consequence, Hsp90 has emerged as a promising target for the development of cancer chemotherapeutics. The first-in-class Hsp90 inhibitor, 17-AAG, showed proof-of-concept efficacy at doses well tolerated by patients and exhibited clear therapeutic benefits in clinical trials.[37-40] Along with combinatorial depletion of oncogenic proteins, the therapeutic selectivity of Hsp90 inhibitors is believed to arise from the altered states of Hsp90.[38] In cancer cells, Hsp90 exists in a super-chaperone complex that is more active than in normal cells, in which case Hsp90 predominantly exists as dormant homodimer and is largely unaffected by inhibitors.[41,42] In support of these findings, 17-AAG was found to accumulate in tumors at higher concentrations than in normal tissues.[43,44] These fundamental differences can partition between normal and tumor tissues and further support Hsp90 as a therapeutic target with the potential for high differential selectivity. In fact, there are currently 16 Hsp90 small molecular inhibitors under clinical evaluation, all of which target the Hsp90 N-terminal ATP-binding site,[45] many of which have been discussed in prior chapters. However, as a master regulator of HSF-1,[26] Hsp90 inhibition by these inhibitors also results in HSF-1 dissociation, which ultimately activates the synthesis of the pro-survival heat-shock response that leads to cell survival. Recent studies suggest that extracellular Hsp90α is associated with invasion by the chaperoning of matrix metalloprotease II.[46] Therefore, Hsp90 up-regulation may promote undesired metastatic activity. Hence, the heat-shock response associated with Hsp90 N-terminal inhibition is a concern and may be detrimental. In addition, distinct from heat-shock response, HSF-1 was shown to drive a transcriptional program that is specific for highly malignant cells.[47] Although detriments associated with HSF-1 dissociation and the subsequent heat-shock response following the administration of N-terminal inhibitors may compromise their anticancer potential, this up-regulation of the heat-shock proteins may have potential to treat neurodegenerative diseases that result from the accumulation of misfolded proteins.[48-52]

Neurodegenerative diseases are among the leading causes of death and disabilities among senior populations around the world.[53] As the global population ages, especially in developed countries, these diseases are expected to become more frequent and more costly. Unfortunately, there is no disease-modifying therapy that exists and therefore, this unmet medical need has promoted the search for new therapeutic paradigms.[54-57] Many neurodegenerative diseases are characterized by the accumulation of aggregates that result from misfolded proteins within or outside the brain or spinal cord neurons and glia.[7] In Alzheimer's disease (AD), the distinct senile plaques are primarily composed of aggregates of a 40–42-amino acid peptide (Aβ) and neurofibrillary tangles (NFT) composed of fibrils of hyper-phosphorylated Tau protein, respectively, in the extra and intra space of neurons and synapses.[58,59] In Parkinson's disease, the development of Lewy bodies composed primarily of fibrillar α-synuclein is believed to be the disease-causing factor.[60] Similarly, aggregates of the poly-glutamine rich protein, huntingtin, in Huntington's disease,[61] inclusions of superoxide dismutase1 (SOD1)[62] and the RNA/DNA binding protein TDP-43 in amyotrophic lateral sclerosis (ALS)[63,64] and the prion protein aggregates in spongiform encephalopathies[65] have been identified as equally debilitating nervous system disorders. Although proteins associated with these neurodegenerative diseases do not share homology in sequence or structure, extensive research on the properties manifested by these proteins has revealed that these monomers commonly undergo conversion from α-helical structures into misfolded β-sheets that exhibit a tendency to self-aggregate and become pathogenic.[7] In other words, these neurodegenerative diseases appear to manifest as a consequence of protein misfolding. As stated earlier in this chapter, Hsp90 is responsible for the refolding of misfolded proteins into their biologically active conformation; however, upon aging, this protein-folding machinery appears less efficient and/or impaired, or aberrant proteins have accumulated beyond the buffering capacity of the cell. Consequently, long-term and progressive abnormal protein accumulation in neurons may result in neurodegenerative disease.[7] It has been hypothesized that induction of the heat-shock response may reverse protein misfolding and facilitate the clearance of toxic aggregates that are responsible for neurodegenerative diseases. Therefore, heat-shock induction resulting from Hsp90 inhibition has produced tremendous enthusiasm for the application of Hsp90 inhibitors for the treatment of neurodegenerative diseases.[7,48,66,67] In fact, it has been demonstrated that the Hsp90 N-terminal inhibitors in the clinic are capable of inducing a robust heat-shock response that provides a beneficial effect on reducing the "proteotoxicity burden" acquired in these diseases. In fact, many Hsp90 N-terminal inhibitors, including PU derivatives, exhibit neuroprotective efficacy, good therapeutic indices and the ability to cross the blood-brain barrier.[68,69] Similar to Hsp90 inhibitors in cancer, the selectivity for such compounds in pathologic and normal brain tissues has been observed.[69] Unfortunately, the concentration of these Hsp90 N-terminal inhibitors needed to induce the beneficial heat-shock response is similar to that needed for induction of cytotoxic client protein degradation; thus, detrimental side effects are likely and, therefore, the therapeutic window for the treatment of neurodegenerative diseases may be low.

If separated, these contradictory mechanisms for client protein degradation and the pro-survival heat-shock induction upon pharmacologic Hsp90 inhibition could have potential applications for the treatment of cancer and neurodegenerative diseases, respectively. Recent studies have shown that small molecules that target the Hsp90 C-terminus can induce the pro-survival heat-shock response at concentrations far below that needed to induce the degradation of client proteins.[70] In contrast, other C-terminal inhibitors have been shown to produce the opposite effect and induce client protein degradation without a concomitant induction of the pro-survival heat-shock response.[71] Consequently, the discovery of efficacious Hsp90 C-terminal inhibitors is expected to provide an alternative paradigm for the treatment of cancer and neurodegenerative diseases.[48,72,73]

10.1.3 The Hsp90 C-terminus: A Second Site to Modulate Hsp90 Function

It is known that in humans, Hsp90 exists as a homodimer and that each monomer contains three highly conserved domains (Figure 10.3):[4] The 25 kDa N-terminus hosts an ATP-binding pocket that was regarded as the most druggable domain,[45] the 35 kDa middle domain facilitates N-terminal substrate recognition and binds to co-chaperones such as Aha1,[74] and the 12 kDa C-terminus is responsible for Hsp90 homodimerization and is the focus of this chapter. The MEEVD domain located at the C-terminus is responsible for mediating interactions with co-chaperones (*e.g.* HOP and CHIP) and immunophilins (*e.g.* FKBP51 and FKBP52) that contain a TPR-recognition sequence,[17,75] and to facilitate client protein loading onto Hsp90 for maturation/degradation. Most interestingly, this C-terminal domain appears to contain a second nucleotide-binding domain that allosterically regulates N-terminal ATPase activity, although the C-terminus itself does not possess ATPase activity.[76–78] Occupation of this domain has been shown to disrupt dimerization, induce a change in Hsp90 conformation, dissociate bound client and partner proteins and prevent ligands from binding to the N-terminus.[79–82]

Novobiocin, chlorobiocin and coumermycin A1 (Figure 10.4) are antibiotics that can be isolated from several *Streptomyces* strains and are known to exhibit antimicrobial activity by binding to the DNA gyrase ATP-binding pocket,[83] a unique nucleotide-binding motif shared only by members of the GHKL superfamily, which includes DNA gyrase, Hsp90, histidine kinases and MutL.[84] Due to the anticancer activities manifested by novobiocin and the similar bent conformation exhibited by ADP bound to both DNA gyrase and the Hsp90 N-terminal domain, Neckers and co-workers hypothesized that

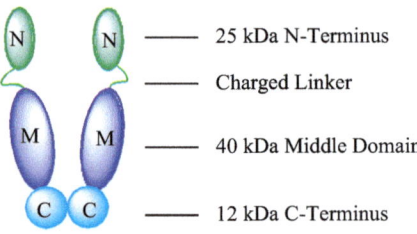

Figure 10.3 Hsp90 domains.

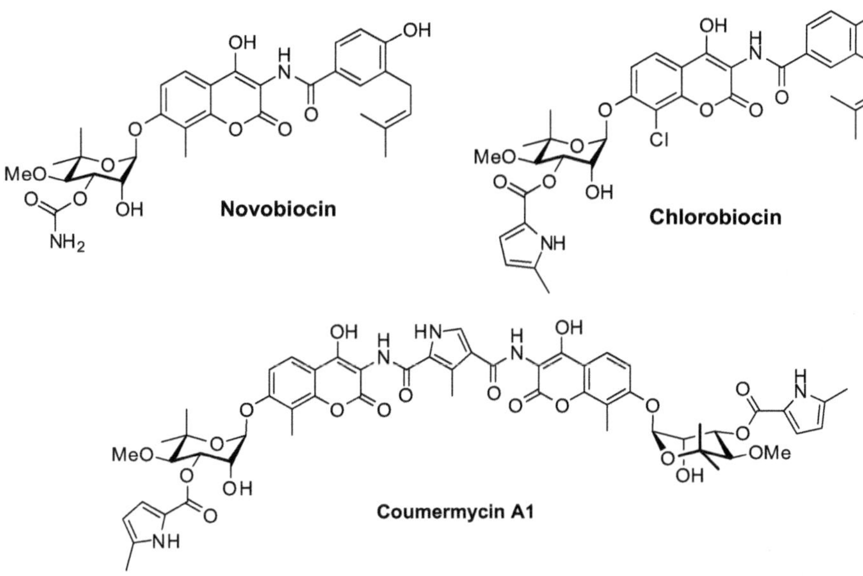

Figure 10.4 Novobiocin and related coumarin antibiotics.

novobiocin may manifest its anticancer activity through binding Hsp90's N-terminal ATP binding pocket.[77] In fact, upon the administration of NB, various Hsp90 client proteins including Raf-1, mutated p53, v-Src and HER2 underwent degradation in a manner similar to that observed upon treatment with GDA or radicicol, suggesting novobiocin binds Hsp90. However, subsequent affinity binding experiments concluded that it bound to a previously unrecognized C-terminal region, instead of the well-recognized N-terminal domain, albeit with low efficiency ($\sim 700\,\mu M$ in SKBr3 cells).[76] Consistent with these studies, Csermely and co-workers demonstrated the existence of a second ATP binding site at the Hsp90 C-terminus.[81,85] More recently, Ovsenek and co-workers compared the effect of novobiocin to geldenamycin on HSF-1 activity in Xenopus oocytes.[86] They demonstrated that oocytes treated with novobiocin followed by heat shock decreased HSF-1 DNA-binding and transcriptional activity in a dose-dependent manner while upon geldenamycin administration, a dose-dependent increase in unbound HSF-1 was observed following sub-maximal heat shock. Co-immunoprecipitation analyses showed that in the presence of novobiocin, Hsp90 associated with both monomeric and trimerized HSF-1, suggesting that novobiocin may not cause the dissociation of HSF-1 from Hsp90, which may avoid the detrimental heat-shock response for the treatment of cancer.[87]

Although small molecules that bind to the N- or C-terminus of Hsp90 both can alter its function as a molecular chaperone, current inhibitor development has focused primarily on the N-terminal domain, whereas significantly less investigation has been pursued on the C-terminus. Novobiocin was the first Hsp90 C-terminal inhibitor identified; however, other small molecules that target the C-terminus have since been discovered. In this chapter, the structure-activity relationships and the development of novobiocin analogues will be discussed.

10.2 Hsp90 C-terminal Inhibitors Derived from Novobiocin

10.2.1 Cytoprotective Hsp90 C-terminal Inhibitors

Although the efficacy ($\sim$700 µM in SKBr3 cells) exhibited by novobiocin is far from that desired for a lead compound, its ability to pharmacologically disrupt the Hsp90 C-terminal domain and cause similar downstream effects to Hsp90 N-terminal inhibitors was worthy of investigation. Extensive structural modifications to the first identified Hsp90 C-terminal inhibitor were expected to produce molecules that manifest increased potency. However, since no co-crystal structure was available, novobiocin was assumed to bind Hsp90 in a similar mode to that observed for binding to DNA gyrase, in which the carbamate of novobiocin formed a key hydrogen-bond network with aspartate 73 and the coumarin lactone produced complementary binding interactions with arginine 136 (Figure 10.5). The prenylated benzamide side-chain sat upon the rim of the binding pocket with a *cis*-amide exposed to the solvent and the prenyl side-chain rotated back into the binding pocket to form hydrophobic interactions.[88]

Initial studies on the novobiocin scaffold as described by Yu and co-workers proposed to maintain these critical interactions and the library of compounds investigated contained these presumably important structural features.[89] As shown in Figure 10.6, these novobiocin analogues contained a shortened amide side-chain that lacked the 4-hydroxy substituent (**A**), coumarin rings that did not contain a 4-hydroxy or amide linker (**B**), scaffolds that contained steric replacements of both the 4-hydroxy and benzamide side-chain (**C**) and 1,2-positional isomers of the noviosyl linkage (**D** and **E**). Lewis acid-catalyzed

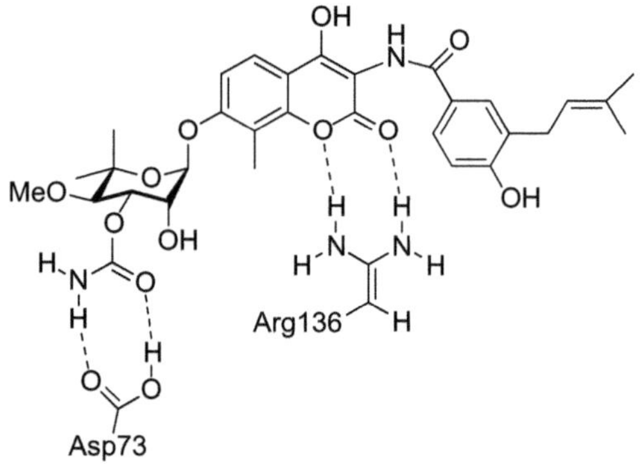

Figure 10.5 Key interactions between DNA gyrase and novobiocin.

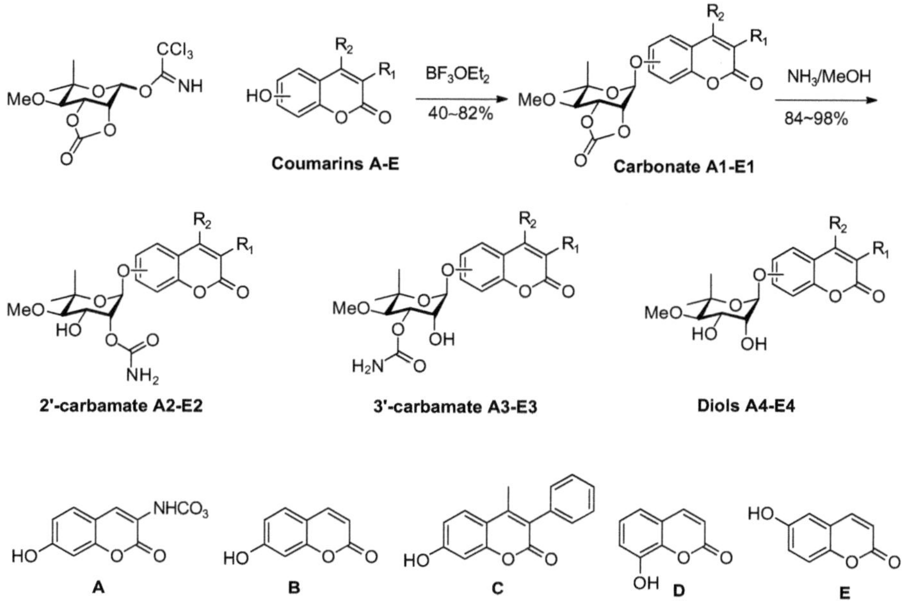

Inhibitors of the Hsp90 C-terminus

Figure 10.6 The first focused novobiocin library targeting Hsp90 C-terminus.

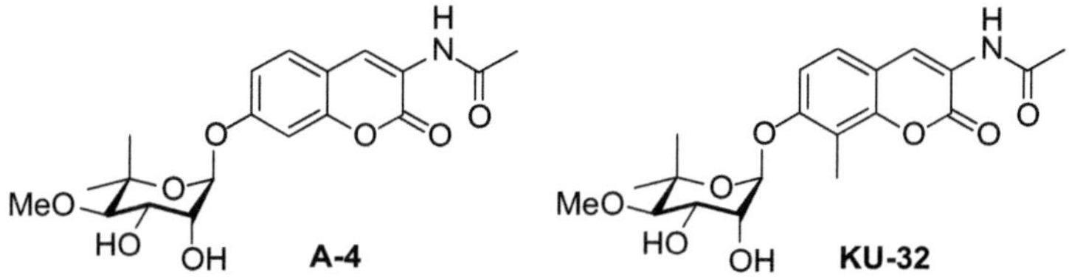

Figure 10.7 Structures of neuroprotective agents A4 and KU-32.

glycosidation of coumarin **A–E** with the trichloroacetimidate of noviose carbonate provided cyclic carbamates **A1–E1**, which upon treatment with methanolic ammonia provided 2′-carbamates **A2–E2**, 3′-carbamates **A3–E3** and the corresponding diols, **A4–E4**.

In an Hsp90 client protein degradation assay using phospho-Akt as an Hsp90 client protein, the results suggested that the attachment of noviose to the 7-position and an amide linker at the 3-position of the coumarin ring were critical for Hsp90 inhibitory activity, whereas modifications to the sugar were detrimental, including carbamoylation. This study not only established the first structure-activity relationships for novobiocin as an Hsp90 inhibitor, but also discovered **A4** (Figure 10.7), a novobiocin analogue that contains a shortened N-acyl side-chain and lacks the 4-deshydroxyl and 3′-descarbamoyl substituents. **A4** dramatically induced the degradation of other Hsp90-dependent client proteins such as Akt, AR and Hif-α at $\sim 10\,\mu M$ in the LNCaP cell line. Most intriguing was the observation that a molecule containing a shortened N-acyl side-chain in lieu of the prenylated benzamide moiety induced the heat-shock response at concentrations ~1000-fold lower than that required for client protein degradation, a phenomenon that had not been observed with N-terminal inhibitors. Due to its strong heat-shock induction activity and its non-toxic nature against neuronal cells, **A4** was evaluated as a neuroprotective agent. Pre-treatment of embryonic primary neurons with **A4** significantly reduced Aβ-induced toxicity in a dose-dependent manner, with an EC_{50} of ~6 nM.[90] Expression of Hsp90 and Hsp70 paralleled neuroprotective activity in this Aβ-induced toxicity model for AD. Furthermore, **A4** demonstrated a time-dependent linear transport across a brain micro-vessel endothelial layer and was determined not to be a substrate for the P-glycoprotein pump, indicating its potential to cross the blood-brain barrier with potential application for the treatment of neurodegenerative diseases, such as Alzheimer's disease.[70]

KU-32 (Figure 10.7), a structural variant of **A4**, manifests ~3-fold increase activity over **A4** in neuroprotection assays using SH-SY5Y cells. KU-32 manifests no toxicity against a wide range of cancer cells even at concentrations as high as 100 µM.[90] Recently, Menchen and Michaelis demonstrated that KU-32 exhibited robust protection in cellular and animal models of AD.[91] KU-32 protected primary cortical neurons against various cellular toxins such as $Aβ_{1-42}$, $Aβ_{25-35}$ and thapsigargin. Pharmacokinetic studies indicated that KU-32 is orally bioavailable and permeable across the blood-brain barrier. *In vivo* studies in two AD mouse models, JNPL3 and rTg4510, both of which contain the P301L Tau mutation and develop aberrant Tau tangles as well as cognitive impairments, showed that chronic KU-32 treatment significantly decreased CP-13 and AT8-labeled Tau in the premotor cortex. In the rTg4510 mouse model, which demonstrated neurodegenerative qualities in the brain, three different markers of neuroprotection, synaptophysin for synaptic damage, MAP2 labeling for neurotic dystrophy and NeuN for neuronal loss, were significantly reduced in the premotor cortex and hippocampus region upon KU-32 administration. At a molecular level, time- and region-specific and statistically significant increases in the somatosensory cortex of Hsp70 were observed; however, no significant changes in Hsp70 or HSF-1 levels were shown in lysate or brain homogenates.

However, utilizing KU-32 as a probe, Urban and co-workers demonstrated that KU-32 induced Hsp70 levels that are sufficient to improve several indices of neuronal function in an animal model of diabetic peripheral neuropathy (DPN).[92] This protection was not observed in Hsp70.1 and Hsp70.3 knockout mice, suggesting that the protective effect of KU-32 is manifested through Hsp70 up-regulation. KU-32 also protects against glucose-induced death of embryonic dorsal root ganglia (DRG) neurons *in vitro* and significantly decreases neuregulin 1 (NRG1)-induced degeneration of myelinated Schwann cell DRG neuron co-cultures that were obtained from wild-type mice. Mechanistically, the over-expression of Hsp70 upon KU-32 administration in myelinated DRG explants prepared from WT and Hsp70 mice were shown to be sufficient to antagonize the induction of c-jun, a negative regulator of myelination, and prevent the loss of myelin segments induced by NRG1.[93]

Moreover, KU-32 is readily bioavailable and when administered at a 20 mg/kg dosage once a week for 6 weeks, it reversed the NCV and sensory deficits in a streptozotocin-induced diabetic mouse model, demonstrating that upregulation of molecular chaperones reverses the sensory hypoalgesia associated with DPN. KU-32 significantly reversed the loss of intra-epidermal nerve fibers (iENFs), a physiological and morphologic marker of degenerative neuropathy.[94] The beneficial effects provided by KU-32 may be related to decreased hyperglycemia-induced oxidative stress and improved mitochondrial bioenergetics in sensory neurons.[95] Taken together, these studies provide biological and clinical rationale to support the hypothesis that modulation of molecular chaperones is a viable approach for the treatment of DPN.

10.2.2 Hsp90 C-terminal Inhibitors for the Treatment of Cancer

10.2.2.1 Transforming a DNA Gyrase Inhibitor into a Selective Hsp90 Inhibitor

Compounds **A4** and **KU-32** suggested that attachment of the noviose appendage to the 7-position of the coumarin ring and an amide linker at the 3-position are critical, whereas the 4-hydroxy substituent and the 3′-carbamate are detrimental. In addition, the 4-hydroxy appears critical for DNA gyrase inhibition due to its role in isomerization of the amide bond upon binding DNA gyrase. To confirm this SAR trend and to pursue more efficacious inhibitors, Burlison and co-workers developed two natural product analogues, 4-deshydroxy novobiocin (DHN1, Figure 10.8) and 3′-descarbamoyl-4-deshydroxy novobiocin (DHN2).[96] In an Hsp90 client protein degradation assay in SKBr3 cells, DHN1 induced both ErB2 and p53 degradation at concentrations between 5 and 10 µM, whereas DHN2 induced degradation

	DNA gyrase activity	Hsp90 activity
Novobiocin	0.9 µM	700 µM
DHN1	250 µM	~7.5 µM
DHN2	500 µM	~0.5 µM

Figure 10.8 Structures of cytotoxic agents DHN1 and DHN2.

between 0.1 and 1.0 µM, suggesting that DHN2 is more efficacious than DHN1. This study confirmed that the 4-hydroxyl and the 3′-carbamate are detrimental to Hsp90 inhibitory activity. In a DNA gyrase inhibitory assay, DHN1 manifested 50% inhibition at ~250 µM, which is in sharp contrast to novobiocin (~0.9 µM), thus confirming that the 4-hydroxyl is critical for DNA gyrase inhibition. DHN2 exhibited no activity in this surpercoiling assay even at the highest concentration tested (500 µM), consistent with the observation that 3′-carbamate maintains a key interaction with Asp73 in the DNA gyrase ATP binding pocket. Furthermore, these studies suggested that the original model for novobiocin binding to DNA gyrase does not accurately reflect the Hsp90 C-terminal binding pocket. Nonetheless, this study produced the first selective inhibitors of the Hsp90 C-terminus and transformed a clinically useful DNA gyrase inhibitor into a selective Hsp90 inhibitor.

10.2.2.2 SAR Studies on the Benzamide Side-chain

In an effort to develop SAR for the benzamide side-chain and to identify more efficacious Hsp90 C-terminal inhibitors, a series of novobiocin analogues with diverse structural features on the benzamide side-chain were synthesized and evaluated against a panel of cancer cell lines.[97] The non-substituted benzamide **1** manifested anti-proliferative activity against breast and prostate cancer cell lines (Figure 10.9). Most of the novobiocin analogues containing a monosubstituted phenyl ring manifested increased anti-proliferative activity. It fact, a *p*-hydrogen-bond acceptor and an *m*-aryl substituent on the benzamide were found most effective. Replacing the amide with a sulfonamide (**8**) resulted in diminished inhibitory activity, suggesting the importance of the amide functionality. Compounds **10** and **11** suggested that the hydrophobic pocket into which the benzamide side-chain projects may accommodate larger aromatic systems that may produce increased activity. Furthermore, increasing flexibility of the side-chain appeared to be well tolerated, and compounds with a rigid two carbon spacer between the amide and phenyl ring (**12**) manifested the greatest potency.

Subsequent optimization of the benzamide side-chain to contain a *p*-methoxy and various *m*-phenyl substituents (**13–18**, Figure 10.10) suggested that incorporation of polar substituents onto the second phenyl ring led to increased activity, including a phenol that produced the most potent inhibitors (**16–18**). In addition, replacement of the phenyl ring with an indole moiety led to increased anti-proliferative activity, suggesting that an H-bond donor is favored in this region. Collectively, the first set of SAR for the amide side-chain demonstrated that the amide is critical for anti-proliferative activity, and novobiocin analogues incorporating a biaryl or heterocyclic ring on the amide side-chain produces molecules that exhibit increased anti-proliferative activity.

10.2.2.3 SAR Studies on the Coumarin Ring

To elucidate structure-activity relationships for the novobiocin coumarin ring, analogues were designed to probe interactions that are manifested by

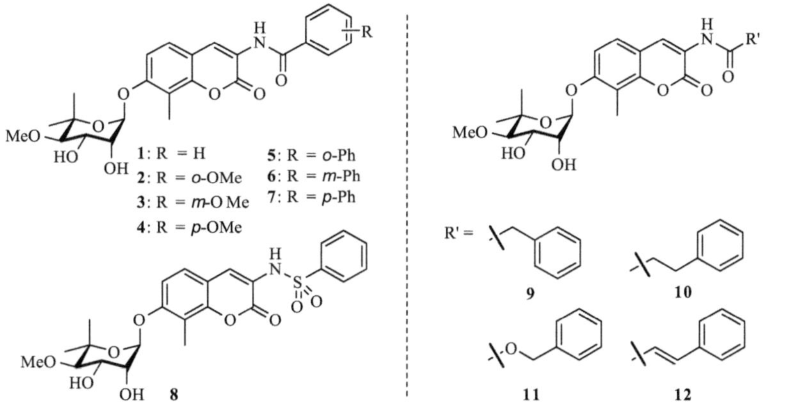

Inhibitors of the Hsp90 C-terminus

Entry	SkBr3	MCF-7	HCT-116	PL45	LNCaP	PC-3
1	21.5 ± 1.4	20.6 ± 0.4	13.0 ± 2.1	3.4 ± 0.6	72.0 ± 4.0	67.6 ± 9.7
2	>100	5.3 ± 1.3	>100	35.8 ± 3.4	N/T	9.1 ± 0.5
3	>100	5.6 ± 2.5	1.9 ± 0.6	2.8 ± 0.8	21.7 ± 2.0	N/T
4	15.6 ± 4.2	10.3 ± 0.9	15.9 ± 1.9	5.9 ± 2.1	7.3 ± 0.9	17.1 ± 4.3
5	39.1 ± 4.1	18.9 ± 7.0	32.7 ± 1.6	14.4 ± 2.4	17.3 ± 5.2	65.3 ± 5.6
6	13.0 ± 1.4	18.0 ± 3.8	12.8 ± 2.3	1.6 ± 0.2	1.6 ± 0.5	11.6 ± 1.4
7	16.3 ± 1.6	8.1 ± 6.0	3.6 ± 2.0	1.6 ± 0.2	44.9 ± 31.6	19.3 ± 5.1
8	>100	>100	>100	>100	>100	>100
9	21.4 ± 2.2	16.4 ± 0.4	13.2 ± 0.6	8.5 ± 1.7	10.4 ± 0.2	43.3 ± 13.4
10	10.2 ± 2.3	6.9 ± 0.3	5.4 ± 0.6	9.8 ± 0.2	92.8 ± 0.9	22.0 ± 5.8
11	17.8 ± 2.4	12.1 ± 0.1	13.0 ± 0.2	11.4 ± 0.6	N/T	N/T
12	2.6 ± 0.6	4.0 ± 0.3	3.2 ± 0.5	3.0 ± 1.0	4.5 ± 0.7	3.9 ± 0.05

Figure 10.9 Representative compounds identified in the benzamide side-chain SAR study and their anti-proliferative activity (IC$_{50}$, μM).

nucleotide bases. While it has been reported that the nucleotide binding site located in Hsp90 N-terminus exhibits selectivity for guanosine, the C-terminal region has been shown to bind both purines and pyrimidines. Therefore, mimics of the guanosine nucleus were chosen to take advantage of this apparent selectivity. The identification of additional interactions within the C-terminal nucleotide-binding domain that are typically manifested by neucleotide bases were hypothesized to produce compounds that would exhibit enhanced inhibitory affinity.[98] Therefore, hydrogen-bond acceptors were placed at the 5-, 6- and 8-positions of the coumarin ring to mimic those at the 6-, 7- and 3-positions of guanine, respectively (Figure 10.11). Since there is limited knowledge of the binding pocket due to the inability to obtain an Hsp90 co-crystal structure bound to C-terminal inhibitors, alkyl and aryl groups of

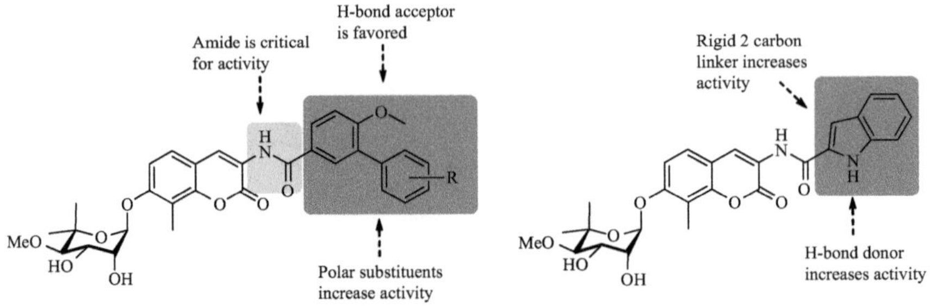

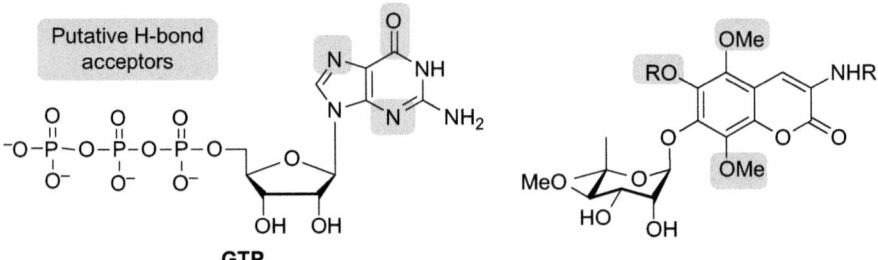

13–18: Biaryl derivatives

19: 2-Indole derivative

Entry	R	SkBr3	MCF-7	HCT-116	PL45	LNCaP	PC-3
13	o-OMe	7.1 ± 0.6	15.6 ± 6.3	5.2 ± 2.1	3.8 ± 1.7	6.0 ± 1.0	52.3 ± 34.7
14	m-OMe	7.5 ± 1.0	18.7 ± 1.8	5.1 ± 1.1	2.0 ± 0.6	2.1 ± 0.3	53.3 ± 4.5
15	p-OMe	7.8 ± 0.5	37.9 ± 2.3	24.0 ± 0.3	2.6 ± 0.5	3.3 ± 0.6	44.0 ± 20.3
16	o-OH	1.5 ± 0.1	1.5 ± 0.1	4.7 ± 1.4	1.4 ± 0.2	2.6 ± 0.6	16.6 ± 4.4
17	m-OH	2.9 ± 1.2	5.3 ± 1.5	4.5 ± 1.2	1.1 ± 0.0	2.5 ± 1.2	14.7 ± 2.4
18	p-OH	1.6 ± 0.2	2.3 ± 0.8	1.4 ± 0.1	1.9 ± 0.8	2.6 ± 0.5	22.3 ± 3.6
19	–	0.37 ± 0.06	0.57 ± 0.07	0.17 ± 0.01	0.47 ± 0.34	12.2 ± 0.0	22.3 ± 10.1

Figure 10.10 Representative compounds identified in the further benzamide side-chain SAR study and their anti-proliferative activity (IC$_{50}$, µM).

Figure 10.11 Design of coumarin analogues that mimic quinine interactions.

variable size were attached to the 5-, 6- and 8-positions of the coumarin ring to probe the size and dimension of the pocket.

The anti-proliferative activity manifested by these compounds was evaluated against both breast and prostate cancer cell lines (Figure 10.12), which demonstrated that the 5- and 6-substituted analogues that contain the biaryl side-chain were less active than analogues containing a hydrogen atom at these positions.[98] 8-Substituted biaryl analogues (**20** and **21**) manifested activity against all of these cell lines and was found to be more active against prostate cancer cells. Analogues containing the 2-indole side-chain exhibited increased activity over the corresponding biaryl derivatives, which corresponds with

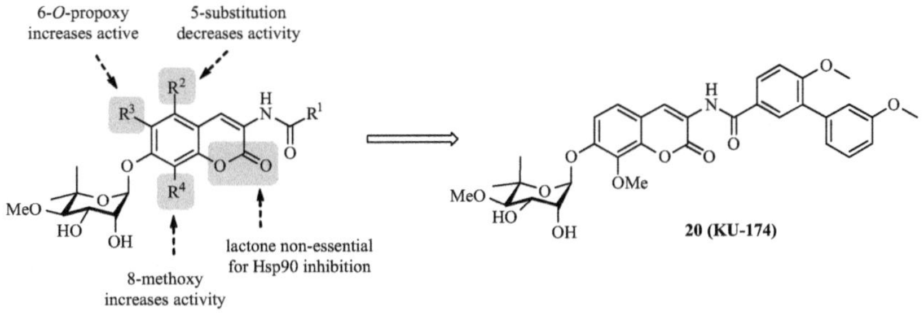

Figure 10.12 Representative compounds identified in the coumarin modification (IC$_{50}$, μM).

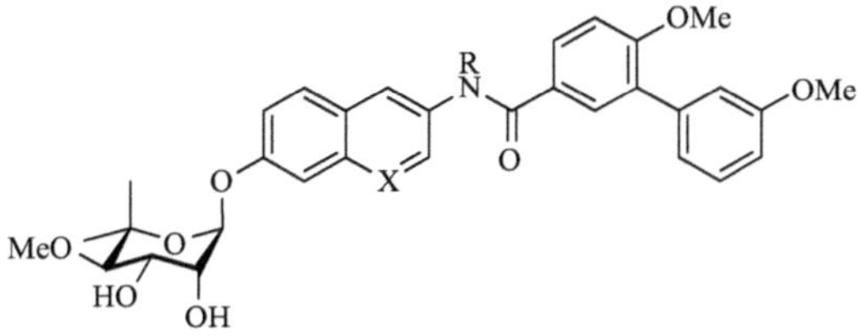

Entry	R^1	R^2	R^3	R^4	MCF-7	SkBr3	PC-3	LNCaP
20	biaryl	H	H	OMe	9.0 ± 5.4	13.9 ± 1.2	2.3 ± 2.9	1.1 ± 0.1
21	biaryl	H	H	Et	41.7 ± 4.0	28.6 ± 1.1	1.8 ± 0.6	1.6 ± 0.3
22	2-indole	H	OPr	Me	2.1 ± 0.1	2.1 ± 0.8	6.2 ± 1.8	1.8 ± 0.7
23	2-indole	H	OMe	Me	24.4 ± 1.2	25.1 ± 7.7	20.2 ± 9.8	10.5 ± 0.3
24	2-indole	OMe	H	Me	6.1 ± 1.7	9.0 ± 0.8	11.8 ± 1.3	12.9 ± 4.4

Entry	X	R^1	MCF-7	SkBr3	PC-3	LNCaP
25	N	Biaryl	13.1 ± 4.1	16.5 ± 6.2	17.6 ± 4.6	14.2 ± 0.4
26	CH	Biaryl	46.4 ± 5.3	38.9 ± 2.4	10.9 ± 0.7	19.6 ± 1.6

Figure 10.13 Structure and anti-proliferative activity of analogues with a coumarin replacement (IC$_{50}$, μM).

previously observed trends. An analogue containing a 6-propoxy-coumarin (**22**) was the most potent derivative produced from this library, which exhibited anti-proliferative activity against breast and prostate cancer cell lines. By comparison, methoxy substitution at the 5-position (**24**) and 6-position (**23**) were detrimental to inhibitory activity.

In addition, analogues bearing replacements for the coumarin lactone were synthesized to probe the importance of hydrogen bond donors/acceptors in this region. As shown in Figure 10.13, coumarin replacements manifested low micromolar anti-proliferative activity against breast and prostate cancer cell lines, with the quinoline analogue (**25**) maintaining better activity than the naphthalene analogue (**26**). Structure-activity relationships manifested by these compounds suggest the lactone moiety to provide beneficial hydrogen-bonding interactions with the novobiocin binding pocket; however, these interactions are not required for anti-proliferative activity. More importantly, these results indicate that further optimization of the coumarin ring may produce compounds that exhibit improved anti-proliferative activity.

Compound **20** (**KU-174**) was identified as one of the most interesting novobiocin analogues from these SAR studies.[71] The NCI60 screen determined that **KU-174** exhibited broad activity against multiple cancer cell lines, in particular, it was active against melanoma cell lines and exhibited cytotoxicity against the multi-drug-resistant ovarian adenocarcinoma cell line. In PC-3 and DU145 prostate cancer cell lines, KU-174 manifested cytostatic activity at a single dose of 10 μM. Notably, in the androgen-dependent prostate cancer cell line, LNCaP-LN3, **KU-174** exhibited anti-proliferative activity with a GI_{50} of 128 nM. **KU-174** also reduced PC3-MM2 cell viability in a dose-dependent manner at lower concentrations (0.1 ~ 1 μM) and manifested cytotoxic activity at higher concentration (10 ~ 50 μM). It appears that within 6 hours, **KU-174** manifests optimal cytotoxicity. In contrast, normal human renal proximal tubule epithelial cells dosed with **KU-174** for 6 hours showed no loss in viability, suggesting that KU-174 is relatively selective for prostate cancer cell lines. Consistent with novobiocin, **KU-174** induced a dose-dependent degradation of client proteins, including survivin, Akt, HER2, nestin, CXCR4 and caspase-3, in PC3-MM2 and LNCap-LN3 cells, supporting Hsp90 inhibition as the mechanism for cell death. A concomitant increase in heat-shock protein expression was not observed, which is consistent with C-terminal inhibition. *In vivo* efficacy studies in a mouse xenograft model of the PC3-MM2 tumor, **KU-174** significantly reduced tumor volume without apparent toxicity.

10.2.2.4 *Click Chemistry for Investigation of the Amide*

Preliminary SAR studies conducted by modification of the benzamide sidechain and coumarin core suggested that both hydrogen bonding and geometry of the amide bond are important for novobiocin binding. Due to similarities of the electronic nature and structural characteristics between the amide and 1,2,3-triazole, it was proposed that inclusion of a 1,2,3-triazole could function as a bioisosteric replacement for the amide.[99] In contrast to amides, the 1,2,3-triazole ring is stable to metabolic hydrolysis and can be easily accessed by well-established click-chemistry.[100,101] In order to investigate the electronic nature of the aryl side-chain, analogues containing neutral (**28**, Figure 10.14), electron-withdrawing (**29**) and electron-donating (**30, 31**) phenyl rings, as identified during prior studies, along with sterically demanding side-chains (**32–34**) were prepared. Although substituted aryl compounds (**28–31**) showed minimal activity against both SKBr3 and MCF-7 cell lines, suggesting *para*-substitution of the aryl ring is not well tolerated, analogues bearing sterically demanding biaryl (**32**), homologated aryl groups (**33**) or indole (**34**) side-chains exhibited

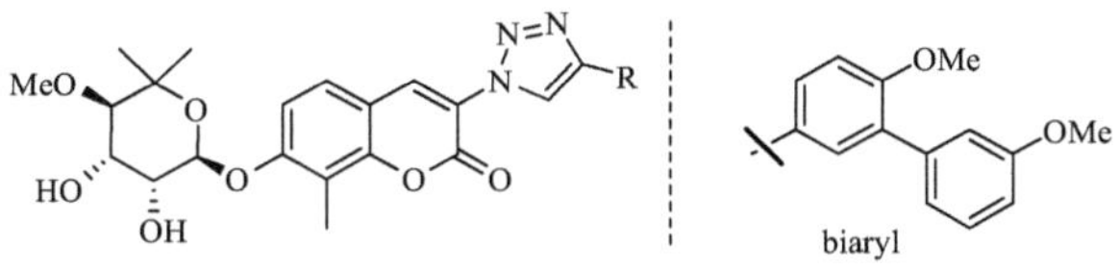

Inhibitors of the Hsp90 C-terminus

Entry	R	MCF-7	SkBr3
27	Phenyl	64.4 ± 5.8	>100
28	4-Methylphenyl	>100	>100
29	2,4-Difluorophenyl	>100	>100
30	4-Methoxylphenyl	>100	>100
31	4-(N,N-Dimethyl)aminophenyl	>100	>100
32	Biaryl	13.2 ± 3.9	21.2 ± 6.0
33	E-styryl	14.6 ± 0.4	51.9 ± 3.0
34	3-Indonyl	18.3 ± 4.7	8.2 ± 0.1

Figure 10.14 Structure and anti-proliferative activity of analogues with a triazole as amide replacement (IC$_{50}$, µM).

comparable activity to their amide counterparts. These data suggest that the C-terminal binding pocket contains a large hydrophobic pocket that can accommodate a variety of substituents. In addition, these results also suggest the importance of a hydrogen bond donor in the form of an amide linker and the steric bulk of the side-chain. Introduction of the triazole moiety in lieu of amide resulted in a loss of hydrogen bonding interactions and loss of activity. However, this loss could be compensated by the introduction of steric bulk into the amide side-chain. A similar steric bulk preference was previously reported with the related novobiocin analogues, and appears to be a contributing factor for anti-proliferative activity.[97]

10.2.2.5 SAR Studies on the Noviose Sugar and Sugar Mimics

Noviose is the sugar attached to novobiocin at the 7-position and is prepared *via* extensive efforts. Even the most efficient synthesis of noviose requires more than 10 steps and the overall yields are less than optimal.[102,103] During SAR studies on the coumarin core, compounds **35** and **36** (**KU-135**, Figure 10.15), novobiocin analogues that contain an acyl ester in lieu of noviose, were shown to manifest more potent anti-proliferative activity than their noviosylated counterparts.[104] In particular, **KU-135** exhibited better inhibitory activity against human leukemic cell Jurkat T-lymphocytes than the Hsp90 N-terminal inhibitor, 17-AAG (0.42 µM *vs.* 1.3 µM). **KU-135** causes G2/M arrest of these cells and induced mitochondria-mediated apoptosis. In addition, **KU-135**

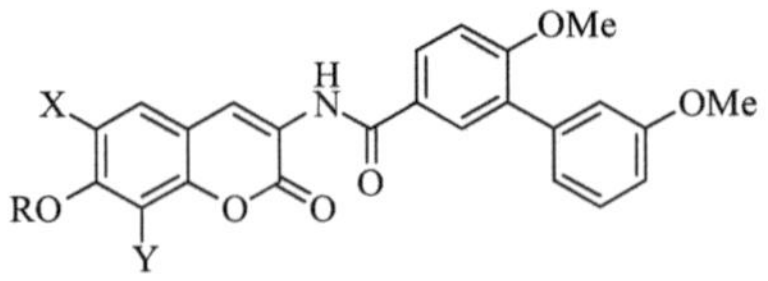

Entry	X	Y	R	SkBr3	MCF-7	LNCaP	PC-3
35	H	CH_3	Ac	0.98 ± 0.02	1.40	1.50 ± 1.00	2.85 ± 1.66
36	OCH_3	CH_3	Ac	5.72 ± 0.02	1.50 ± 0.24	1.05 ± 0.59	1.69 ± 2.05

Figure 10.15 Structure and anti-proliferative activity of analogues without noviose (IC_{50}, μM).

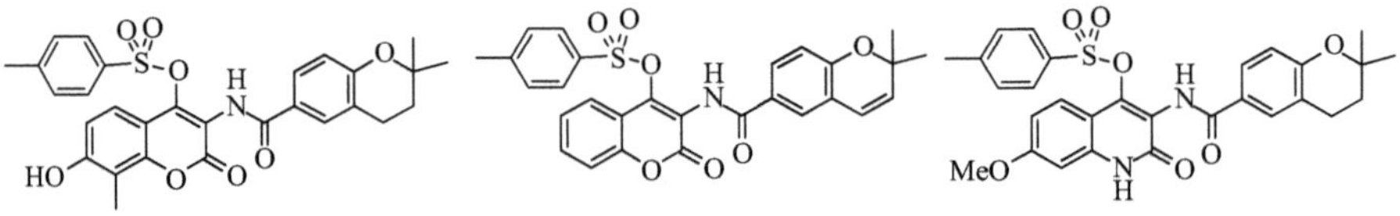

4TCNA ~50 μM (MCF-7)   **4DHTCNA** ~35 μM (MCF-7)   **4TCCQ** ~8 μM (MCF-7)

Figure 10.16 Structure and anti-proliferative activity of denoviose 4-tosyl-analogues.

induced Hsp90-dependent client protein (*e.g. p*-Akt) degradation in a time-dependent manner. Three complementary approaches were used to confirm Hsp90 inhibitory activity and they included a novobiocin-binding assay, proteolytic finger-printing of Hsp90 and surface plasmon resonance spectroscopy. The attachment of an acetyl group in lieu of noviose resulted in a 10-fold improvement in Hsp90 inhibitory activity, and suggested that while the sugar moiety may play a role in maintaining binding interactions with Hsp90, not all of these functionalities are necessary.

Consistent with our studies, Alami and co-workers demonstrated that denoviosylated novobiocin analogues containing a tosyl group at the C-4 position of the coumarin ring (such as 4TCNA and 4DHTCNA, Figure 10.16), exhibited mid-micromolar Hsp90 inhibition.[105] Replacement of the coumarin ring with a 2-quinolone moiety produced analogues (4TCCQ) with enhanced cytotoxicity against the MCF-7 cancer cell line, as compared to novobiocin.[106] However, the modest activity associated with these 4-tosyl analogues indicated that modification of the 4-position may improve binding affinity, but it is mutually exclusive with regards to modification at the 7-position.

Encouraged by the enhanced anti-proliferative activity manifested by denoviosylated analogues, surrogates of noviose were investigated, including 5-, 6- and 7-membered sugar derivatives, as well as cyclohexyl-derived mimics for the replacement of noviose (Figure 10.17).[107] Analogue **38** did not show any activity even at the highest concentration tested (100 μM), suggesting that the 3′-hydroxy is important. Compounds **37** and **39** exhibited similar activity to **14**, suggesting that the 2′-hydroxy is dispensable, whereas the

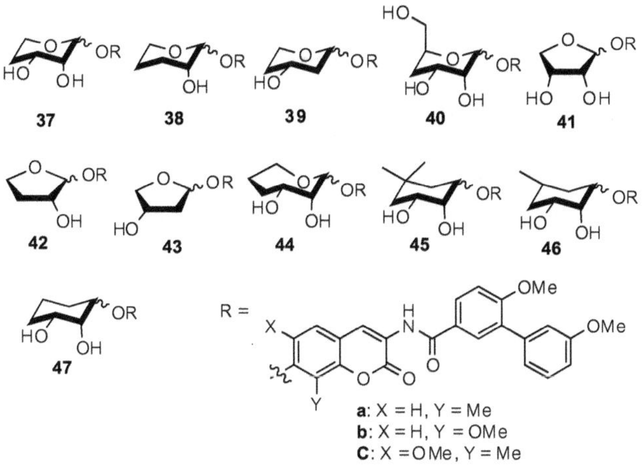

Entry	Anomer	SKBr3	MCF-7	LNCAP-LN3	PC3-MM2
14		7.5 ± 1.0	18.7 ± 1.8	2.1 ± 0.3	53.3 ± 4.5
37a	β	6.23 ± 0.52	2.56 ± 0.08	3.59 ± 3.40	NT
37a	α	6.71 ± 0.80	42.56 ± 2.48	1.20 ± 0.83	12.88 ± 9.30
39a	α	9.28 ± 0.04	11.20 ± 0.49	8.56 ± 9.00	NT
39b	β	8.75 ± 0.49	95.77 ± 3.21	11.50 ± 5.66	NT
39c	α	1.37 ± 0.14	>100	3.27 ± 0.09	2.35 ± 1.11
40a	β	9.59 ± 0.42	11.07 ± 0.47	3.23 ± 3.76	4.58 ± 2.78
41a	β	12.46 ± 0.52	37.17 ± 1.68	3.65 ± 2.83	NT
41b	β	7.57 ± 1.05	11.73 ± 0.78	2.60 ± 1.51	10.64 ± 20.83
41c	β	9.45 ± 0.19	3.37 ± 0.07	4.72 ± 1.48	9.56 ± 2.58
43a	β	35.24 ± 1.76	>100	3.72	10.05 ± 6.75
43a	α	>100	>100	4.96 ± 10.58	26.79 ± 58.24
44a	β	2.38 ± 0.06	>100	>100	NT
45b	–	3.45 ± 1.73	1.56 ± 0.03	NT	NT
46b	–	6.38 ± 0.71	8.52 ± 0.36	NT	NT
47b	–	7.44 ± 0.36	5.46 ± 0.36	NT	NT

Figure 10.17 Structures and anti-proliferative activities of analogues with simplified sugar replacement (IC_{50}, μM).

$6'$-gem-dimethyl and the $5'$-methoxy are not essential for inhibitory activity. Similar trends were observed for analogues that contain 5-membered mimics (**41–43**). Compound **44** was completely inactive against both breast cancer cell lines and prevented further exploration of ring-expanded analogues. It was interesting to note that when the 8-position of the coumarin ring contained a methoxy group (**45b–47b**), the sugar was not necessary.

Initial SAR studies (Figure 10.18) indicated that the sugar portion may play a significant role towards solubilization of the aglycon. Since N-heterocycles are found in a variety of biologically active compounds, and impart excellent solubility, the insertion of nitrogen into the diol-containing ring system was pursued (Figure 10.19). In addition, compounds that lack the diol altogether

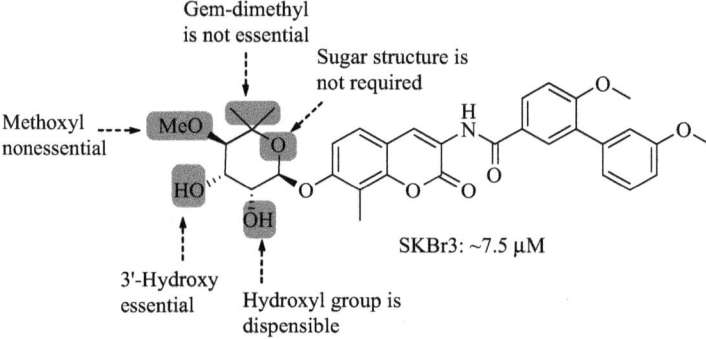

Figure 10.18 SAR observations for noviose surrogates.

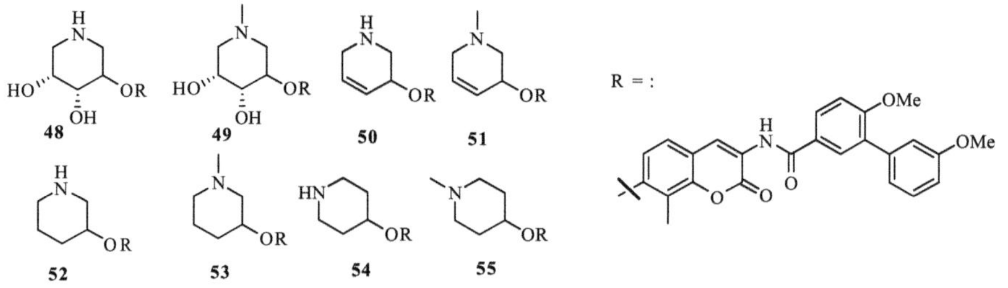

Entry	SkBr3	MCF-7	LNCaP	PC-3
48	2.91 ± 0.90	2.07 ± 0.86	4.90 ± 9.16	6.33 ± 4.12
49	3.92 ± 0.32	1.85 ± 0.02	0.59 ± 0.54	2.99 ± 1.27
50	1.61 ± 0.05	1.73	4.27 ± 0.05	4.40 ± 0.06
51	2.92 ± 1.33	5.29 ± 0.23	1.22	1.73 ± 1.80
52	1.16 ± 0.16	1.63 ± 0.28	3.02 ± 0.97	2.57 ± 1.13
53	2.06 ± 0.57	5.04 ± 0.02	1.22 ± 0.17	4.23 ± 1.68
54	1.19 ± 0.06	1.47 ± 0.02	3.38 ± 1.25	4.12 ± 1.29
55	1.34 ± 0.18	1.51 ± 0.24	4.12 ± 0.16	3.13 ± 0.67

Figure 10.19 Structures and anti-proliferative activities of analogues with N-heterocycle as noviose replacement (IC_{50}, μM).

and instead contained nitrogen at various locations within the heterocycle (**52–55**) were investigated. Anti-proliferative activity against breast cancer cell lines determined that the majority of these secondary and tertiary amines exhibited activity between 1 and 3 µM, making them ∼700-fold more efficacious than novobiocin. Notably, these data showed that dihydroxylation of the piperidine ring is not required for the manifestation of anti-proliferative activity. In fact, analogues containing a saturated or unsaturated piperidine ring were effective inhibitors. Surprisingly, there was little effect upon transposition of the nitrogen within the piperidine ring, as all of the analogues exhibited similar activity. The secondary and tertiary amines exhibited similar activity as well. Overall, these azasugar analogues consistently exhibited low micromolar anti-proliferative activity, and represent an improved sugar mimic that can be readily prepared or purchased, which overcomes the burden of noviose preparation.

The promising anti-proliferative activity and solubility of these azasugar-containing novobiocin analogues inspired the construction of ring-opened amine-containing surrogates to investigate the significance of the constrained ring system (Figure 10.20). Interestingly, the two-carbon tertiary amine analogue (**57**) demonstrated notable anti-proliferative activity at ∼1 µM, while the analogue containing a three methylene linker (**58**) manifested anti-proliferative activity in the mid-nanomolar range. In addition to the aliphatic amino analogues, several simplified non-sugar molecules, such as glycerol (**59**), carbamates (**60–62**), phosphate (**63**) and sulfonic esters (**64** and **65**), were also appended to the 7-position *in lieu* of noviose in an effort to further mimic the functionality present in **KU-135**. The carbamate (**60**) and methylated carbamate (**61**) produced low micromolar activity while the phosphate ester (**63**) and sulfonic esters (**64** and **65**) manifested little or no activity, suggesting that hydrogen bonding is important and that hydrophobic bulk is required for inhibition.

10.2.2.6 Summary of Noviose SAR

Based on SAR studies of the noviose sugar, *N*-methyl-4-hydroxy-piperidine and 3-(dimethylamino)propan-1-ol were identified as optimal replacements. These surrogates were coupled with previously optimized coumarin scaffolds that contain either the 2-indole, or prenylated/biarylphenol benzamide side-chain, and, as shown in Figure 10.21, these analogues exhibited efficacious activity against the breast cell lines, SKBr3 and MCF-7. Most of these compounds manifested submicromolar activity against the HER2 over-expressing SKBr3 cell line. When compared to their biaryl counterparts, the piperidine-containing analogues that contain the indole side-chain were shown to exhibit increased anti-proliferative activity when attached to the 8-methyl and 6-methoxy coumarin (**66** and **68**). Notably, **68** (**KU-363**) was 10-fold more active against SKBr3 cells, compared to its biaryl counterpart. In the case of the prenylated benzamide side-chain (**72–84**), the acetylated phenols demonstrated comparable activity to the hydrolyzed, free phenol derivatives. Attachment of

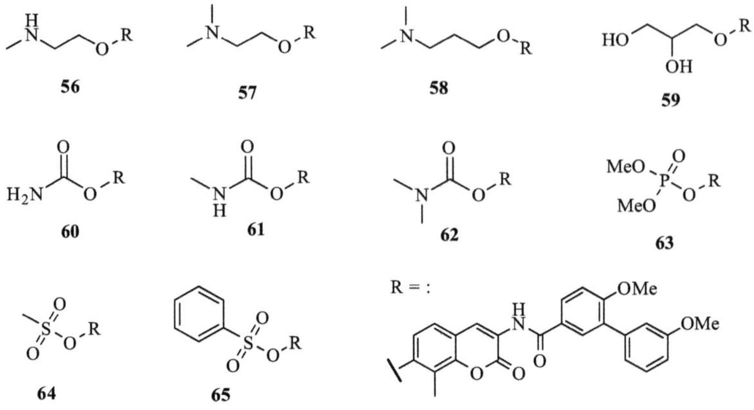

Figure 10.20 Structures and anti-proliferative activities of analogues with linear noviose replacement (IC$_{50}$, μM).

Entry	SkBr3	MCF-7	LNCaP	PC-3
56	5.36 ± 0.08	9.80 ± 0.11	>100	14.01 ± 0.28
57	1.02 ± 0.13	1.46 ± 0.08	6.65 ± 12.43	4.17 ± 19.64
58	0.60 ± 0.01	0.50 ± 0.03	37.5 ± 54.2	12.9 ± 11.4
60	3.02 ± 0.56	1.16 ± 0.08	2.61 ± 1.09	6.37 ± 4.31
61	2.40 ± 0.16	1.72 ± 0.16	3.75 ± 0.76	5.22 ± 2.16
62	39.85 ± 0.48	74.35 ± 3.92	5.40 ± 6.58	>100
63	>100	>100	3.65 ± 5.73	>100
64	7.33 ± 0.96	8.85 ± 0.88	19.96 ± 42.67	>100
65	>100	>100	0.58 ± 1.34	>100

the azasugar increased activity against both cell lines, and the 6-methoxy coumarin manifested the most efficacious inhibitory activity (**74** and **77**, **80** and **83**). Although data from the biaryl containing analogues suggested that attachment of the aliphatic amine would result in increased activity (**55** *vs.* **58**), the anti-proliferative activity of these indole analogues (**69–71**) was similar.

It is worth noting that **68** was identified as the most active compound against the SKBr3 cancer cell line. When evaluated against head and neck squamous cell cancer cell lines (HNSCC) including MDA-1986, JMAR, UM-SCC-2, JUH-011 and MRC-5, it exhibited an IC$_{50}$ value comparable to 17-AAG.[108] It was shown to induce apoptosis at ~1 μM through the cleavage of PARP and activation of caspase 3 within 24 hours. The dose-dependent degradation of Hsp90 client proteins, including Akt and Raf-1, were also observed in the MDA-1986 lysate upon administration of **68**. In an *in vivo* mouse model, high

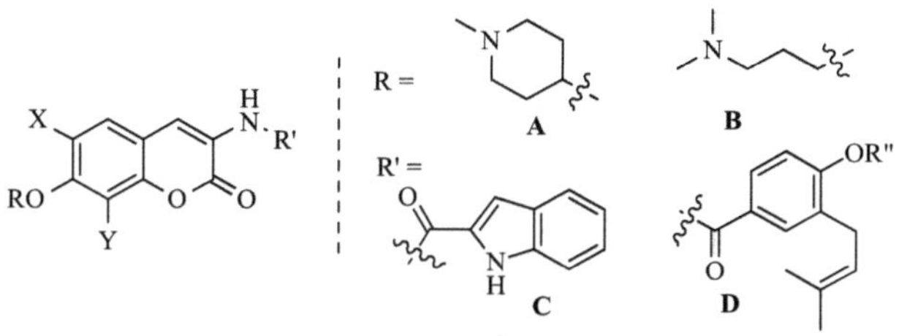

Inhibitors of the Hsp90 C-terminus 281

Entry	R	X	Y	R'	R"	SKBr3	MCF-7	PC3-MM2
66	A	H	CH$_3$	C	–	0.48 ± 0.09	0.57 ± 0.03	11.40 ± 5.25
67	A	H	OCH$_3$	C	–	2.58 ± 0.28	1.86 ± 0.08	7.93 ± 4.18
68	A	OCH$_3$	CH$_3$	C	–	0.11 ± 0.01	0.52 ± 0.04	0.87 ± 0.46
69	B	H	CH$_3$	C	–	1.13 ± 0.01	5.23 ± 0.22	13.69 ± 0.18
70	B	H	OCH$_3$	C	–	1.50 ± 0.13	1.41 ± 0.09	8.95 ± 6.11
71	B	OCH$_3$	CH$_3$	C	–	0.57 ± 0.09	0.56	2.58 ± 4.47
72	A	H	CH$_3$	D	Ac	0.58 ± 0.04	1.18 ± 0.16	2.12 ± 3.32
73	A	H	OCH$_3$	D	Ac	1.07 ± 0.14	1.64 ± 0.24	3.98 ± 0.06
74	A	OCH$_3$	CH$_3$	D	Ac	0.42 ± 0.01	0.58 ± 0.02	1.41 ± 0.04
75	A	H	CH$_3$	D	H	0.76 ± 0.14	1.09 ± 0.08	1.37 ± 1.42
76	A	H	OCH$_3$	D	H	0.92 ± 0.01	1.54 ± 0.21	3.53 ± 0.01
77	A	OCH$_3$	CH$_3$	D	H	0.42 ± 0.01	0.54 ± 0.02	2.26 ± 1.43
78	B	H	CH$_3$	D	Ac	0.46 ± 0.15	1.18 ± 0.02	1.42 ± 0.05
79	B	H	OCH$_3$	D	Ac	0.78 ± 0.17	2.14 ± 0.22	4.59 ± 4.23
80	B	OCH$_3$	CH$_3$	D	Ac	0.36 ± 0.03	0.70 ± 0.03	1.46 ± 0.03
81	B	H	CH$_3$	D	H	0.44 ± 0.02	1.35 ± 0.30	1.81 ± 1.22
82	B	H	OCH$_3$	D	H	0.77 ± 0.08	3.26 ± 0.26	9.24 ± 17.79
83	B	OCH$_3$	CH$_3$	D	H	0.39 ± 0.06	0.80 ± 0.07	1.38 ± 0.02

Figure 10.21 Structures and anti-proliferative activities of analogues with optimized structural features (IC$_{50}$, μM).

doses of KU363 showed an 88% animal response with the low dose producing an astonishing 75% response rate after three weeks of treatment. In parallel, 100% of control animals had progressive disease (PD), while 100% of cisplatin animals showed some response. Although cisplatin showed an overall better efficacy than **KU-363**, it manifested a worse toxicity profile. All the mice treated with cisplatin reported a 20% or greater weight loss after a three-week treatment. In comparison, weight loss was observed in one animal within each of the low-dose and high-dose **68** groups.

10.2.2.7 Novobiocin Analogues with Second Generation Noviose Surrogates

Considering the synthetic feasibility and anti-proliferative activity of **75** and **78**, which showed lead-like properties, including increased anti-proliferative activity and solubility, Huang and co-workers pursued a three-dimensional quantitative structure-activity relationship (3D-QSAR) model and suggested that modifications to the amine may produce analogues with improved activity.[109] Consequently, a second generation of amino-analogues that focused around **75** and **78** were designed, synthesized and evaluated.[110]

Sugar SAR studies suggested that a three-carbon spacer linking the 7-coumarin phenol with the amine was optimal.[107] Although **75** exhibited slightly increased anti-proliferative activity compared to **78**, increasing the number of rotatable bonds is unfavored due to the entropic penalties. Therefore, novobiocin analogues containing rigid heterocyclic piperidine and pyrrolidine derivatives were proposed. As shown in Figure 10.22, C-linked heterocyclic analogues (**84–87**) manifested similar anti-proliferative activities to **75**, suggesting that various ring structures are readily accommodated. Interestingly, the secondary amino analogues exhibited greater anti-proliferative activity than the corresponding tertiary amines (**86** vs. **85**), which was further exacerbated against MCF-7 cells. Although piperizine analogues (**88** and **89**) produced

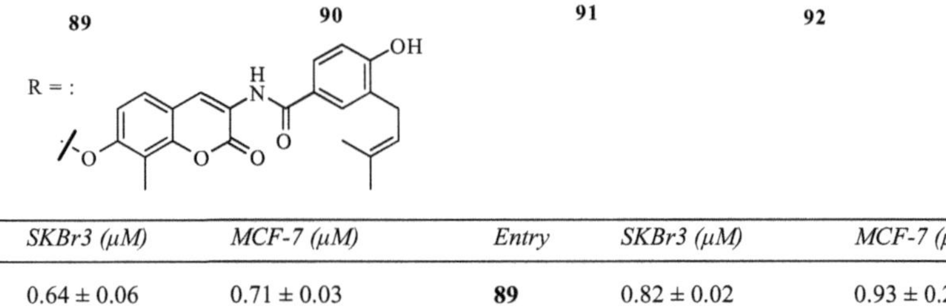

Entry	SKBr3 (µM)	MCF-7 (µM)	Entry	SKBr3 (µM)	MCF-7 (µM)
84	0.64 ± 0.06	0.71 ± 0.03	89	0.82 ± 0.02	0.93 ± 0.21
85	1.30 ± 0.63	2.64 ± 0.45	90	>50	>50
86	0.60 ± 0.01	0.79 ± 0.10	91	>50	>50
87	0.92 ± 0.56	1.55 ± 0.04	92	0.72 ± 0.01	0.99 ± 0.04
88	1.01 ± 0.03	1.20 ± 0.12			

Figure 10.22 Structures and anti-proliferative activities of heterocyclic analogues (IC_{50}, µM).

comparable inhibitory activity to **75**, compounds **90** and **91** were inactive. However, deleting one carbon atom (**92**) restored activity, suggesting that hydrophobicity is limited in this region and that the incorporation of heteroatoms is necessary for further expansion.

The 3D-QSAR studies suggested that sterically bulky N-substitution of **78** could lead to compounds that exhibit increased anti-proliferative activity.[109] Therefore, analogues containing both linear and branched N-alkyl substitutions were pursued in an effort to maintain hydrophobicity while allowing for H-bond donors. As shown in Figure 10.23, a linear (**93, 94** and **96**) or branched (**98–100, 102** and **103**) chain was shown to be appropriate for binding. The diethylamino (**95**) and diisopropylamino (**101**) analogues produced decreased anti-proliferative activity, while the dipropylamino analogue manifested no activity (**97**), indicating a limit to the amount of available space. Glycine ester analogue **104** and amide analogue **106** exhibited decreased anti-proliferative activity, while alcohol **107** retained activity against both cancer cell lines, supporting further extension into the binding site requires the incorporation of a hydrogen bond donor.

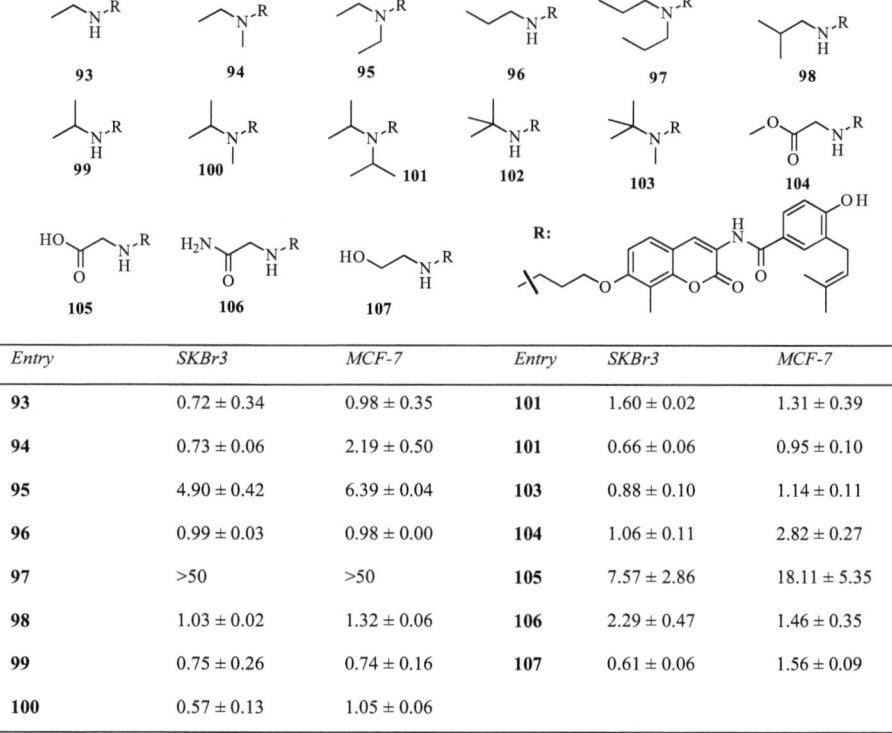

Entry	SKBr3	MCF-7	Entry	SKBr3	MCF-7
93	0.72 ± 0.34	0.98 ± 0.35	101	1.60 ± 0.02	1.31 ± 0.39
94	0.73 ± 0.06	2.19 ± 0.50	101	0.66 ± 0.06	0.95 ± 0.10
95	4.90 ± 0.42	6.39 ± 0.04	103	0.88 ± 0.10	1.14 ± 0.11
96	0.99 ± 0.03	0.98 ± 0.00	104	1.06 ± 0.11	2.82 ± 0.27
97	>50	>50	105	7.57 ± 2.86	18.11 ± 5.35
98	1.03 ± 0.02	1.32 ± 0.06	106	2.29 ± 0.47	1.46 ± 0.35
99	0.75 ± 0.26	0.74 ± 0.16	107	0.61 ± 0.06	1.56 ± 0.09
100	0.57 ± 0.13	1.05 ± 0.06			

Figure 10.23 Structures and anti-proliferative activities of acyclic amino-analogues (IC_{50}, μM).

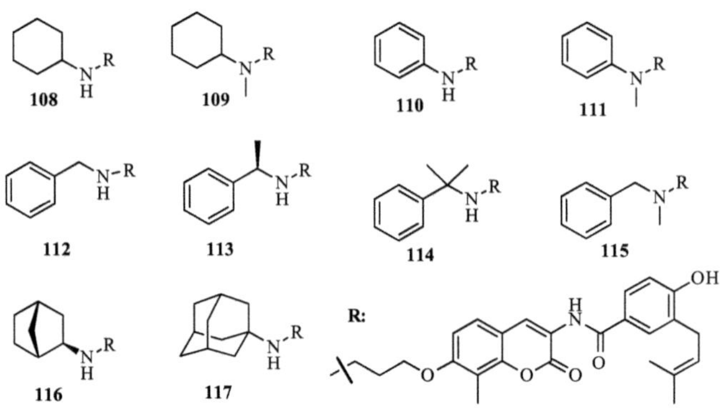

Entry	SKBr3	MCF-7	Entry	SKBr3	MCF-7
108	0.66 ± 0.12	1.07 ± 0.35	114	0.96 ± 0.01	1.07 ± 0.01
109	0.97 ± 0.03	1.08 ± 0.07	115	10.73 ± 0.49	11.52 ± 4.16
112	0.47 ± 0.16	0.91 ± 0.01	116	0.77 ± 0.13	1.03 ± 0.12
113	1.00 ± 0.31	1.45 ± 0.20	117	0.31 ± 0.04	0.32 ± 0.03

Figure 10.24 Structures and anti-proliferative activities of analogues with bulky acyclic amine (IC_{50}, μM).

Although no improved anti-proliferative activity was observed amongst the first two sets of compounds, steric bulk appears to be tolerated in the binding site. In fact, subsequent modifications showed that the cyclohexyl (**108** and **109**, Figure 10.24), benzyl (**112**), substituted benzyl (**113** and **114**) and bicyclic alkyl (**116** and **117**) analogues to possess inhibitory activity. However, aniline-containing analogues (**110** and **111**) produced no activity, even at a concentration of 50 μM, potentially a result of its altered hydrogen bonding properties. The *N*-adamantyl-containing analogue, **117**, manifested the most potent anti-proliferative activity in this series, suggesting that significant hydrophobic space is available. In addition, **117** exhibited much lower toxicity against normal cells (MRC5, 5.42 μM; HMLE, 15.05 μM), and therefore, provides a potential probe for further development.

10.2.2.8 3D-QSAR-Assisted Benzamide Phenyl Optimization

In an effort to gain better understanding of the determinants produced upon novobiocin binding to Hsp90 and to produce a set of rational, quantitative rules for the development of new analogues, a three-dimensional quantitative structure-activity model (3D-QSAR) of the most potent novobiocin-analogues was generated.[111] The model suggested that the noviose sugar should be replaced by an amine, the amide side-chain should exhibit hydrophobic character and the *para*-oxygen atom on the amide side-chain should be incorporated for the maintenance of a hydrogen bond donor.

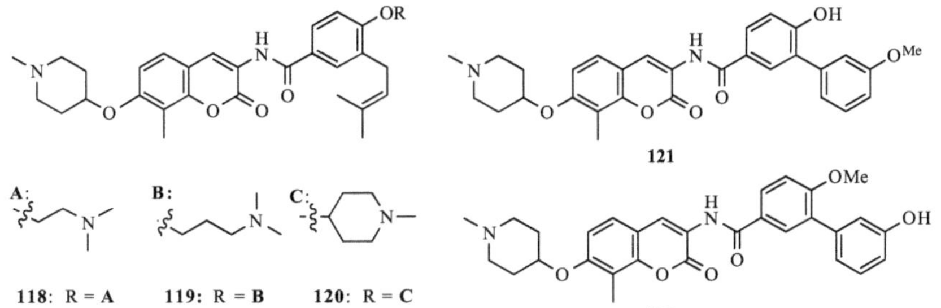

Entry	SKBr3 (µM)	MCF-7 (µM)
118	0.33 ± 0.00	0.72 ± 0.08
119	0.25 ± 0.02	0.46 ± 0.05
120	0.21 ± 0.00	0.47 ± 0.081
121	0.86 ± 0.16	0.81 ± 0.14
122	0.31 ± 0.04	0.38 ± 0.08

Figure 10.25 Structure and anti-proliferative activity of novobiocin analogues containing a modified benzamide side-chain (IC$_{50}$, µM).

Since the 3D-QSAR model suggests that the presence of a hydrogen bond donor and a hydrogen bond acceptor on the side-chain can increase activity, two approaches were undertaken. The first approach required alkylation of the 4′-phenol with an alkyl tertiary amine, in which the 4′-oxygen serves as a hydrogen acceptor and the protonated amine served as a hydrogen donor. Three amines (2-(dimethylamino)ethanol, 3-(dimethylamino)-propan-1-ol and 1-methylpiperidin-4-ol) were attached to the 4′-phenol benzamide side-chain of **75**, a lead compound identified from these studies. Evaluation of these benzamide analogues against breast cancer cell lines demonstrated that attachment of the phenol with alkyl amines (**118–120**, Figure 10.25) led to molecules that exhibit increased anti-proliferative activity. For biaryl analogues (**121** and **122**), replacement of the prenyl group with a substituted phenyl ring resulted in similar inhibitory activity (**121**) to **75**; however, switching the phenol from the first to the second benzene ring increased activity almost three-fold (**122**). This latter result suggests that the incorporation of a hydrogen bond donor into the second ring is favorable. Overall, these results support the use of quantitative structure-based rational studies to investigate the chemical space of molecules targeting the Hsp90 C-terminus.

10.2.2.9 Dimeric Inhibitors

Coumermycin A1 (Figure 10.4), a dimeric natural product, manifests ten-fold greater inhibitory activity than novobiocin (70 µM vs. 700 µM) against the

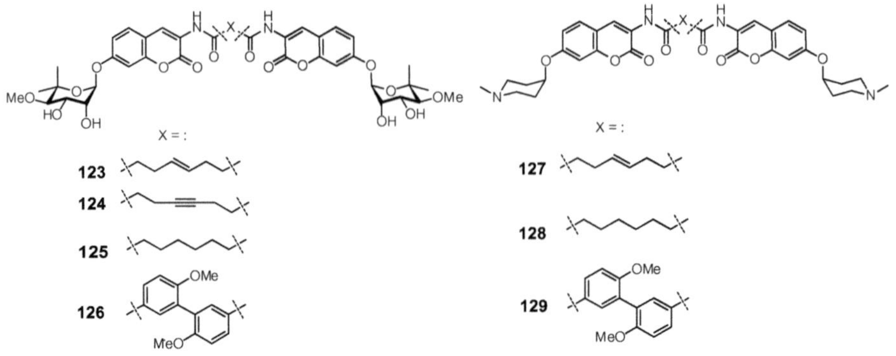

Figure 10.26 Structures and anti-proliferative activities of dimeric novobiocin analogues (IC$_{50}$, μM).

Entry	SKBr3 (μM)	MCF-7 (μM)	Entry	SKBr3 (μM)	MCF-7 (μM)
123	1.5 ± 0.1	3.9 ± 0.7	127	0.15 ± 0.01	0.27 ± 0.02
124	82.2 ± 0.7	16.2 ± 0.2	128	1.2 ± 0.3	2.8 ± 0.3
125	1.9 ± 0.2	2.7 ± 1.0	129	5.0 ± 0.7	14.2 ± 2.3
126	0.11 ± 0.05	0.72 ± 0.21			

Hsp90 protein folding machinery. This increase may be the consequence of the homodimeric nature of Hsp90, wherein each side of the inhibitor could bind to each monomeric unit of Hsp90, producing a "bidentate" inhibitory effect. Based on this rationale, the dimerization of A4 was pursued in an effort to generate compounds that inhibit Hsp90 function more effectively than coumermycin A1. Two approaches were taken towards this objective:[112] the first involved the preparation of A4 dimers linked through *para*- or *meta*-phenyl rings in lieu of the pyrrole bridge present in coumermycin A1. Unfortunately, neither of these dimers manifested activity at the highest concentration tested. The second approach utilized a cross metathesis reaction of two olefins to generate dimers that contain flexible methylene spacers. The anti-proliferative activities manifested by these dimeric compounds indicated that a tether of six carbon linkers (**123**, Figure 10.25) was optimal, suggesting a fixed distance between the two nucleotide binding sites in the Hsp90 monomers. Replacing the alkene with alkyne (**124**) greatly decreased activity, while saturation of the double bond (**125**) retained activity, which suggests that flexibility is important in this region for alignment of the two A4 coumarin rings. The dose-dependent client protein degradation produced by these Hsp90 inhibitors paralleled the anti-proliferative activity observed upon the administration of **125**. Replacement of noviose with a piperidine ring gave mixed results.[113] The A4 dimer containing an alkene linker produced a ten-fold increase (**127**) as expected; however, the A4 dimer with a saturated linker maintained similar activity (**128**) to the monomeric species.

10.3 Additional Hsp90 C-terminal Inhibitors that Induce the Heat-shock Response

10.3.1 Novologues

As discussed above, novobiocin analogues containing a benzamide side-chain exhibit anti-proliferative activities (*e.g.* **KU-174**), whereas analogues containing an acetamide (*e.g.* **KU-32**) do not, but instead produce neuroprotective activity. Although extensive studies have sought to elucidate structure–activity relationships for novobiocin analogues as anticancer agents, little has been done to explore the chemical attributes necessary for enhancing neuroprotective activity. Unfortunately, the lack of a co-crystal structure with a ligand bound to the Hsp90 C-terminus makes the rational design of new analogues difficult. However, utilizing the molecular model produced upon photoaffinity labeling of Hsp90, an approach towards rational design was pursued. Docking studies suggested that **KU-32** binds to the C-terminal region and appears to maintain binding interactions with both the protein backbone and the amino acid side-chains, similar to those manifested by novobiocin.[114] In this model, the amide side-chain appears to project into a large hydrophobic pocket that could tolerate more flexible linkers. Therefore, novologues containing a B-ring (Figure 10.27) that can project into this area of the Lys539-residing pocket were designed as a new scaffold for diversification. To validate this scaffold for use as a neuroprotective agent, a library was prepared to contain complementary interactions with Lys539. As shown in Figure 10.26, the compounds contained noviose with the requisite 2′,3′-diol; a derivatized 3-aryl substituent (B-ring) for probing hydrophobic and hydrogen bonding interactions that complement Lys539, and a 4-ethyl acetamide to occupy the binding pocket.

The neuroprotective activity of these compounds against glucose-induced toxicity of embryonic dorsal root ganglion (DRG) sensory neuron cultures was determined at 1 µM concentration (Figure 10.28). The results demonstrated

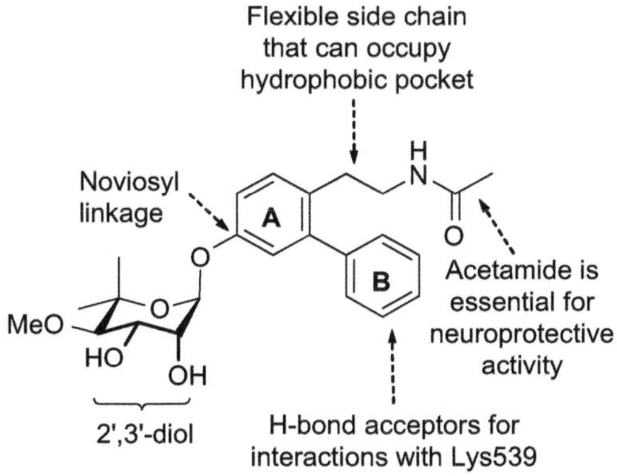

Figure 10.27 Structure of novologue and its attributes.

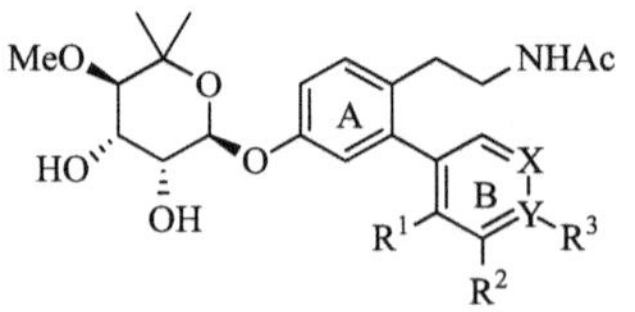

Figure 10.28 Structures and cell viability data of ethyl acetamide side-chain novologues.

Entry	R^1	R^2	R^3	X	Y	% of cell viability
130	H	H	H	C	C	76% ± 11
131	H	F	H	C	C	95% ± 14
132	H	H	F	C	C	75% ± 27
133	Cl	H	H	C	C	71% ± 21
134	H	Cl	H	C	C	90% ± 23
135	H	CF_3	H	C	C	83% ± 16
136	H	H	CF_3	C	C	74% ± 19
137	SMe	H	H	C	C	83% ± 40
138	OMe	H	H	C	C	92% ± 10
139	H	OMe	H	C	C	78% ± 34
140	H	Me	H	C	C	82% ± 30
141	H	CH_2-N-morpholine	H	C	C	83% ± 26
142	H	H	OH	C	C	67% ± 10
143	H	-OCH_2O-		C	C	83%±18
144	H	H	H	N	C	61% ± 7
145	H	H	H	C	N	81% ± 12

that the inclusion of *meta*-substituted acetamide novologues (**131, 134** and **135**) manifested comparable protection against glucotoxicity as KU-32, while the *ortho*- and *para*-substituted (**132, 133** and **136**) derivatives were less effective. However, methoxy-substituted analogues **78** (*ortho*-OMe) and **139** (*meta*-OMe) exhibited an opposing trend. Greater cytoprotective activity was observed for analogues that contain electronegative atoms at the *meta*-position (F, Cl, CF_3), which is believed to result from favorable interactions with Lys539 in the Hsp90 C-terminal binding pocket. Increasing the size of the electronegative atom at the *meta*-position (F to Cl to CF3) resulted in decreased neuroprotective activity. Hydrogen bond donors at the *para*-position appeared detrimental, as **142** (*para*-OH) did not provide significant protection against glucotoxicity. The 4-pyridine analogue (**144**) was unable to protect against glucose-induced toxicity and was less protective than the corresponding 3-pyridine analogue, **145**, suggesting that the 4-pyridine analogue maintains hydrogen bond interaction with Lys 539 more efficiently.

Further evaluation demonstrated that the most efficacious novologue was **131**, which is approximately 14-fold more efficacious than KU-32 (13.0 ± 3.6 nM vs. 240.2 ± 42.5 nM). In a luciferase assay, **131** was more effective than KU-32 at activation of the Hsp70 promoter. In addition, it also increased Hsp70 expression at lower concentrations than KU-32.

Systematic investigation of the substituents on the novologue B-ring provided a new scaffold for the development of Hsp90 C-terminal inhibitors and identified several efficacious Hsp90 modulators. This study also demonstrated that a rationally designed scaffold could be developed, which provides a platform upon which the development of new neuroprotective agents can be pursued.

10.3.2 AEG3482

The interruption of pathways that lead to neuronal cell death may delay or halt the progression of neurodegenerative diseases.[115] During neuronal apoptosis, activation of the N-terminal kinase (JNK) signaling pathway occurs in both mouse models and pathological specimens from AD brain. It appears that *c-jun* induction upon JNK activation is a contributing factor to the pathophysiology of neurodegenerative diseases, including Alzheimer's disease.[116] It has been shown that JNK suppression occurs upon Hsp70 binding and can reduce the number of neuronal cells that undergo apoptosis.[117] AEG3482 (Figure 10.29), an Hsp90 inhibitor, was demonstrated to dissociate HSF-1 from the Hsp90 complex, and result in the transcriptional activation of Hsp70, which blocked apoptosis.[118] Akt, a pro-survival kinase, was not affected upon AEG3482 treatment. Similar to many novobiocin family analogues, AEG3482 exhibits no effect on Hsp90 N-terminal ATPase activity, but geldanamycin has not been shown to compete with AEG3482 for Hsp90 binding. It is likely that AEG3482 binds to the Hsp90 C-terminus and manifests its activity similar to KU-32.

10.3.3 ITZ-1

In a drug screening program aimed at the discovery of anti-osteoarthritis agents, ITZ-1 (Figure 10.30) was identified as a chondroprotective agent.[119] It inhibited interleukin (IL)-1β-induced proteoglycan and collagen release from bovine nasal cartilage *in vitro* and protected rat knee joints from intra-articular infusion of IL-1β-induced cartilage proteogycan degradation. In addition, ITZ-1 reduced nitric oxide-mediated death of human articular chondrocyte by

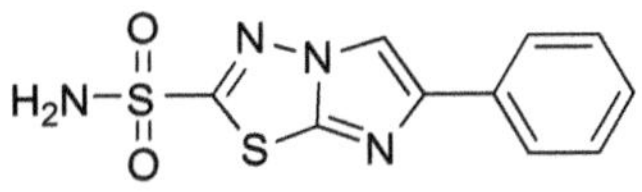

AEG3482

Figure 10.29 Structure of AEG3482.

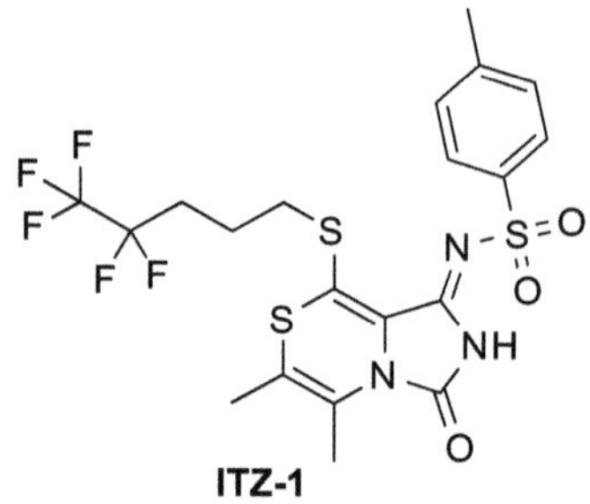

Figure 10.30 Structure of ITZ-1.

selectively inhibiting IL-1β induced ERK activation without affecting p38 kinase and JNK activation. Subsequent studies to elucidate the mechanism by which ITZ-1 manifests its activity demonstrated it to bind the Hsp90 C-terminal region.[120] Similar to other known Hsp90 C-terminal inhibitors, ITZ-1 did not disrupt Hsp90 ATPase activity. In the presence of IL-1β as co-stimulator, ITZ-1 increased mRNA and protein levels for Hsp70 and Hsp90 in human articular chondrocytes, while in the absence of IL-1β such increases were not observed. Co-immunoprecipitation studies showed that ITZ-1 significantly decreased the amount of HSF-1 co-precipitated with Hsp90. Although mild Raf-1 degradation upon ITZ-1 administration was observed, its ability to induce the degradation of other Hsp90 client proteins, such as the glucocorticoid receptor (GR), Akt, epidermal growth factor receptor (EGFR) and receptor interacting proteins (RIP)-1, was not observed until significantly higher concentrations were reached. Although research on ITZ-1 has focused on its chondroprotective properties, its strong HSF-1 induction, minimal client-protein degradation and low cytotoxicity indicate ITZ-1 may have potential for the treatment of neurodegenerative diseases.

10.4 Other Hsp90 C-terminal Inhibitors that Induce Client Protein Degradation

10.4.1 EGCG

EGCG (Figure 10.31) is one of the most abundant polyphenolic constituents in green tea. As a potent antioxidant against reactive oxygen species, its anticancer potential has attracted considerable attention and the mechanism by which its activity is manifested has been extensively studied.[121–124] Evidence demonstrates EGCG to affect multiple molecular targets, including growth factors signaling (which involves epidermal and vascular endothelial growth factors) and transcription factors such as AP-1 and NF-κB. EGCG is known to induce the degradation of many Hsp90 client proteins, including telomerase, diverse kinases and the aryl hydrocarbon receptor (AhR).[125] In addition, EGCG also disrupts association between Hsp90 and co-chaperones, such as

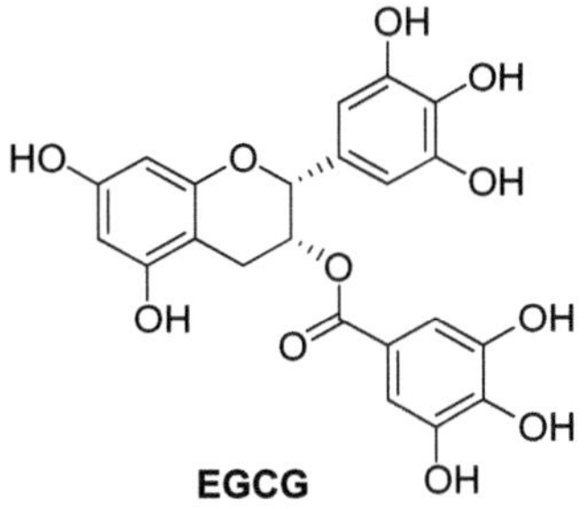

Figure 10.31 Structure of EGCG.

p23 and Hsc70.[126] In fact, EGCG was shown to bind Hsp90, specifically to amino acids 538–728 in the C-terminal region, as demonstrated by proteolytic footprinting, immunoprecipitation and an ATP-agarose pull-down assay.[127] Interestingly, against the MCF-7 cell line, EGCG was shown to reduce the expression of Hsp70 and Hsp90 without affecting other Hsps,[128] which is similar to the activity manifested by novobiocin.

Alongside its anticancer activity, the neuroprotective role of EGCG, which appears unrelated to molecular chaperone modulation, has been increasingly recognized as a potential mechanism for the treatment of Alzheimer's disease.[129–131] EGCG alters APP processing by enhancing non-amyloidgenic α-secretase cleavage without affecting the competing β- and γ-secretase cleavage that lead to amyloidogenic Aβ peptides.[132] Evidence suggests that Aβ oligomers might be the most toxic form of this peptide and that EGCG can act like a small molecule chaperone by directly binding to unfolded polypeptides and to shift their aggregation away from amyloid oligomers and fibrils, and to redirect them towards unstructured, non-toxic spherical oligomers.[133]

10.4.2 Silybin

Similar to EGCG, silybin is also a naturally occurring polyphenol that constitutes the major component of silymarin, a biologically active flavonolignan extract from the seed of milk thistle (*Silybum marianum*).[134,135] It can be resolved into two diastereoisomers, A and B, in a 1 : 1 ratio (Figure 10.32). As a traditional herb medicine, silybin (silymarin) has been used for the treatment of liver and gallbladder disorders.[136] At present, silybin is clinically used as an antihepatotoxic agent against chronic liver disease and is marketed as a nutritional supplement to protect the liver from diseases resulting from alcohol consumption and exposure to chemical and/or environmental toxins.[137,138] Recent studies have shown that silybin inhibits the proliferation of human breast, lung, colon and prostate cancer cells and manifests both prevention and anticancer activities in various tumor models.[136] In an effort to elucidate the mechanism by which silybin exhibits anticancer activity, an Hsp90-dependent firefly luciferase refolding and Hsp90-dependent heme-regulated eIF2a kinase

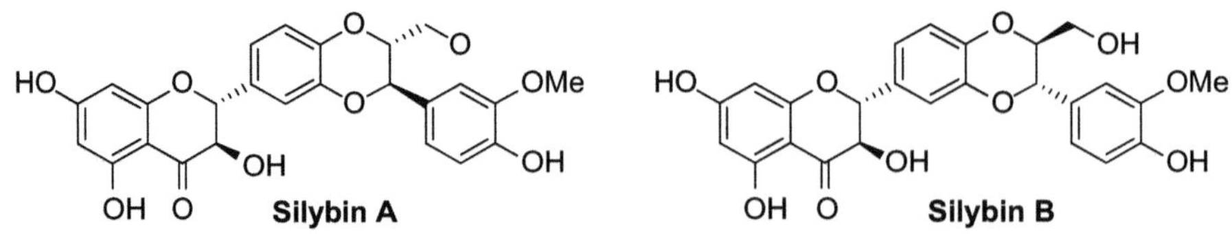

Figure 10.32 Structures of silybin A and silybin B.

(HRI) activation assay were performed, which identified silybin as an Hsp90 inhibitor.[139] In MCF-7 cancer cells, Hsp90 client proteins, including HER2, Raf and Akt, were degraded in a concentration-dependent manner upon the administration of silybin. In addition, Hsp90 levels remained constant at all concentrations tested, suggesting that, like novobiocin, silybin may also be an Hsp90 C-terminal inhibitor. Subsequent SAR studies identified essential and non-essential functionalities that are present on the silybin scaffold and required for Hsp90 inhibition. As shown in Figure 10.33, silybin analogues containing modifications to the A-ring exhibited improved anti-proliferative activity. These results indicated that while the phenol on the A-ring is not necessary (since compound **148** shows comparable activity to **146**, **147** and **149**), one phenol can be tolerated whereas the pattern of substitution is not important. It appears that the C-3 hydroxyl group is not essential, since removal of the C-3 hydroxyl group resulted in only a ~2-fold increase in anti-proliferative activity. Although removal of the C-3 hydroxyl group on compounds **148** and **149** resulted in similar activity (compound **153** and **154**), removal of the C-3 hydroxyl group on compounds **146** and **147** resulted in decreased activity (**151** and **152**).

Similar to EGCG, and due to its excellent antioxidant and anti-inflammatory activities, the potential application of silybin for the treatment of Alzheimer's disease has been evaluated. Silybin was shown to suppress nitrotyrosine levels and inhibit the over-expression of iNOS and TNF-α mRNA, which is induced by $A\beta_{25-35}$ in the hippocampus and amygdale.[140] In addition, silybin alleviated memory deficits resulting from exposure to $A\beta_{25-35}$ and $A\beta_{1-42}$ in several AD mouse models.[141,142]

10.4.3 Hybrids of Novobiocin and Silybin

In an effort to develop new Hsp90 inhibitors, silybin and novobiocin were overlayed to produce small molecules that were identified as Hsp90 C-terminal inhibitors.[143] It appears that the coumarin and flavonone ring systems can be interchanged, and that the corresponding 3-arylcoumarin derivatives target the Hsp90 protein folding machinery. Compound **155**, the first hybridized molecule that was designed to maintain the coumarin ring while incorporating the 3-aryl appendage of silybin, manifested anti-proliferative activity at low micromolar concentrations against both SKBr3 and MCF-7 cells. Western blot analyses of

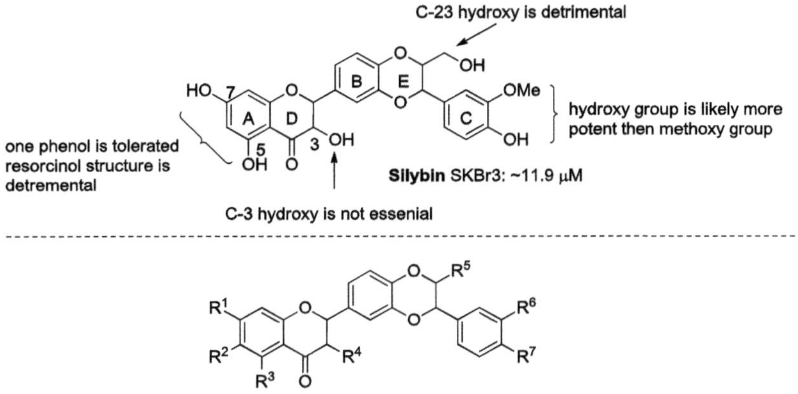

Inhibitors of the Hsp90 C-terminus

C-23 hydroxy is detrimental

one phenol is tolerated resorcinol structure is detremental

hydroxy group is likely more potent then methoxy group

Silybin SKBr3: ~11.9 µM

C-3 hydroxy is not essenial

Entry	R^1	R^2	R^3	R^4	R^5	R^6	R^7	SKBR3	MCF-7
Silybin	OH	H	OH	OH	CH$_2$OH	OMe	OH	197.0 ± 45.3	222.8 ± 3.6
146	OH	H	H	OH	CH$_2$OH	OMe	OH	16.04 ± 2.23	11.92 ± 0.64
147	H	H	OH	OH	CH$_2$OH	OMe	OH	11.90 ± 2.43	17.66 ± 6.55
148	H	H	H	OH	CH$_2$OH	OMe	OH	11.12 ± 1.42	13.42 ± 2.56
149	H	OH	H	OH	CH$_2$OH	OMe	OH	13.41 ± 1.32	17.80 ± 7.38
150	OH	H	OH	H	CH$_2$OH	OMe	OH	101.2 ± 2.50	104.2 ± 5.44
151	OH	H	H	H	CH$_2$OH	OMe	OH	41.91 ± 2.96	41.58 ± 3.68
152	OH	H	OH	H	CH$_2$OH	OMe	OH	47.73 ± 1.51	50.32 ± 2.83
153	H	H	H	H	CH$_2$OH	OMe	OH	16.25 ± 0.63	13.66 ± 2.58
154	H	OH	H	H	CH$_2$OH	OMe	OH	15.63 ± 4.35	15.62 ± 0.50

Figure 10.33 SAR and silybin analogues with improved anti-proliferative activity (IC$_{50}$, µM).

MCF-7 cell lysates upon the administration of **155** produced a similar result to that determined with **75** and silybin, suggesting that 3-arylcoumarin derivatives represent a new class of Hsp90 inhibitors.

Further SAR studies suggested that substitution at the 8-position of the coumarin ring is important, since the 8-methyl-3-arylcoumarin derivatives manifested improved anti-proliferative activity when compared to the 8-desmethyl analogues (**155–159** vs. **160–169**, Figure 10.34), particularly against the MCF-7 breast cancer cell line. Any substitution on the phenyl ring was deemed beneficial, but it appears that sterically bulky groups are favored (**158** vs. **157**, **164** vs. **166** vs. **163**). Bulky substituents (**168**) or a phenoxy group (**167**) at the *para* position can also lead to compounds that exhibit increased anti-proliferative activity. In addition, electron-deficient groups were more efficacious than those bearing electron-rich substitutions (**157/164** vs. **161**).

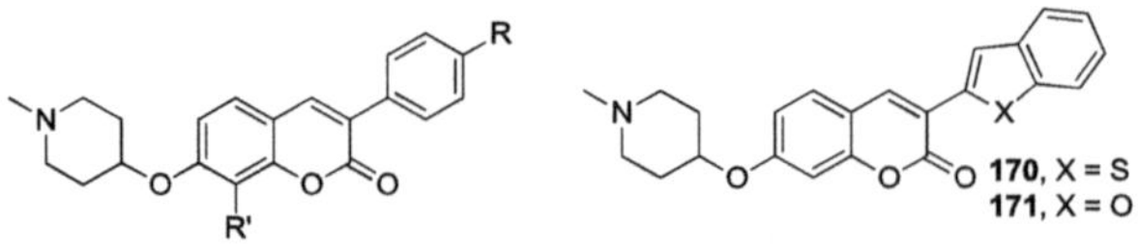

Figure 10.34 Structures and anti-proliferative activity for 8-methyl-3-arylcoumarins (IC$_{50}$, μM).

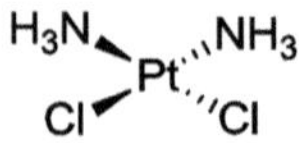

Entry	R	R	SKBR3	MCF-7
155	3,4-methylenedioxyl	CH$_3$	5.21±1.14	3.71±0.11
156	p-OCH$_3$	CH$_3$	5.52±1.44	1.97±0.11
157	p-Cl	CH$_3$	4.94±0.03	1.24±0.06
158	p-F	CH$_3$	7.38±0.23	3.83±0.00
159	p-OCF$_3$	CH$_3$	4.51±0.42	1.65±0.16
160	H	H	38.41±2.31	28.98±14.96
161	3,4-methylenedioxyl	H	14.80±0.01	8.79±1.32
162	p-OCH$_3$	H	29.93±15.44	16.92±2.34
163	p-Cl	H	8.24±1.59	3.89±0.61
164	p-F	H	43.85±0.32	10.50±0.57
165	p-OCF$_3$	H	5.78±0.59	4.12±0.37
166	p-CH$_3$	H	11.23±1.11	8.60±0.25
167	p-phenoxy	H	4.43±0.14	1.45±0.11
168	p-t-butyl	H	4.42±1.75	2.36±0.30
169	p-phenyl	H	3.33±0.69	3.08±0.09
170	--	H	0.98±0.01	0.81±0.02
171	--	H	5.84±0.06	4.93±0.06

Figure 10.35 Structure of cisplatin.

Interestingly, the benzo[b]thiophene derivative **170** exhibited activity against both cell lines at concentrations below 1 μM, and may serve as a viable lead compound for further investigation.

10.4.4 Cisplatin

Cisplatin (Figure 10.35) is a platinum-containing chemotherapeutic agent that is clinically used for the treatment of various cancers, including testicular, ovarian, bladder and small cell lung cancer. It has been well recognized that cisplatin coordinates and cross-links DNA, and therefore prevents DNA replication in rapidly dividing cells. Itoh and co-workers showed that, in addition to DNA targeting, cisplatin inhibits Hsp90 chaperone activity. By applying bovine brain cytosol to a cisplatin-affinity column, followed by elution with cisplatin, Hsp90 was eluted in a concentration-dependent manner, suggesting that cisplatin has a high affinity for Hsp90. Subsequent affinity purification experiments demonstrated that cisplatin binds to the Hsp90 C-terminal region. Csermely and co-workers determined that the cisplatin biding site is located proximal to C-terminal nucleotide binding site and that it can bind selectively to this region. Moreover, Rosenhagen and co-workers showed that upon the administration of cisplatin in neuroblastoma cells, degradation of the androgen and glucocorticoid receptors was observed. However, there was no effect on other Hsp90 clients, such as Raf-1, Lck and c-Src, indicating selective client protein degradation may be achievable with Hsp90 C-terminal inhibitors.

10.5 Conclusion

Discovery of the Hsp90 C-terminal nucleotide-binding site provides a new paradigm for modulation of the Hsp90 protein folding machinery. Considerable progress has been made during the past decade, and it is now clear that small molecules that bind to the C-terminus can segregate the pro-apoptotic client protein degradation activity from the pro-survival heat-shock response. The properties possessed by C-terminal inhibitors may have promising potential for the treatment of cancer and/or neurodegenerative disorders (Figure 10.36), which appear to be readily separable by individual scaffolds.

Despite these promising features, the Hsp90 C-terminal binding pocket remains elusive by co-crystal structures and inhibitor binding has not been confirmed. The development of new drug candidates that modulate the Hsp90 C-terminal domain will be aided greatly by solution of the co-crystal structure of Hsp90 C-terminus bound to these inhibitors, which represents the greatest obstacle to be overcome.

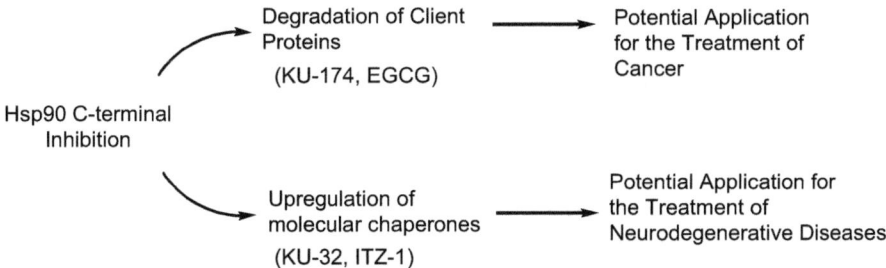

Figure 10.36 The effects of Hsp90 C-terminal inhibition and potential applications.

References

1. F. Ritossa, *Cell Stress Chaperones*, 1996, **1**, 97.
2. A. De Maio, M. G. Santoro, R. M. Tanguay and L. E. Hightower, *Cell Stress Chaperones*, 2012, **17**, 139.
3. M. Taipale, D. F. Jarosz and S. Lindquist, *Nat. Rev. Mol. Cell Biol.*, 2010, **11**, 515.
4. B. S. J. Blagg and T. A. Kerr, *Med. Res. Rev.*, 2006, **26**, 310.
5. B. T. Lai, N. W. Chin, A. E. Stanek, W. Keh and K. W. Lanks, *Mol. Cell. Biol.*, 1984, **4**, 2802.
6. M. P. Goetz, D. O. Toft, M. M. Ames and C. Erlichman, *Ann. Oncol.*, 2003, **14**, 1169.
7. H. Zhao, M. L. Michaelis and B. S. J. Blagg, *Adv. Pharmacol.*, 2012, **64**, 1.
8. E. Ruckova, P. Muller, R. Nenutil and B. Vojtesek, *Cellular & Molecular Biology Letters*, 2012, **17**, 446.
9. H. Pearl Laurence, C. Prodromou and P. Workman, *Biochem. J.*, 2008, **410**, 439.
10. K. A. Krukenberg, T. O. Street, L. A. Lavery and D. A. Agard, *Q. Rev. Biophys.*, 2011, **44**, 229.
11. L. Neckers, M. Mollapour and S. Tsutsumi, *Trends Biochem. Sci.*, 2009, **34**, 223.
12. A. Bracher and F. U. Hartl, *Nat. Struct. Mol. Biol.*, 2006, **13**, 478.
13. K. Richter and J. Buchner, *Cell*, 2006, **127**, 251.
14. L. Whitesell and S. L. Lindquist, *Nat. Rev. Cancer*, 2005, **5**, 761.
15. J. Frydman, *Annu. Rev. Biochem.*, 2001, **70**, 603.
16. A. B. Schmid, S. Lagleder, M. A. Grawert, A. Rohl, F. Hagn, S. K. Wandinger, M. B. Cox, O. Demmer, K. Richter, M. Groll, H. Kessler and J. Buchner, *EMBO J.*, 2012, **31**, 1506.
17. S. Y. Chen, W. P. Sullivan, D. O. Toft and D. F. Smith, *Cell Stress Chaperones*, 1998, **3**, 118.
18. C. Prodromou, B. Panaretou, S. Chohan, G. Siligardi, R. O'Brien, J. E. Ladbury, S. M. Roe, P. W. Piper and L. H. Pearl, *EMBO J.*, 2000, **19**, 4383.
19. S. J. Felts and D. O. Toft, *Cell Stress Chaperones*, 2003, **8**, 108.
20. C. A. Ballinger, P. Connell, Y. X. Wu, Z. Y. Hu, L. J. Thompson, L. Y. Yin and C. Patterson, *Mol. Cell. Biol.*, 1999, **19**, 4535.
21. P. Connell, C. A. Ballinger, J. H. Jiang, Y. X. Wu, L. J. Thompson, J. Hohfeld and C. Patterson, *Nat. Cell Biol.*, 2001, **3**, 93.
22. S. Murata, Y. Minami, M. Minami, T. Chiba and K. Tanaka, *EMBO Rep.*, 2001, **2**, 1133.
23. W. P. Xu, M. Marcu, X. T. Yuan, E. Mimnaugh, C. Patterson and L. Neckers, *Proc. Natl Acad. Sci. USA*, 2002, **99**, 12847.
24. P. Workman, F. Burrows, L. Neckers and N. Rosen, *Ann. NY Acad. Sci.*, 2007, **1113**, 202.
25. H. Zhang and F. Burrows, *J. Mol. Med.*, 2004, **82**, 488.

26. J. Y. Zou, Y. L. Guo, T. Guettouche, D. F. Smith and R. Voellmy, *Cell*, 1998, **94**, 471.
27. Y. H. Shi, D. D. Mosser and R. I. Morimoto, *Gene. Dev.*, 1998, **12**, 654.
28. N. D. Trinklein, J. I. Murray, S. J. Hartman, D. Botstein and R. M. Myers, *Mol. Biol. Cell*, 2004, **15**, 1254.
29. D. B. Solit and G. Chiosis, *Drug Discov. Today*, 2008, **13**, 38.
30. G. E. L. Brandt and B. S. J. Blagg, *Curr. Top. Med. Chem.*, 2009, **9**, 1447.
31. M. B. Almeida, J. L. M. do Nascimento, A. M. Herculano and M. E. Crespo-Lopez, *Biomed. Pharmacother.*, 2011, **65**, 239.
32. W. P. Sullivan and D. O. Toft, *J. Biol. Chem.*, 1993, **268**, 20373.
33. D. F. Nathan and S. Lindquist, *Mol. Cell. Biol.*, 1995, **15**, 3917.
34. D. Picard, http://www.picard.ch/downloads/Hsp90facts.pdf.
35. W. Xu and L. Neckers, *Clin. Cancer Res.*, 2007, **13**, 1625.
36. D. Hanahan and R. A. Weinberg, *Cell*, 2011, **144**, 646.
37. U. Banerji, *Proc. Am. Assoc. Cancer Res.*, 2003, **44**, 677.
38. P. Workman, *Trends Mol. Med.*, 2004, **10**, 47.
39. E. A. Sausville, J. E. Tomaszewski and P. Ivy, *Current Cancer Drug Targets*, 2003, **3**, 377.
40. S. Pacey, M. Gore, D. Chao, U. Banerji, J. Larkin, S. Sarker, K. Owen, Y. Asad, F. Raynaud, M. Walton, I. Judson, P. Workman and T. Eisen, *Invest. New Drugs*, 2012, **30**, 341.
41. A. Kamal, L. Thao, J. Sensintaffar, L. Zhang, M. F. Boehm, L. C. Fritz and F. J. Burrows, *Nature*, 2003, **425**, 407.
42. A. Kamal, L. Thao, J. Sensintaffar, L. Zhang, M. F. Boehm, L. C. Fritz and F. J. Burrows, *Clin. Cancer Res.*, 2003, **9**, 6126S.
43. P. Workman, *Mol. Canc. Therapeut.*, 2003, **2**, 131.
44. G. Chiosis, H. Huezo, N. Rosen, E. Mimnaugh, L. Whitesell and L. Neckers, *Mol. Canc. Therapeut.*, 2003, **2**, 123.
45. K. Jhaveri, T. Taldone, S. Modi and G. Chiosis, *Biochim. Biophys. Acta Mol. Cell Res.*, 2012, **1823**, 742.
46. B. K. Eustace, T. Sakurai, J. K. Stewart, D. Yimlamai, C. Unger, C. Zehetmeier, B. Lain, C. Torella, S. W. Henning, G. Beste, B. T. Scroggins, L. Neckers, L. L. Ilag and D. G. Jay, *J. Pharmacol. Sci.*, 2005, **97**, 90P.
47. M. L. Mendillo, S. Santagata, M. Koeva, G. W. Bell, R. Hu, R. M. Tamimi, E. Fraenkel, T. A. Ince, L. Whitesell and S. Lindquist, *Cell*, 2012, **150**, 549.
48. W. J. Luo, A. Rodina and G. Chiosis, *BMC Neurosci.*, 2008, **9**(Suppl 2), S7.
49. H. Adachi, M. Katsuno, M. Waza, M. Minamiyama, F. Tanaka and G. Sobue, *Int. J. Hyperther.*, 2009, **25**, 647.
50. A. Klettner, *Drug News Perspect.*, 2004, **17**, 299.
51. B. De Paepe, K. K. Creus, J. Weis and J. L. De Bleecker, *Neuromuscul. Disord.*, 2012, **22**, 26.
52. M. E. Jackrel and J. Shorter, *J. Clin. Invest.*, 2011, **121**, 2972.
53. A. I. Kaplin and M. Williams, *Neurology*, 2007, **69**, 410.

54. H. Harris and D. C. Rubinsztein, *Nat. Rev. Neurol.*, 2012, **8**, 108.
55. M. M. Bednar and A. Perry, *Neurol. Res.*, 2012, **34**, 129.
56. Y. W. Jung, E. Hysolli, K. Y. Kim, Y. Tanaka and I. H. Park, *Curr. Opin. Neurol.*, 2012, **25**, 125.
57. J. E. Gestwicki and D. Garza, in *Molecular Biology of Neurodegenerative Diseases*, D. B. Teplow, Academic Press Publication, Elsevier Inc. San Diego, Calif, 2012, vol. 107, p. 327.
58. G. G. Glenner and C. W. Wong, *Biochem. Biophys. Res. Commun.*, 1984, **120**, 885.
59. I. Grundkeiqbal, K. Iqbal, Y. C. Tung, M. Quinlan, H. M. Wisniewski and L. I. Binder, *Proc. Natl Acad. Sci. USA*, 1986, **83**, 4913.
60. M. G. Spillantini and M. Goedert, *Trends Neurosci.*, 1998, **21**, 428.
61. M. DiFiglia, E. Sapp, K. O. Chase, S. W. Davies, G. P. Bates, J. P. Vonsattel and N. Aronin, *Science*, 1997, **277**, 1990.
62. L. I. Bruijn, M. K. Houseweart, S. Kato, K. L. Anderson, S. D. Anderson, E. Ohama, A. G. Reaume, R. W. Scott and D. W. Cleveland, *Science*, 1998, **281**, 1851.
63. T. Arai, M. Hasegawa, H. Akiyama, K. Ikeda, T. Nonaka, H. Mori, D. Mann, K. Tsuchiya, M. Yoshida, Y. Hashizume and T. Oda, *Biochem. Biophys. Res. Comm.*, 2006, **351**, 602.
64. M. Neumann, D. M. Sampathu, L. K. Kwong, A. C. Truax, M. C. Micsenyi, T. T. Chou, J. Bruce, T. Schuck, M. Grossman, C. M. Clark, L. F. McCluskey, B. L. Miller, E. Masliah, I. R. Mackenzie, H. Feldman, W. Feiden, H. A. Kretzschmar, J. Q. Trojanowski and V. M. Y. Lee, *Science*, 2006, **314**, 130.
65. D. C. Bolton, M. P. McKinley and S. B. Prusiner, *Science*, 1982, **218**, 1309.
66. W. J. Luo, W. L. Sun, T. Taldone, A. Rodina and G. Chiosis, *Mol. Neurodegener.*, 2010, **5**, 24.
67. L. B. Peterson and B. S. J. Blagg, *Future Med. Chem.*, 2009, **1**, 267.
68. T. Taldone and G. Chiosis, *Curr. Top. Med. Chem.*, 2009, **9**, 1436.
69. C. A. Dickey, A. Kamal, K. Lundgren, N. Klosak, R. M. Bailey, J. Dunmore, P. Ash, S. Shoraka, J. Zlatkovic, C. B. Eckman, C. Patterson, D. W. Dickson, N. S. Nahman, M. Hutton, F. Burrows and L. Petrucelli, *J. Clin. Invest.*, 2007, **117**, 648.
70. S. Ansar, J. A. Burlison, M. K. Hadden, X. M. Yu, K. E. Desino, J. Bean, L. Neckers, K. L. Audus, M. L. Michaelis and B. S. J. Blagg, *Bioorg. Med. Chem. Lett.*, 2007, **17**, 1984.
71. J. D. Eskew, T. Sadikot, P. Morales, A. Duren, I. Dunwiddie, M. Swink, X. Y. Zhang, S. Hembruff, A. Donnelly, R. A. Rajewski, B. S. J. Blagg, J. R. Manjarrez, R. L. Matts, J. M. Holzbeierlein and G. A. Vielhauer, *BMC Cancer*, 2011, **11**, 468.
72. M. G. Marcu and L. M. Neckers, *Curr. Cancer Drug Target*, 2003, **3**, 343.
73. A. Donnelly and B. S. J. Blagg, *Curr. Med. Chem.*, 2008, **15**, 2702.
74. G. P. Lotz, H. Y. Lin, A. Harst and W. M. J. Obermann, *J. Biol. Chem.*, 2003, **278**, 17228.

75. F. Pirkl and J. Buchner, *J. Mol. Biol.*, 2001, **308**, 795.
76. M. G. Marcu, A. Chadli, I. Bouhouche, M. Catelli and L. M. Neckers, *J. Biol. Chem.*, 2000, **275**, 37181.
77. M. G. Marcu, T. W. Schulte and L. Neckers, *J. Natl. Canc. Inst.*, 2000, **92**, 242.
78. B. A. L. Owen, W. P. Sullivan, S. J. Felts and D. O. Toft, *J. Biol. Chem.*, 2002, **277**, 7086.
79. B. G. Yun, W. J. Huang, N. Leach, S. D. Hartson and R. L. Matts, *Biochemistry*, 2004, **43**, 8217.
80. R. K. Allan, D. Mok, B. K. Ward and T. Ratajczak, *J. Biol. Chem.*, 2006, **281**, 7161.
81. C. Soti, A. Racz and P. Csermely, *J. Biol. Chem.*, 2002, **277**, 7066.
82. R. L. Matts, G. E. L. Brandt, Y. M. Lu, A. Dixit, M. Mollapour, S. Q. Wang, A. C. Donnelly, L. Neckers, G. Verkhivker and B. S. J. Blagg, *Bioorg. Med. Chem.*, 2011, **19**, 684.
83. D. C. Hooper, J. S. Wolfson, G. L. McHugh, M. B. Winters and M. N. Swartz, *Antimicrob. Agents Chemother.*, 1982, **22**, 662.
84. R. Dutta and M. Inouye, *Trends Biochem. Sci.*, 2000, **25**, 24.
85. C. Soti, A. Vermes, T. A. J. Haystead and P. Csermely, *Eur. J. Biochem.*, 2003, **270**, 2421.
86. R. Conde, Z. R. Belak, M. Nair, R. F. O'Carroll and N. Ovsenek, *Biochem. Cell Biol.*, 2009, **87**, 845.
87. L. Neckers and P. Workman, *Clin. Canc. Res.*, 2012, **18**, 64.
88. R. J. Lewis, O. M. P. Singh, C. V. Smith, T. Skarzynski, A. Maxwell, A. J. Wonacott and D. B. Wigley, *EMBO J.*, 1996, **15**, 1412.
89. X. M. Yu, G. Shen, L. Neckers, H. Blake, J. Holzbeierlein, B. Cronk and B. S. J. Blagg, *J. Am. Chem. Soc.*, 2005, **127**, 12778.
90. Y. Lu, S. Ansar, M. L. Michaelis and B. S. J. Blagg, *Bioorg. Med. Chem.*, 2009, **17**, 1709.
91. H. Menchen, PhD Dissertation, The University of Kansas, 2012, 199 pages, http://kuscholarworks.ku.edu/dspace/handle/1808/10216.
92. M. J. Urban, C. Y. Li, C. J. Yu, Y. M. Lu, J. M. Krise, M. P. McIntosh, R. A. Rajewski, B. S. J. Blagg and R. T. Dobrowsky, *ASN Neuro*, 2010, **2**, e00040.
93. C. Li, J. Ma, H. Zhao, B. S. J. Blagg and R. T. Dobrowsky, *ASN Neuro*, 2012, **4**, e00102.
94. M. J. Urban, P. Pan, K. L. Farmer, H. P. Zhao, B. S. J. Blagg and R. T. Dobrowsky, *Exp. Neurol.*, 2012, **235**, 388.
95. L. Zhang, H. P. Zhao, B. S. J. Blagg and R. T. Dobrowsky, *J. Proteome Res.*, 2012, **11**, 2581.
96. J. A. Burlison, L. Neckers, A. B. Smith, A. Maxwell and B. S. J. Blagg, *J. Am. Chem. Soc.*, 2006, **128**, 15529.
97. J. A. Burlison, C. Avila, G. Vielhauer, D. J. Lubbers, J. Holzbeierlein and B. S. J. Blagg, *J. Org. Chem.*, 2008, **73**, 2130.
98. A. C. Donnelly, J. R. Mays, J. A. Burlison, J. T. Nelson, G. Vielhauer, J. Holzbeierlein and B. S. J. Blagg, *J. Org. Chem.*, 2008, **73**, 8901.

99. L. B. Peterson and B. S. J. Blagg, *Bioorg. Med. Chem. Lett.*, 2010, **20**, 3957.
100. H. C. Kolb, M. G. Finn and K. B. Sharpless, *Angew. Chem. Int. Ed.*, 2001, **40**, 2004.
101. J. L. Hou, X. F. Liu, J. Shen, G. L. Zhao and P. G. Wang, *Exp. Opin. Drug Discov.*, 2012, **7**, 489.
102. X. M. Yu, G. Shen and B. S. J. Blagg, *J. Org. Chem.*, 2004, **69**, 7375.
103. W. M. Pankau and W. Kreiser, *Tetrahedron Lett.*, 1998, **39**, 2089.
104. S. N. Shelton, M. E. Shawgo, S. B. Matthews, Y. Lu, A. C. Donnelly, K. Szabla, M. Tanol, G. A. Vielhauer, R. A. Rajewski, R. L. Matts, B. S. J. Blagg and J. D. Robertson, *Mol. Pharmacol.*, 2009, **76**, 1314.
105. G. Le Bras, C. Radanyi, J. F. Peyrat, J. D. Brion, M. Alami, V. Marsaud, B. Stella and J. M. Renoir, *J. Med. Chem.*, 2007, **50**, 6189.
106. D. Audisio, S. Messaoudi, L. Cegielkowski, J. F. Peyrat, J. D. Brion, D. Methy-Gonnot, C. Radanyi, J. M. Renoir and M. Alami, *ChemMedChem*, 2011, **6**, 804.
107. H. P. Zhao, A. C. Donnelly, B. R. Kusuma, G. E. L. Brandt, D. Brown, R. A. Rajewski, G. Vielhauer, J. Holzbeierlein, M. S. Cohen and B. S. J. Blagg, *J. Med. Chem.*, 2011, **54**, 3839.
108. S. M. Cohen, R. Mukerji, A. K. Samadi, X. Zhang, H. Zhao, B. S. J. Blagg and M. S. Cohen, *Ann. Surg. Oncol.*, 2012, **19**(3), S483.
109. X. Y. Huang, Z. J. Shan, H. L. Zhai, L. N. Li and X. Y. Zhang, *J. Chem. Inform. Model.*, 2011, **51**, 1999.
110. H. P. Zhao and B. S. J. Blagg, *Bioorg. Med. Chem. Lett.*, 2013, **23**, 552.
111. H. P. Zhao, E. Moroni, B. Yan, G. Colombo and B. S. J. Blagg, *ACS Med. Chem. Lett.*, 2013, DOI: 10.1021/ml300275g.
112. J. A. Burlison and B. S. J. Blagg, *Org. Lett.*, 2006, **8**, 4855.
113. B. R. Kusuma, L. B. Peterson, H. P. Zhao, G. Vielhauer, J. Holzbeierlein and B. S. J. Blagg, *J. Med. Chem.*, 2011, **54**, 6234.
114. B. R. Kusuma, L. Zhang, T. Sundstrom, L. B. Peterson, R. T. Dobrowsky and B. S. J. Blagg, *J. Med. Chem.*, 2012, **55**, 5797.
115. K. A. Gallo, *Chem. Biol.*, 2006, **13**, 115.
116. L. H. Wang, C. G. Besirli and E. M. Johnson, *Annu. Rev. Pharmacol. Toxicol.*, 2004, **44**, 451.
117. M. Y. Sherman, V. L. Gabai, A. B. Meriin and J. A. Yaglom, *Cell Stress Chaperones*, 2000, **5**, 487.
118. A. H. Salehi, S. J. Morris, W. C. Ho, K. M. Dickson, G. Doucet, S. Milutinovic, J. Durkin, J. W. Gillard and P. A. Barker, *Chem. Biol.*, 2006, **13**, 213.
119. H. Kimura, H. Yukitake, H. Suzuki, Y. Tajima, K. Gomaibashi, S. Morimoto, Y. Funabashi, K. Yamada and M. Takizawa, *J. Pharmacol. Sci.*, 2009, **110**, 201.
120. H. Kimura, H. Yukitake, Y. Tajima, H. Suzuki, T. Chikatsu, S. Morimoto, Y. Funabashi, H. Omae, T. Ito, Y. Yoneda and M. Takizawa, *Chem. Biol.*, 2010, **17**, 18.

121. R. J. Rosengren, *Handbook of Green Tea and Health Research*, Nova Science Publishers, Inc. Hauppauge, N. Y., 2009, p. 301.
122. N. Khan and H. Mukhtar, *Cancer Lett.*, 2008, **269**, 269.
123. G. Zhang, Y. Wang, Y. Zhang, X. Wan, J. Li, K. Liu, F. Wang, Q. Liu, C. Yang, P. Yu, Y. Huang, S. Wang, P. Jiang, Z. Qu, J. Luan, H. Duan, L. Zhang, A. Hou, S. Jin, T. C. Hsieh and E. Wu, *Curr. Mol. Med.*, 2012, **12**, 163.
124. B. L. Queen and T. O. Tollefsbol, *Curr. Aging Sci.*, 2010, **3**, 34.
125. C. M. Palermo, C. A. Westlake and T. A. Gasiewicz, *Biochemistry*, 2005, **44**, 5041.
126. Y. Y. Li, T. Zhang, Y. Q. Jiang, H. F. Lee, S. J. Schwartz and D. X. Sun, *Mol. Pharm.*, 2009, **6**, 1152.
127. Z. Y. Yin, E. C. Henry and T. A. Gasiewicz, *Biochemistry*, 2009, **48**, 336.
128. P. Tran, S. A. Kim, H. S. Choi, J. H. Yoon and S. G. Ahn, *BMC Cancer*, 2010, **10**, 276.
129. M. J. R. Howes and E. Perry, *Drugs Aging*, 2011, **28**, 439.
130. J. Kim, H. J. Lee and K. W. Lee, *J. Neurochem.*, 2010, **112**, 1415.
131. M. Singh, M. Arseneault, T. Sanderson, V. Murthy and C. Ramassamy, *J. Agric. Food Chem.*, 2008, **56**, 4855.
132. J. W. Fernandez, K. Rezai-Zadeh, D. Obregon and J. Tan, *FEBS Lett.*, 2010, **584**, 4259.
133. D. E. Ehrnhoefer, J. Bieschke, A. Boeddrich, M. Herbst, L. Masino, R. Lurz, S. Engemann, A. Pastore and E. E. Wanker, *Nat. Struct. Mol. Biol.*, 2008, **15**, 558.
134. L. Abenavoli, R. Capasso, N. Milic and F. Capasso, *Phytother. Res.*, 2010, **24**, 1423.
135. T. Kumar, Y. K. Larokar, S. K. Iyer, A. Kumar and D. K. Tripathi, *Int. J. Pharm. Phytopharm. Res.*, 2011, **1**, 124.
136. R. Gazak, D. Walterova and V. Kren, *Curr. Med. Chem.*, 2007, **14**, 315.
137. C. Loguercio and D. Festi, *World J. Gastroenterol.*, 2011, **17**, 2288.
138. R. Gazak, D. Walterova and V. Kren, *Curr. Med. Chem.*, 2007, **14**, 315.
139. H. P. Zhao, G. E. Brandt, L. Galam, R. L. Matts and B. S. J. Blagg, *Bioorg. Med. Chem. Lett.*, 2011, **21**, 2659.
140. P. Lu, T. Mamiya, L. L. Lu, A. Mouri, L. B. Zou, T. Nagai, M. Hiramatsu, T. Ikejima and T. Nabeshima, *Br. J. Pharmacol.*, 2009, **157**, 1270.
141. P. Lu, T. Mamiya, L. L. Lu, A. Mouri, M. Niwa, M. Hiramatsu, L. B. Zou, T. Nagai, T. Ikejima and T. Nabeshima, *J. Pharmacol. Exp. Ther.*, 2009, **331**, 319.
142. F. Yin, J. H. Liu, X. H. Ji, Y. W. Wang, J. Zidichouski and J. Z. Zhang, *Neurochem. Int.*, 2011, **58**, 399.
143. H. P. Zhao, B. Yan, L. B. Peterson and B. S. J. Blagg, *ACS Med. Chem. Lett.*, 2012, **3**, 327.

CHAPTER 11

Hsp70 Inhibitors

YAOYU CHEN* AND WENLAI ZHOU*

Oncology, Novartis Institutes for Biomedical Research, Cambridge, MA, USA
*Email: Wenlai.Zhou@novartis.com; Yaoyu.Chen@novartis.com

11.1 Hsp70 Chaperone Function

Hsp70 proteins as ATP-dependent molecular chaperones are involved in folding of newly synthesized polypeptides, the assembly of multi-protein complexes and the transport of proteins across cellular membranes.[1] Hsp70 family members can be divided into two major categories: the stress-inducible Hsp70 (such as Hsp72) and the constitutively expressed Hsc70 (such as Hspa8).[2] Hsp70 proteins contain two distinct functional regions: a peptide binding domain (PBD) and the amino-terminal ATPase domain (ABD). The PBD is responsible for substrate binding and refolding of the protein upon substrate binding, while the ABD facilitates the release of the client protein after ATP hydrolysis.

Hsp70 co-chaperones bind to Hsp70 and regulate its chaperone function. Those co-chaperones can be classified into three groups: 1. The J-domain co-chaperones (JDCs), such as Hsp40, are a relatively large group and bind to the Hsp70 ABD and stimulate the ATPase activity. 2. The nucleotide exchange factor co-chaperones (NEFCs) catalyze the release of ADP, which is required for completion of Hsp70 ATPase cycle. NEFCs include a few members like Bag-1, Hsp110 or Hspbp1. 3. The tetratricopeptide repeat (TPR) domain co-chaperones bind to the C-terminal EEVD motif of Hsp70 or Hsp90. The typical members of this group include Hop and carboxyl terminus of Hsc70 interacting

RSC Drug Discovery Series No. 37
Inhibitors of Molecular Chaperones as Therapeutic Agents
Edited by Timothy Machajewski and Zhenhai Gao
© The Royal Society of Chemistry 2014
Published by the Royal Society of Chemistry, www.rsc.org

protein (CHIP). They possess ubiquitin ligase activities and participate in the ubiquitination of some Hsp client proteins.

11.2 How to Target Hsp70?

A lot of efforts have been made to develop effective Hsp70 inhibitors in the past decade. So far, most of the compounds developed target the distinct domains of Hsp70, including the nucleotide binding domain (NBD), the substrate binding domain (SBD) and Hsp70 co-chaperones binding sites. The Hsp70 function cycle includes ATP binding, nucleotide hydrolysis, allostery and substrate binding (Figure 11.1). Firstly, when ATP binds to the NBD of Hsp70, SBD and NBD it shows coupled motion, which suggested their tight association. SBD binding to the substrate, coupled with J-domain interactions in the NBD, promote ATP hydrolysis. Secondly, ATP hydrolysis leads to a significant increase of the affinity of substrate to SBD (high affinity). SBD is also released by conformation change. Lastly, the interaction between nucleotide exchange factor (NEF) and NBD releases ADP and further releases folded substrates from the Hsp70 binding site. Those folded substrates are involved in a broad array of critical cellular functions in regard to protein homeostasis.[3,4] Thus, a couple of regulatory interactions are thought to be the key processes for targeting Hsp70.

Different platforms and assays have been developed to identify the small molecules that inhibit the activity of Hsp70. One platform is based on physical interaction between certain molecules and the Hsp70 protein. The pull-down assay of Hsp70 showed that the polyamine 15-deoxyspergualin (DSG) binds to Hsp70 and enhances the steady state ATPase activity of Hsp70.[5] Based on the structure of DSG, NSC630668-R/1, which efficiently inhibits ATPase activity and blocks Hsp70-mediated trafficking of polypeptides, was identified as an Hsp70 inhibitor through a small-scale screening.[6] Lately, a collection of about 30 dihydropyrimidines related to R/1 were screened using single-turnover ATP hydrolysis reactions and several novel compounds were found to modulate Hsp70 ATPase activity.[7] The classes of those chaperone inhibitors can be further defined: some directly inhibited Hsp70 ATPase activity; others blocked J domain proteins, which can enhance the ATPase activity of Hsp70.[7]

Another platform employs the inorganic phosphate chelator malachite green (MG) to monitor the ATPase activity of Hsp70.[8] MG assay is colorimetric and

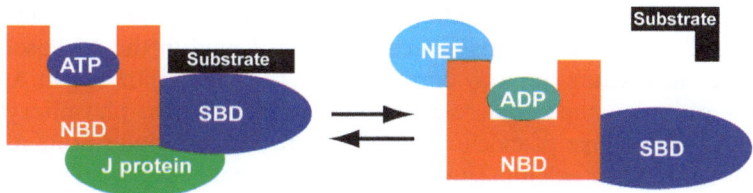

Figure 11.1 Schematic of ATPase cycle of Hsp70.

has the advantages of being robust, cost effective and suitable for automated screening, which led to compounds being identified that alter the ATPase activity of Hsp70. Using this assay, seven new inhibitors of Hsp70 were identified from a collection of 204 dihydropyrimidines, which suggests that the MG assay may be a useful platform for finding a small-molecule inhibitor of Hsp70.[8]

Fluorescence polarization (FP) is an assay with wide application in the discovery of novel protein modulators and in measuring real-time interactions between proteins or proteins and their ligands.[9] When the compound is bound to a molecule with greater mass, the complex tumbles much slower and the light emitted is polarized.[9] One FP assay was also developed through measuring the Hsp70 binding competition with a fluorescence labeled ATP-based probe.[10] Adenosine, a fragment of ATP, was chosen as a start point for initial investigation, and commercially available adenosine analogues were screened in the FP assay.[10] By using this platform, VER-155008 was identified to target the ATPase binding domain of Hsp70.[11]

The property of Hsp70 that restores the enzyme activity of denatured luciferase was used to develop a high-throughput screening (HTS) assay.[12] In this assay, denatured firefly luciferase was treated with a mixture of Hsp70 and prospective chemical modulators and the luminescence of refolded luciferase was used to follow the reaction progress, and counterscreens were used to exclude compounds that target luciferase.[13] Therefore, the compound hits identified from the screen may modify protein folding *via* their effects on the Hsp70 chaperone activity. By using this platform, a chemical library was screened and five new inhibitors of Hsp70 were identified.[13]

By taking advantage of the platform mentioned above, several compounds have been selected and applied as chemical genetic tools to elucidate the role of Hsp70 in different diseases. We will discuss them in detail in the following sections.

11.3 Hsp70 Inhibitors

11.3.1 Targeting the Nucleotide Binding Domain

A 44 kDa nucleotide-binding domain locates at the N-terminus of Hsp70 protein and possesses ATPase activity, which is essential for controlling its structure and function. Structural studies have revealed that the Hsp70 NBD (ATPase fragment) is composed of two lobes (I and II), which form a clamp-like structure with a large cleft between them. ATP is found to bind at the base of the cleft and the nucleotide-binding part of the ATPase fragment is similar to that of hexokinase.[14] There are two calcium ions found in the ATPase structure. While one calcium ion may correspond to the magnesium site in ATPase and likely plays an important role for ATP hydrolysis and phosphorylation *in vitro* and the change of calcium ion structure may facilitate protein phosphorylation, the other calcium site represents a new calcium-binding motif that may be important in maintaining the protein stabilization.[15] Recently, the crystal structures of NBD in four human Hsp70 isoforms were solved and all those proteins crystallized in a closed cleft conformation, indicating that the NBDs of Hsp70 function are conserved.[16] NMR data suggest that the NBDs of Hsp70 are allosterically regulated by binding either ADP or ATP in a cleft and that the orientation of two lobes of the Hsc70 NBD in solution deviates up to 10 degrees from their positions in 14 superimposing X-ray structures.[17]

Compounds that target the nucleotide binding domain are expected to compete with ATP for binding to Hsp70, which should have a significant impact on chaperone function. Most Hsp90 inhibitors developed so far are also ATP-competitive.[18] Interestingly, Hsp90 inhibitors that bind Hsp90's ATP-binding sites are not cross-reactive with Hsp70, because the Hsp70 chaperone binds to nucleotide using the actin-like fold, while the Hsp90 chaperone binds to the ATP site directly.[18] Several chemical structures have been reported to have the capability to target the NBD, including (−)-epigallocatechin gallate (EGCG), MKT-077, sulfoglycolipids, NSC 630668-R/1, VER-155008 and apoptozole.[6,11,19–22] Most of those compounds were identified from the ATP domain binding screen and were validated to bind indeed to the ATP binding site of Hsp70 to inhibit its function. Those compounds share little structural similarity as they bind to the different sites of NBD. In addition, those compounds have shown different IC_{50}s on inhibition of ATP activities of Hsp70 in a biochemical assay and have different biological activities in terms of impact on tumorigenesis (Table 11.1).

11.3.1.1 (−)-Epigallocatechin Gallate (EGCG)

(−)-epigallocatechin gallate (EGCG) is a major component in green tea and it has been shown to have an inhibitory activity against tumorigenesis.[23–25] However, the underlying mechanism is not clear. Recently, some studies suggest that EGCG interacts with Hsp70 at the ATP-binding site of the protein and inhibits Hsp70's function by competing with ATP binding.[19] After binding to Hsp70, the EGCG switches Hsp70 from its active monomer to the inactive dimer and oligomer forms.[19] EGCG may also prevent the anti-apoptotic effect of Hsp70, which usually suppresses the caspase-mediated cell death pathways in drug-treated cancer cells, contributing to the development of drug resistance.[19] Another flavonoid myricetin has also been reported as an inhibitor of Hsp70 and this compound inhibits the ATPase activity of Hsp70 ($IC_{50} \sim 10\,\mu M$) and leads to rapid proteasome-dependent tau degradation in a cell-based model.[25]

11.3.1.2 MKT-077

MKT-077 (formerly known as FJ-776 (1-ethyl-2-[[3-ethyl-5-(3-methylbenzothiazolin-2-yliden)]-4-oxothiazolidin-2-ylidenemethyl] pyridinium chloride)) is a lipophilic cation/rhodacyanine analogue (related to rhodamine 123) dye.[20] It is a highly water-soluble (>200 mg/ml) lipophilic compound and displays significant anticancer activity both *in vitro* and *in vivo*.[26,27] MKT-077 is an "allosteric inhibitor", because it plays a role through a differential interaction

Table 11.1 Summary of miscellaneous Hsp70 inhibitors

Compound	Chemical Structure	Mechanism of Action	IC50	Biological activity
EGCG		Inhibit ATP binding site	10 μM	Against tumorigenesis
MKT-077		Binds to the NBD	7 μM	Inhibit the growth of human cancer cell lines
AdaSGC		Binds to ATPase domain	50 μM	Not clear

Hsp70 Inhibitors

Compound	Mechanism	Concentration	Effect
NSC 630668-R/1	Altering the oligomeric state of Hsc70	10–200 μM	Against tumorigenesis
VER-155008	Binds to ATPase domain	0.5 μM	Against tumorigenesis
Apoptozole	Binds to ATPase domain	0.25 μM	Induce apoptotic cell death and against tumorigenesis

Table 11.1 (Continued)

Compound	Chemical Structure	Mechanism of Action	IC50	Biological activity
Acyl Benzamides		Blocks APiase sit	2.7–9.5 µM	AntiEcoli
DSG (Gusperimus)		Binds to EEVD domain	158–501 µM	Rheumatoid arthritis, Crohn's disease, lupus erythematosus, and the prevention and therapy of transplant rejection or graft-versus-host disease.
PES		Binds to the SBD	20 µM	Inhibit the growth of human cancer cell line

with Hsp70 allosteric states.[28] MKT-077 was shown to bind to the nucleotide-binding domain of Hsp70, leading to a tertiary structural change of Hsp70 protein, thus inactivates its chaperone function and induces senescence in human tumor cell lines.[29] MKT-077 was known to bind to several Hsp70 family members, such as mortalin or Hsc70, and abrogate the interaction between Hsp70 family proteins and the tumor suppressor protein, p53.[20] In cancer cells, upon MKT-077 treatment, p53 is released from p53-Hsp70 complexes and its transcriptional function is rescued.[20] MKT-077 selectively inhibits the proliferation of cancer cells through its function of regulating oncogenic Ras mutants such as v-Ha-Ras.[30] MKT-077 binds actins directly, bundles actin filaments by cross-linking and blocks membrane ruffling, which suppresses Ras transformation.[30] Cancer cells with elevated level of Hsp70 expression are more sensitive to MKT-077 and MKT-077 treatment leads cancer cells into senescence, which is similar to the effect of ECEG on cancer cells.[29] MKT-077 was shown to inhibit the growth of at least seven human cancer cell lines, including colon carcinoma CX-1, breast carcinoma MCF-7, pancreatic carcinoma CRL 1420, bladder transitional cell carcinoma EJ, melanoma LOX, human erythroleukemia K562 and colorectal carcinoma HCT116.[26,31]

11.3.1.3 Sulfoglycolipids

Sulfogalactoglycerolipid (SGG) and sulfogalactosyl ceramide (SGC) isolated from the mammalian reproductive tract were shown to be immunologically related to the Hsp70 family of stress proteins.[32] The N-terminal ATPase domain of Hsp70 binds to sulfogalactosyl ceramide and sulfogalactosyl glycerolipid specifically while the C-terminal domain does not.[21] Mutagenesis analysis suggests that the conserved residues within the ATPase domain of Hsp70, Arg(342) and Phe(198), are crucial residues for this binding activity.[21] SGC can be coupled to an alpha-adamantane or a norbornane rigid hydrophobic frame and these conjugates inhibited the specific binding of bovine brain Hsc70 within the range of 100–300 mM.[33] A dose of inhibition of bovine brain Hsc70 ATPase activity by SGC was observed within a similar range.[33] Kinetic analysis indicated that SGC was a non-competitive inhibitor of Hsp70 ATPase activity.[33] However, the dependence of ATPase inhibition on the rate of hydrolysis indicates that SGC binding occurs at a certain stage of the ATPase cycle.[33] Furthermore, phospholipids were also shown to bind to Hsp70, but the complex formation with the acidic glyco- and phospholipids involves different functions of Hsp70.[34] The acidic lipid/Hsp70 complexes are categorized into two groups. The first group is the sulfatide-induced large-sized complex and only the N-terminal ATPase domain is responsible for the complex formation. The second group is the ganglioside-induced complex and both the N-terminal ATPase and the C-terminal peptide-binding domains are involved in the complex formation.[34] It is not clear whether there is an effect of sulfoglycolipids on the proliferation of cancer cells.

11.3.1.4 NSC 630668-R/1 (Dihydropyrimidines)

Through an analysis of compounds with some structural similarities to 15-deoxyspergualin (DSG), NSC 630668-R/1 was identified as a novel inhibitor of Hsc70 activity.[6] R/1 inhibits the endogenous and DnaJ-stimulated ATPase activity of Hsc70 and blocks the Hsc70-mediated protein translocation.[6] Biochemical studies demonstrate that R/1 most likely exerts those functions by altering the oligomeric state of Hsc70.[6] Furthermore, a pool of small molecules with structural similarity to 15-deoxyspergualin and NSC 630668-R/1were further tested for their effects on endogenous and Hsp40-stimulated Hsp70 ATPase activity.[7] Some compounds were found to enhance the rate of Hsp70 ATP hydrolysis by ≥2.5-fold at 0.3 mm, for example, MAL3-38 and MAL3-90 increase the ATP turnover rate by 3- and 5-fold, respectively.[9] Others like MAL3-55 exhibited concentration-dependent effects on Hsc70-mediated ATP hydrolysis.[9] MAL3-39 and MAL3-101 inhibit J-chaperone-stimulated Hsp70 ATPase activity by 6–8-fold.[7] MAL3-101 interfering with stimulation of Hsp70's ATPase activity was also shown to have antitumorigenic effect.[35] MAL3-101 induces apoptosis in the SK-BR-3 breast cancer cell line with an apparent ED50 of <10 µM, although this compound shows high molecular weight, low solubility and high lipophilicity.[35]

11.3.1.5 VER-155008

In a fluorescence polarization (FP) assay, VER-155008 was shown to potently compete with ATP for the binding to Hsp70 with an IC_{50} of 0.5 µM.[10,11] In addition, VER-155008 binds to Hsc70 and the endoplasmic reticulum Hsp70 family member Grp78 and inhibits the ATP turnover of Hsc70 by 24.5% at 30 µM and 48.7% at 100 µM, which suggests that it is an ATP competitive inhibitor.[11] The ATPase domain of Hsc70 was further determined as the binding site of VER-155008.[11] VER-155008 inhibits proliferation of multiple human tumor cell lines including HCT116 and HT29 *in vitro* and induces the degradation of Hsp90 client proteins such as HER2 and Raf-1 in HCT116 and BT474 cells.[11] VER-155008 was also able to induce apoptosis in BT474 human breast tumor cell line.[11] VER-155008 has limited oral absorption and is therefore administered *via* the i.v. route.[11] Following a single 25 mg/kg i.v. dose to female BALB/c mice, VER-155008 showed a plasma clearance of 49 ml/min/kg, which was reflected in a short half-life of 0.6 h.[11] In HCT116 tumor-bearing nude mice, VER-155008 was shown to distribute to the tumor tissue with a longer tumor half-life than the plasma (1.5 *vs.* 0.5 h).[11]

11.3.1.6 Apoptozole

Apoptozole is a novel apoptosis-inducing small molecule that interacts with Hsc70 and Hsp70 and was discovered in a cell-based screen using an imidazole compound library.[22] A ligand-directed protein labeling and molecular modeling study further suggest the binding of apoptozole to an ATP-binding pocket.[36] By binding to ATPase domain of Hsc70, apoptozole inhibits the ATPase activity of Hsc70 and induces apoptosis in the human embryonic cancer cell line.[22] However, the precise molecular mode of action and the selectivity of apoptozole call for further studies.

11.3.2 Targeting the Substrate Binding Domain

A 15 kDa substrate binding domain (SBD) contains a groove with an affinity for neutral, hydrophobic amino acid residues. The groove interacts with peptides, which is up to seven residues in length.[16] SBDs typically consist of a β-sandwich subdomain, which comprises two 4-stranded β-sheets and is followed by an α-helical lid that folds back over the β-sandwich. Two loops extend from the β-sandwich and form the substrate-binding site.[37] The client polypeptide substrate binds in an extended conformation and an alpha-helical domain, which doesn't contact with peptide directly, is found to stabilize the whole complex.[37,38]

A couple of known peptides have been shown to bind to the substrate-binding cleft of the SBD and the C-terminal "lid" domain.[39] The interaction between peptide and Hsp70 inhibits the ATP turnover and folding of model substrates.[40] Therefore, targeting the substrate-binding pocket is also an effective strategy for inhibiting Hsp70 function. Acyl benzamides and 15-deoxyspergualin (DSG) were shown to bind to the SBD.[41,42] 15-deoxyspergualin (DSG), known as Gusperimus, was first approved by FDA as an immunosuppressive in clinic. Later, DSG was shown to bind to Hsc70 and inhibits its function.[42]

11.3.2.1 Acyl Benzamides

Hsp70 might help protein folding by catalyzing the *cis/trans* isomerization of secondary amide peptide bonds in unfolded or partially folded proteins, which opened up the new possibility of compounds for inhibiting this essential DnaK function. A series of fatty acylated benzamido inhibitors of the *cis/trans* isomerase activity of Hsp70 were developed and tested for their effects.[41] The fatty acylated benzamido derivatives such as Nα-[fatty-acyl-(4-aminoalkylbenzoyl)]-L-α-amino acids reversibly block the ATPase site of Hsp70 with an IC_{50} value of 2.7 μm and decrease the yield of native protein from the guanidinium chloride-denatured firefly luciferase in a DnaK/DnaJ/GrpE-assisted refolding assay with an IC_{50} value of 9.5 μm.[41] The compounds did not influence the ATPase activity of Hsp70, but located on the peptide binding cleft of the Hsp70 protein.[41] Hsp70 is known to play a critical role in regulating heat-shock response by modulating the activity, stability and amounts of free δ^{32} and an elevated level of Hsp70 represses the synthesis of δ^{32}-regulated heat-shock proteins.[43] Fatty acylated benzamido enhances transcriptional activity of δ^{32}-regulated genes, which suggests that this effect may be *via* the inhibition of APIase activity of Hsp70.[41]

11.3.2.2 15-deoxyspergualin (DSG)

15-deoxyspergualin, also known as Gusperimus, is an immunosuppressive drug and is shown to be efficacious in transplant rejection and autoimmune disease.[42,44] It is a derivative of the antitumor antibiotic spergualin, and inhibits the interleukin-2-stimulated maturation of T-cells to the S and G2/M phases and the polarization of the T-cells into IFN-gamma-secreting Th1 effector T-cells, resulting in the inhibition of growth of activated naïve CD4 T-cells.[42,44] DSG also interacts with Hsc70 specifically and this interaction is characterized through affinity capillary electrophoresis and the binding affinity of purified Hsc70 is around 4 µM.[5,45] Further studies revealed that DSG and its analogs stimulate the ATPase activity of Hsc70 (2-fold; Km = 3 µM), but have no effect on stimulation by J-domain co-chaperones or on the release of substrates.[5,45] To further understand how DSG binds to Hsc70, ^{14}C-DSG was cross-linked to purified Hsc70 by adding EDC (1-ethyl-3-(3-dimethoxy-aminopropyl)-carbodiimide) and mass spectrometry analysis of the peptide fragments was performed and the results suggest that the specific binding site of DSG to Hsc70 was through four amino acids, 647EEVD650, which locates in the C-terminal of Hsc70, the same binding site of the TPR-domain co-chaperones.[5,45] The specific binding of DSG to four amino acids 647EEVD650 of Hsc70 regulates the ATPase activity of Hsc70.[5,45]

So far, DSG is one of the most clearly characterized Hsp70 inhibitors in terms of its clinical utility. DSG (Gusperimus) has been used in a few common diseases and conditions such as rheumatoid arthritis, Crohn's disease, lupus erythematosus and the prevention and therapy of transplant rejection or graft-*versus*-host disease.[46–49] DSG has also been tested in organ transplantation and other autoimmune diseases in clinic as well as being a potential antibacterial compound.[50] In a graft-*versus*-host disease rat model, DSG is able to reduce GVHD-associated morbidity.[50] After the DSG treatment, Hsp70 level was reduced in spleen and lymph nodes in rats. Therefore, anti-Hsp70 antibody production and the serum levels of IL-2, IFN-gamma, TNF-alpha and IL-10 were diminished.[50] The mechanism of DSG regulating immunosuppressive effect in GVHD rats may be through the binding of DSG to Hsp70 to reduce the levels of anti-Hsp70 antibodies and other cytokines.[50] Up-regulation of Hsc70 was also observed in human CML cells and inhibition of CML cell viability by DSG may be through the inhibition of Hsc70 activities.[51]

11.3.3 Targeting the Co-chaperones

Hsp70 chaperones help protein folding through ATP-controlled cycles of substrate binding and release. ATP hydrolysis is the rate-limiting step of locking in of substrates into the substrate-binding cavity of Hsp70. This process is stimulated by a large class of Hsp70-associated co-chaperones containing the N-terminal J-domain. There are six of these proteins in *E. coli*, 20 in *Saccharomyces cerevisiae* and 40 in *human*.[52] Those Hsp70-associated co-chaperones are characterized by containing a conserved 70 amino acid J-domain, which is first named after the founding of DnaJ gene.[52,53] The major function of J-domain is to increase the ATPase activity of Hsp70.[54] J-domain interacts with NBD of Hsp70 and the interaction increases ATPase activity more than 5- to 10-fold, which enhanced the substrate affinity.[54] The C-terminal domains of those Hsp70-associated co-chaperones are used for classification.[55] For example, the class I and II proteins contain domains involved in dimerization and substrate binding.[55]

Hsp70 chaperones help Hsp70s function through protein folding. J-domain co-chaperones modulate the function of Hsp70 by delivering substrate peptides to Hsp70 and stimulating ATP turnover. D-peptides were shown to inhibit DnaJ stimulated ATPase activity of Hsp70 and these peptides are thought to compete with binding of substrates to DnaJ.[56] Therefore, a novel approach of targeting Hsp70 has been proposed to inhibit Hsp70 functions through modulating the essential co-chaperones instead of Hsp70 proteins themselves.

11.3.3.1 2-phenylethynesulfonamide (PES)

A small-molecule 2-phenylethynesulfonamide (PES), also called phenylacetylenylsulfonamide or pifithrin-m (PFTm), interacts selectively with the stress-inducible Hsp70 protein and inhibits its functions.[57] PES was first identified in a screen of the ChemBridge DIVERSet library of small molecules, having the capability to inhibit the p53-mediated apoptosis.[58] Furthermore, PES was found to prevent the accumulation of p53 and inhibit caspase cleavage in a cisplatin-treated human cancer cell line through interfering with the chaperone function of Hsp70 protein.[57] PES interacts with in Hsp70 at carboxyterminal substrate-binding domain (amino acids 386–641), but not with Hsp90.[57] PES-treated cells contain less Hsp70 in association with Chip, Hsp40 and Apaf1.[57] The disruption of Hsp70 functions by PES influences a number of cell signaling pathways such as P53 and NFκB.[57] PES treatment leads to cancer cell death, which is associated with protein aggregation, impaired autophagy and inhibition of lysosomal function.[57] PES-mediated inhibition of Hsp70 family proteins in cancer cells impairs two major protein degradation systems: the autophagy-lysosome system and the proteasome pathway.[59] As a consequence of the disruption in the Hsp70 chaperone system caused by PES, many cellular proteins, including Hsp70/Hsp90 substrates, accumulate in detergent-insoluble cell fractions, indicative of aggregation and functional inactivation.[59] PES blocks the trafficking of Arf by inhibiting Hsp70 mediated mitochondrial localization, which leads to cancer cell autophagy.[60] Interestingly, cancer cells expressing high levels of Arf are more sensitive to PES than counterparts with Arf silenced.[60] Lastly, PES was also shown to suppress tumor development and enhance survival rate in a Myc-induced lymphomagenesis mouse model.[57]

11.3.3.2 D-peptides

The co-chaperone DnaJ has been reported to bind native and denatured proteins as well as peptides.[61] The normal all-D and retro all-D peptides don't bind to Hsp70, but they bind to DnaJ with K_d values of 6.8 mM and 0.9 mM, respectively. The emission spectrum of the DnaJ-bound peptides suggests that DnaJ bound both D-peptides with the same main chain direction as L-peptides.[61] Binding of the normal all-D and all-L peptides inhibited the DnaJ-induced stimulation of DnaK ATPase.[61] D-peptides have also been shown to efficiently inhibit the refolding of denatured luciferase in the Hsp70/DnaJ/GrpE chaperone system (EC_{50} = 1–2 mM).[56] The inhibition of the chaperone action is due to the binding of D-peptide to DnaJ (K_d = 1–2 mM), which seems to preclude DnaJ from forming Hsp70/substrate/DnaJ complexes.[56]

11.4 The Prospectus of Hsp70 Inhibitor

Although Hsp70 has been proposed as a promising pharmacological target for many years, several factors make the drug discovery process complicated and challenging. Firstly, Hsp70 has multiple isoforms. Thirteen Hsp70 isoforms are found in human and most of them are found in all the major subcellular compartments. Among them, inducible Hsp70 is the most relevant to cancer. High Hsp70 expression is associated with poor prognosis and drug resistance of many cancer types, such as breast, endometrial, oral colorectal and prostate cancers.[18] Importantly, Hsp70 inhibition *via* antisense, siRNA/shRNA and small-molecule inhibitors (affecting ATPase, substrate or co-chaperone binding) demonstrated antitumor activities.[18,62] However, the human Hsc70 and Hsp70 isoforms are 86% identical to each other.[63] Can we develop the compounds that selectively inhibit a subset of Hsp70 isoforms (*e.g.* inhibiting the stress-inducible Hsp70 instead of the constitutively expressed Hsc70 protein)? Secondly, Hsp70s are involved in almost all cellular functions including protein folding, degradation, trafficking, remodeling, apoptosis, autophage and cell signaling. It is difficult to determine the cellular functions governing the cellular sensitivity to Hsp70 inhibition. Moreover, the roles of measurable Hsp70 functions (*e.g.* ATPase, refolding *etc.*) in controlling its biology in cell remain unclear. Lastly, Hsp70 binds very tightly to ATP (*E. coli* Dnak binding with a KD of 1 nM).[64] It will be extremely challenging to develop an ATP competitive inhibitor. We believe that more specific inhibitors of Hsp70 isoforms will be identified to effectively target different diseases with the development of novel HTS platforms in the near future.

References

1. C. A. Ballinger, P. Connell, Y. Wu, Z. Hu, L. J. Thompson, L. Y. Yin and C. Patterson, *Mol. Cell. Biol.*, 1999, **19**, 4535–4545.
2. H. Wegele, S. K. Wandinger, A. B. Schmid, J. Reinstein and J. Buchner, *J. Mol. Biol.*, 2006, **356**, 802–811.
3. M. Vogel, B. Bukau and M. P. Mayer, *Mol. Cell*, 2006, **21**, 359–367.
4. J. F. Swain, G. Dinler, R. Sivendran, D. L. Montgomery, M. Stotz and L. M. Gierasch, *Mol. Cell*, 2007, **26**, 27–39.

5. S. G. Nadler, M. A. Tepper, B. Schacter and C. E. Mazzucco, *Science*, 1992, **258**, 484–486.
6. S. W. Fewell, B. W. Day and J. L. Brodsky, *J. Biol. Chem.*, 2001, **276**, 910–914.
7. S. W. Fewell, C. M. Smith, M. A. Lyon, T. P. Dumitrescu, P. Wipf, B. W. Day and J. L. Brodsky, *J. Biol. Chem.*, 2004, **279**, 51131–51140.
8. L. Chang, E. B. Bertelsen, S. Wisen, E. M. Larsen, E. R. Zuiderweg and J. E. Gestwicki, *Anal. Biochem.*, 2008, **372**, 167–176.
9. Y. Kang, T. Taldone, C. C. Clement, S. W. Fewell, J. Aguirre, J. L. Brodsky and G. Chiosis, *Bioorg. Med. Chem. Lett.*, 2008, **18**, 3749–3751.
10. D. S. Williamson, J. Borgognoni, A. Clay, Z. Daniels, P. Dokurno, M. J. Drysdale, N. Foloppe, G. L. Francis, C. J. Graham, R. Howes, A. T. Macias, J. B. Murray, R. Parsons, T. Shaw, A. E. Surgenor, L. Terry, Y. Wang, M. Wood and A. J. Massey, *J. Med. Chem.*, 2009, **52**, 1510–1513.
11. A. J. Massey, D. S. Williamson, H. Browne, J. B. Murray, P. Dokurno, T. Shaw, A. T. Macias, Z. Daniels, S. Geoffroy, M. Dopson, P. Lavan, N. Matassova, G. L. Francis, C. J. Graham, R. Parsons, Y. Wang, A. Padfield, M. Comer, M. J. Drysdale and M. Wood, *Cancer Chemother. Pharmacol.*, 2010, **66**, 535–545.
12. L. Galam, M. K. Hadden, Z. Ma, Q. Z. Ye, B. G. Yun, B. S. Blagg and R. L. Matts, *Bioorg. Med. Chem.*, 2007, **15**, 1939–1946.
13. S. Wisen and J. E. Gestwicki, *Anal. Biochem.*, 2008, **374**, 371–377.
14. K. M. Flaherty, C. DeLuca-Flaherty and D. B. McKay, *Nature*, 1990, **346**, 623–628.
15. M. Sriram, J. Osipiuk, B. Freeman, R. Morimoto and A. Joachimiak, *Structure*, 1997, **5**, 403–414.
16. M. Wisniewska, T. Karlberg, L. Lehtio, I. Johansson, T. Kotenyova, M. Moche and H. Schuler, *PLoS One*, 2010, **5**, e8625.
17. Y. Zhang and E. R. Zuiderweg, *Proc. Natl Acad. Sci. USA*, 2004, **101**, 10272–10277.
18. A. R. Goloudina, O. N. Demidov and C. Garrido, *Cancer Lett.*, 2012, **325**, 117–124.
19. S. P. Ermakova, B. S. Kang, B. Y. Choi, H. S. Choi, T. F. Schuster, W. Y. Ma, A. M. Bode and Z. Dong, *Cancer Res.*, 2006, **66**, 9260–9269.
20. R. Wadhwa, T. Sugihara, A. Yoshida, H. Nomura, R. R. Reddel, R. Simpson, H. Maruta and S. C. Kaul, *Cancer Res.*, 2000, **60**, 6818–6821.
21. D. Mamelak and C. Lingwood, *J. Biol. Chem.*, 2001, **276**, 449–456.
22. D. R. Williams, S. K. Ko, S. Park, M. R. Lee and I. Shin, *Angew. Chem. Int. Ed. Engl.*, 2008, **47**, 7466–7469.
23. A. M. Bode and Z. Dong, *Mutat. Res.*, 2004, **555**, 33–51.
24. O. J. Park and Y. J. Surh, *Toxicol. Lett.*, 2004, **150**, 43–56.
25. J. D. Lambert and C. S. Yang, *J. Nutr.*, 2003, **133**, 3262S–3267S.
26. K. Koya, Y. Li, H. Wang, T. Ukai, N. Tatsuta, M. Kawakami, Shishido and L. B. Chen, *Cancer Res.*, 1996, **56**, 538–543.

27. J. S. Modica-Napolitano, K. Koya, E. Weisberg, B. T. Brunelli, Y. Li and L. B. Chen, *Cancer Res.*, 1996, **56**, 544–550.
28. A. Rousaki, Y. Miyata, U. K. Jinwal, C. A. Dickey, J. E. Gestwicki and E. R. Zuiderweg, *J. Mol. Biol.*, 2011, **411**, 614–632.
29. C. C. Deocaris, N. Widodo, B. G. Shrestha, K. Kaur, M. Ohtaka, K. Yamasaki, S. C. Kaul and R. Wadhwa, *Cancer Lett.*, 2007, **252**, 259–269.
30. A. Tikoo, R. Shakri, L. Connolly, Y. Hirokawa, T. Shishido, B. Bowers, L. H. Ye, K. Kohama, R. J. Simpson and H. Maruta, *Cancer J.*, 2000, **6**, 162–168.
31. D. Pilzer, M. Saar, K. Koya and Z. Fishelson, *Int. J. Cancer*, 2010, **126**, 1428–1435.
32. J. Boulanger, D. Faulds, E. M. Eddy and C. A. Lingwood, *J. Cell. Physiol.*, 1995, **165**, 7–17.
33. H. Whetstone and C. Lingwood, *Biochemistry*, 2003, **42**, 1611–1617.
34. Y. Harada, C. Sato and K. Kitajima, *Biochem. Biophys. Res. Commun.*, 2007, **353**, 655–660.
35. C. M. Wright, R. J. Chovatiya, N. E. Jameson, D. M. Turner, G. Zhu, S. Werner, D. M. Huryn, J. M. Pipas, B. W. Day, P. Wipf and J. L. Brodsky, *Bioorg. Med. Chem.*, 2008, **16**, 3291–3301.
36. H. J. Cho, H. Y. Gee, K. H. Baek, S. K. Ko, J. M. Park, H. Lee, N. D. Kim, M. G. Lee and I. Shin, *J. Am. Chem. Soc.*, 2011, **133**, 20267–20276.
37. X. Zhu, X. Zhao, W. F. Burkholder, A. Gragerov, C. M. Ogata, M. E. Gottesman and W. A. Hendrickson, *Science*, 1996, **272**, 1606–1614.
38. R. C. Morshauser, W. Hu, H. Wang, Y. Pang, G. C. Flynn and E. R. Zuiderweg, *J. Mol. Biol.*, 1999, **289**, 1387–1403.
39. L. Otvos, Jr., I. O, M. E. Rogers, P. J. Consolvo, B. A. Condie, S. Lovas, P. Bulet and M. Blaszczyk-Thurin, *Biochemistry*, 2000, **39**, 14150–14159.
40. G. Kragol, S. Lovas, G. Varadi, B. A. Condie, R. Hoffmann and L. Otvos, Jr., *Biochemistry*, 2001, **40**, 3016–3026.
41. M. Liebscher, G. Jahreis, C. Lucke, S. Grabley, S. Raina and C. Schiene-Fischer, *J. Biol. Chem.*, 2007, **282**, 4437–4446.
42. J. Lee, M. S. Kim, E. Y. Kim, H. J. Park, C. Y. Chang, D. Y. Jung, C. H. Kwon, J. W. Joh and S. J. Kim, *Int. Immunopharmacol.*, 2007, **7**, 1003–1012.
43. B. Bukau, *Mol. Microbiol.*, 1993, **9**, 671–680.
44. E. L. Ramos, S. G. Nadler, D. M. Grasela and S. L. Kelley, *Transplant. Proc.*, 1996, **28**, 873–875.
45. S. G. Nadler, D. D. Dischino, A. R. Malacko, J. S. Cleaveland, S. M. Fujihara and H. Marquardt, *Biochem. Biophys. Res. Commun.*, 1998, **253**, 176–180.
46. H. M. Lorenz, W. H. Schmitt, V. Tesar, U. Muller-Ladner, I. Tarner, I. A. Hauser, F. Hiepe, T. Alexander, H. Woehling, K. Nemoto and P. A. Heinzel, *Arthritis Res. Ther.*, 2011, **13**, R36.
47. T. Shimizu, H. Ishida, H. Shirakawa, K. Omoto, K. Tsunoyama, T. Tokumoto and K. Tanabe, *Clin. Transplant.*, 2010, **24**(22), 22–26.

48. O. Flossmann and D. R. Jayne, *Rheumatology (Oxford)*, 2010, **49**, 556–562.
49. H. Imai, O. Hotta, M. Yoshimura, T. Konta, Y. Tsubakihara, M. Miyazaki, C. Tomida, M. Kobayashi, S. Suzuki, H. Shiiki, A. Yamauchi, H. Yokoyama and M. Nose, *Clin. Exp. Nephrol.*, 2006, **10**, 40–54.
50. J. Goral, H. L. Mathews, S. G. Nadler and J. Clancy, *Immunobiology*, 2000, **202**, 254–266.
51. E. S. Jose-Eneriz, J. Roman-Gomez, L. Cordeu, E. Ballestar, L. Garate, E. J. Andreu, I. Isidro, E. Guruceaga, A. Jimenez-Velasco, A. Heiniger, A. Torres, M. J. Calasanz, M. Esteller, N. C. Gutierrez, A. Rubio, I. Perez-Roger, X. Agirre and F. Prosper, *Br. J. Haematol.*, 2008, **142**, 571–582.
52. X. B. Qiu, Y. M. Shao, S. Miao and L. Wang, *Cell. Mol. Life Sci.*, 2006, **63**, 2560–2570.
53. M. J. Vos, J. Hageman, S. Carra and H. H. Kampinga, *Biochemistry*, 2008, **47**, 7001–7011.
54. C. S. Gassler, A. Buchberger, T. Laufen, M. P. Mayer, H. Schroder, A. Valencia and B. Bukau, *Proc. Natl Acad. Sci. USA*, 1998, **95**, 15229–15234.
55. A. Szabo, R. Korszun, F. U. Hartl and J. Flanagan, *EMBO J.*, 1996, **15**, 408–417.
56. P. Bischofberger, W. Han, B. Feifel, H. J. Schonfeld and P. Christen, *J. Biol. Chem.*, 2003, **278**, 19044–19047.
57. J. I. Leu, J. Pimkina, A. Frank, M. E. Murphy and D. L. George, *Mol. Cell*, 2009, **36**, 15–27.
58. E. Strom, S. Sathe, P. G. Komarov, O. B. Chernova, I. Pavlovska, I. Shyshynova, D. A. Bosykh, L. G. Burdelya, R. M. Macklis, R. Skaliter, E. A. Komarova and A. V. Gudkov, *Nat. Chem. Biol.*, 2006, **2**, 474–479.
59. J. I. Leu, J. Pimkina, P. Pandey, M. E. Murphy and D. L. George, *Mol. Cancer Res.*, 2011, **9**, 936–947.
60. J. Pimkina and M. E. Murphy, *Cancer Biol. Ther.*, 2011, **12**, 503–509.
61. B. Feifel, H. J. Schonfeld and P. Christen, *J. Biol. Chem.*, 1998, **273**, 11999–12002.
62. P. P. Mehta, P. P. Kung, S. Yamazaki, M. Walls, A. Shen, L. Nguyen, M. R. Gehring, G. Los, T. Smeal and M. J. Yin, *Cancer Lett.*, 2011, **300**, 30–39.
63. C. G. Evans, L. Chang and J. E. Gestwicki, *J. Med. Chem.*, 2010, **53**, 4585–4602.
64. R. Russell, R. Jordan and R. McMacken, *Biochemistry*, 1998, **37**, 596–607.

CHAPTER 12

The Cancer Super-chaperone Hsp90: Drug Targeting and Post-translational Regulation

ANNERLEIM WALTON-DIAZ, SAHAR KHAN,
JANE B. TREPEL, MEHDI MOLLAPOUR AND
LEN NECKERS*

Urologic Oncology Branch and Medical Oncology Branch, Center for Cancer Research, National Cancer Institute, 9000 Rockville Pike, Bethesda, MD 20892-1107, USA
*Email: neckersl@mail.nih.gov

12.1 Background and Biology of Hsp90

During the past decade much has been learned about the nature and function of heat-shock protein 90 (Hsp90), especially concerning its association with several molecules and pathways important in cancer. Hsp90 is an abundant (2–5% of total cellular protein) molecular chaperone whose homeostatic functions include stabilization and modulation of a number of proteins (clients) that comprise various cell signaling nodes, and in fostering cellular responses to environmental stress.[1–3] The chaperone has been hijacked by, and represents a non-oncogene addiction of, cancer cells, where its expression is further elevated above that of untransformed cells.[4] Hsp90 is found in all kingdoms except *Archaea*.[5] In humans, as in other eukaryotes, there are two Hsp90 isoforms: stress-inducible Hsp90α and constitutively expressed Hsp90β.[6] These isoforms, although highly homologous, do not fully complement each other. Hsp90β knockout is embryologically lethal, while mice lacking Hsp90α are viable but sterile.[7] Eukaryotes also express organelle specific Hsp90 paralogs: glucose-regulated protein 94 (Grp94), also known as Hsp90B1, is found in the endoplasmic reticulum, where it participates in folding proteins destined for secretion.[8,9] Hsp75, also known as TNF receptor-associated protein 1 (TRAP1), is a mitochondrial paralog that provides protection from proteotoxic stress and may impact mitochondrial metabolism.[10,11] Hsp90 is a member of the ATPase/kinase GHKL (DNA Gyrase, Hsp90, Histidine Kinase, MutL) superfamily – a small group of proteins that are characterized by a unique ATP binding cleft.[2,12] The N-terminal domain of the chaperone contains an ATP-binding site, which is also the target for Hsp90 inhibitors now in clinical trial. The middle (M) domain has binding sites for clients and co-chaperones, and the C-terminal domain contains a dimerization motif and binding sites for other co-chaperones. Connecting the N and M domains are a number of charged amino acids. This unstructured region is referred to as the "charged linker" and plays an important role in Hsp90 chaperone function.[13–15]

Hsp90 is a dynamic protein that undergoes a conformational cycle determined in part by ATP binding/hydrolysis and co-chaperone binding (see Figure 12.1). However, the Hsp90-directed chaperone cycle is complex and its regulation is also impacted by numerous post-translational modifications to Hsp90 and the various co-chaperones. Clinically evaluated Hsp90 inhibitors disrupt the chaperone cycle by occupying the ATP binding pocket. Co-chaperones interact with distinct Hsp90 conformational states (see Figure 12.2). Presumably, these interactions lower the energy barrier between certain conformations, thus providing directionality to the Hsp90 cycle.[16,17] Further, certain co-chaperones, such as Hop and Cdc37, assist in delivery of distinct sets of client proteins (steroid hormone receptors and kinases, respectively) to the Hsp90 chaperone machine. The complex and highly regulated conformational dynamics allow Hsp90 to bind, chaperone and release client proteins. During this process, the conformation and activity/stability of the client protein is altered.[16–20] As stated above, Hsp90 inhibitors currently in clinical trial replace ATP in the N-domain nucleotide-binding pocket, thereby preventing the chaperone cycle from progressing. As a result, Hsp90-dependent client proteins are ubiquitinated and degraded in the proteasome, in a process involving Hsp70 and several chaperone-interacting E3 ubiquitin ligases.[4,12,21]

12.2 Post-translational Modifications of Hsp90

In eukaryotes, numerous post-translational modifications (PTMs) contribute to the regulation of the Hsp90 chaperone cycle (see Figure 12.1).[22] These PTMs seem likely to be an evolutionarily acquired characteristic since bacterial Hsp90 lacks any known post-translational modification. Further, the extent of Hsp90 post-translational modification is greater in metazoans compared to single cell eukaryotes, suggesting that increased use of PTMs provides the possibility for more complex regulation of Hsp90 activity as its client repertoire has increased during evolution. For a detailed description of Hsp90 post-translational

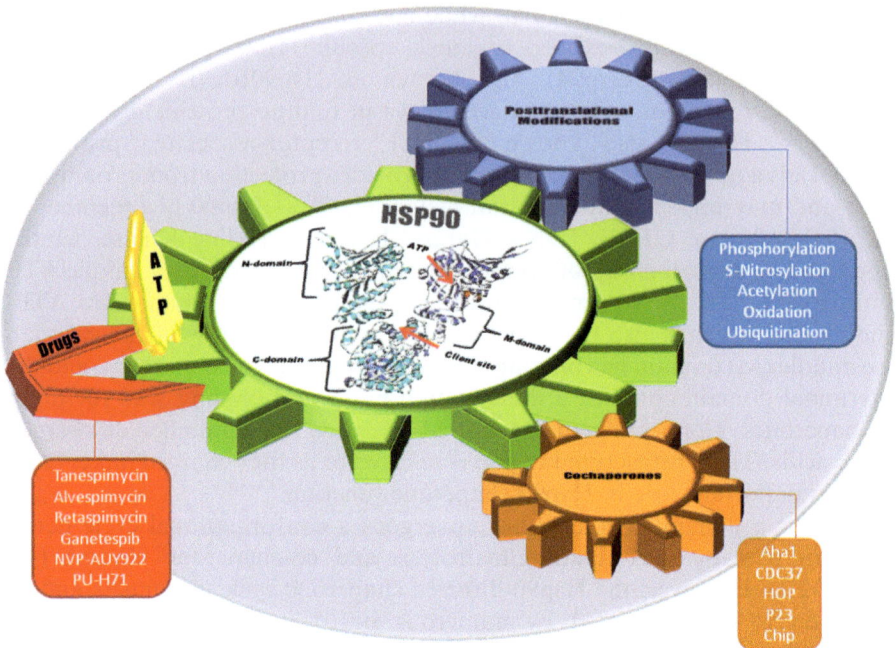

Figure 12.1 The Hsp90 chaperone cycle is regulated by the interplay between ATP binding and the regulated association/dissociation of various co-chaperones. These co-chaperones have distinct activities, including delivering client proteins (HOP, CDC37), stimulating or inhibiting Hsp90 ATPase activity (Aha1, p23) or promoting ubiquitination of Hsp90 clients (CHIP). The Hsp90 chaperone machine is also regulated by a number of diverse posttranslational modifications, including phosphorylation, S-nitrosylation, oxidation, acetylation and ubiquitination. Hsp90 inhibitors that are currently in the clinic displace ATP from its binding pocket in the N-domain of Hsp90 and thus prevent chaperone cycling.

modification sites, the interested reader is directed to two curated websites: PhosphoSitePlus (www.phosphosite.org/protein) and PhosphoPep (www.phosphopep.org/index.php). In the following sections, we will briefly discuss several recently reported Hsp90 PTMs.

12.2.1 Phosphorylation

Phosphorylation of Hsp90 was first described in the 1980s.[23] More recently, Muller *et al.* reported that C-terminal phosphorylation of Hsp90 determines co-chaperone binding. These authors showed that phosphorylation of Hsp90 and Hsp70 prevents binding of CHIP (a co-chaperone with ubiquitin ligase activity), while simultaneously enhancing the binding of the co-chaperone p60^{HOP}. Since both co-chaperones compete for the same binding site on Hsp90 but have opposing activities (CHIP promotes client degradation while p60^{HOP} is involved in client folding), these findings demonstrate how specific

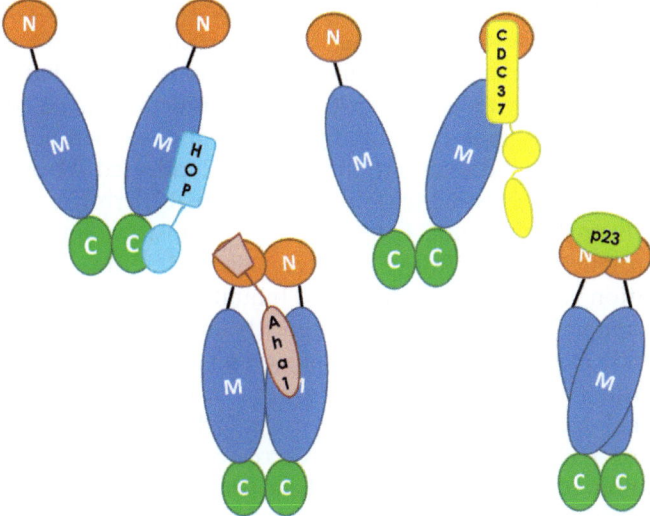

Figure 12.2 Hsp90 conformational dynamics are potentiated by co-chaperones with distinct functions. HOP delivers certain client proteins, including steroid hormone receptors, to Hsp90 and competes with the ubiquitin ligase CHIP for chaperone binding. At the same time, HOP inhibits Hsp90 ATPase to facilitate client protein loading onto the "open" conformation of Hsp90. CDC37 likewise inhibits ATP hydrolysis and primarily delivers kinase clientele to Hsp90. Aha1 and p23 stabilize distinct conformational states of the chaperone and either stimulate (Aha1) or pause (p23) Hsp90-mediated ATP hydrolysis.

phosphorylation events determine whether an Hsp90-dependent client is more likely to be degraded or properly folded.[24] Further, the authors show that cancer cells are characterized by excessive phosphorylation of Hsp90 in this region and this is coincident with increased binding of p60[HOP] at the expense of CHIP, suggesting that this PTM could be targeted in certain malignancies.

CK2 is a ubiquitous Ser/Thr kinase whose activity depends on Hsp90 function.[25–27] CK2 phosphorylates two serine residues, Ser-231 and Ser-263, in the charged-linker of Hsp90α.[28] Equivalent residues in Hsp90β (Ser-226 and Ser-255) are also phosphorylated in untransformed cells but not in leukemic cells.[29] The leukemogenic tyrosine kinases, Bcr-Abl, FLT3/D835Y and Tel-PDGFRβ, all suppress constitutive phosphorylation of Hsp90β at these sites, and this leads to inhibition of apoptosome function. This is achieved by stabilization of a strong interaction between Hsp90β and apoptotic peptidase activating factor 1 (Apaf-1), which prevents cytochrome c-induced Apaf-1 oligomerization and caspase-9 recruitment. Stabilization of the Hsp90β-apoptosome interaction by suppression of Ser-226 and Ser-255 phosphorylation may contribute to chemoresistance in leukemias.[29]

CK2 also phosphorylates a conserved threonine residue (Thr-22) in the N-domain of yeast Hsp90 both *in vitro* and *in vivo*. Thr-22 is the only threonine residue in the N-domain targeted by CK2. We recently showed that ATP binding to the Hsp90 N-domain is necessary for CK2-mediated phosphorylation of Thr-22, suggesting that this PTM participates in the ATP-driven chaperone cycle.[30,31] In support of this hypothesis, we found that mutation of this residue significantly reduced Hsp90 interaction with Aha1, a co-chaperone that up-regulates the rate of ATP hydrolysis and chaperone activity. As expected from these data, mutation of Thr-22 affected Hsp90-dependent chaperoning of kinase (v-Src, Mpk1/Slt2, Raf-1, HER2/ErbB2 and CDK4) and non-kinase (heat-shock factor 1, cystic fibrosis transmembrane conductance regulator protein, glucocorticoid receptor) clients.[30,31] In addition, Thr-22 phosphorylation status also affects Hsp90 inhibitor sensitivity.[31]

Other phosphorylations of Hsp90 also directly impact the interaction of Aha1, consistent with recent structural data suggesting that interaction of this co-chaperone with Hsp90 is necessary to lower the energy barrier of a critical conformational step in the chaperone cycle.[32] Thus, cSrc-dependent phosphorylation of Y301 in Hsp90β leads to decreased Aha1 interaction,[33,34] while phosphorylation of Hsp90α Y313 has been shown to dramatically enhance Hsp90 binding to Aha1.[35] More recently, interaction of another co-chaperone, CDC37, was also shown to be regulated by Hsp90 phosphorylation.[35] Taken together, this growing body of data suggests that eukaryotic cells utilize a diverse set of phosphorylation events to regulate co-chaperone interactions with Hsp90.

12.2.2 Nitrosylation/Oxidation

Hsp90 also undergoes S-nitrosylation. Nitric oxide promotes S-nitrosylation of Hsp90 Cys-597 in endothelial cells and this inhibits Hsp90 chaperone activity.[36] Since endothelial nitric oxide synthase (eNOS) is an Hsp90 client, S-nitrosylation provides a way to regulate NO production that relies on direct feedback inhibition of Hsp90 chaperone activity.[36–38] Oxidative stress has also been shown to cause post-translational modification of Hsp90. Tubocapsenolide, a novel withanolide, increases reactive oxygen species, decreases glutathione levels and causes direct thiol oxidation of Hsp90.[39] Likewise, 4-hydroxy-2-nonenal targets Hsp90 Cys-572 and inhibits its ability to chaperone clients.[40]

12.2.3 Acetylation

Acetylation is a reversible PTM that adds acetyl groups to proteins, usually on lysine residues. Histones are a major target for acetylation and historically acetylating and deacetylating enzymes were termed histone acetylases (HATs) and histone deacetylases (HDACs), respectively. However, these enzymes are now known to be capable of modifying numerous non-histone proteins, including Hsp90. Hsp90 acetylation in response to HDAC inhibitors (HDACi) was first reported by Yu and colleagues. These investigators showed that the HDACi depsipeptide (Romidepsin) increased steady state acetylation of Hsp90 while simultaneously destabilizing Hsp90 interaction with several client proteins, including ErbB2, Raf-1 and mutant p53.[41] Others have shown that additional HDACi also cause Hsp90 hyperacetylation.[42] p300 and HDAC6 promote acetylation and deacetylation of Hsp90, respectively.[43–46] Acetylated Hsp90 levels are increased in HDAC6-deficient mouse embryonic fibroblasts and glucocorticoid receptor function in these cells is compromised.[47] Androgen receptor, an Hsp90 client, is also down-regulated upon HDAC6 inhibition.[48] Reduction in *HDAC6* expression also promotes destabilization of another Hsp90 client protein, the hypoxia-inducible transcription factor HIF-1α.[49] HDAC6 and HDAC10 have been shown to regulate Hsp90-mediated VEGF receptor regulation.[50] Although the impact of HDAC6 on Hsp90 acetylation has been extensively studied, other HDACs also are able to deacetylate the chaperone. HDAC1 has been reported to deacetylate Hsp90 in the nucleus of human breast cancer cells,[51] and both HDAC1 and HDAC10 inhibit the productive Hsp90 chaperoning of VEGF receptor proteins.[50]

Treating SKBr3 breast cancer cells with the pan HDACi trichostatin A (TSA) caused hyperacetylation of Hsp90α at lysine 294.[52] Interestingly, K294 acetylation can be detected even in the absence of HDACi, suggesting that a pool of Hsp90α may be constitutively acetylated on this lysine residue.[52] K294Q or K294A Hsp90α mutants (acetylated lysine mimics) displayed reduced interaction with numerous client proteins (ErbB2, p60$^{v\text{-}Src}$, mutant p53, androgen receptor, Raf-1, HIF1α), and they failed to associate with the co-chaperones Aha1, CHIP and FKBP52. Conversely, the non-acetylatable mutant K294R displayed an equivalent or stronger interaction with these co-chaperone and client proteins compared to wild-type Hsp90.[52] Treating human embryonic kidney 293 cells (HEK293) with the pan-HDACi Panobinostat (LBH589) led to identification of seven additional acetylated lysine residues in Hsp90. Mutation of these lysine residues to glutamine affected the binding of Hsp90α to several co-chaperones, including CHIP, Hsp70 and p23, inhibited ATP binding and inhibited Hsp90 chaperoning of Raf-1.[53] Taken together, these data suggest that reversible acetylation of Hsp90 at multiple sites is a dynamic, tightly regulated process impacting Hsp90 function.

12.2.4 Client-dependent Regulation of Hsp90 PTMs: a Novel Approach to Therapy?

Hsp90 is frequently modified, and its activity impacted, by its own clients. An illustrative example is the tyrosine kinase Wee1. Wee1 is involved in the regulation of the G2/M cell cycle checkpoint and is an Hsp90 client. Wee1 phosphorylates Hsp90 on a conserved tyrosine residue (Y38 on Hsp90α) in the N-domain. Phosphorylation of Hsp90 at this site increases its ability to chaperone several cancer-related kinases including HER2/ErbB2, Src, C-Raf, Cdk4 and Wee1 itself. These data suggest that Wee1 inhibitors might potentiate or synergize with Hsp90 inhibitors. In support of this hypothesis, we recently reported that inhibition of Wee1 sensitizes prostate and cervical cancer cells to Hsp90 inhibitors and causes activation of the intrinsic apoptotic pathway. Dual Wee1 and Hsp90 inhibition in prostate cancer causes down-regulation of Wee1 and Survivin, a suppressor of apoptosis.[54] Combined inhibition of Wee1 and

Hsp90, using drug concentrations that individually were ineffective *in vivo*, caused significant inhibition of tumor growth and led to prolonged survival. These data suggest a novel strategy to enhance the efficacy of Hsp90 inhibitors by combining Hsp90 inhibitors with inhibitors of certain Hsp90 clients that themselves modify Hsp90.

12.3 Targeting Hsp90: Implications and Best Clinical Outcomes

The concept of targeting Hsp90 for cancer therapy was initially viewed with some skepticism because its high expression in non-transformed cells suggested the significant possibility of generating unwanted toxicity. This concern has proven not to be relevant, however, for reasons that are still the subject of intense investigation. Surprisingly, Hsp90 inhibitors *in vivo* tend to concentrate and persist in tumors while being more rapidly cleared from blood and normal tissues.[4] Thus, perhaps even more relevant than with most drugs, Hsp90 inhibitor dose and schedule of administration are important. Careful consideration of these parameters, together with an appreciation of the specific role of Hsp90 relevant to the cancer in question, will be key to designing the most informative clinical trials going forward.

Although the first Hsp90 inhibitors to be identified, the natural products geldanamycin and radicicol, proved too toxic for clinical use, these agents served as invaluable chemical tools for probing the biology of Hsp90 and validating this molecular chaperone as a therapeutic target. These compounds also provided the foundation for development of future clinically well-tolerated drugs. Of these, the first drug to progress to clinical trials was the geldanamycin derivative tanespimycin (17-allylamino-17-demethoxygeldanamycin, 17-AAG). This drug demonstrated clinical activity (as defined by RECIST criteria) in HER2/ErbB2-positive, trastuzumab-resistant breast cancer patients.[55]

Other examples of first-generation Hsp90 inhibitors that have been clinically evaluated include a second geldanamycin derivative, alvespimycin (17-dimethylaminoethylamino-17-demethoxygeldanamycin, 17-DMAG) in castrate-refractory prostate cancer, chondrosarcoma and renal cancer[56] and retaspimycin (IPI-504, a soluble, stable hydroquinone form of 17-AAG), which remains in clinical development.[57] The trailblazing proof-of-concept work with geldanamycin analogues stimulated the race to discover synthetic small-molecule Hsp90 inhibitors that would overcome some or all of the limitations of this class, and would provide the ability to use doses and schedules capable of insuring sufficiently sustained client depletion while sparing liver toxicity, and a large number of synthetic Hsp90 inhibitors are now in clinical development (see Figure 12.3 and Table 12.1; as of this writing, 9 drugs are being evaluated in a total of 44 clinical trials).

As more is learned about the role of Hsp90 in modulating signaling networks in cancer (and normal) cells, and about the sensitivity of various client proteins to Hsp90 inhibition (and the importance of these sensitive clients for tumor

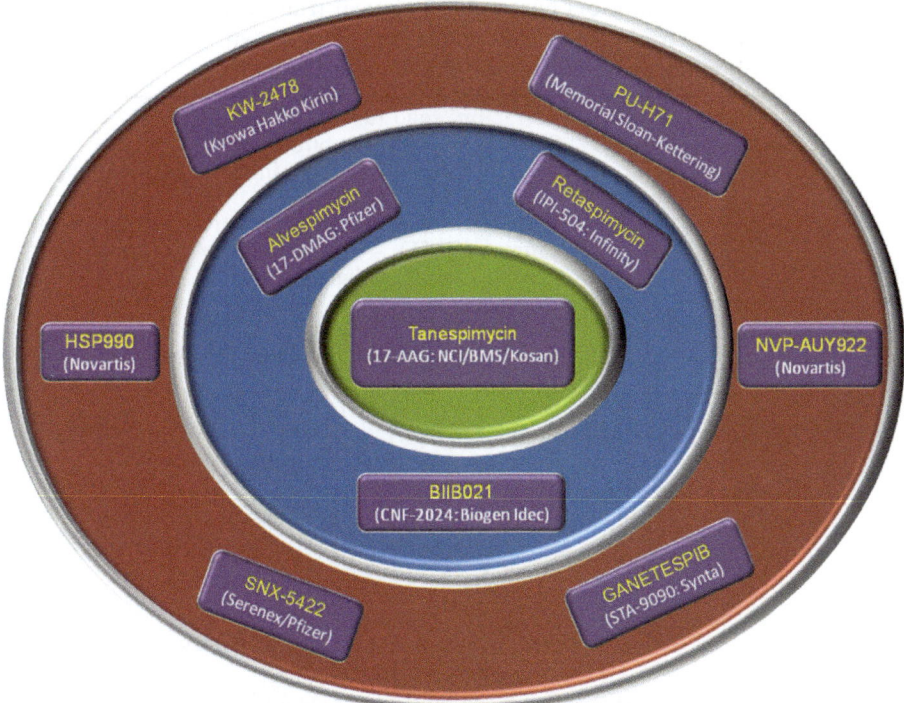

Figure 12.3 A selection of N-domain Hsp90 inhibitors. Tanespimycin (17-AAG) was the first Hsp90 inhibitor to be extensively clinically evaluated, with other agents following into clinical trial during the last several years. Drugs in the outer circle represent a selection of the most recently evaluated (clinically or pre-clinically) N-domain-targeting Hsp90 inhibitors.

survival), we will be better able to predict the patient population most likely to benefit from Hsp90 inhibition, either as a single treatment strategy or in combination with other therapies. As of this writing, several selection criteria already seem apparent and will be briefly discussed (see Figure 12.4). These include (1) targeting cancers that depend on highly sensitive Hsp90 clients (*e.g.* HER2/ErbB2 positive breast cancer, ALK positive non-small-cell lung cancer); (2) targeting cancers that are characterized by constitutive proteotoxic stress (*e.g.* multiple myeloma and K-Ras-driven tumors); (3) employing Hsp90 inhibitors to prevent or combat tumor escape from molecularly targeted therapy (*e.g.* escape from tyrosine kinase inhibitors).

12.3.1 Targeting Sensitive Clients that are Also Tumor Drivers

HER2/ErbB2 is one of the most sensitive clients of Hsp90. In breast cancer the role of HER2 as an important driver oncoprotein has been well established in the literature.[58] In 2011, Scaltriti *et al.* assessed the antitumor activity of retaspimycin in trastuzumab-resistant HER2 + breast cancer cells. They tested

Table 12.1 Hsp90 inhibitors currently in clinical trial (as of February, 2013). Drug, trial phase and indication are shown. Information was obtained from www.cancer.gov/clinicaltrials.

AT13387	I	Refractory solid tumors
(Astex Pharma)	I, II	Single agent or with abiraterone
	I, II	With or without crizotinib in non-small-cell lung cancer (NSCLC)
	II	With or without imatinib in gastrointestinal stromal tumor (GIST)
AUY922	I	Advanced solid malignancies
(Novartis)	I, II	Undesignated
	II	Undesignated
	I	With BYL719 (PI3K inhibitor) in advanced/metastatic gastric cancer
	I, II	With erlotinib (NSCLC)
	II	Refractory GIST
	II	Refractory metastatic pancreatic cancer
	II	Myelofibrosis
	II	NSCLC with EGFR mutations
	Ib	With LDK378 in ALK-rearranged NSCLC
	II	GIST
	II	NSCLC after 2 lines of prior chemotherapy
	II	Advanced ALK + NSCLC
	I	With cetuximab in metastatic KRAS mutant metastatic colorectal cancer
Ganetespib	II	Small cell lung cancer (SCLC)
(Synta Pharma)	II	NSCLC
	II	Prostate cancer after docetaxel
	II	Metastatic pancreatic cancer (2nd or 3rd line)
	II	Metastatic ocular melanoma
	II	With fulvestrant in hormone receptor positive breast cancer
	II	Unresectable melanoma (stage III or IV)
	I	Advance hepatocellular carcinoma
	II	ALK-positive NSCLC
		With or without bortezomib in relapsed/refractory multiple myeloma
	I	Refractory solid tumors
	II	Metastatic HER2 + or triple negative breast cancer
	I	With capecitabine and radiation in rectal cancer
	II/III	With docetaxel in advanced NSCLC
	I	With docetaxel in solid tumors
	I, II	With crizotinib in ALK + lung cancers
	I, II	With pemetrexed-cisplatin in malignant pleural mesothelioma
SNX-5422	I	Refractory solid tumors
(Esanex)		Hematologic cancers
PU-H71	I	Advanced malignancies
(Samus Therapeutics)	I	Solid tumors/low-grade non-Hodgkin's lymphoma not responding to treatment PET imaging of tumors
XL-888	I	With vemurafenib in unresectable BRAF-mutated stage III/IV melanoma

Table 12.1 (*Continued*)

DS-2248 (Daiichi Sankyo)	I	Advanced solid tumors
Debio 0932 (Debiopharm)	I	With standard care in NSCLC
IPI-504 (Infinity Pharma)	I, II	With everolimus in KRAS mutant NSCLC

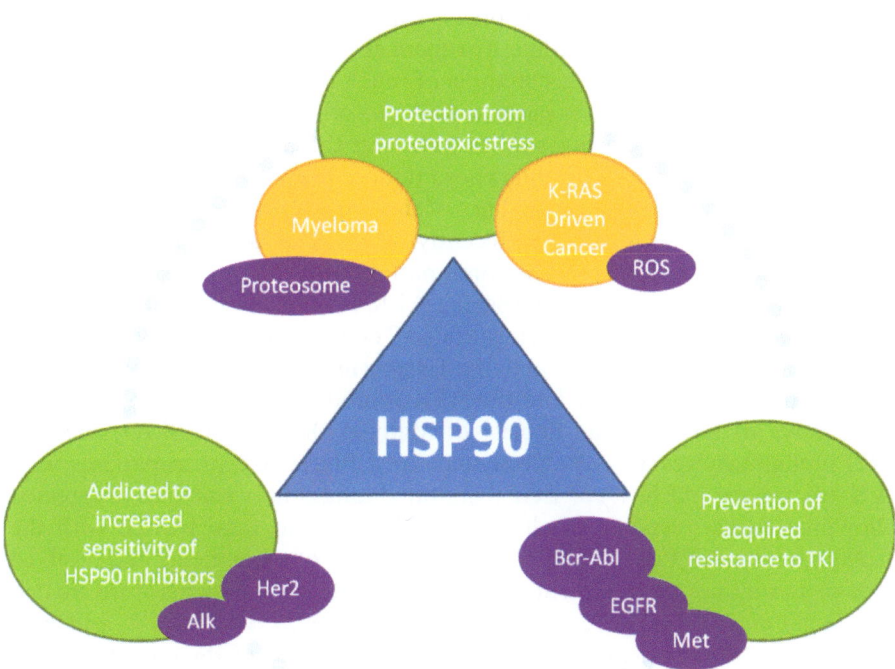

Figure 12.4 Hsp90 inhibitors will likely be most useful as anticancer drugs in situations where a tumor driver protein is also a highly dependent Hsp90 client (*e.g.* ALK fusion proteins and HER2). Additionally, Hsp90 inhibitors may be efficacious for cancers that are exposed to constitutively elevated levels of proteotoxic stress and so depend on the cellular chaperone network to buffer this stress (*e.g.* multiple myeloma and K-Ras-driven cancers). In this case, Hsp90 inhibitors are not aimed at inhibiting a particular, dominant Hsp90 client but in abrogating the tumor's ability to cope with excessive stress. Finally, Hsp90 inhibitors are likely to prevent or significantly retard the development of resistance to targeted therapy that arises as a result of secondary kinase mutation or oncogene switching.

trastuzumab, retaspimycin and a combination of both agents *in vitro* and in xenograft models. Retaspimycin was able to reduce total levels of HER2 equally in trastuzumab-sensitive and -resistant cells. The Hsp90 inhibitor was also able to inhibit tumor growth in xenografts when used as a single agent or in combination with trastuzumab.[59]

As mentioned above, the closely related Hsp90 inhibitor tanespimycin has shown clinical activity in Phase I and II trials in HER2+, trastuzumab-refractory breast cancer. RECIST responses were seen with a weekly schedule of 450 mg/m^2. The overall response rate was 22% and the clinical benefit rate (complete response + partial response + stable disease) was 59%.[60,61]

Another ongoing trial is the ENCHANT Trial, evaluating the Hsp90 inhibitor ganetespib (STA-9090) in breast cancer. This is a Phase II trial for evaluating ganetespib as front-line treatment for HER2+ and triple-negative metastatic breast cancer.[62] This study is of much interest since it is evaluating the use of Hsp90 inhibitors in patients naïve to other lines of therapy.

There have been promising studies assessing the activity of ganetespib in non-small-cell lung cancer patients whose tumors harbor ALK fusion proteins and who have progressed on other lines of therapy. In these studies, ganetespib showed response rates of 50%.[63] Recently, Proia *et al.* also evaluated the activity of ganetespib alone and in combination with crizotinib, in crizotinib-sensitive and -resistant cancers harboring ALK fusions, and in cells expressing amplified ALK or ROS1 translocations. They reported that single agent ganetespib displayed antitumor activity. Furthermore, there was strong synergy between this Hsp90 inhibitor and the ALK inhibitor crizotinib.[63]

Recently, at the European Society of Medical Oncology (ESMO) Meeting 2012, preliminary results were presented for a Phase II trial in patients with ALK-rearranged (ALK+) or EGFR-mutated lung cancer who had progressed following at least one line of chemotherapy and who were treated with the Hsp90 inhibitor NVP-AUY922. In this study, 61% of the patients had received three or more lines of treatment. Up to April 2012, 121 patients were treated with 70 mg/m^2 NVP-AUY922 once weekly. Of these, 29% of ALK+ and 20% of EGFR-mutated patients had partial responses. Four of the six ALK+ responders were naïve to crizotinib. The median progression-free survival rate at 18 weeks was 42% in ALK+ and 34% in EGFR-mutated patients.[64]

Ganetespib was also studied by Kau *et al.*, who reported a Phase I and pharmacokinetic study of multiple schedules of ganetespib in combination with docetaxel for patients with advanced solid tumors. This combination was well tolerated at doses of 75 mg/m^2 for docetaxel and 150 mg/m^2 for ganetespib. As of this writing, a randomized Phase IIb/III study with a regimen including docetaxel on day 1 and ganetespib on days 1 and 15 is ongoing for advanced lung cancer.[65,66]

12.3.2 Hsp90 and Proteotoxic Stress

In some cancers, there is a delicate balance between proteotoxic stress and cancer cell survival. For example, the proteasome machinery is used to maximum capacity in highly secretory multiple myeloma cells and the proteasome is a validated molecular target in this cancer.[67] Hsp90 inhibitors (which redirect many Hsp90-dependent clients to the proteasome) have been shown to synergize with proteasome inhibitors in multiple myeloma and other cancers.[68] A Phase I/II trial was undertaken to evaluate combination of the Hsp90 inhibitor tanespimycin and the proteasome inhibitor bortezomib. Among evaluable patients, 3% had a complete response and 12% had a partial response. The objective response rate was 27%. Of note, the highest response rates were observed in patients naïve to bortezomib.[69,70] Unfortunately, the development of tanespimycin was discontinued, due in part to patent issues (www.myelomabeacon.com/news/2010//tanespymicin-development-halted). However, given this promising activity, additional Hsp90 inhibitors are being evaluated in combination with proteasome inhibitors in multiple myeloma.[71] For example, a current Phase I study is comparing ganetespib alone *versus* ganetespib plus bortezomib in patients with relapsed and/or refractory multiple myeloma. Although results are not available as of this writing (http://clinicaltrials.gov/ct2/show/NCT01485835?term = ganetespib&rank = 2), Hsp90 inhibitors are likely to be efficacious in such a setting because they interfere with the machinery, already running at maximum capacity, used by myeloma cells to maintain cellular homeostasis.

K-Ras is mutated in 25% of non-small-cell lung cancer and is recognized as an important oncogenic driver. Patients whose cancers harbor this mutation respond poorly to existing therapies. Hes *et al.* recently published a study evaluating ganetespib in patients with K-Ras mutations. They observed single agent activity and even better outcomes when ganetespib was combined with MEK or PI3K/mTOR inhibitors. Of interest, these investigators reported that ganetespib sensitized mutant K-Ras cells to standard of care therapies.[72]

Mutant K-Ras-driven tumor cells depend on optimal function of the cellular stress response machinery necessary to cope with constitutively elevated reactive oxygen species (ROS). Moderate ROS levels are needed by K-Ras-driven tumor cells to regulate important signal transduction pathways on which they depend for proliferation.[73] However, excessive oxidative stress can overwhelm the stress response machinery, damaging proteins and leading to cell death. For this reason, K-Ras-driven tumors rely on an active chaperone network and on mTOR to maintain sufficient levels of reduced glutathione needed to modulate oxidative stress. In this context, Hsp90 inhibitors help to collapse this safety net and, when combined with mTOR inhibition, this strategy provides a novel therapeutic approach to attack such cancers.[74] A Ib/II clinical trial evaluating the safety and efficacy of retaspimycin plus the mTOR inhibitor everolimus in patients with K-Ras mutant non-small-cell lung cancer is ongoing (http://www.infi.com/product-candidates-pipeline-IPI-504.asp). In this case, as with proteasome inhibitor combination in multiple myeloma, Hsp90 inhibition is not aimed at inactivating a specific tumor-driver, but at compromising the cancer cell's ability to cope with persistent environmental stress.

12.3.3 Hsp90 Inhibitors and TKIs

Chronic myeloid leukemia (CML) is a hematologic disorder characterized by a translocation between chromosomes 9 and 22.[75] This produces the constitutively active chimeric oncogenic tyrosine kinase BCR-ABL. Tyrosine kinase inhibitors (TKIs), such as imatinib mesylate (Gleevec) have been an important tool for treatment of these malignancies since BCR-ABL is necessary to maintain their deregulated growth.[76] Although responses to Gleevec are dramatic, this TKI is not able to completely eradicate BCR-ABL + tumor cells, and resistance eventually develops. BCR-ABL kinase domain mutations account for up to 90% of secondary resistance mechanisms in CML.[77] Among the strategies explored to combat this resistance is the use of Hsp90 inhibitors, since Hsp90 is known to chaperone and promote the stability of BCR-ABL.[78] Investigators have shown that both Gleevec-sensitive and -resistant BCR-ABL retain dependence on Hsp90 and sensitivity to Hsp90 inhibitors.[76] Peng et al. evaluated the combination of retaspimycin and Gleevec in mice with leukemia harboring Gleevec-resistant BCR-ABL. These investigators reported that this combination was more effective than either drug alone and was able to significantly prolong survival in mice.[79] These studies provide evidence that Hsp90 inhibitors might abrogate or significantly delay escape of certain kinase-driven cancers from TKIs.

Hsp90 inhibition has also been studied in relation to tyrosine kinase KIT-driven gastrointestinal stromal tumors (GIST). Investigators at Memorial Sloan-Kettering Cancer Center reported a Phase II trial using the Hsp90 inhibitor BIIB021 in which they observed a metabolic response. This was assessed by fluorodeoxyglucose positron emission tomography (FDG-PET). The study included 23 patients with GIST refractory to Gleevec and sunitinib (Sutent). The investigators reported a partial response in three patients treated with a dose of 600 mg bi-weekly, and in two patients treated with a dose of 400 mg three times a week, with an overall response rate of 22%.[80]

12.4 Concluding Remarks

As described in this chapter, ongoing efforts are being made to better understand the structure and functional biology of Hsp90 in cancer and in normal cells, as well as to identify sensitive tumor-driving clients and novel approaches to sensitize cancer cells to Hsp90 inhibitors. We continue to learn about additional post-translational modifications of Hsp90 that affect not only the chaperone itself but its interaction with numerous client proteins and affect its sensitivity to Hsp90 inhibitors. However, one should not lose sight of the fact that Hsp90 is also highly expressed in normal cells where it contributes towards maintaining protein homeostasis. This needs to be considered in trial design so as to take advantage of the beneficial, if not completely understood, property of current clinically evaluated Hsp90 inhibitors to persist preferentially in tumors and not in normal tissues. Further, predicting those cancers most likely to respond to Hsp90 inhibition is a work in progress. It is likely that initial success will be best achieved by using these inhibitors to treat cancers that are addicted to particular amplified, mutated or translocated driver oncogenes, which are themselves highly dependent Hsp90 clients, such as HER2/ErrbB2 and ALK. Use of Hsp90 inhibitors to interfere with cancer cells' ability to cope with persistent proteotoxic or other environmental stresses is likely to represent an additional paradigm, as is their use to combat development of TKI resistance or escape. Going forward, realization of the full therapeutic potential of inhibiting Hsp90 will certainly also benefit from a more complete genetic profiling of tumors.

References

1. S. K. Wandinger, K. Richter and J. Buchner, *J. Biol. Chem.*, 2008, **283**(27), 18473–18477.
2. D. Picard, *Cell. Mol. Life Sci.*, 2002, **59**(10), 1640–1648.
3. M. Taipale, D. F. Jarosz and S. Lindquist, *Nat. Rev. Mol. Cell Biol.*, 2010, **11**(7), 515–528.
4. J. Trepel, M. Mollapour, G. Giaccone and L. Neckers, *Nat. Rev. Cancer*, 2010, **10**(8), 537–549.
5. A. T. Large, M. D. Goldberg and P. A. Lund, *Biochem. Soc. Trans.*, 2009, **37**(Pt 1), 46–51.
6. I. Grad, C. R. Cederroth, J. Walicki, C. Grey, S. Barluenga, N. Winssinger, B. De Massy, S. Nef and Didier Picard, *PLoS One*, 2010, **5**(12), e15770.
7. A. K. Voss, T. Thomas and P. Gruss, *Development*, 2000, **127**(1), 1–11.
8. D. E. Dollins, J. J. Warren, R. M. Immormino and D. T. Gewirth, *Mol. Cell*, 2007, **28**(1), 41–56.
9. S. Frey, A. Leskovar, J Reinstein and J. Buchner, *J. Biol. Chem.*, 2007, **282**(49), 35612–35620.
10. S. J. Felts, B. A. Owen, P. Nguyen, J. Trepel, D. B. Donner and D. O. Toft, *J. Biol. Chem.*, 2000, **275**(5), 3305–3312.
11. A. Leskovar, H. Wegele, N. D. Werbeck, J. Buchner and J. Reinstein, *J. Biol. Chem.*, 2008, **283**(17), 11677–11688.
12. L. H. Pearl and C. Prodromou, *Annu. Rev. Biochem.*, 2006, **75**, 271–294.
13. S. Tsutsumi, M. Mollapour, C. Prodromou, C. T. Lee, B. Panaretou, S. Yoshida, M. P. Mayer and L. M. Neckers, *Proc. Natl Acad. Sci. USA*, 2012, **109**(8), 2937–2942.
14. S. Tsutsumi, M. Mollapour, C. Graf, C. T. Lee, B. T. Scroggins, W. Xu, L. Haslerova, M. Hessling, A. A. Konstantinova, J. B. Trepel, B. Panaretou, J. Buchner, M. P. Mayer, C. Prodromou and L. Neckers, *Nat. Struct. Mol. Biol.*, 2009, **16**(11), 1141–1147.
15. O. Hainzl, M. C. Lapina, J. Buchner and K. Richter, *J. Biol. Chem.*, 2009, **284**(34), 22559–22567.
16. M. Hessling, K. Richter and J. Buchner, *Nat. Struct. Mol. Biol.*, 2009, **16**(3), 287–293.
17. M. Mickler, M. Hessling, C. Ratzke, J. Buchner and T. Hugel, *Nat. Struct. Mol. Biol.*, 2009, **16**(3), 281–286.
18. C. K. Vaughan, U. Gohlke, F. Sobott, V. M. Good, M. M. Ali, C. Prodromou, C. V. Robinson, H. R. Saibil and L. H. Pearl, *Mol. Cell*, 2006, **23**(5), 697–707.

19. S. H. McLaughlin, L. A. Ventouras, B. Lobbezoo and S. E. Jackson, *J. Mol. Biol.*, 2004, **344**(3), 813–826.
20. A. K. Shiau, S. F. Harris, D. R. Southworth and D. A. Agard, *Cell*, 2006, **127**(2), 329–340.
21. L. Neckers, *Curr. Top. Med. Chem.*, 2006, **6**(11), 1163–1171.
22. M. Mollapour and L. Neckers, *Biochim. Biophys. Acta*, 2012, **1823**(3), 648–655.
23. J. J. Dougherty, R. K. Puri and D. O. Toft, *J. Biol. Chem.*, 1982, **257**(23), 14226–14230.
24. P. Muller, E. Ruckova, P. Halada, P. J. Coates, R. Hrstka, D. P. Lane and B. Vojtesek, *Oncogene*, 2013, in press.
25. M. Ruzzene, G. Di Maira, K. Tosoni and L. A. Pinna, *Methods Enzymol.*, 2010, **484**, 495–514.
26. Y. Miyata, *Cell. Mol. Life Sci.*, 2009, **66**(11–12), 1840–1849.
27. J. J. Dougherty, D. A. Rabideau, A. M. Iannotti, W. P. Sullivan and D.O. Toft, *Biochim. Biophys. Acta*, 1987, **927**(1), 74–80.
28. S. P. Lees-Miller and C. W. Anderson, *J. Biol. Chem.*, 1989, **264**(5), 2431–2437.
29. M. Kurokawa, C. Zhao, T. Reya and S. Kornbluth, *Mol. Cell. Biol.*, 2008, **28**(17), 5494–5506.
30. M. Mollapour, S. Tsutsumi, Y. S. Kim, J. Trepel and L. Neckers, *Oncotarget*, 2011, **2**(5), 407–417.
31. M. Mollapour, S. Tsutsumi, A. W. Truman, W. Xu, C. K. Vaughan, K. Beebe, A. Konstantinova, S. Vourganti, B. Panaretou, P. W. Piper, J. B. Trepel, C. Prodromou, L. H. Pearl and L. Neckers, *Mol. Cell*, 2011, **41**(6), 672–681.
32. M. Retzlaff, F. Hagn, L. Mitschke, M. Hessling, F. Gugel, H. Kessler, K. Richter and J. Buchner, *Mol. Cell*, 2010, **37**(3), 344–354.
33. F. Desjardins, C. Delisle and J. P. Gratton, *Arterioscler. Thromb. Vasc. Biol.*, 2012, **32**(10), 2484–2492.
34. M. Duval, F. Le Boeuf, J. Huot and J. P. Gratton, *Mol. Biol. Cell*, 2007, **18**(11), 4659–4668.
35. W. Xu, M. Mollapour, C. Prodromou, S. Wang, B. T. Scroggins, Z. Palchick, K. Beebe, M. Siderius, M. J. Lee, A. Couvillon, J. B. Trepel, Y. Miyata, R. Matts and L. Neckers, *Mol. Cell*, 2012, **47**(3), 434–443.
36. A. Martinez-Ruiz, L. Villanueva, C. Gonzalez de Orduna, D. López-Ferrer, J. Vázquez and S. Lamas, *Proc. Natl Acad. Sci. USA*, 2005, **102**(24), 8525–8530.
37. M. Retzlaff, M. Stahl, H. C. Eberl, S. Lagleder, J. Beck, H. Kessler and J Buchner, *EMBO Rep.*, 2009, **10**(10), 1147–1153.
38. G. Garcia-Cardena, R. Fan, V. Shah, R. Sorrentino, G. Cirino, A. Papapetropoulos and W. C. Sessa, *Nature*, 1998, **392**(6678), 821–824.
39. W. Y. Chen, F. R. Chang, Z. Y. Huang, J. H. Chen, Y. C. Wu and C. C. Wu, *J. Biol. Chem.*, 2008, **283**(25), 17184–17193.
40. D. L. Carbone, J. A. Doorn, Z. Kiebler, B. R. Ickes and D. R. Petersen, *J. Pharmacol. Exp. Ther.*, 2005, **315**(1), 8–15.

41. X. Yu, Z. S. Guo, M. G. Marcu, L. Neckers, D. M. Nguyen, G. A. Chen and D. S. Schrump, *J. Natl Cancer Inst.*, 2002, **94**(7), 504–513.
42. R. Nimmanapalli, L. Fuino, P. Bali, M. Gasparetto, M. Glozak, J. Tao, L. Moscinski, C. Smith, J. Wu, R. Jove, P. Atadja and K. Bhalla, *Cancer Res.*, 2003, **63**(16), 5126–5135.
43. P. Bali, M. Pranpat, J. Bradner, M. Balasis, W. Fiskus, F. Guo, K. Rocha, S. Kumaraswamy, S. Boyapalle, P. Atadja, E. Seto and K. Bhalla, *J. Biol. Chem.*, 2005, **280**(29), 26729–26734.
44. J. J. Kovacs, P. J. Murphy, S. Gaillard, X. Zhao, J. T. Wu, C. V. Nicchitta, M. Yoshida, D. O. Toft, W. B. Pratt and T. P. Yao, *Mol. Cell*, 2005, **18**(5), 601–607.
45. V. D. Kekatpure, A. J. Dannenberg and K. Subbaramaiah, *J. Biol. Chem.*, 2009, **284**(12), 7436–7445.
46. P. J. Murphy, Y. Morishima, J. J. Kovacs, T. P. Yao and W. B. Pratt, *J. Biol. Chem.*, 2005, **280**(40), 33792–33799.
47. Y. Zhang, S. Kwon, T. Yamaguchi, F. Cubizolles, S. Rousseaux, M. Kneissel, C. Cao, N. Li, H. L. Cheng, K. Chua, D. Lombard, A. Mizeracki, G. Matthias, F. W. Alt, S. Khochbin and P. Matthias, *Mol. Cell. Biol.*, 2008, **28**(5), 1688–1701.
48. J. Ai, Y. Wang, J. A. Dar, J. Liu, L. Liu, J. B. Nelson and Z Wang, *Mol. Endocrinol.*, 2009, **23**(12), 1963–1972.
49. Q. Zhang and D. L. Denlinger, *J. Insect Physiol.*, 2010, **56**(2), 138–150.
50. J. H. Park, S. H. Kim, M. C. Choi, J. Lee, D. Y. Oh, S. A. Im, Y. J. Bang and T. Y. Kim, *Biochem. Biophys. Res. Commun.*, 2008, **368**(2), 318–322.
51. Q. Zhou, A. T. Agoston, P. Atadja, W. G. Nelson and N. E. Davidson, *Mol. Cancer Res.*, 2008, **6**(5), 873–883.
52. B. T. Scroggins, K. Robzyk, D. Wang, M. G. Marcu, S. Tsutsumi, K. Beebe, R. J. Cotter, S. Felts, D. Toft, L. Karnitz, N. Rosen and L. Neckers, *Mol. Cell*, 2007, **25**(1), 151–159.
53. Y. Yang, R. Rao, J. Shen, Y. Tang, W. Fiskus, J. Nechtman, P. Atadja and K. Bhalla, *Cancer Res.*, 2008, **68**(12), 4833–4842.
54. A. Iwai, D. Bourboulia, M. Mollapour, S. Jensen-Taubman, S. Lee, A. C. Donnelly, S. Yoshida, N. Miyajima, S. Tsutsumi, A. K. Smith, D. Sun, X. Wu, B. S. Blagg, J. B. Trepel, W. G. Stetler-Stevenson and L. Neckers, *Cell Cycle*, 2012, **11**(19), 3649–3655.
55. S. Modi, S. Sugarman, A. Stopeck, H. Linden, W. Ma, K. Kerscy, R. G. Johnson, N. Rosen, A. L. Hannah and C. A. Hudis, *J. Clin. Oncol.*, 2008, **26**(15S), 1027.
56. S. Pacey, R. H. Wilson, M. Walton, M. M. Eatock, A. Hardcastle, A. Zetterlund, H. T. Arkenau, J. Moreno-Farre, U. Banerji, B. Roels, H. Peachey, W. Aherne, J. S. de Bono, F. Raynaud, P. Workman and I. Judson, *Clin. Cancer Res.*, 2011, **17**(6), 1561–1570.
57. L. V. Sequist, S. Gettinger, N. N. Senzer, R. G. Martins, P. A. Jänne, R. Lilenbaum, J. E. Gray, A. J. Iafrate, R. Katayama, N. Hafeez, J. Sweeney, J. R. Walker, C. Fritz, R. W. Ross, D. Grayzel,

J. A. Engelman, D. R. Borger, G. Paez and R. Natale, *J. Clin. Oncol.*, 2010, **28**(33), 4953–4960.
58. S. Loi, E. de Azambuja, L. Pugliano, C. Sotiriou and M. J. Piccart, *Curr. Opin. Oncol.*, 2011, **23**(6), 547–558.
59. M. Scaltriti, V. Serra, E. Normant, M. Guzman, O. Rodriguez, A. R. Lim, K. L. Slocum, K. A. West, V. Rodriguez, L. Prudkin, J. Jimenez, C. Aura and J. Baselga, *Mol. Cancer Ther.*, 2011, **10**(5), 817–824.
60. S. Modi, A. Stopeck, H. Linden, D. Solit, S. Chandarlapaty, N. Rosen, G. D'Andrea, M. Dickler, M. E. Moynahan, S. Sugarman, W. Ma, S. Patil, L. Norton, A. L. Hannah and C. Hudis, *Clin. Cancer Res.*, 2011, **17**(15), 5132–5139.
61. S. Modi, A. T. Stopeck, M. S. Gordon, D. Mendelson, D. B. Solit, R. Bagatell, W. Ma, J. Wheler, N. Rosen, L. Norton, G. F. Cropp, R. G. Johnson, A. L. Hannah and C. A. Hudis, *J. Clin. Oncol.*, 2007, **25**(34), 5410–5417.
62. D. Cameron, M. S. Mano, V. Vukovic, F. Teofilovici, R. Bradley and A. Awada, *European Society for Medical Oncology, Vienna, Austria*, 2012, Abstract 347P.
63. D. A. Proia, J. Acquaviva, Q. Jiang, L. Xue, D. Smith, J. C. Friedland, S. He, S. Morris and Y. Wada, *J. Clin. Oncol.*, 2012, **30**, 3090.
64. E. Felip, E. Carcereny, F. Barlesi, L. Gandhi, L. V. Sequist, S.-W. Kim, H. J. M. Groen, B. Besse, D.-W. Kim, E. F. Smit, M. Akimov, E. Avsar, S. Bailey, W. Ofosu-Appiah and E. B. Garon, *European Society for Medical Oncology, Vienna, Austria*, 2012, Abstract 4380.
65. J. S. Kauh, T. K. Owonikoko, B. F. El-Rayes, D. M. Shin, S. Murali, C. M. Lewis, M. D. Karol, F. Teofilovici, Y. Du, H. Fu, F. R. Khuri and S. S. Ramalingam, *J. Clin. Oncol.*, 2012, **30**, 3094.
66. S. S. Ramalingam, B. Zaric, G. Goss, C. Manegold, R. Rosell, V. Vukovic, I. El-Hairy, F. Teofilovici, A. Enke and D. Fennell, *European Society for Medical Oncology, Vienna, Austria*, 2012, Abstract 1248P-PR.
67. P. Lawasut, D. Chauhan, J. Laubach, C. Hayes, C. Fabre, M. Maglio, C. Mitsiades, T. Hideshima, K. C. Anderson and P. G. Richardson, *Curr. Hematol. Malig. Rep.*, 2012, **7**(4), 258–266.
68. E. G. Mimnaugh, W. Xu, M. Vos, X. Yuan, J. S. Isaacs, K. S. Bisht, D. Gius and L. Neckers, *Mol. Cancer Ther.*, 2004, **3**(5), 551–566.
69. P. G. Richardson, A. A. Chanan-Khan, S. Lonial, A. Y. Krishnan, M. P. Carroll, M. Alsina, M. Albitar, D. Berman, M. Messina and K. C. Anderson, *Br. J. Haematol.*, 2011, **153**(6), 729–740.
70. P. G. Richardson, C. S. Mitsiades, J. P. Laubach, S. Lonial, A. A. Chanan-Khan and K. C. Anderson, *Br. J. Haematol.*, 2011, **152**(4), 367–379.
71. M. Ri, S. Iida, T. Nakashima, H. Miyazaki, F. Mori, A. Ito, A. Inagaki, S. Kusumoto, T. Ishida, H. Komatsu, Y. Shiotsu and R. Ueda, *Leukemia*, 2010, **24**(8), 1506–1512.
72. J. Acquaviva, D. L. Smith, J. Sang, J. C. Friedland, S. He, M. Sequeira, C. Zhang, Y. Wada and D. A. Proia, *Mol. Cancer Ther.*, 2012, **11**(12), 2633–2643.

73. W. Xu, J. Trepel and L. Neckers, *Cancer Cell*, 2011, **20**(3), 281–282.
74. T. De Raedt, Z. Walton, J. L. Yecies, D. Li, Y. Chen, C. F. Malone, O. Maertens, S. M. Jeong, R. T. Bronson, V. Lebleu, R. Kalluri, E. Normant, M. C. Haigis, B. D. Manning, K. K. Wong, K. F. Macleod and K. Cichowski, *Cancer Cell*, 2011, **20**(3), 400–413.
75. C. R. Bartram, A. de Klein, A. Hagemeijer, T. van Agthoven, A. G. van Kessel, D. Bootsma, G. Grosveld, M. A. Ferguson-Smith, T. Davies, M. Stone, N. Heisterkamp, J. R. Stephenson and J. Groffen, *Nature*, 1983, **306**(5940), 277–280.
76. M. E. Gorre, K. Ellwood-Yen, G. Chiosis, N. Rosen and C. L. Sawyers, *Blood*, 2002, **100**(8), 3041–3044.
77. E. Weisberg and J. D. Griffin, *Blood*, 2000, **95**(11), 3498–3505.
78. M. E. Gorre, M. Mohammed, K. Ellwood, N. Hsu, R. Paquette, P. N. Rao and C. L. Sawyers, *Science*, 2001, **293**(5531), 876–880.
79. C. Peng, J. Brain, Y. Hu, A. Goodrich, L. Kong, D. Grayzel, R. Pak, M. Read and S. Li, *Blood*, 2007, **110**(2), 678–685.
80. M. A. Dickson, S. H. Okuno, M. L. Keohan, R. G. Maki, D. R. D'Adamo, T. J. Akhurst, C. R. Antonescu and G. K. Schwartz, *Ann. Oncol.*, 2013, **24**(1), 252–257.

CHAPTER 13

Hsp90 Inhibitors in Clinic

EMIN AVSAR

Novartis Pharmaceuticals, Oncology Translational Medicine, 1 Health Plaza, New Jersey, USA
Email: emin.avsar@novartis.com

13.1 Introduction

The quest of finding Hsp90 inhibitors started when two early naturally occurring compounds, geldanamycin and radicicol, were found to inhibit Hsp90.[1] Radicicol was first discovered in 1953 as an antifungal agent.[2] Geldanamycin was discovered in 1970 while screening for antibiotics against protozoa in the fermented cultures of *Streptomycis Hygroscopieus*.[3] It was not discovered until 1985 that both agents might have antineoplastic effects. Uehara and colleagues showed that geldanamycin and other ansamycins were able to reverse the oncogenic transformation of a rat kidney cell line model based on heat activated rous sarcoma virus (RSV).[4] For almost a decade the mechanism for the antineoplastic effects were unknown, and different ideas were postulated (for a nice review see ref. 5), until Whitesell and colleagues showed that geldanamycin did not reverse the src mediated formation by inhibiting src phosphorylation *per se*, but by disrupting the src-hsp90 complex formation by binding to Hsp90.[6] Subsequent to this finding, two research groups simultaneously showed that geldanamycin binds to the ATP pocket present in the N-terminal of Hsp90.[7,8] This finding opened the possibilities for Hsp90 as a viable drug target in oncogenesis, and led to further studies to show many oncogenic proteins to be Hsp90 clients.[1,9,10] The pre-clinical findings are summarized by my colleagues in recent chapters. Below I will try to summarize the clinical findings thus far and future development for Hsp90 inhibitors.

RSC Drug Discovery Series No. 37
Inhibitors of Molecular Chaperones as Therapeutic Agents
Edited by Timothy Machajewski and Zhenhai Gao
© The Royal Society of Chemistry 2014
Published by the Royal Society of Chemistry, www.rsc.org

13.2 Geldanamycin Analogues

13.2.1 17-AAG (Tanespimycin)

Geldanamycin and radicicol never entered clinic due to the hepatoxicity seen in toxicology studies, and the poor solubility and stability issues associated with both compounds.[11] The hepatoxicity has been attributed to the quinone ring, and to overcome the solubility and toxicity issues a non-essential carboxyl in the 17 position of the benzoquinone ring was replaced with amines to form less toxic, and more stable analogues.[1] One such analogue was 17-allyamino-17-demethoxygeldanamycin (17-AAG), which showed increased biologic activity and antitumor properties in rat embryo cell lines transfected with human HER2.[12] Later, it was shown that 17-AAG, tanespimycin, also binds to Hsp90, and destabilizes Raf-1 and mutated p53.[13] The encouraging pre-clinical results of tanespimycin led to the first clinical trials conducted in both the US and the UK supported by the National Cancer Institute (NCI). These studies tried multiple dosing schemes. Banerji and colleagues evaluated tanespimycin, given intravenously (i.v.), in a weekly dosing schedule in patients with advanced solid malignancies,[14] whereas other studies have evaluated different schedules, such as daily dosing with two-day rest and weekly dosing with one-week rest.[15] The once weekly schedule started at the low dose of $10\,mg/m^2$ and was escalated up to the maximum tolerated dose (MTD) of $450\,mg/m^2$. A total of 31 patients were dosed. The dose was doubled from $10\,mg/m^2$ in subsequent cohorts until the $320\,mg/m^2$ dose, where this dose cohort was expanded to six patients as Hsp70 induction was seen in peripheral blood leucocytes.[14] One dose limiting toxicity (DLT), grade 3 diarrhea, was observed in the expanded cohort. Due to this observation the next dose cohort was escalated in a 40% incremental increase to $450\,mg/m^2$, as defined per the study protocol. At this dose cohort one patient out of the first three patients experienced a DLT in the form of grade 3 diarrhea, grade 3 ALT and grade 4 AST elevations. The cohort was expanded to six patients and no further DLTs were observed. An additional three patients were enrolled at the $450\,mg/m^2$ dose in order to further evaluate the pharmacological changes. One out of these patients experienced a DLT, grade 3 diarrhea, bringing the total to two patients out of nine to have a DLT at $450\,mg/m^2$. Although per protocol definition MTD had not been reached, the investigators did not attempt to test higher dose levels because PK and PD objectives of the trial had been met, and the dose administration required 200 minutes of infusion time and over 40 ml of DMSO as solvent. Other notable observed toxicities were nausea, fatigue and vomiting. Three patients experienced grade 3 and grade 2 hypersensitivity, but these were not classified as DLTs as they were attributed to the egg phospholipid used to reconstitute tanespimycin. No partial or completed responses were noticed; however, several patients experienced disease stabilization. Two patients with metastatic melanoma remained receiving treatment for 15 and 35 months, respectively, and one patient at the time of publication was still on treatment for 41 months.[14]

A study that assessed the daily dosing of tanespimycin for 5 days with two-day rest every three weeks was initiated by Dr. Grem and colleagues at the NCI.[15] The starting dose was also 10 mg/m^2 and dose escalation was conducted in 40% increments following an accelerated titration dose escalation schema. This schema was followed until one patient experience a DLT, or two patients experience grade 2 events. Once any such event was experienced, the escalation design was reversed to traditional 3 plus 3, Fibonacci escalation design. The accelerated escalation was employed until 28 mg/m^2 and the first DLT, grade 3 hepatoxicity, was experienced at 56 mg/m^2. Other notable toxicities experienced were nausea, emesis, anemia and fever. There were no radiological responses observed but Hsp90 modulation was observed in peripheral blood mono nuclear cells (PBMC) at 40 mg/m^2 by measuring Hsp70 induction.

Solit and colleagues also evaluated tanespimycin in advanced solid tumors by using the same dosing schedule (daily 5-day dosing in 21-day cycle), and an accelerated titration dose escalation schema.[16] The starting dose was 5 mg/m^2 and the first DLT, grade 3 diarrhea/AST elevation in one patient and grade 3 thrombocytopenia in another patient, was also experienced at 56 mg/m^2. Protocol amendment was introduced to evaluate other dosing schemes such as daily 3-day dosing in 14-day cycles. This schema was started at 80 mg/m^2 and the first DLT, grade 3 emesis, was observed at 112 mg/m^2. At 157 mg/m^2 two additional DLTs, grade 3 nausea and grade 3 dyspnea, were observed. The MTD was determined at 112 mg/m^2 for the daily 3-day dosing. The study team also evaluated another dosing schema, twice weekly dosing with one week treatment break in a 21-day dosing cycle. The starting dose for this schema was 112 mg/m^2 and at 307 mg/m^2 two DLTs were experienced. One patient had a seizure and another patient was hospitalized with grade 3 abdominal pain, nausea and fever. Following the DLTs the study continued to treat patients at 220 mg/m^2, where one DLT, grade 2 hepatitis, was observed. Based on this finding the MTD for twice weekly dosing was determined to be 220 mg/m^2. Lastly, a continuous twice weekly dosing schedule was evaluated. Starting dose for this schedule was 150 mg/m^2. At this dose level, and the next evaluated dose level of 210 mg/m^2, all patients experienced delayed grade 3 liver toxicity, hence it was concluded that this dosing schedule was too toxic to evaluate further. Other notable adverse events experienced in all patients were fatigue (91%), constipation (52%), myalgia (57%) and anemia (76%). Of note, none of the patients experienced QTc prolongations. Two patients experienced asymptomatic second-degree atrioventricular block that resolved without intervention. No radiological responses were observed. Some patients observed prolonged stable disease, for example a patient with prostate cancer remained on continuous twice weekly dosing for 11 cycles with a 25% decline of PSA. Although no objective responses were seen radiologically, PMBC samples pre- and post-tanespimycin dosing showed reduction in phospho Akt levels in some patients, suggesting early pharmacodynamic signals.

Tanespimycin was acquired by Bristol Myers and Squibb pharmaceuticals from Kosan pharmaceuticals after early studies showed promise. However, its development was later terminated.[1]

13.2.2 17-DMAG (Alvespimycin)

Due to the poor solubility of tanespimycin, the observed liver toxicity and the adverse events caused by the solvent DMSO, different alternatives of the geldanamycin scaffold were undertaken. A viable alternate structure was discovered by substituting C 17 methoxy group of geldanamycin with N,N-dimethlyethylamine 17-desmethoxy-17-N-N-dimethylethylamine to create 17-DMAG, alvespimycin, an analogue with increased water solubility and better oral bioavailability than tanespimycin. Different schedules of alvespimycin have been tested in clinic. The twice weekly continuing dosing schedule, where drug administration was separated by 72 h, was assessed in 30 patients with advanced solid tumor malignancies between December 2004 and July 2007.[17] The primary objective of the Phase I study was to determine MTD for this continuous dosing schedule. Similar to earlier studies with tanespimycin, Hsp70 induction was assessed in PMBC as a pharmacodynamics marker to assess Hsp90 inhibition. The study utilized an accelerated titration design with intrapatient dose escalation allowed. The starting dose was 1 mg/m^2 and MTD was reached at 21 mg/m^2. At the MTD dose DLTs, grade 3 renal failure and grade 3 motor neuropathy were experienced in two patients out of three. An additional 20 patients were enrolled at the MTD dose. Notable adverse events included grade 1 or 2 fatigue, anorexia, blurred vision and musculoskeletal pain. The observed visual disturbances were not associated with inflammation of the eyes, and the disturbances spontaneously improved in all patients. Notably, all patients experienced musculoskeletal pain starting at 12 mg/m^2. The muscular events were not associated with elevation of creatine kinase, or observation of muscle weakness. No incidences of cardiac toxicity, including QTcF prolongation, were reported. No radiological responses were observed; however, some patients experienced prolonged stable disease, such as one peritoneal mesothelioma patient, who remained on treatment for 22 months. The median duration on treatment was 4 months, notable for a heavily pre-treated Phase I population with advanced disease stage. Hsp70 induction was observed in the MTD cohort. To limit clinic visits for infusions and reduce toxicity, especially for elderly patients with more comorbidities, a weekly dosing schedule of alvespimycin was also evaluated.[18] This study also utilized an accelerated titration design and enrolled 25 patients with advanced solid malignancies between February 2006 and April 2008. The primary objective of the study was to determine MTD of alvespimycin when given weekly i.v. in 4-week cycles without a treatment break. When feasible, the study collected pre- and post-tumor biopsies from patients to assess pharmacodynamic changes. The starting dose was 2.5 mg/m^2, and MTD was determined at 80 mg/m^2. DLTs were experienced in two patients out of four at the 106 mg/m^2 dose. These events were in the form of grade 3 fatigue and hypoalbunemia in one patient, and grade 4 AST elevation with grade 3 diarrhea within 24 hours of treatment in another patient. The latter patient's condition worsened with grade 4 hypotension and grade 3 dehydration, and by day 4 the patient had anuric renal failure. The patient died on day 5 and death was attributed to alvespimycin treatment. In total eight patients were treated at 80 mg/m^2 and from five patients pre- and post-treatment biopsies were collected. No DLTs were observed at 80 mg/m^2 and this dose was declared as MTD for weekly alvespimycin treatment. In general, treatment was tolerated well with most AEs occurring at 80 mg/m^2. Common AEs with weekly treatment comprised nausea, vomiting, fatigue and liver enzyme disturbances. Of note, four patients experienced ocular adverse events in the form of blurred vision (three patients), dry eye (three patients), keratitis (two patients) and conjunctivitis (two patients). The majority of these events occurred at 80 mg/m^2 and were grade 2 or less. For the first time radiological responses were observed in two patients. One patient, with castrate resistant prostate cancer (CPRC), had a complete response by both CT and PSA measurement. This patient stayed on treatment for 124 weeks before progressing. Another melanoma patient had a partial response (PR) and remained on treatment for 159 weeks before progressing. Prolonged disease stabilization was observed in three patients (chondrosarcoma, clear cell renal carcinoma and CRPC) who remained on treatment for more than 6 months. Hsp70 induction was clearly shown in PBMCs and the effect was sustained post 96 h of treatment. Interestingly, the two patients that experienced DLTs at 106 mg/m^2 had the highest Hsp70 induction (1250 and 5610 fmol/ml *vs.* average 86 fmol for the rest of the patients). However, results on client protein degradation in tumor tissue were mixed with three samples out of five showing depletion in CDK4 or HER2. A more frequent dosing schedule has been evaluated for the oral form of alvespimycin. In their Phase I study, Flaherty and colleagues evaluated both every other day (q.o.d.) and daily (q.d.) dosing of alvespimycin for 4 weeks with a 2-week treatment break. Interim results of their study have been published in abstract form.[19] Both treatment schedules were reported to be well tolerated with no treatment-related deaths observed. DLTs were experienced at 50 mg q.o.d. in three out of five treated patients in the form of thrombocytopenia (two patients) and hemorrhagic colitis. At 30 mg q.d. two patients out of four experienced DLTs in the form of nephrodic syndrome and dehydration. Recommended Phase II dose was 40 mg for q.o.d schedule and 20 mg for q.d. schedule. The most common observed adverse event was fatigue (43%), followed by nausea (28%), anorexia (21%), diarrhea (19%), peripheral edema (19%) and arthralgia (15%). Interestingly, no visual adverse events were reported but this may be due to 15% cutoff for their reporting. However, it is still notable that the adverse event profile for their study differs from the studies using the intravenous formulation suggesting differences between oral and i.v. dosing of alvespimycin. No radiological responses were reported; however, prolonged stable disease of more than 12 weeks was observed in 11 patients. A melanoma patient remained on study for 60 weeks.

Development of alvespimycin was also discontinued by Bristol Myers and Squibb despite early observed efficacy signals.

13.2.3 17-Allyamino-17-demethoxygeldanamycin Hydroquinone Hydrochloride (IPI-504; Retaspimycin)

In an attempt to find a better geldanamycin-based Hsp90 inhibitor, Infinity pharmaceuticals further tweaked the structure of tanespimycin to create the water-soluble hydrochloride form IPI-504,[20] which appears to be the most potent form of ansamycin-based Hsp90 inhibitors. Once *in vivo*, IPI-504, retaspimycin, is in equilibrium with tanespimycin, which results from the oxidization of the hydroquinone ring of retaspimycin and subsequent enzymatic reduction of the quinone ring of tanespimycin.[21] This may still explain hepatotoxic AEs seen with retaspimycin in clinic. Phase I studies with IPI-504 were initiated in advanced NSCLC, refractory GIST and relapsed multiple myeloma.[22] These studies have both evaluated twice weekly intermittent dosing, and twice weekly dosing for 14 days followed with 10 days rest (days 1, 4, 8, 11 in a 21-day cycle.) In the Phase I study assessing tolerability in advanced NSCLC patients, the MTD for intermittent dosing was established at $225\,mg/m^2$. In the Phase I studies with refractory multiple myeloma and refractory GIST, MTD for twice weekly dosing with 10-day treatment break was established at $400\,mg/m^2$.[22] DLTs for the continuous twice weekly dosing were observed in two patients out of five treated at $300\,mg/m^2$ in the form of grade 3 AST elevations and grade 3 fatigue. Both events were reversible. Other notable AEs included grade 2 fatigue (36%), grade 1 back pain (36%), grade 1 diarrhea (27%), grade 1 headache (27%), grade 1 infusion site pain (27%), grade 1 infusion reaction (18%), grade 1 peripheral sensory neuropathy (18%) and grade 1 myalgia (18%). In the 76-patient Phase II study in advanced NSCLC patients, assessing efficacy of the twice weekly dosing schedule with treatment break at $400\,mg/m^2$, three patients died while on study.[23] Two of these patients died of complications from pneumonia, including sepsis and respiratory distress, which were considered possibly related to retaspimycin. Grade 3 diarrhea was reported in 11% of patients, whereas grade 3 fatigue, nausea and vomiting were reported in 8% of patients. Of note 70% of patients had abnormality in their liver function tests with varying degrees, and 9% of patients experienced grade 3 AST increases. In general, the most commonly observed AEs, regardless of severity and causality, were alkaline phosphatase increases (62%), fatigue (58%), nausea (57%), diarrhea (53%), AST increases (49%), ALT increases (41%), vomiting (37%), cough (32%), abnormal urine color (29%), anorexia (25%), arthralgia (25%), myalgia (25%) and headache (25%). Of note 15% of patients reported blurred vision while on treatment. Objective radiological responses were seen in five patients (7%) and median PFS was 2.86 months.

13.3 Resorcinol Derivates

13.3.1 NVP-AUY922

NVP-AUY922 is a resorcinylic isoxazole amide Hsp90 inhibitor that was discovered by Vernalis and which is currently being developed by Novartis

Pharmaceuticals. NVP-AUY922 is the most potent Hsp90 inhibitor currently being evaluated in clinic,[24] with several Phase II studies initiated in various solid malignancies. NVP-AUY922 entered clinic in 2008 in a Phase I/II study designed to evaluate MTD when NVP-AUY922 was given i.v. once weekly in patients with advanced solid malignancies. The Phase II extension part was to be initiated at the MTD to evaluate efficacy of NVP-AUY922 in HER2+ or ER+ metastatic breast cancer patients who had failed standard treatment. The dose escalation in the Phase I component was conducted by utilizing an adaptive Bayesian logistics regression model (BLRM) with overdose control.[25] One hundred and one patients were evaluated in nine dose cohorts: $2\,mg/m^2$ (n = 3), $4\,mg/m^2$ (n = 4), $8\,mg/m^2$ (n = 4), $16\,mg/m^2$ (n = 7), $22\,mg/m^2$ (n = 12), $28\,mg/m^2$ (n = 8), $40\,mg/m^2$ (n = 16), $54\,mg/m^2$ (n = 19) and $70\,mg/m^2$ (n = 28). The most common adverse events were diarrhea (68.3%), nausea (53.5%), fatigue (41.6%), vomiting (33.7%), decreased appetite (31.7%), asthenia (30.7%), abdominal pain and anemia (each of 23.8%). At the dose level of $40\,mg/m^2$ the first emergence of visual disturbances were noted. The incidence of the visual AEs increased per dose level, where 25% of patients experienced some sort of visual AE at $40\,mg/m^2$, and 89% of patients experienced visual AEs at $70\,mg/m^2$. The reported events were in the form of night blindness (22.8%), photopsia (13.9%), blurred vision (12.9%) and visual impairment (11.9%.) Several DLTs were also noted at various dose levels. At $22\,mg/m^2$ one patient with liposarcoma experienced grade 3 atrial flutter. At $40\,mg/m^2$ one patient with breast cancer experienced grade 3 diarrhea and another patient with colon cancer experienced grade 3 anorexia and grade 3 fatigue. At $54\,mg/m^2$ a colon cancer patient experienced grade 3 asthenia and an ovarian cancer patient experienced grade 3 diarrhea. At the highest dose of $70\,mg/m^2$, two patients with esophageal cancer experienced grade 3 visual AEs. At the same dose another patient with colon cancer experienced grade 3 diarrhea. PK profiling showed no drug accumulation following weekly i.v. doses, a dose proportional increase in C_{max} and less than dose proportional increase in AUC_t. The terminal half-life of NVP-AUY922 was 120 h at $70\,mg/m^2$. Hsp70 induction was also observed in PBMCs. There were no objective responses in the dose escalation component of the trial; however, metabolic responses were observed by FDG–PET at the dose levels of $54\,mg/m^2$ and $70\,mg/m^2$. Especially, three out of four patients with NSCLC had partial metabolic responses at the dose level of $70\,mg/m^2$.

Although DLTs were observed at $70\,mg/m^2$ per the Bayesian model $70\,mg/m^2$ were not the determined MTD, and further dose escalation was warranted. However, considering two patients experienced grade 3 visual AEs study investigators and the Novartis clinical trial team jointly decided that $70\,mg/m^2$ was a safe and efficacious dose for the Phase II studies, and it was not in the best interest of patients to try to titer the MTD by evaluating higher dose levels. The Phase II component of the study enrolled 16 patients who completed enrolment in May 2012. The safety profile was similar to the Phase I component where the most common AEs were diarrhea (94%), night blindness (69%), blurred vision (50%), photopsia (44%), nausea (38%) and fatigue

(31%). Two patients with HER2+ metastatic breast cancer that had progressed on trastuzumab treatment experienced partial radiological responses, which were also associated with a decrease in HER2+ serum levels.[26]

13.3.2 NVP-HSP990

NVP-HSP990 is a small-molecule inhibitor of Hsp90 that is being developed as a follow-up compound to NVP-AUY922 by Novartis pharmaceuticals. NVP-HSP990 is orally bioavailable and crosses the blood-brain barrier making it an interesting back-up compound. As NVP-AUY922, NVP-HSP990 binds competitively to the N-terminal ATP docking site of Hsp90, with an IC_{50} at 13 nanomolar. NVP-HSP990 has a broad range of activity in various xenograph models, including the trastuzumab resistant, HER2+ over-expressing, BT474 breast cancer cell lines, the Kras mutated NSCLC cell line A549 and the double EGFR mutant (L858R and T790M), erlotinib and gefitinib resistant, NSCLC cell line H-1975.

NVP-HSP990 has been evaluated in a Phase I dose escalation study using the BLRM in patients with advanced solid tumors refractory to standard treatment. The primary purpose of the study was to determine MTD when NVP-HSP990 was given once weekly in a 28-day treatment cycle. Sixty-four patients were evaluated at seven dose cohorts: 2.5 mg (n = 3), 5 mg (n = 5), 10 mg (n = 7), 20 mg (n = 6), 30 mg (n = 5), 50 mg (n = 22) and 60 mg (n = 5). The MTD was determined at 50 mg. An additional 11 patients were evaluated on a twice-weekly schedule at 25 mg. The most frequent AEs, suspected to be related to NVP-HSP990, were diarrhea (73.4%), asthenia (40.6%), decreased appetite (25.0%) and insomnia (25.0%).

Seven DLTs (per protocol definition events occurring in cycle 1) were observed. One patient experienced grade 3 myoclonic hand movements at 50 mg. At the 60 mg cohort, one patient experienced grade 3 tremor and another patient experienced multiple grade 1 events, which led to a dose reduction and therefore constituted a DLT per the protocol definition. In the 25 mg twice-weekly dosing schedule, one patient experienced several AEs during cycle 1 that resulted in treatment discontinuation and was reported as a DLT. The events consisted of dysmetria, extremities tremors, superior limb extrapyramidal hypertrophia (all grade 1), and ataxia, confusion and visual hallucination (all grade 2).

PK analysis from the blood samples collected of the 64 patients showed that drug concentrations reached their peak 3 h post dose and declined mono- or biexponentially with a terminal half-life of 20 hours. No objective radiological responses were observed in the study. However, 25 patients (39%) achieved stable disease. Of note, disease stabilization for longer than four months was observed in seven patients, with one patient remaining on treatment for longer than one year. Additionally, FDG-PET results showed 11 patients to achieve a partial metabolic response.

It is noteworthy that, as the only blood-brain-barrier-crossing Hsp90 inhibitor, NVP-HSP990 also showed neurotoxic AEs that were dose limiting. The most common neurological events that were deemed related to NVP-HSP990 treatment were dizziness (18.8%), tremor (17.2%), ataxia (9.4%), cerebellar syndrome (7.8%), balance disorder and headache (both of 6.3%). Severe neurological events, either grade 3 or grade 4, were noted at the higher dose cohorts of 50 mg and 60 mg and included syncope, myoclonus, neurotoxicity, presyncope (each 1/22 [4.5%] in the 50 mg cohort) and tremor (1/5 patients [20.0%)] in the 60 mg cohort). No neurological events of grade 3 or 4 were reported for patients in the 25 mg twice-weekly cohort. Although the prospect of an Hsp90 inhibitor that crosses the blood-brain barrier was exciting, given the fact that development of brain metastasis is usually a sign of resistance to targeted agents and most brain tumors remain untreatable by chemotherapy, the involvement of Hsp90 in synapse formation and maintenance may have a limiting effect in the development of such agents.

13.3.3 STA9090 (Ganetespib)

Like NVP-AUY922, STA 9090, ganetespib, is a synthetic resorcinylic-based triazalone compound with substantial pre-clinical activity in various cell lines.[27–29] Ganetespib is currently being evaluated in Phase II studies, primarily in NSCLC and metastatic breast cancer, either as a single agent or in combination treatment.[30] In clinic, both once-weekly and twice-weekly dosing schedules have been evaluated and Synta pharmaceuticals, developer of ganetespib, has decided on the once weekly dosing schedule for the ongoing Phase II studies. Interim results for the twice-weekly dosing schedule were presented at ASCO 2010.[31] This was a dose escalation study using the standard 3+3 Fibonacci design with the primary objective to evaluate MTD when ganetespib is given i.v. twice weekly for three weeks in a 28-day cycle. At the time of the presentation the MTD had not yet been reached. Data were presented up to the dose level of 50 mg/m^2. The most frequent AEs at 50 mg/m^2 were diarrhea (67%), anemia (67%), vomiting (33%), nausea (33%), fatigue (33%), AST increases (33%), hypomagnesemia (33%) and ALT increases (17%). One partial response was documented in a melanoma patient.

Once weekly continuous dosing was also evaluated in a Phase I study using standard Fibonacci design with the primary objective to establish MTD.[32] Fifty-three patients were evaluated in multiple dose cohorts up to 259 mg/m^2. Three patients experienced DLTs, grade 3 diarrhea or grade 4 asthenia, at the highest dose level of 259 mg/m^2. The MTD was determined to be 216 mg/m^2. The most common AEs at the higher dose levels of 180–216 mg/m^2 were diarrhea (93%), fatigue (73%), abdominal pain (47%), nausea (40%), anemia (33%), elevated AST (27%), decreased appetite (27%), hypokalemia (20%) and elevated alkaline phosphate (20%). One patient with colorectal carcinoma had a PR.

13.3.4 AT13387

Developed by Astex pharmaceuticals AT13387 is a resorcinol-based synthetic Hsp90 inhibitor administered i.v., currently evaluated in Phase II studies in multiple indications. A Phase I study was initiated to explore the MTD of AT13387, when given either once weekly or twice weekly for three weeks in a 28-day cycle, in patients with advanced solid tumors. Results from the twice-weekly dosing schedule were presented by Dr. Geoffrey Shapiro of the Dana Farber Cancer Institute at ASCO 2010.[33] Dose levels from 10 mg/m^2 up to 120 mg/m^2 were evaluated in 26 patients using a modified Fibonacci dose escalation design. AT13387 was well tolerated with only AEs observed at the highest dose levels of 80 mg/m^2 and 120 mg/m^2. There were no DLTs. However, dose levels higher than 120 mg/m^2 were not evaluated due to the accumulation of low-grade toxicities such as visual disturbances. The most commonly observed AEs were diarrhea, reversible visual changes, hypotension during AT13387 infusion and dry skin/mouth. The severity for the observed AEs was grade 1 or 2. The visual changes that emerged at the highest dose of 120 mg/m^2 were usually observed around day 11 and consisted of delayed light/dark accommodation, blurred vision and flashes. No radiological responses were observed; however, two patients remained on treatment for six months. Pharmacological changes were observed in collected tumor samples. This included Hsp70 induction, and caspase 3 elevation that is indicative of apoptosis.

Final results for the once-weekly and twice-weekly dosing schedule were presented at ASCO 2012.[34] MTD for once-weekly schedule was determined at 260 mg/m^2. The safety profile in 53 patients that included both schedules was diarrhea (62%), fatigue (40%), visual disturbances (36%), nausea (26%), injection site events (pain or inflammation) (25%), dry mouth (23%) and anemia (21%). From the 53 patients, one patient with imatinib-resistant GIST (exon 17 c-kit secondary mutation) had a partial response that lasted eight months. Another GIST patient with a metabolic response assessed by PET remained on treatment for seven months. PK profiling revealed that AUC and C_{max} increased dose proportionally and the elimination half-life for once-weekly dosing at the MTD 260 mg/m^2 was 6.5–9.1 h.

13.4 Purine Scaffold-based Inhibitors

Due to the limitation of ansamycin-based scaffold proteins, several companies have tried to develop the second-generation synthetic inhibitors by taking advantage of the unique folding properties of the Hsp90 in comparison to other enzymes that bind ATP.[35] Working on a new structure Chiosis and colleagues at MSKCC were the first to develop a purine scaffold-based Hsp90 inhibitor PU-3 with double the potency of geldanamycin.[36] This led to several other compounds being developed that have entered clinic. All of these compounds have a common pattern in the form of an NH$_2$ group attached to a purine scaffold core and an aryl moiety separated by 5 Å to establish Hsp90 binding.[1]

13.4.1 Debio 0932/CUDC-305

First discovered by Curis pharmaceuticals as CUDC-305, Debio 0932 is currently being developed by the Debiopharm group in Switzerland. Debio 0932 is a purine-based structure that is orally administered. A Phase I study with Debio 0932 has been completed, and Debiopharm has started a Phase I/II combination treatment study with pemetrexed/cisplatin in patients with advanced non-small-cell lung cancer. The Phase I dose determination study enrolled 50 patients with solid malignancies with the primary objective of determining MTD in two different dose schedules, once daily (q.d.) and once every two days dosing (q.2d.), by utilizing traditional Fibonacci 3+3 dose escalation design.[37] Twenty-three patients received Debio 0932 once every two days up to 1600 mg. One DLT, grade 3 febrile neutropenia, was reported at 1600 mg. However, further dose escalation was not possible due to the excessive number of 100 mg capsules that was needed to be taken. Twenty-eight patients were treated at the q.d. schedule. At 1600 mg two DLTs, diarrhea and asthenia, were observed. The MTD for the q.d. dosing schedule was determined to be 1000 mg. Other common AEs were constipation, asthenia, decreased appetite, diarrhea, nausea and vomiting. No ocular toxicities or cardiac toxicities were observed. Two patients in the q.2d. schedule had a partial response. One of these patients was a 63-year-old male with Kras mutated adenocarcinoma. The other patient had breast cancer but the type of breast cancer was not specified.[37,38]

13.4.2 MPC-3100

MPC-3100 is a fully synthetic, orally bioavailable, purine-based Hsp90 inhibitor being developed by Myrexis Pharmaceuticals. A recently completed Phase I study enrolled 26 patients with the primary objective of determining MTD when MPC-3100 was given q.d. for 21 days in a 28-day cycle.[39] Dose escalation was conducted utilizing the modified 3+3 Fibonacci with accelerated titration design. One DLT, grade 3 supraventricular tachycardia, was observed at 245 mg/m^2. Another patient treated at the same dose level had respiratory failure with fatal outcome. After no DLTs were observed at the next dose level of 340 mg/m^2 dosing was switched to twice daily (bid) without treatment break due to high pill burden. The two following dose levels were 480 mg and 640 mg, irrespective of patient body surface area. No other DLTs were observed. The most common AEs were diarrhea, nausea, vomiting and fatigue. Liver or cardiac toxicity were not observed, but visual disturbances were noted at the higher dose cohorts in three patients. The events, all grade 1 or 2 in severity, consisted of bilateral white ocular crescents, intermittent light to dark accommodation difficulties, dimming vision, blurry vision and visual floaters, suggesting an effect on the photoreceptor layers of the retina. Opthalmaloscopic exams did not reveal morphological changes and all events were reversible upon drug withdrawal. Although MTD was not reached the recommended Phase II dose was established as 240 mg b.i.d. per the study investigators. No objective responses were observed; however, Hsp70 induction was observed in PBMCs.

Myrexis has also developed a follow-up compound MPC-0767 that is an L-alanine ester pro-drug of MPC-3100 with better solubility and bioavailability. It does not appear that an IND has been activated for MPC-0767 and there are currently no active Phase II studies evaluating the efficacy of MPC-3100.[30]

13.4.3 BIIB021/CNF 2024

CNF 2024 was first discovered by Conforma Pharmaceuticals, but is now being developed by Biogen Idec under the compound name BIIB021. CNF2024 is given by oral administration, and both daily and twice-daily schedules have been evaluated.[30] There are currently no active studies evaluating CNF 2024/BIIB021 and there are scarce publicly available data on the completed Phase I dose finding studies. Elfiky and colleagues presented Phase I data at ASCO 2008 from 23 patients with solid malignancies or chronic lymphocytic leukemia (CLL).[40] BIIB021 was given once daily continuously for three weeks in a 28-day cycle in CLL patients and twice weekly for three weeks in a 28-day cycle in patients with solid malignancies. Two DLTs, grade 3 dizziness and grade 3 syncope, were noted in patients receiving 800 mg twice weekly. Other notable toxicities in patients with solid malignancies were fatigue, hyponatremia or hypoglycemia that were grade 3 or grade 4 in severity. One patient with CLL experienced abnormal grade 3 liver enzyme elevation. 800 mg was determined as the MTD for the twice-weekly dosing schedule. No MTD was reported for the once-daily schedule. Except for one CLL patient that had a 39% reduction in a lymph node, no objective responses were reported. In patients with solid malignancies, 11 patients out of 16 had stable disease at the first disease assessment. However, some early PD effects were observed. These included Hsp70 induction and decreases of serum HER2. It appears Biogen Idec have terminated the development of BIIB021 as the compound is no longer listed in the pipeline for the company.[41]

13.5 Aminobenzamide Derivates

13.5.1 SNX-5422/PF-04928473

SNX-5422 is a novel class of Hsp90 inhibitor that is a 2-aminobenzamide derivate. SNX-5455 was constructed by scientists at Serenex. It is orally bioavailable and has shown promising pre-clinical efficacy in a broad range of cell lines.[42] Later with Pfizer's acquisition of Serenex, SNX-5422 was renamed with the Pfizer investigational compound label PF-04928473. SNX-5422/PF-04928473 entered clinic in March 2008 in a Phase I study conducted at the NCI.[43] The primary objective of the study was to find the MTD, safety and toxicity of SNX-5422 when administered twice weekly in 28-day cycles in patients with advanced solid or lymphoid (lymphoma or CLL) malignancies. Standard Fibonacci dose escalation design was utilized and intradose patient escalation was allowed. In total, 33 patients were enrolled. The starting dose was $4\,\text{mg/m}^2$ and dose escalation continued up to $177\,\text{mg/m}^2$. SNX-5422 was well tolerated with only one DLT, grade 3 non-specific arthritis, observed at a dose level of $18\,\text{mg/m}^2$. Most of the observed AEs were in the form of grade 1 or grade 2 events. The most common adverse events attributed to SNX-5422 were nausea (24%), ALT elevation (15%), diarrhea (12%), thrombocytopenia (12%) and fatigue (12%). Grade 3 events were limited and were reported at the

Table 13.1 Hsp90 inhibitors that have entered clinic.

Compound	Structure	Adverse events	Administration Route	Development phase	Key indication
Tanespimycin (17-AAG)		Diarrhea, nausea, fatigue, emesis, liver enzyme increases, hypersensitivity, anemia, constipation	i.v.	Discontinued	NA
Alvespimycin (17-DMAG)		Diarrhea, nausea, fatigue, emesis, liver enzyme increases, visual disturbances, motor neuropathy, musculoskeletal pain	Oral	Discontinued	NA

Compound	Structure	Side effects	Route	Phase	Cancer type
Retaspimycin (IPI-504)		Diarrhea, nausea, emesis, fatigue, liver enzyme increases, anorexia, arthralgia, myalgia, headache, abnormal urine color, visual disturbances	i.v.	II	NSCLC
NVP-AUY922		Diarrhea, visual disturbances, nausea, fatigue, emesis, anorexia, abdominal pain, anemia, myalgia	i.v	II	NSCLC, breast, GIST, gastric
NVP-HSP990		Diarrhea, fatigue, anorexia, insomnia, neurotoxicities	Oral	I	NSCLC, glioblastoma

Table 13.1 (*Continued*)

Compound	Structure	Adverse events	Administration Route	Development phase	Key indication
Ganetespib (STA-9090)		Diarrhea, fatigue, nausea, emesis, constipation, back pain, anorexia, liver enzyme increases, muscular weakness, dizziness, insomnia	Oral	II/III	NSCLC, breast, MM
AT13387		Diarrhea, fatigue, visual disturbances, nausea, hypotension, dry mouth/skin, anemia, injection site inflammation/pain	Oral	II	Prostate, NSCLC, GIST
Debio0932 / CUDC305		Diarrhea, fatigue, constipation, anorexia, nausea, emesis, febrile neutropenia	Oral	II	NSCLC

Hsp90 Inhibitors in Clinic

Compound	Structure	Side effects	Route	Phase	Status	
MPC 3100		Diarrhea, nausea, emesis, fatigue and visual disturbances	Oral	I	?	
BIIB021/ CNF2024		Diarrhea, fatigue, nausea, emesis, visual disturbances, dizziness, syncope	Oral		Discontinued	NA
SNX-5422/ PF-04928473		Diarrhea, emesis, visual disturbances, liver enzyme increases, fatigue, thrombocytopenia	Oral		Discontinued	NA

following frequencies: diarrhea (9%), non-septic arthritis (3%), thrombocytopenia (3%) and AST elevation (3%). However, dose escalation was halted at 177 mg/m^2 due to reports from canine studies that showed SNX-5422 to cause irreversible retinal toxicity even though no ocular adverse events were observed in patients. Once the toxicology data were available, the protocol was amended to include comprehensive eye examinations on all patients. At the time of the amendment only three patients were undergoing study treatment. Two patients who had normal eye exams had been receiving treatment for 8 cycles and 28 cycles, respectively. The eye exam from the third patient, who had been complaining of blurred vision, revealed bilateral cataracts and a slightly prolonged rod to cone break in the right eye. However, the ERG recordings from this patient were normal and it is possible that these events were present at baseline. Even though all three patients did not show any clinical ocular adverse events, it was later decided to discontinue all three patients due to the ocular toxicology findings from the pre-clinical studies. The patients had been on treatment for 28 cycles, 11 cycles and 10 cycles and had stable disease during study discontinuation.

No objective responses were observed. Some patients achieved disease stabilization with a median duration of 16 weeks (3–110 weeks.) The half-life of SNX-5422 did not appear to change between the different dose levels evaluated, and was between 8 and 15 hours. No drug accumulation was noticed due to multiple dosing and a dose proportional increase of AUC was observed. In addition, Hsp70 induction was observed in PBMCs and showed a linear correlation with PK parameters.

SNX 5422 was also evaluated, in parallel run, separate Phase I studies that evaluated the dosing schedule of every other day for 21 days in a 28-day cycle, and every day dosing for 21 days in a 28-day cycle.[43] The results from these studies suggest that the more frequent dosing appears to cause increased AE rate without much added efficacy. The most frequently observed AEs for these schedules were nausea, emesis, fatigue and diarrhea. In addition, 4 patients out of 20 experienced visual AEs in the form of nictalopia and blurred vision at doses of 50 mg/m^2 to 89 mg/m^2. In most cases the onset of the events were after two weeks of treatment, and were reversible within a few days upon discontinuing treatment. Together with these results it appears that Pfizer terminated all ongoing studies due to visual toxicity concerns, and no further dose escalations were attempted in any of these Phase I studies.

13.6 Indications that Appear to Have Promise for Hsp90 Inhibitor Development

13.6.1 Non-small-cell Lung Cancer (NSCLC)

Lung cancer is a common type of cancer that affects men and women around the globe and is the leading cause of cancer deaths in the US with an estimated 156,940 deaths in 2012.[44] NSCLC is the most common type of lung cancer that accounts for roughly 85% of all cases with approximately 70% of these patients presenting with advance disease (Stage IIIB or Stage IV) at the time of diagnosis.[45,46] Currently, there is no curative treatment for advanced NSCLC except for rare cases involving only satellite metastasis that can be surgically removed. The median survival for advanced NSCLC is approximately 8–10 months,[47] unless the patient has specific somatic mutations such as epidermal growth factor receptor (EGFR) mutations or echinoderm microtubule associated protein like 4-anaplastic large cell kinase (EML4-ALK) translocations that now have established active treatment.

13.6.1.1 Epidermal Growth Factor Receptor

The control of cell growth is mediated by a complex network of signaling pathways responsive to external influences, such as growth factors, as well as to internal controls and checks.[48,49] Epidermal growth factor was one of the first growth factors to be described. EGFR over-expression is common in NSCLC and based on the detection threshold can be found in 40–90% of patients.[50] In addition to EGFR over-expression or amplification, mutations on the EGFR itself can cause the receptor to be constitutively active, regardless of the binding of its ligand. This results in phosphorylation of downstream molecular pathways such as RAS/RAF/MEK or c-SRC/AKT that cause proliferation and oncogenesis. The most common EGFR mutations are either deletions in the EGFR gene in exon 19 (E19del) of the tyrosine kinase domain or missense mutations in exon 21 (L858R).[51,52] The frequency of EGFR mutations is more common in patients of Asian descent (around 40%) in comparison to Caucasian patients (around 15%).[53]

The recent discovery that the response to EGFR tyrosine kinase inhibitors (EGFR TKIs) is linked to aberrant EGFR tyrosine kinase signaling has created a new paradigm for the treatment of NSCLC. Emerging data from the literature suggest that a new form of malignancy within NSCLC, driven by EGFR mutation, is a distinct disease compared to other forms of metastatic NSCLC.[54–56] Retrospective analyses from earlier studies conducted with EGFR TKIs have shown that responses to EGFR TKI are independently associated with EGFR mutation.[57,58] Recently completed Phase III studies have showed a PFS of 9 to 13 months when patients with activating EGFR mutations in advance NSCLC stage receive EGFR TKIs as first line treatment.[59–62] The overall response rate (ORR) in patients receiving EGFR TKI ranged from 71% to 82%. The mature OS data from these studies have still not been reported but are expected to be up to 30 months for patients receiving EGFR TKI.

Although, remarkable response rates are seen in patients with EGFR mutations that are treated with EGFR TKIs ($\sim$80%), the responses are of limited duration because eventually patients develop "acquired resistance". Several resistance mechanisms have been identified, the most common being a secondary mutation at the threonine gatekeeper residue at position 790, T790M.[63,64] Other secondary mutations in exon 19 or exon 21 have also been associated with acquired resistance to EGFR TKI,[65–67] but their prevalence is much lower than T790M and pre-clinical work suggests that T790M has the highest degree of EGFR TKI resistance.[68] The other commonly seen reason for resistance is the amplification of the Met oncogene that causes proliferation through the HER3 pathway by activating AKT downstream and accounts for roughly 20% of cases.[64,69] These mutations can occur in the presence or absence of T790M.[70] However, T790M and Met amplification account for around 60% of cases for acquired resistance, while tumors from approximately 40% of remaining patients confer resistance through mechanisms yet to be explained.[64]

Interestingly, T790M, as well as the mutated EGFR, is a highly sensitive client protein of Hsp90. This was first shown pre-clinically with geldanamycin[71] and has been also shown for other Hsp90 inhibitors such as ganetespib.[28] This sensitivity is also evident by the fact that most Hsp90 inhibitors, such as NVP-AUY922, IPI-504, MPC-3100, *etc.*, appear to be efficacious in xenograph model H-1972 that has T790M. Based on these pre-clinical results several Hsp90 inhibitors were evaluated in NSCLC in small Phase II studies that were enriching for patients with EGFR activating mutations that had been previously treated with EGFR TKIs.

The initial results in patients with activating EGFR mutations treated with Hsp90 inhibitors were largely disappointing. The first study to evaluate an Hsp90 inhibitor in this patient population was retaspimycin.[23] In this study patients with advanced NSCLC that had received prior EGFR TKI (gefitinib or erlotinib) and had available tumor tissue for mutational analysis were treated with retaspimycin on days 1, 4, 8 and 11 of a 21-day cycle. Seventy-six patients were enrolled in the study. Patients were treated at $400\,mg/m^2$, which later was reduced to $225\,mg/m^2$ due to emerging hepatotoxicity signals in a Phase III GIST study. At the time of dose reduction 75 patients had been treated at $400\,mg/m^2$ and 19 patients that were continuing treatment continued at the lower dose. Of the 76 patients 68 patients had enough archival tissue for EGFR mutation analysis and 28 of these patients were found to have EGFR activating mutations. Two patients had T790M; however, it should be noted that most samples analyzed were from tissue taken prior to EGFR TKI treatment start, and the T790M incidence in this study would be expected to be much higher. Only one of these 28 patients, with an L858R exon 21 mutation, had a partial response to retaspimycin treatment. The overall response rate for the whole study population was 7% with four more partial responses observed in the EGFR wild-type group of patients, two of whom were later found to have ALK rearrangement. Similar disappointing results have been seen with ganetespib. In a similarly designed Phase II study, ganetespib was evaluated in molecularly defined NSCLC patient population.[72] Ninety-five patients were enrolled into different strata defined as EGFR mut, Kras mut and patients that were wild-type for EGFR or Kras. Ganetespib was given $200\,mg/m^2$ weekly for 21 days in a 28-day cycle. No objective responses were seen in the 16 patients with activating EGFR mutations.

In contrast, activity has been seen with NVP-AUY922 in this hard-to-treat patient population. NVP-AUY922 was evaluated in a similar molecularly defined Phase II study that assessed efficacy when given weekly at $70\,mg/m^2$.

Patients were stratified based on whether they had EGFR mutations, KRAS mutations and EML4-ALK translocations.[73,74] Of the 121 patients that were enrolled, 35 patients were assigned to the EGFR mutant stratum. These patients were heavily pre-treated where 37% had two prior lines of treatment and 54% had three lines or more. Objective PRs were seen in 7 patients (20%) and 13 patients (37%) had stable disease. Disease control rate was achieved in 57% of EGFR mutant patients. The PFS rate at 18 weeks for the 35 patients was 35.2% [95% Cl 18.7, 52.2]. Interestingly, a subanalysis showed that in patients who had received an EGFR TKI as the last regimen prior to receiving AUY922, 5 patients (26%) out of 19 had a partial response. DCR for this set of patients was 67% and the PFS rate at 18 weeks was 46.6% [95% Cl 21.6, 68.4]. Although the sample size was small, and these results may simply reflect that these patients showed better performance than the rest when they entered the study, it is encouraging to think that Hsp90 inhibitors are especially effective in patients that are "truly" refractory to EGFR TKI treatment. For instance, it has been shown that tumor growth in EGFR mutant patients can be driven by different factors depending on treatment.[75] Unfortunately, baseline biopsies were not mandated in the study, and the resistance mechanism for these patients is not known. Future studies in this patient population will have to address this. After these encouraging results the study was amended to include an additional 35 patients with EGFR mutations. The study is ongoing and updated results are expected in 2013.

13.6.1.2 EML4-ALK

In a small subset of NSCLC patients, an inversion within chromosome 2 p results in the fusion of the anaplastic large cell kinase (ALK) with echinoderm microtubule associated protein like 4 (EML4) to create the EML4-ALK fusion gene product.[76] The frequency of this rearrangement appears to be more common in patients with Asian ethnicity than Caucasian, where the frequency for the rearrangement is around 6–7% for Asian patients, and less than 5% for Caucasian patients.[77] These patients, like those with EGFR activating mutations, tend to be young and never have smoked.[78] Patients with EML4-ALK rearrangements respond very well to ALK/MET/ROS-1 inhibitor crizotinib, where response rate is 61% and median PFS is around 10 months.[79] However, the resistance mechanism develops during crizotinib treatment with secondary mutations in the ALK domain such as L1196M, or other oncogenic mutations such as activating EGFR mutations, or up-regulation of other proliferative pathways.[80–82]

EML4-ALK transfusion gene product is a sensitive client protein,[83] probably more so than the HER2 receptor. In addition, pre-clinical results have shown secondary ALK mutations such as L1196M and others to be sensitive to Hsp90 inhibition.[84] Given that the explained resistance mechanisms, such as the EGFR or the RAS pathway, consist of several client proteins of Hsp90, several studies have now been initiated in EML4-ALK patients that are either naïve or crizotinib pre-treated.

IPI-504 was the first compound to show efficacy in patients with EML4-ALK translocations. In the molecularly driven Phase II study explained above, Sequist and colleagues noticed that 4 out of 40 patients (10%) who were EGFR wild-type had a partial response to IPI-504 treatment.[23] When they assessed tumor samples from these patients they noticed that two out of the four patients with a response had EML4-ALK translocation. All these patients had not received crizotinib because crizotinib Phase I studies in NSCLC had not started at the time of enrollment. There were only enough tumor samples from 15 patients to test for ALK rearrangements and 3 patients were positive (4%). All three patients remained on study for approximately 7 months. The estimated median PFS for all patients in the study was 2.86 months.

In the molecularly defined Phase II study with ganetespib described above, 4 patients out of 62 patients had a radiological partial response.[72] When these patients were further characterized it was found that all four had EML4-ALK rearrangements. Four more patients were found to have EML4-ALK mutations and three of these patients had disease stabilization that lasted for 16 weeks or longer. Testing for EML4-ALK was not done in all 62 patients, only for 23 patients, and it is possible that some EML4-ALK patients did not respond to ganetespib; for instance two patients that had progressed on crizotinib, who are presumably ALK positive, did also progress on ganetespib. However, these early results mirror those of retaspimycin and suggest that EML4-ALK patients, especially patients that are naïve to crizotinib treatment, do well on Hsp90 inhibitor treatment.

The other compound being evaluated in this patient population is NVP-AUY922. In the Phase II study assessing efficacy of NVP-AUY922 in molecularly defined NSCLC, 22 patients with EML4-ALK translocations were enrolled.[73] Eight patients were crizotinib treatment naïve. Seven patients out of the 22 (32%) had partial radiological responses.[74] Disease control was achieved in 59% of the patients, and the PFS rate at 18 weeks was 35.8% [95% CI 16.8, 55.3]. Response rates for crizotinib naïve patients were much higher. Four patients (50%) had partial radiologic responses and disease control was achieved in all patients (100%). The PFS rate at 18 weeks was 62.5% [95% CI 22.9, 86.1]. Although the sample size was small these initial results are similar to response rates seen with crizotinib.

Different variants of EML4-ALK depending on the break apart region of ALK domain have been described.[85] Emerging data have suggested that ALK and Hsp90 inhibitors may have different sensitivity to the different variants, with especially variant 3 being insensitive.[86] This may to a degree explain why responses to crizotinib in ALK rearrangement patients are less than the response rates achieved with EGFR TKIs in patients with EGFR activating mutations. However, pre-clinical data have shown combining Hsp90 inhibitors with ALK inhibitors may overcome this insensitivity.[86] Such trials are now being evaluated.[30]

13.6.2 Breast Cancer

Breast cancer is the most common form of cancer in women with an estimation of 226,870 new cases in 2012.[44] Due to the emergence of several new treatment options, and better screening, breast cancer mortality rates have been steadily declining; however, it is still estimated that 39,510 women will have succumbed to this disease in 2012.[44] Breast cancer can be broadly divided into types that are mainly androgen driven, such Endocrine receptor (ER) positive breast cancer, epidermal growth factor receptor (EGFR) driven, such as HER2 positive breast cancer, and forms such as triple negative breast cancer, where the oncogenic driver is neither the androgenic pathways, nor the growth factor pathways.

13.6.2.1 HER2+ Breast Cancer

It was first reported by Slamon and colleagues that the HER2 receptor was amplified in 30% of analyzed breast cancer samples, which correlated to poor prognosis.[87] The subsequent development of HER2 directed monoclonal antibody trastuzumab has vastly improved the outcomes in both the early and metastatic stages of HER2 driven breast cancer.[88,89] However, as with most solid tumors, resistance sooner or later develops to HER2 targeted therapy. Common resistance mechanisms include the PI3K pathway activation through either PTEN loss or PI3K mutations and amplifications.[90] Perhaps because many kinases in the PI3K pathway are known client proteins, small Phase II studies are evaluating multiple Hsp90 inhibitors, either as single agent or in combination with trastuzumab, in patients expressing HER2 that have progressed on trastuzumab containing regimen.

Due to the strong pre-clinical data seen in breast cancer cell lines several Hsp90 inhibitors have been tried in small Phase II studies in patients with metastatic breast cancer. Tanespimycin was evaluated in 11 patients with metastatic or locally advanced breast cancer.[91] Patients were not further selected based on molecular oncogenes. Six of the patients were triple negative. Tanespimycin was given on days 1, 4, 8, 11 in a 21-day cycle at a dose of 220 mg/m^2. No responses were seen and only three patients had stable disease at first staging.

In another Phase II study, tanespimycin was combined with HER2 receptor antagonist Trastuzumab.[92] Thirty-one patients with HER2 over-expressed metastatic breast cancer, who had progressed on prior trastuzumab containing treatment, received tanespimycin weekly at 450 mg/m^2 and trastuzumab at the conventional dose of 2 mg/m^2. HER2 over-expression was determined as +3 staining using immunohistochemistry or FISH with a ratio ≥ 2. The majority of patients had received at least one prior line of chemotherapy in addition to trastuzumab treatment, and had a PS of 0–1. The majority of the patients (84%) were receiving trastuzumab at the time of study entry. Of the 31 patients enrolled, 27 were evaluable for efficacy per protocol based criteria, meaning patients had to have received more than one dose of treatment or not more than one prior trastuzumab-based treatment. Out of the 27 evaluable patients 6 patients had confirmed partial responses, which were evaluated by an independent radiologist. In addition, ten patients had stable disease at first tumor staging. These responses were durable with a median of 147 days

(109–203 days). The median PFS for the 27 patients was 6 months (4–9 months) and median overall survival was 17 months (16–26 months). The combination treatment was tolerated relatively well for the majority of patients; however, five patients discontinued treatment due to AEs. These events included fatigue, decline in ejection fraction, depression and elevated liver enzymes, all grade 3 in severity. The patient that experienced decline in left ventricular ejection fraction (LVEF) was asymptomatic and this event developed after 14 months of treatment. This patient had a similar episode when receiving their prior treatment, which was a combination of paclitaxel and trastuzumab, hence this event was not judged to be related to the tanespimycin trastuzumab combination treatment.

Ganetespib is also being evaluated in HER2+ and triple negative metastatic breast cancer. In a recent Phase II study conducted at MSKCC 22 patients with metastatic breast cancer received 200 mg/m^2 ganetespib weekly for 21 days in a 28-day cycle. All patients had to have received prior trastuzumab containing treatment. A Simon two-stage design was applied and, if 3 responses were observed, additional patients up to a study total of 40 were to be recruited. Of the 22 patients, 10 patients were HER2+ and ER+, 3 patients were HER2+, 6 patients were ER+ and 3 patients were triple negative. Radiological responses were seen in two patients who were HER2+ (however, it was not indicated if these patients were also ER+). The duration of the responses was 5 months. Stable disease at the first staging was seen in one triple negative patient and 6 HER2+ patients. All 6 patients that were ER+ had progressive disease. Median PFS was 7 weeks (7–19) and median OS was 28 weeks. The response rate of 9% was below the pre-specified criteria and enrollment was stopped. Another Phase II study evaluating ganetespib in combination with trastuzumab has been initiated in patients that have triple negative or HER2+ metastatic breast cancer.

NVP-AUY922 has been evaluated as both single agent and in combination with trastuzumab. In the first-in-man Phase I study, an extension arm evaluated either HER2 over-expressing or ER over-expressing locally advanced or metastatic breast cancer that was refractory to standard treatment at the Phase II recommended dose, 70 mg/m^2. The safety profile was similar to what was seen in the dose escalation phase with diarrhea (100%), fatigue (81.3%), night blindness (68.8%), nausea (62.5%), blurred vision (50%), headache (43.8%), photopsia (43.8%), decreased appetite (31.3%) and emesis (31.3%) being the most frequent adverse events. Encouraging early efficacy signals were seen in both patient groups with one HER2+ patient achieving partial response and one ER+ patient achieving disease stabilization of over 6 months.[93]

Phase Ib/II combination study with trastuzumab has also been conducted in advanced or metastatic HER2 over-expressing breast cancer patients that had progressed on trastuzumab containing treatment. The Phase Ib part determined the MTD for AUY922 and trastuzumab using Bayesian logistic regression model with over dose control for dose escalation. AUY922 was given weekly and trastuzumab was administered weekly at the standard dose of 2 mg/kg. The starting dose for AUY922 was 54 mg/m^2. A total of 13 patients were evaluated at the dose escalation phase and the recommended Phase II dose for the combination was 70 mg/m^2 for NVP-AUY922 and 2 mg/kg for trastuzumab. The combination was tolerated well with only one DLT reported for grade 3 diarrhea observed at the 70 mg/m^2 cohort. The most frequent AEs were diarrhea (83.7%), nausea (37.2%), fatigue and headache (both at 32.6%), photopsia and blurred vision (both at 30.2%), visual impairment (27.9%), night blindness (25.6%), muscle spasm and vomiting (both at 23.3%). The extension phase of the study is ongoing and an additional 40 patients are expected to be enrolled. Preliminary data from 31 patients have shown a clear efficacy signal for the combination treatment with 7 patients achieving a partial tumor response and 16 patients (52%) having disease stabilization at the first scan assessment.[94]

Collectively, these results show that further confirmatory studies with Hsp90 inhibitors should be conducted in combination with trastuzumab in patients expressing HER2+ metastatic breast cancer.

13.6.3 Prostate Cancer

Prostate cancer is after lung cancer the deadliest cancer form in man. An estimated 241,740 new cases will emerge in the US and 28,170 patients are expected to have succumbed to their disease in 2012.[44] Prostate cancer, especially in advance castrate resistant stage, is largely driven by the androgen receptor,[95] which is a known client protein for Hsp90.[96] In addition to androgen signaling, other pathways including Pi3K/Akt/mTOR, the EGFR signaling pathway and HER2 amplification have also been described to play a role in prostate cancer proliferation.[97,98] HER2, EGFR and Akt are all known client proteins of Hsp90.[10,71] Pre-clinical work has yielded promising results in prostate cancer where tanespimycin and analogues have shown dose-dependent growth inhibition in prostate cancer cell lines and xenograph models.[21,99–101] However, these promising pre-clinical results have unfortunately not been replicated in clinic. Retaspimycin was tested in 19 castrate resistant prostate cancer patients in a twice-weekly dosing one week rest cycle.[101] Retaspimycin was given on days 1, 4, 8, 11 in a 21-day cycle at 400 mg/m^2, which was the determined Phase I dose.[16] The study followed response assessments as defined per NCI PSA working group, meaning a response was defined as 50% or greater reduction in PSA levels compared to pre-treatment baseline PSA levels, which had to be confirmed in two serial PSA measurements at least 28 days apart. Progression was defined as rising PSA levels 25% or greater than nadir or 50% or greater than baseline PSA levels. The study looked at two subtypes of patients. Group 1 patients included chemo naïve patients and group 2 enrolled patients that had progressed on docetaxel based treatment. A Simon two-stage statistical design was utilized, where the primary end point of the study was response rate. The model predicted 5% response or less to be unacceptable for continuation and 25% or greater response rate a desirable outcome for the study. Two subgroups were defined and target enrollment for each was 15 patients. If one patient responded, an additional 10 patients were planned to be recruited.

The trial was prematurely stopped after 19 patients had been accrued. The reason for the termination was that two treatment-related deaths were experienced in group 1 patients while an unplanned interim analysis provided no clinically meaningful antitumor activity in the 19 patients. The deaths were due to hepatic failure in a 57-year-old patient who died during cycle 1 and a 60-year-old patient who died in cycle 2 due to insulin-resistant hyperglycemia with ketoacidosis and multiple organ failure. Both patients had grade 3 and grade 4 elevations in liver enzymes.

AEs were noted in more than 25% of patients, and included nausea (47%), diarrhea (42%), fatigue (32%), anorexia (26%) and arthralgia (26%). No patient responses were observed. One patient in group 1 received 9 cycles of treatment with a PSA decrease of 48%. The authors concluded that further expansion of the trial was not warranted given the two treatment deaths and lack of meaningful clinical response.

Tanespimycin was also evaluated in a Phase II multi-center study.[102] The study enrolled male patients with histologically confirmed metastatic prostate adenocarcinoma that had radiologic disease progression, or rising PSA levels, despite androgen deprivation treatment and anti-androgen withdrawal. The study enrolled 17 patients; however, one patient never started treatment due to high potassium levels and another was deemed not eligible due to receiving contra-indicated medication. Tanespimycin was given on days 1, 8 and 15 at $300\,mg/m^2$ in a 28-day cycle. The median age of the 15 patients was 68 (52–78), median PSA number was 261 (46–1705) and median gleason score was 8 (5–10). Twenty-seven percent of the patients had a PS of 0, 67% had a PS of 1 and 6% had a PS of 2. Sixty percent of the patients were receiving zolendronic acid for bone metastasis. A Simon two-stage design was utilized, where 16 patients would be enrolled in the first stage and after a 6-month observation an interim analysis was planned to be conducted. If one or more patients had a clinical response than an additional 9 patients were planned to be accrued. The primary end point of the study was to assess PSA response per PSA working group guidelines. The secondary end point of the study was PFS, OS and radiological response rate on patients with measurable disease. Changes in IL-6, IL-8 and maspin levels were also measured as an exploratory objective.

The study was terminated at the interim analysis due to lack of any PSA response. The majority of patients, 13 in total (87%), discontinued treatment due to disease progression. Nine patients (60%) experienced grade 3 adverse events; no grade 4 events or treatment-related deaths were observed. The most common grade 3 event was fatigue (4 patients, 27%.) In addition to PSA levels, no significant changes were observed in IL-6, IL-8 or maspin levels.

The lack of any meaningful clinical response in advanced prostate cancer may be due to the fact that AR appears to be more dependent on the co-chaperone p23 rather than Hsp90 for stability.[103] However, pre-clinical data in the second-generation inhibitors such as ganetespib[104] or NVP-AUY922 still warrant further testing of these agents in both AR-dependent and AR-independent prostate cancer.

13.6.4 Gastrointestinal Stromal Cancers (GIST)

Gastrointestinal stromal cancers (GIST) are the most common mesenchymal neoplasms of the gastrointestinal system.[105] One of the key features of GIST is activating mutations in either c-KIT gene or the platelet derived growth factor α (PDGFRA) gene, which accounts for 90% of GIST cases.[105,106] Hence, GIST tumors respond very well to either c-KIT inhibitors such as imatinib, or to kinase inhibitors with a broad range of action such as sunitinib, which blocks PDGFRA mediated signaling. Despite these advances neither imatinib nor sunitinib are curative in the advance setting, and resistance to both tyrosine kinase inhibitors develops. These include secondary mutations in the c-KIT or PDGFRA domains or activation of alternate oncogenic pathways caused by secondary mutations in KRAS, Braf or Nras,[107] all of which are known client proteins of Hsp90. Pre-clinical studies have shown Hsp90 inhibitors to be very effective in GIST cell lines resistant to imatinib.[108–111]

Given the encouraging pre-clinical results seen in GIST cell lines, retaspimycin entered clinic to evaluate patients with advanced GIST that was imatinib and sunitinib resistant.[112] Retaspimycin was given on days 1, 4, 8, 11 of a 21-day cycle at five dose levels ranging from $90\,mg/m^2$ to $400\,mg/m^2$. Twenty-one patients were evaluated. Although no radiological responses were observed per RECIST, FDG PET imaging revealed partial metabolic activities in seven patients. These responses were seen in all dose cohorts. Interestingly, re-activation of the metabolic signaling was seen during the 10-day treatment break, validating earlier observations that the effect of tanespimycin on the tumor pathways is short lived. In addition, seven patients also achieved stable disease and continued treatment past four cycles. In general, the regimen was tolerated well in this patient population with mostly grade 1–2 observed adverse events that included alkaline phosphatase increases, fatigue and headache.

The results from the Phase I study with evidence of early efficacy signals captured by FDG PET imaging encouraged Infinity pharmaceuticals to start a Phase III double-blinded placebo-controlled randomized study. It was planned that a total of 195 patients with metastatic or unresectable GIST that had previously progressed on imatinib and sunitinib would be randomized in a 2:1 fashion to receive either retaspimycin at $400\,mg/m^2$ twice weekly for two weeks followed with a week rest or placebo. Unfortunately, the study was terminated early after retaspimycin-related deaths were experienced.[112] A small Phase II study with BIIB021 was also undertaken in patients with advanced GIST refractory to imatinib and sunitinib.[113] The study enrolled 23 patients and assessed two dosing schedules: 12 patients received BIIB021 twice a week at 600 mg and 11 patients received the drug three times a week at 400 mg. The primary end point of the study was metabolic partial response, assessed by FDG-PET. Mutation status was determined in nine patients, where seven patients had exon 11 KIT mutation, one patient had an exon 9 KIT mutation and one patient had no PDGFR or KIT mutation detected. Treatment was well tolerated and most AEs were grade 1 and grade 2. Commonly observed AEs consisted of increased liver enzymes, dizziness, diarrhea, nausea, fatigue, emesis, constipation and night blindness. Partial metabolic responses were observed in three patients receiving 600 mg dose and two patients receiving 400 mg dose. These responses were seen in patients with exon 11 KIT mutations, those with exon 9 KIT mutations and in patients with unknown mutational status. No radiological responses per RECIST criteria were observed. The median duration on study was 35 days (8–138 days). Although several patients achieved metabolic responses, target inhibition appeared short lived. In 64% of the patients there were substantial increases in FDG update between day 5 and day 8, suggesting a more frequent dosing such as daily dosing may be beneficial for achieving durable responses. Currently, second-generation non-ansamycin Hsp90 inhibitors such as NVP-AUY922, and ganetespib are being evaluated in investigator initiated small Phase II studies. Interim results from the Phase II study with single ganetespib in imatinib- and sunitinib-resistant patients were recently revealed at ASCO.[114] The weekly ganetespib treatment for three weeks with one-week treatment break appears to be well tolerated. Although no objective responses were yet seen, metabolic PET responses were seen in several patients. In addition, similar to what was observed with BIIB021, pre- and post-biopsies showed c-kit degradation, which was short lived. Complete results from these studies are eagerly waited. Other possibilities for future studies include combining Hsp90 inhibitors with imatinib.

13.6.5 Gastric Cancer

Although not among the higher incidence rates of cancer types in the US and Europe, gastric cancer is the second deadliest form of cancer after lung cancer with 866,000 deaths globally in 2007.[115] Gastric cancer is common in Eastern Europe, Latin America and Asia and the high incidence rates in these regions are probably due to lifestyles. It is known that certain factors such as *H. pylori* infection, low socioeconomic status, smoking, use of alcohol, intake of salty and smoked food and low consumption of fruits and vegetables are all associated with gastric cancer.[116] Approximately 84% of patients with gastric cancer will be diagnosed in the advance setting where treatment is not curative.[117] Chemotherapy still remains the cornerstone of treatment for gastric cancer. 5-Flurouracil (5-FU) in combination with cisplatin and epirubicin achieves a median overall survival (OS) of approximately 10 months, and similar OS rates are achieved in various combinations of 5-FU, platinum-containing agents and taxanes.[118] Recently, based on the positive results from the TOGA study,[119] trastuzumab has been approved in the US, in combination with cisplatin and capecitabine or 5-FU, for the treatment of patients with HER2 over-expressing advanced gastric cancer who are treatment naïve for metastatic disease.

Recent studies have shown a number of client proteins of Hsp90 to be associated with gastric cancer. Around 15–45% of gastric cancers over-express HER2.[119,120] In addition, c-MET over-expression or activating mutations have also been reported in gastric cancer tissues.[121] The vascular endothelial growth factor receptor (VEGFR) is also involved in the tumor genesis of gastric cancer. It has been shown that serum levels of VEGF-A and VEGFR-2 are elevated in patients and high levels of VEGFR-1 have been associated with poor overall survival.[122] Both VEGFR-1 and VEGFR-2 have been shown to be down-regulated by Hsp90 inhibitors,[123,124] through a mechanism involving the proteasome.[125]

NVP-AUY922 has been shown to down-regulate HER2 and pAKT in gastric cancer cell lines.[126] These results have been replicated in xenograft models including HER2 over-expressing N87 cell lines and c-MET over-expressing GTL-16 gastric tumors (internal NVP data). Considering the strong pre-clinical signals seen with NVP-AUY922, Novartis initiated two Phase II studies in advanced gastric cancer patients.

The first study initiated randomized advanced gastric cancer patients who had progressed after first-line chemotherapy treatment to either receive NVP-AUY922 or the comparator drug irinotecan or docetaxel (based on first-line treatment; non-taxane: docetaxel, taxane: irinotecan). NVP-AUY922 was given weekly at $70\,mg/m^2$, irinotecan at $350\,mg/m^2$ and docetaxel at $75\,mg/m^2$. Randomization was done 1:1. The primary objective of the study was to assess PFS. Assessment of OS was a secondary objective of the study, hence crossover between treatment arms were not allowed. Target recruitment for the study was 100 patients and an interim analysis was planned to be conducted after at least 50 patients had been randomized to the study.

Unfortunately the study was terminated after 65 patients were randomized as the interim analysis did not show any superiority of NVP-AUY922 treatment to chemotherapy. At the time of the analysis 33 patients had been assigned to the NVP-AUY922 treatment arm, 25 patients to the docetaxel treatment arm and 7 patients to the irinotecan treatment arm. One patient randomized to docetaxel never started treatment. The safety profile was similar in this patient population receiving NVP-AUY922 with diarrhea (57.6%) being the most commonly observed AE. Other frequently seen AEs were nausea (51.5%), anemia (42.4%), decreased appetite (39.4%), abdominal pain (36.4%), emesis (33.3%) and constipation (30.3%). Visual AEs were observed in 66.7% of patients. These events consisted of blurred vision (24.2%), photopsia (21.2%) and visual impairment (15.2%). No objective responses were seen with NVP-AUY922. Median PFS was 1.6 months for NVP-AUY922 and 4.4 months for the comparator arm.

In a second Phase II study, NVP-AUY922 was explored in combination with trastuzumab in patients with HER2+ over-expressing advanced gastric cancer who had progressed on trastuzumab based first-line treatment. Target recruitment for this study was 48 patients and the primary objective was to determine the overall response rate to the combination treatment. NVP-AUY922 was given weekly at $70\,mg/m^2$ and trastuzumab was given standard dose, loading dose 8 mg/kg and maintenance dose 6 mg/kg, every three weeks. This study was also prematurely terminated as only one unconfirmed partial response was observed in 21 patients. However, the combination treatment with trastuzumab was tolerated well and the most commonly observed AEs were diarrhea, visual disturbances, hypertension, nausea and increases in liver enzymes. There were no reported DLTs.

Given the strong pre-clinical signals it was disappointing to see both studies not showing any meaningful response in this hard-to-treat indication. Although, our second study focused on HER2+ population it did not factor in what resistance mechanisms patients had towards HER2 directed treatment. Future studies with more selected patient population may be warranted to see whether Hsp90 inhibitors have a role in the treatment of gastric cancer.

13.6.6 Multiple Myeloma

Multiple myeloma (MM) is one of the common hematological malignancies with an estimated 21,700 newly diagnosed cases in the US in 2012.[44] Multiple myeloma (MM) is a slowly progressing malignancy characterized by proliferation of clonal plasmocytes within the bone marrow. In the advanced stages MM usually manifests itself with osteolytic lesions, renal failure due to high levels of circulating immunoglobins and bone marrow failure leading to anemia and infections.[127] Treatment for MM comprises multi-modal treatment combining chemotherapy, bone marrow transplant and more recently approved targeted agents such as dexamethasone, melphalan, thalidomide, lenalidomide or bortezomib. Despite all the therapeutic options, MM is an incurable disease. Median survival for patients that have relapsed on thalidomide/lenalidomide and bortezomib is approximately 9 months.[128] Hence, newer treatment alternatives are needed.

The accumulation of unfolded and misfolded proteins in the endoplasmic reticulum (ER) activates intracellular pathways to resolve the protein folding defect, a response that is termed unfolded protein response (UPR).[129] This response helps increase the ER protein folding and modification, increases the cellular protein synthesis and activates ER-mediated protein degradation.[129] However, if the ER is taxed too much by the imbalance between accumulation of proteins and clearance by UPR than apoptotic signaling pathways are triggered.[130] In MM, which is characterized by the synthesis of a large number of immunoglobins, the UPR pathway is highly active.[128,131] Several components of the UPR pathway are known Hsp90 client proteins and Hsp90 inhibition induces apoptosis in MM cell lines.[132,133]

Tanespimycin was the first Hsp90 inhibitor to be assessed in patients with multiple myeloma both as single agent and in combination with the proteasome inhibitor bortezomib. In a dose escalation Phase I study tanespimycin was evaluated in 29 patients with advanced multiple myeloma that had progressed on multiple lines of treatment.[134] The primary end point of the study was to evaluate MTD of tanespimycin when given twice weekly with a one-week break in a 21-day cycle by using standard Fibonacci dose escalation scheme. Plasma PK and preliminary efficacy was also evaluated. The evaluated doses included 150, 220, 275, 340, 420 and 525 mg/m^2. MTD was not determined per protocol established criteria. Five patients experienced DLTs. One patient at 150 mg/m^2 and another patient in the 340 mg/m^2 dose cohort experienced grade 4 thrombocytopenia. One patient at $220\,mg/m^2$ experienced grade 3 ALT elevation. One patient at $340\,mg/m^2$ experienced grade 4 AST and Grade 3 ALT elevation. One patient at $525\,mg/m^2$ experienced grade 3 cardiac ischemia. In general, tanespimycin was tolerated well in this patient population and only three patients discontinued treatment due to unacceptable toxicity. The most common observed non-hematological AEs were diarrhea (59%), fatigue (38%), back pain (35%) and nausea (35%). Seven patients experienced elevations in ALT, and nine cases of AST elevations were observed. The most common hematological AE was anemia (24%) and thrombocytopenia (21%). Three patients experienced grade 3 anemia, and two cases of grade 4 thrombocytopenia was observed. There was no incidence of neutropenia. Of note, no QTcF prolongations or changes in LVEF were observed.

Early efficacy signals were detected in this study. Although no complete or partial responses were observed, one patient at the $150\,mg/m^2$ cohort experienced a minor response, which is defined in MM as at least 25% reduction in M protein urine and plasma levels, 25% reduction in plasmacytomas and absence of an increase in lytic bone lesions. This patient had a 41% reduction in urine M protein and a 33% decrease in serum M protein and achieved disease stabilization for 3 months. An additional 15 patients were deemed stable disease during the first efficacy evaluation, and had a median PFS of 2.1 months (95% CL 1.48–2.60). Five of these patients had serum M protein reductions ranging from 7 to 38% and urine M protein reductions from 6% to 96%.

The rationale for combining Hsp90 inhibitors with proteasome inhibitors has been strong, as cells rely on the proteasome to recycle accumulated proteins. This becomes important especially in MM where large numbers of proteins, such as immunoglobins, are accumulated. Proteasome inhibitors such as bortezomib trigger apoptosis by inhibiting the proteasome activity.[135] Hsp90 inhibitors have been shown pre-clinically to combine well with the proteasome inhibitor bortezomib.[136–138] Based on the encouraging early signals of efficacy in single agent tanespimycin, Richardson and colleagues initiated a Phase I/II study in combination with bortezomib.[137] The study enrolled 72 heavily pretreated patients with MM. The combination treatment was well tolerated and responses were noted. Overall response rate in bortezomib naïve patients was 48%, and 13% in bortezomib refractory patients. The median duration of response was 12 months (CL 95% 5–19). Due to these encouraging results the study was expanded, but slight changes were made to the patient population, namely bortezomib naïve patients were excluded.[137] The study, which was conducted in the US, compared $1.3\,mg/m^2$ bortezomib to three different doses of tanespimycin: $50\,mg/m^2$, $175\,mg/m^2$ and $340\,mg/m^2$. Unfortunately, this study was terminated early for reasons that were only related to resources and had nothing to do with lack of efficacy or intolerable safety.

Twenty-two patients participated in the study. Eight patients in the $340\,mg/m^2$ cohort, eight in the $175\,mg/m^2$ cohort and six patients in the $50\,mg/m^2$ group. The most common experienced AEs were fatigue (73%), nausea (68%), diarrhea (64%), constipation (50%) and vomiting (45%). The most common grade 3 and grade 4 events included fatigue (32%), thrombocytopenia (27%), neutropenia (18%) and abdominal pain (9%). Categories of AEs that led to discontinuation of treatment were gastrointestinal disturbances (18%) and musculoskeletal and connective tissue disorders (14%). Liver toxicities were noticed in four patients. One patient treated at 340 mg/m^2 cohort experienced grade 4 AST and grade 2 elevations. Another patient, treated at 340 mg/m^2, experienced grade 4 liver toxicity, which resolved within two weeks of tanespimycin withdrawal. The patient was rechallenged with tanespimycin at 275 mg/m^2 and subsequently developed grade 3 fatigue and was discontinued from treatment. Thrombocytopenia was the most common hematological AE reported. Two patients treated in the 340 mg/m^2 experienced grade 4 thrombocytopenia, which led to discontinuation for one patient. Treatment-related neuropathy, a common side effect of bortezomib treatment, was reported in only six patients, all of whom had prior treatment-related peripheral neuropathy.

One MR response was noted in the 340 mg/m^2 cohort and two PRs were reported in the 175 mg/m^2 cohort. Ten patients achieved SD. Of note one patient remained on study for 22 cycles and another patient received treatment for 19 cycles, with a significant reduction of plasma cells in their bone marrow (60% at baseline down to 5% at end of study). Despite these promising early results no future studies with tanespimycin in MM have been undertaken.

In a similar Phase I study retaspimycin was also evaluated in heavily pretreated MM patients.[128] Retaspimycin was given twice weekly for two weeks with a ten-day treatment break. The primary objective of the study was to evaluate MTD using a Fibonacci dose escalation scheme with two phases, an accelerated escalation phase enrolling one patient per dose cohort and the standard phase enrolling three patients per cohort. The study was designed to enroll 34 patients during the escalation phase and an additional 14 patients at the MTD dose level. Starting dose was 22.5 mg/m^2 and doses were escalated in the accelerated phase up to 150 mg/m^2. The standard escalation schema evaluated doses of 150 mg/m^2, 225 mg/m^2, 300 mg/m^2 and 400 mg/m^2. The study was terminated early at the 400 mg/m^2 dose and enrolled 18 patients. The reason for early termination was the lack of any responses during the escalation phase.

Retaspimycin was well tolerated and no DLTs were observed. The most common AEs were fatigue (44%), anemia and diarrhea (33%), back pain (28%), increased lactate blood dehydrogenase (28%) and nausea (28%). The most common grade 3 and grade 4 AEs were anemia, neutropenia and thrombocytopenia, which were reported in two patients and were most likely related to disease-infiltrated bone marrow. No cardiac toxicity was reported. There was only one hepatic toxicity reported, which consisted of reversible grade 1 alanine aminotransferase. There was only one incidence of ocular toxicity, grade 1 dry eye, reported. The lack of any responses seen in this study was probably due to the fact that higher doses of retaspimycin were never tested.

NVP-AUY22 was also evaluated in patients with relapsed or refractory MM. In a rather complex Phase I-Ib/II trial, NVP-AUY922 was evaluated first as single agent and then in combination with bortezomib with our without dexamethasone. A Bayesian logistics regression model with overdose control was utilized both in the single agent and combination parts of the study. The primary purpose of the study was to determine MTD for the single and combination treatments. NVP-AUY922 was administered once weekly in a 21-day treatment cycle. The study was terminated early due to slow recruitment.

The single agent escalation part enrolled 24 patients that were evaluated in six cohorts; $8\,mg/m^2$ (n = 3), $16\,mg/m^2$ (n = 3), $30\,mg/m^2$ (n = 3), $45\,mg/m^2$ (n = 5), $60\,mg/m^2$ (n = 3) and $70\,mg/m^2$ (n = 7). The Phase Ib part enrolled five patients who received $50\,mg/m^2$ NVP-AUY22 and $1.3\,mg/m^2$ bortezomib.

In the single agent dose escalation part, the most commonly observed AEs were diarrhea (66.7%), pyrexia (37.5%), anemia and nausea (both of 29.2%), thrombocytopenia (25%), fatigue (20.8%), QT prolongation, hypertension, night blindness, photopsia, upper respiratory tract infection, visual impairment and emesis (each of 16.7%). A total of nine patients (37.5%) experienced ocular toxicity including night blindness, photopsia and visual impairment (all of 16.7%). The ocular toxicities were transient and resolved upon drug discontinuation and in general were grade 1 or grade 2 events. However, more severe ocular toxicity was noted at the $45\,mg/m^2$ and $70\,mg/m^2$ dose cohorts, where a total of four patients (16.7%) reported CTCAE grade 3 or 4 visual events, including cataracts, night blindness, retinopathy, blurred vision and diminished visual acuity. One of the ocular events, grade 3 ocular toxicity, seen in a patient treated at $70\,mg/m^2$, was deemed a DLT.

In the combination part with bortezomib the most commonly observed AEs were diarrhea that was observed in four patients (80%), musculoskeletal pain, nasopharyngitis and thrombocytopenia observed each in three patients (60%), abdominal pain, C-reactive protein increase, fatigue, leukopenia, nausea, neutropenia, night blindness, night sweats, photopsia, pyrexia, visual impairment and vomiting each seen in two patients (40%). The combination of NVP-AUY922 and bortezomib was clearly more toxic than each single agent alone and three patients experienced four cases of DLTs at the $50\,mg/m^2$ cohort. These events included two cases of grade 3 musculoskeletal pain, one case of grade 3 non-cardiac chest pain, and one case of grade 3 diarrhea.

Although no objective responses were seen in the single agent escalation portion of the study, NVP-AUY922 single agent treatment achieved disease stabilization in 67% of patients in this heavily pre-treated population. One partial response was observed in the combination phase of the study and an additional three patients had stable disease.[139] These results are in line with earlier tanespimycin studies that Hsp90 inhibitors, especially in combination with proteasome inhibitors, may yield benefit in refractory MM patients.

13.7 Conclusions

Since the discovery in 1985 that Hsp90 plays a role in oncogenic proliferation a lot of ground has been covered, with several exciting Hsp90 inhibitors now being evaluated in different stages of clinical trials (Table 13.1). All Hsp90 inhibitors have a distinct safety profile and it appears that GI-related toxicities are the common treatment-related AE. These usually consist of diarrhea, nausea and to a lesser extent emesis, all of which can be managed by antidiarrheal and anti-emesis treatment in clinic and these events seldom require treatment discontinuation.

One of the biggest concerns with Hsp90 inhibitors has been the potential for causing cardiac toxicity. *In vitro* experiments have shown interference with the human ether-a-go-go-related gene (hERG)-related proteins that are involved in the proper functioning of potassium ion channels in heart tissue.[140,141] Hence, all clinical protocols with Hsp90 inhibitors have developed strict ECG monitoring and patients with minor baseline cardiac toxicities have been excluded. So far none of the Hsp90 inhibitors in clinic have shown clear cardiac toxicity. It is possible that this is due to the carefully selected patient population, as all studies have very stringent cardiac exclusion criteria. QTc-PK analysis we have performed with NVP-AUY922 has shown some changes in QTcF parameters at 70 mg/m^2. We observed that there was a mean decrease of –8.3 ms 8 hours post-dose in cycle 1 day 1 and a mean increase of +13.4 ms 24 hours post-dose (C1D2). However, all other timepoints, including pre- and post-infusion in subsequent treatment days, did not show any changes. It is possible that this transient effect is seen with the other Hsp90 inhibitors, but so far no detailed reports have been published. It is probably still prudent to use caution and monitor patients closely for QTcF changes.

The ansamycin first-generation compounds all show liver toxicity and this unfortunately has hampered the development of these agents, despite early efficacy signals seen in myeloma, GIST, NSCLC and breast cancer. Liver toxicity has not been a limiting issue for the non-ansamycin second-generation inhibitors, suggesting that the removal of the quinine ring has been beneficial. However, grade 3 or grade 4 events have still been reported in some patients receiving ganetespib. Caution should be exercised, especially when treating patients with liver metastasis.

Ocular toxicity is seen at various frequencies, to a lesser degree with alvespimycin, retaspimycin, BIIB021, MPC3100 and more frequently with AT13387, SNX-5455 and NVP-AUY922, among Hsp90 inhibitors suggesting a class effect. Many proteins that are involved in the functioning of the photoreceptors in the retina are Hsp90 client proteins.[142] This is in line with the ocular AEs usually described by patients, which involves delayed light to dark adaptation difficulties (night blindness), photopsias, peripheral vision loss and in rare cases color blindness, all suggesting an effect on the photoreceptor layer. Electroretinogram recordings from patients and animals confirm this with significant reductions in a-wave and b-wave amplitudes. However, especially in the case of NVP-AUY922, these events are reversible in humans and animals, and dose reduction and interruption appears to be helpful with very few patients having to discontinue treatment. What is surprising is that ocular toxicity does not seem to be seen at the same frequency among all Hsp90 inhibitors. The structural differences between Hsp90 inhibitors may affect their retinal elimination rate, which could be a decisive factor for ocular toxicity. For instance, NVP-AUY922 has a slow elimination rate, whereas ganetespib has a very rapid elimination rate.[143] In addition, the half-life of ganetespib is much shorter than that of NVP-AUY922, raising the question that from an efficacy stand-point ocular toxicity may be a sort of response sign akin to rash and EGFR TKIs. For instance, the effect on client proteins appears to be short lived for ganetespib and tanespimycin.[114,144] It is interesting that NVP-AUY922 has a clear response in EGFR mutant patients, whereas this has only been seen in one patient treated with retaspimycin and none with ganetespib. Tanespimycin and alvespimycin, which has a similar structure to retaspimycin, have slower retinal elimination rates than ganetespib.

Except for myalgias and muscular pain reported with some Hsp90 inhibitors, effects on the central nervous system (CNS) seem to be limited to compounds that cross the blood-brain barrier such as NVP-HSP990. At the high dose levels there was a clear effect on CNS and grade 1, 2 and 3 neurological symptoms such as dizziness, tremors, ataxia and myoclonic movements in limbs were noted. This is not surprising given the role Hsp90 plays in the synapse.[145] Neuro-toxicities were observed at the high dose levels of 50 mg and 60 mg in the weekly dosing schedule. The events were milder in the schedule using 25 mg twice weekly dosing, which may be more suitable for these types of compounds. However, the question arises whether more frequent dosing with lower doses is enough for client protein degradation. Further studies, with a mandate of collecting tumor biopsies to assess target pathway modulation, will have to be conducted.

The big question is still what oncogenic client proteins are the most susceptible to Hsp90 inhibition, and hold promise for the first indication where an Hsp90 inhibitor will be granted marketing authorization. From the conducted studies so far it appears that the EML4-ALK, mutant EGFR or HER2+ expressing cancers hold the most promise. ROS-1 translocations also seem to be a prospective target,[84] but clinical trials in this patient population still have to be conducted. Therefore, it appears likely that either NSCLC or breast cancer will be one of the indications having an Hsp90 inhibitor as part of standard treatment. The only question that remains is when this will happen.

References

1. K. Jhaveri, T. Taldone, S. Modi and G. Chiosis, *Biochim. Biophys. Acta*, 2012, **1823**, 742.
2. P. Delmotte and J. Delmotte-Plaque, *Nature*, 1953, **171**, 344.
3. C. Deboer, P. A. Meulman, R. J. Wnuk and D. H. Peterson, *J. Antibiot. (Tokyo)*, 1970, **23**, 442.
4. Y. Uehara, M. Hori, T. Takeuchi and H. Umezawa, *Mol. Cell Biol.*, 1986, **6**, 2198.
5. Y. Uehara, *Curr. Cancer Drug Targets*, 2003, **3**, 325.
6. L. Whitesell, E. G. Mimnaugh, C. B. De, C. E. Myers and L. M. Neckers, *Proc. Natl Acad. Sci. USA*, 1994, **91**, 8324.
7. J. P. Grenert, W. P. Sullivan, P. Fadden, T. A. Haystead, J. Clark, E. Mimnaugh, H. Krutzsch, H. J. Ochel, T. W. Schulte, E. Sausville, L. M. Neckers and D. O. Toft, *J. Biol. Chem.*, 1997, **272**, 23843.

8. C. Prodromou, S. M. Roe, R. O'Brien, J. E. Ladbury, P. W. Piper and L. H. Pearl, *Cell*, 1997, **90**, 65.
9. L. Neckers and P. Workman, *Clin. Cancer Res.*, 2012, **18**, 64.
10. P. Workman, F. Burrows, L. Neckers and N. Rosen, *Ann. NY Acad. Sci.*, 2007, **1113**, 202.
11. L. Neckers, *Curr. Top. Med. Chem.*, 2006, **6**, 1163.
12. R. C. Schnur, M. L. Corman, R. J. Gallaschun, B. A. Cooper, M. F. Dee, J. L. Doty, M. L. Muzzi, C. I. DiOrio, E. G. Barbacci, P. E. Miller, A. T. O'Brien, M. J. Morin, B. A. Foster, V. A. Pollack, D. M. Savage, D. E. Sloan, L. R. Pustilnik and M. P. Moyer, *J. Med. Chem.*, 1995, **38**, 3813.
13. T. W. Schulte and L. M. Neckers, *Cancer Chemother. Pharmacol.*, 1998, **42**, 273.
14. U. Banerji, A. O'Donnell, M. Scurr, S. Pacey, S. Stapleton, Y. Asad, L. Simmons, A. Maloney, F. Raynaud, M. Campbell, M. Walton, S. Lakhani, S. Kaye, P. Workman and I. Judson, *J. Clin. Oncol.*, 2005, **23**, 4152.
15. E. A. Sausville, J. E. Tomaszewski and P. Ivy, *Curr. Cancer Drug Targets*, 2003, **3**, 377.
16. D. B. Solit, S. P. Ivy, C. Kopil, R. Sikorski, M. J. Morris, S. F. Slovin, W. K. Kelly, A. DeLaCruz, T. Curley, G. Heller, S. Larson, L. Schwartz, M. J. Egorin, N. Rosen and H. I. Scher, *Clin. Cancer Res.*, 2007, **13**, 1775.
17. S. Kummar, M. E. Gutierrez, E. R. Gardner, X. Chen, W. D. Figg, M. Zajac-Kaye, M. Chen, S. M. Steinberg, C. A. Muir, M. A. Yancey, Y. R. Horneffer, L. Juwara, G. Melillo, S. P. Ivy, M. Merino, L. Neckers, P. S. Steeg, B. A. Conley, G. Giaccone, J. H. Doroshow and A. J. Murgo, *Eur. J. Cancer*, 2010, **46**, 340.
18. S. Pacey, R. H. Wilson, M. Walton, M. M. Eatock, A. Hardcastle, A. Zetterlund, H. T. Arkenau, J. Moreno-Farre, U. Banerji, B. Roels, H. Peachey, W. Aherne, J. S. de Bono, F. Raynaud, P. Workman and I. Judson, *Clin. Cancer Res.*, 2011, **17**, 1561.
19. K. T. Flaherty, L. Gore, A. Avadhani, S. Leong, K. Harlacker, Z. Zhong, R. G. Johnson, A. L. Hannah, P. O'Dywer and S. G. Eckhardt, *J. Clin. Oncol.*, 2007, **25**(18S), 14059.
20. J. Ge, E. Normant, J. R. Porter, J. A. Ali, M. S. Dembski, Y. Gao, A. T. Georges, L. Grenier, R. H. Pak, J. Patterson, J. R. Sydor, T. T. Tibbitts, J. K. Tong, J. Adams and V. J. Palombella, *J. Med. Chem.*, 2006, **49**, 4606.
21. J. R. Sydor, E. Normant, C. S. Pien, J. R. Porter, J. Ge, L. Grenier, R. H. Pak, J. A. Ali, M. S. Dembski, J. Hudak, J. Patterson, C. Penders, M. Pink, M. A. Read, J. Sang, C. Woodward, Y. Zhang, D. S. Grayzel, J. Wright, J. A. Barrett, V. J. Palombella, J. Adams and J. K. Tong, *Proc. Natl Acad. Sci. USA*, 2006, **103**, 17408.
22. L. V. Sequist, P. A. Janne, J. Sweeney, J. R. Walker, D. Grayzel and T. J. Lynch, *AACR-NCI-EORTC International Conferance on Molecular Targets and Cancer Therapuetics*, San Francisco, USA, 2007.

23. L. V. Sequist, S. Gettinger, N. N. Senzer, R. G. Martins, P. A. Janne, R. Lilenbaum, J. E. Gray, A. J. Iafrate, R. Katayama, N. Hafeez, J. Sweeney, J. R. Walker, C. Fritz, R. W. Ross, D. Grayzel, J. A. Engelman, D. R. Borger, G. Paez and R. Natale, *J. Clin. Oncol.*, 2010, **28**, 4953.
24. P. A. Brough, W. Aherne, X. Barril, J. Borgognoni, K. Boxall, J. E. Cansfield, K. M. Cheung, I. Collins, N. G. Davies, M. J. Drysdale, B. Dymock, S. A. Eccles, H. Finch, A. Fink, A. Hayes, R. Howes, R. E. Hubbard, K. James, A. M. Jordan, A. Lockie, V. Martins, A. Massey, T. P. Matthews, E. McDonald, C. J. Northfield, L. H. Pearl, C. Prodromou, S. Ray, F. I. Raynaud, S. D. Roughley, S. Y. Sharp, A. Surgenor, D. L. Walmsley, P. Webb, M. Wood, P. Workman and L. Wright, *J. Med. Chem.*, 2008, **51**, 196.
25. S. Bailey, B. Neuenschwander, G. Laird and M. Branson, *J. Biopharm. Stat.*, 2009, **19**, 469.
26. C. P. Schroeder, J. V. Pedersen, S. Chua, C. Swanton, M. Akimov, S. Ide, C. Fernandez-Ibarra, A. Dzik-Jurasz, E. De Vries, S. B. Gaykema and U. Banerji, *J. Clin. Oncol.*, 2011, **29**(suppl).
27. T. Y. Lin, M. Bear, Z. Du, K. P. Foley, W. Ying, J. Barsoum and C. London, *Exp. Hematol.*, 2008, **36**, 1266.
28. T. Shimamura, S. A. Perera, K. P. Foley, J. Sang, S. J. Rodig, T. Inoue, L. Chen, D. Li, J. Carretero, Y. C. Li, P. Sinha, C. D. Carey, C. L. Borgman, J. P. Jimenez, M. Meyerson, W. Ying, J. Barsoum, K. K. Wong and G. I. Shapiro, *Clin. Cancer Res.*, 2012, **18**, 4973.
29. W. Ying, Z. Du, L. Sun, K. P. Foley, D. A. Proia, R. K. Blackman, D. Zhou, T. Inoue, N. Tatsuta, J. Sang, S. Ye, J. Acquaviva, L. S. Ogawa, Y. Wada, J. Barsoum and K. Koya, *Mol. Cancer Ther.*, 2012, **11**, 475.
30. www.clinicaltrials.gov.
31. J. M. Cleary, E. I. Heath, E. L. Kwak, B. J. Dezube, L. Gandhi, C. Zack, R. Bradley, V. M. Vukovic, G. Shapiro and P. LoRusso, *J. Clin. Oncol.*, 2010, **28**(15s).
32. J. W. Goldman, R. N. Raju, G. A. Gordon, V. M. Vukovic, R. Bradley and L. S. Rosen, *J. Clin. Oncol.*, 2010, **28**(15s).
33. G. I. Shapiro, E. L. Kwak, B. J. Dezube, D. P. Lawrance, J. M. Cleary, S. Lewis, M. Squires, V. Lock, J. F. Lyons and M. Yule, *J. Clin. Oncol.*, 2010, **28**(15s).
34. D. Mahadevan, G. I. Shapiro, S. E. Kurtin, J. M. Cleary, J. F. Lyons, A. Rodriguez-Lopez, M. Yule, V. Ahanonu, G. S. Choy, M. Noursalehi and M. Azab, *J. Clin. Oncol.*, 2012, **30**, 34.
35. P. Chene, *Nat. Rev. Drug Discov.*, 2002, **1**, 665.
36. G. Chiosis, B. Lucas, A. Shtil, H. Huezo and N. Rosen, *Bioorg. Med. Chem.*, 2002, **10**, 3555.
37. N. Isambert, A. Hollebecque, Y. Berge, H. van Ingen, S. Brienza, A. Estaillats, J. C. Soria, P. Fumoleau and J. P. Delord, *J. Clin. Oncol.*, 2012, **30**(suppl.).

38. http://www.debiopharm.com/media/publications/72-debio-0932/3188-a-phase-i-study-of-debio-0932-an-oral-hsp90-inhibitor-in-patients-with-solid-tumors-.html.
39. W. Samlowski, K. Papadopoulos, A. J. Olszanski, K. Zavitz, D. M. Cimbora, S. Shawbell, A. Balch, G. Mather and A. Beelen, *Mol. Cancer Ther.*, 2011, **10**, 1.
40. A. Elfiky, W. M. Saif, M. Beeram, S. O'Brien, N. Lammanna, J. E. Castro, J. Woodworth, R. Perea, C. Storgard and D. D. Von Hoff, *J. Clin. Oncol.*, 2008, **26**.
41. www.biogenidec.com.
42. K. H. Huang, J. M. Veal, R. P. Fadden, J. W. Rice, J. Eaves, J. P. Strachan, A. F. Barabasz, B. E. Foley, T. E. Barta, W. Ma, M. A. Silinski, M. Hu, J. M. Partridge, A. Scott, L. G. Dubois, T. Freed, P. M. Steed, A. J. Ommen, E. D. Smith, P. F. Hughes, A. R. Woodward, G. J. Hanson, W. S. McCall, C. J. Markworth, L. Hinkley, M. Jenks, L. Geng, M. Lewis, J. Otto, B. Pronk, K. Verleysen and S. E. Hall, *J. Med. Chem.*, 2009, **52**, 4288.
43. A. Rajan, R. J. Kelly, J. B. Trepel, Y. S. Kim, S. V. Alarcon, S. Kummar, M. Gutierrez, S. Crandon, W. M. Zein, L. Jain, B. Mannargudi, W. D. Figg, B. E. Houk, M. Shnaidman, N. Brega and G. Giaccone, *Clin. Cancer Res.*, 2011, **17**, 6831.
44. R. Siegel, D. Naishadham and A. Jemal, *CA Cancer J. Clin.*, 2012, **62**, 10.
45. M. Alvarez, E. Roman, E. S. Santos and L. E. Raez, *Expert. Rev. Anticancer Ther.*, 2007, **7**, 1423.
46. M. I. Gallegos Ruiz, K. Floor, P. Roepman, J. A. Rodriguez, G. A. Meijer, W. J. Mooi, E. Jassem, J. Niklinski, T. Muley, Z. N. van, E. F. Smit, K. Beebe, L. Neckers, B. Ylstra and G. Giaccone, *PLoS.One*, 2008, **3**, e0001722.
47. T. E. Stinchcombe and M. A. Socinski, *Proc. Am. Thorac. Soc.*, 2009, **6**, 233.
48. W. J. Gullick, *Br. Med. Bull.*, 1991, **47**, 87.
49. D. S. Salomon, R. Brandt, F. Ciardiello and N. Normanno, *Crit Rev. Oncol. Hematol.*, 1995, **19**, 183.
50. M. F. Zakowski, S. Hussain, W. Pao, M. Ladanyi, M. S. Ginsberg, R. Heelan, V. A. Miller, V. W. Rusch and M. G. Kris, *Arch. Pathol. Lab Med.*, 2009, **133**, 470.
51. J. G. Paez, P. A. Janne, J. C. Lee, S. Tracy, H. Greulich, S. Gabriel, P. Herman, F. J. Kaye, N. Lindeman, T. J. Boggon, K. Naoki, H. Sasaki, Y. Fujii, M. J. Eck, W. R. Sellers, B. E. Johnson and M. Meyerson, *Science*, 2004, **304**, 1497.
52. T. J. Lynch, D. W. Bell, R. Sordella, S. Gurubhagavatula, R. A. Okimoto, B. W. Brannigan, P. L. Harris, S. M. Haserlat, J. G. Supko, F. G. Haluska, D. N. Louis, D. C. Christiani, J. Settleman and D. A. Haber, *N. Engl. J. Med.*, 2004, **350**, 2129.
53. L. V. Sequist and T. J. Lynch, *Annu. Rev. Med.*, 2008, **59**, 429.

54. T. J. Lynch, D. W. Bell, R. Sordella, S. Gurubhagavatula, R. A. Okimoto, B. W. Brannigan, P. L. Harris, S. M. Haserlat, J. G. Supko, F. G. Haluska, D. N. Louis, D. C. Christiani, J. Settleman and D. A. Haber, *N. Engl. J. Med.*, 2004, **350**, 2129.
55. T. Mitsudomi and Y. Yatabe, *Cancer Sci.*, 2007, **98**, 1817.
56. L. Paz-Ares, D. Soulieres, I. Melezinek, J. Moecks, L. Keil, T. Mok, R. Rosell and B. Klughammer, *J. Cell Mol. Med.*, 2010, **14**, 51.
57. V. A. Miller, G. J. Riely, M. F. Zakowski, A. R. Li, J. D. Patel, R. T. Heelan, M. G. Kris, A. B. Sandler, D. P. Carbone, A. Tsao, R. S. Herbst, G. Heller, M. Ladanyi, W. Pao and D. H. Johnson, *J. Clin. Oncol.*, 2008, **26**, 1472.
58. S. Toyooka, T. Takano, T. Kosaka, K. Hotta, K. Matsuo, S. Ichihara, Y. Fujiwara, J. Soh, H. Otani, K. Kiura, K. Aoe, Y. Yatabe, Y. Ohe, T. Mitsudomi and H. Date, *Cancer Sci.*, 2008, **99**, 303.
59. T. S. Mok, Y. L. Wu, S. Thongprasert, C. H. Yang, D. T. Chu, N. Saijo, P. Sunpaweravong, B. Han, B. Margono, Y. Ichinose, Y. Nishiwaki, Y. Ohe, J. J. Yang, B. Chewaskulyong, H. Jiang, E. L. Duffield, C. L. Watkins, A. A. Armour and M. Fukuoka, *N. Engl. J. Med.*, 2009, **361**, 947.
60. M. Maemondo, A. Inoue, K. Kobayashi, S. Sugawara, S. Oizumi, H. Isobe, A. Gemma, M. Harada, H. Yoshizawa, I. Kinoshita, Y. Fujita, S. Okinaga, H. Hirano, K. Yoshimori, T. Harada, T. Ogura, M. Ando, H. Miyazawa, T. Tanaka, Y. Saijo, K. Hagiwara, S. Morita and T. Nukiwa, *N. Engl. J. Med.*, 2010, **362**, 2380.
61. C. Zhou, Y. L. Wu, G. Chen, J. Feng, X. Q. Liu, C. Wang, S. Zhang, J. Wang, S. Zhou, S. Ren, S. Lu, L. Zhang, C. Hu, C. Hu, Y. Luo, L. Chen, M. Ye, J. Huang, X. Zhi, Y. Zhang, Q. Xiu, J. Ma, L. Zhang and C. You, *Lancet Oncol.*, 2011, **12**, 735.
62. R. Rosell, E. Carcereny, R. Gervais, A. Vergnenegre, B. Massuti, E. Felip, R. Palmero, R. Garcia-Gomez, C. Pallares, J. M. Sanchez, R. Porta, M. Cobo, P. Garrido, F. Longo, T. Moran, A. Insa, M. F. de, R. Corre, I. Bover, A. Illiano, E. Dansin, C. J. de, M. Milella, N. Reguart, G. Altavilla, U. Jimenez, M. Provencio, M. A. Moreno, J. Terrasa, J. Munoz-Langa, J. Valdivia, D. Isla, M. Domine, O. Molinier, J. Mazieres, N. Baize, R. Garcia-Campelo, G. Robinet, D. Rodriguez-Abreu, G. Lopez-Vivanco, V. Gebbia, L. Ferrera-Delgado, P. Bombaron, R. Bernabe, A. Bearz, A. Artal, E. Cortesi, C. Rolfo, M. Sanchez-Ronco, A. Drozdowskyj, C. Queralt, A. de, I, J. L. Ramirez, J. J. Sanchez, M. A. Molina, M. Taron and L. Paz-Ares, *Lancet Oncol.*, 2012, **13**, 239.
63. W. Pao, V. A. Miller, K. A. Politi, G. J. Riely, R. Somwar, M. F. Zakowski, M. G. Kris and H. Varmus, *PLoS Med.*, 2005, **2**, e73.
64. W. Pao and J. Chmielecki, *Nat. Rev. Cancer*, 2010, **10**, 760.
65. D. B. Costa, K. S. Nguyen, B. C. Cho, L. V. Sequist, D. M. Jackman, G. J. Riely, B. Y. Yeap, B. Halmos, J. H. Kim, P. A. Janne, M. S. Huberman, W. Pao, D. G. Tenen and S. Kobayashi, *Clin. Cancer Res.*, 2008, **14**, 7060.

66. M. N. Balak, Y. Gong, G. J. Riely, R. Somwar, A. R. Li, M. F. Zakowski, A. Chiang, G. Yang, O. Ouerfelli, M. G. Kris, M. Ladanyi, V. A. Miller and W. Pao, *Clin. Cancer Res.*, 2006, **12**, 6494.
67. J. Bean, G. J. Riely, M. Balak, J. L. Marks, M. Ladanyi, V. A. Miller and W. Pao, *Clin. Cancer Res.*, 2008, **14**, 7519.
68. E. Avizienyte, R. A. Ward and A. P. Garner, *Biochem. J.*, 2008, **415**, 197.
69. J. A. Engelman, K. Zejnullahu, T. Mitsudomi, Y. Song, C. Hyland, J. O. Park, N. Lindeman, C. M. Gale, X. Zhao, J. Christensen, T. Kosaka, A. J. Holmes, A. M. Rogers, F. Cappuzzo, T. Mok, C. Lee, B. E. Johnson, L. C. Cantley and P. A. Janne, *Science*, 2007, **316**, 1039.
70. J. Bean, C. Brennan, J. Y. Shih, G. Riely, A. Viale, L. Wang, D. Chitale, N. Motoi, J. Szoke, S. Broderick, M. Balak, W. C. Chang, C. J. Yu, A. Gazdar, H. Pass, V. Rusch, W. Gerald, S. F. Huang, P. C. Yang, V. Miller, M. Ladanyi, C. H. Yang and W. Pao, *Proc. Natl Acad. Sci. USA*, 2007, **104**, 20932.
71. T. Shimamura, A. M. Lowell, J. A. Engelman and G. I. Shapiro, *Cancer Res.*, 2005, **65**, 6401.
72. K. Wong, J. Kocztmas, J. Goldman, E. Paschold, L. Horn, J. Lufkin, R. K. Balckman, F. Teofilovici, G. Shapiro and M. A. Socinski, *J. Clin. Oncol.*, 2011, **29**.
73. E. B. Garon, T. Moran, B. Barlesi, L. Gandhi, L. V. Sequist, S. W. Kim, H. J. M. Groen, B. Besse, E. F. Smit, D. W. Kim, M. Akimov, E. Avsar, S. Bailey and E. Felip, *J. Clin. Oncol.*, 2012, **30**(suppl.).
74. E. Felip, E. Carcereny, F. Barlesi, L. Gandhi, L. V. Sequist, S. W. Kim, H. J. M. Groen, B. Besse, D. W. Kim, E. F. Smit, M. Akimov, E. Avsar, S. Bailey, W. Ofosu-Appiah and E. B. Garon, *Annals Oncol.*, 2012, **23**, 9.
75. L. V. Sequist, B. A. Waltman, D. Dias-Santagata, S. Digumarthy, A. B. Turke, P. Fidias, K. Bergethon, A. T. Shaw, S. Gettinger, A. K. Cosper, S. Akhavanfard, R. S. Heist, J. Temel, J. G. Christensen, J. C. Wain, T. J. Lynch, K. Vernovsky, E. J. Mark, M. Lanuti, A. J. Iafrate, M. Mino-Kenudson and J. A. Engelman, *Sci. Transl. Med.*, 2011, **3**, 75ra26.
76. M. Soda, Y. L. Choi, M. Enomoto, S. Takada, Y. Yamashita, S. Ishikawa, S. Fujiwara, H. Watanabe, K. Kurashina, H. Hatanaka, M. Bando, S. Ohno, Y. Ishikawa, H. Aburatani, T. Niki, Y. Sohara, Y. Sugiyama and H. Mano, *Nature*, 2007, **448**, 561.
77. S. Perner, P. L. Wagner, F. Demichelis, R. Mehra, C. J. Lafargue, B. J. Moss, S. Arbogast, A. Soltermann, W. Weder, T. J. Giordano, D. G. Beer, D. S. Rickman, A. M. Chinnaiyan, H. Moch and M. A. Rubin, *Neoplasia*, 2008, **10**, 298.
78. S. J. Rodig, M. Mino-Kenudson, S. Dacic, B. Y. Yeap, A. Shaw, J. A. Barletta, H. Stubbs, K. Law, N. Lindeman, E. Mark, P. A. Janne, T. Lynch, B. E. Johnson, A. J. Iafrate and L. R. Chirieac, *Clin. Cancer Res.*, 2009, **15**, 5216.
79. D. R. Camidge, Y. J. Bang, E. L. Kwak, A. J. Iafrate, M. Varella-Garcia, S. B. Fox, G. J. Riely, B. Solomon, S. H. Ou, D. W. Kim, R. Salgia,

P. Fidias, J. A. Engelman, L. Gandhi, P. A. Janne, D. B. Costa, G. I. Shapiro, P. Lorusso, K. Ruffner, P. Stephenson, Y. Tang, K. Wilner, J. W. Clark and A. T. Shaw, *Lancet Oncol.*, 2012, **13**, 1011.
80. R. C. Doebele, A. B. Pilling, D. L. Aisner, T. G. Kutateladze, A. T. Le, A. J. Weickhardt, K. L. Kondo, D. J. Linderman, L. E. Heasley, W. A. Franklin, M. Varella-Garcia and D. R. Camidge, *Clin. Cancer Res.*, 2012, **18**, 1472.
81. T. Sasaki, J. Koivunen, A. Ogino, M. Yanagita, S. Nikiforow, W. Zheng, C. Lathan, J. P. Marcoux, J. Du, K. Okuda, M. Capelletti, T. Shimamura, D. Ercan, M. Stumpfova, Y. Xiao, S. Weremowicz, M. Butaney, S. Heon, K. Wilner, J. G. Christensen, M. J. Eck, K. K. Wong, N. Lindeman, N. S. Gray, S. J. Rodig and P. A. Janne, *Cancer Res.*, 2011, **71**, 6051.
82. Y. L. Choi, M. Soda, Y. Yamashita, T. Ueno, J. Takashima, T. Nakajima, Y. Yatabe, K. Takeuchi, T. Hamada, H. Haruta, Y. Ishikawa, H. Kimura, T. Mitsudomi, Y. Tanio and H. Mano, *N. Engl. J. Med.*, 2010, **363**, 1734.
83. E. Normant, G. Paez, K. A. West, A. R. Lim, K. L. Slocum, C. Tunkey, J. McDougall, A. A. Wylie, K. Robison, K. Caliri, V. J. Palombella and C. C. Fritz, *Oncogene*, 2011, **30**, 2581.
84. D. A. Proria, J. Acquaviva, Q. Jiang, L. Xue, D. Smith, J. C. Friedland, S. He, J. Sang, S. W. Morris and Y. Wada, *J. Clin. Oncol.*, 2012, **30**(suppl.).
85. T. Sasaki and P. A. Janne, *Clin. Cancer Res.*, 2011, **17**, 7213.
86. J. M. Heuckmann, H. Balke-Want, F. Malchers, M. Peifer, M. L. Sos, M. Koker, L. Meder, C. M. Lovly, L. C. Heukamp, W. Pao, R. Kuppers and R. K. Thomas, *Clin. Cancer Res.*, 2012, **18**, 4682.
87. D. J. Slamon, G. M. Clark, S. G. Wong, W. J. Levin, A. Ullrich and W. L. McGuire, *Science*, 1987, **235**, 177.
88. D. Slamon and M. Pegram, *Semin. Oncol.*, 2001, **28**, 13.
89. M. J. Piccart-Gebhart, M. Procter, B. Leyland-Jones, A. Goldhirsch, M. Untch, I. Smith, L. Gianni, J. Baselga, R. Bell, C. Jackisch, D. Cameron, M. Dowsett, C. H. Barrios, G. Steger, C. S. Huang, M. Andersson, M. Inbar, M. Lichinitser, I. Lang, U. Nitz, H. Iwata, C. Thomssen, C. Lohrisch, T. M. Suter, J. Ruschoff, T. Suto, V. Greatorex, C. Ward, C. Straehle, E. McFadden, M. S. Dolci and R. D. Gelber, *N. Engl. J. Med.*, 2005, **353**, 1659.
90. T. W. Miller, B. N. Rexer, J. T. Garrett and C. L. Arteaga, *Breast Cancer Res.*, 2011, **13**, 224.
91. E. M. Gartner, P. Silverman, M. Simon, L. Flaherty, J. Abrams, P. Ivy and P. M. Lorusso, *Breast Cancer Res. Treat.*, 2012, **131**, 933.
92. S. Modi, A. Stopeck, H. Linden, D. Solit, S. Chandarlapaty, N. Rosen, G. D'Andrea, M. Dickler, M. E. Moynahan, S. Sugarman, W. Ma, S. Patil, L. Norton, A. L. Hannah and C. Hudis, *Clin. Cancer Res.*, 2011, **17**, 5132.
93. C. P. Schroder, J. V. Pedersen, S. Chua, C. Swanton, M. Akimov, S. Ide, A. Fernandez-Ibarra, A. Dzik-Jurasz, E. De Vries, S. B. Gaykema and U. Banerji, *J. Clin. Oncol.*, 2011, **29**(suppl.).

94. A. Kong, A. Rea, S. Ahmed, T. Beck, R. L. Lopez, L. Biganzoli, A. Armstrong, M. Aglietta, E. Alba, M. Campone, M. Akimov, A. Matano, C. Lefebvre and S. Lee, *J. Clin. Oncol.*, 2012, **30**(suppl.).
95. M. Stanbrough, G. J. Bubley, K. Ross, T. R. Golub, M. A. Rubin, T. M. Penning, P. G. Febbo and S. P. Balk, *Cancer Res.*, 2006, **66**, 2815.
96. W. B. Pratt and D. O. Toft, *Exp. Biol. Med. (Maywood.)*, 2003, **228**, 111.
97. H. I. Scher, *J. Natl Cancer Inst.*, 2000, **92**, 1866.
98. J. Li, C. Yen, D. Liaw, K. Podsypanina, S. Bose, S. I. Wang, J. Puc, C. Miliaresis, L. Rodgers, R. McCombie, S. H. Bigner, B. C. Giovanella, M. Ittmann, B. Tycko, H. Hibshoosh, M. H. Wigler and R. Parsons, *Science*, 1997, **275**, 1943.
99. D. B. Solit, F. F. Zheng, M. Drobnjak, P. N. Munster, B. Higgins, D. Verbel, G. Heller, W. Tong, C. Cordon-Cardo, D. B. Agus, H. I. Scher and N. Rosen, *Clin. Cancer Res.*, 2002, **8**, 986.
100. B. E. Hanson and D. H. Vesole, *Expert. Opin. Investig. Drugs*, 2009, **18**, 1375.
101. W. K. Oh, M. D. Galsky, W. M. Stadler, S. Srinivas, F. Chu, G. Bubley, J. Goddard, J. Dunbar and R. W. Ross, *Urology*, 2011, **78**, 626.
102. E. I. Heath, D. W. Hillman, U. Vaishampayan, S. Sheng, F. Sarkar, F. Harper, M. Gaskins, H. C. Pitot, W. Tan, S. P. Ivy, R. Pili, M. A. Carducci, C. Erlichman and G. Liu, *Clin. Cancer Res.*, 2008, **14**, 7940.
103. V. Reebye, C. L. Querol, D. N. Lavery, G. N. Brooke, S. M. Powell, D. Chotai, M. M. Walker, H. C. Whitaker, R. Wait, H. C. Hurst and C. L. Bevan, *Mol. Endocrinol.*, 2012, **26**, 1694.
104. S. He, C. Zhang, A. A. Shafi, M. Sequeira, J. Acquaviva, J. C. Friedland, J. Sang, D. L. Smith, N. L. Weigel, Y. Wada and D. A. Proia, *Int. J. Oncol.*, 2013, **42**, 35.
105. B. P. Rubin, S. Singer, C. Tsao, A. Duensing, M. L. Lux, R. Ruiz, M. K. Hibbard, C. J. Chen, S. Xiao, D. A. Tuveson, G. D. Demetri, C. D. Fletcher and J. A. Fletcher, *Cancer Res.*, 2001, **61**, 8118.
106. B. P. Rubin, M. C. Heinrich and C. L. Corless, *Lancet*, 2007, **369**, 1731.
107. J. Y. Blay, C. A. Le, P. A. Cassier and I. L. Ray-Coquard, *Discov. Med.*, 2012, **13**, 357.
108. G. Floris, R. Sciot, A. Wozniak, L. T. Van, J. Wellens, G. Faa, E. Normant, M. Debiec-Rychter and P. Schoffski, *Clin. Cancer Res.*, 2011, **17**, 5604.
109. T. Smyth, L. T. Van, J. E. Curry, A. M. Rodriguez-Lopez, A. Wozniak, M. Zhu, R. Donsky, J. G. Morgan, M. Mayeda, J. A. Fletcher, P. Schoffski, J. Lyons, N. T. Thompson and N. G. Wallis, *Mol. Cancer Ther.*, 2012, **11**, 1799.
110. G. Floris, M. Debiec-Rychter, A. Wozniak, C. Stefan, E. Normant, G. Faa, K. Machiels, U. Vanleeuw, R. Sciot and P. Schoffski, *Mol. Cancer Ther.*, 2011, **10**, 1897.
111. S. Bauer, L. K. Yu, G. D. Demetri and J. A. Fletcher, *Cancer Res.*, 2006, **66**, 9153.

112. G. D. Demetri, C. A. Le, M. Von Mehren, B. Chmielowski, S. Bauer, W. A. Chow, E. Rodenas, K. McKee, D. S. Graysel and Y. Kang, *Final Results from a Phase III Study of IPI-504 (Retaspimycin Hydrochloride) Versus Placebo in Patients (pts) with Gastrointestinal Stromal Tumors (GIST) Following Failure of Kinase Inhibitor Therapies*, 2010, Gastrointestinal cancers symposium, Orlando, FL USA. [Available at: asco.org].
113. M. A. Dickson, S. H. Okuno, M. L. Keohan, R. G. Maki, D. R. D'Adamo, T. J. Akhurst, C. R. Antonescu and G. K. Schwartz, *Ann. Oncol.*, 2013, **24**, 252.
114. G. D. Demetri, M. C. Heinrich, B. Chmielowski, J. A. Morgan, S. George, R. Bradley, R. K. Blackman, F. Teofilovici, J. A. Fletcher, W. D. Tap and M. Von Mehren, *J. Clin. Oncol.*, 2011, **29**(suppl.).
115. World Health Organization Cancer Fact Sheet Number 297, 2013.
116. R. Wesolowski, C. Lee and R. Kim, *Lancet Oncol.*, 2009, **10**, 903.
117. F. Rivera, M. E. Vega-Villegas and M. F. Lopez-Brea, *Cancer Treat. Rev.*, 2007, **33**, 315.
118. J. L. Lee and Y. K. Kang, *Future Oncol.*, 2008, **4**, 179.
119. Y. J. Bang, C. E. Van, A. Feyereislova, H. C. Chung, L. Shen, A. Sawaki, F. Lordick, A. Ohtsu, Y. Omuro, T. Satoh, G. Aprile, E. Kulikov, J. Hill, M. Lehle, J. Ruschoff and Y. K. Kang, *Lancet*, 2010, **376**, 687.
120. J. T. Jorgensen, *Oncology*, 2010, **78**, 26.
121. F. Cecchi, D. C. Rabe and D. P. Bottaro, *Expert. Opin. Ther. Targets*, 2012, **16**, 553.
122. M. S. Al-Moundhri, A. Al-Shukaili, M. Al-Nabhani, B. Al-Bahrani, I. A. Burney, A. Rizivi and S. S. Ganguly, *World J. Gastroenterol.*, 2008, **14**, 3879.
123. W. B. Nagengast, M. A. de Korte, T. H. Oude Munnink, H. Timmer-Bosscha, W. F. den Dunnen, H. Hollema, J. R. de Jong, M. R. Jensen, C. Quadt, C. Garcia-Echeverria, G. A. van Dongen, M. N. Lub-de Hooge, C. P. Schroder and E. G. de Vries, *J. Nucl. Med.*, 2010, **51**, 761.
124. T. H. Oude Munnink, M. A. Korte, W. B. Nagengast, H. Timmer-Bosscha, C. P. Schroder, J. R. Jong, G. A. Dongen, M. R. Jensen, C. Quadt, M. N. Hooge and E. G. Vries, *Eur. J. Cancer*, 2010, **46**, 678.
125. J. H. Park, S. H. Kim, M. C. Choi, J. Lee, D. Y. Oh, S. A. Im, Y. J. Bang and T. Y. Kim, *Biochem. Biophys. Res. Commun.*, 2008, **368**, 318.
126. Z. A. Wainberg, A. Anghel, A. M. Rogers, A. J. Desai, O. Kalous, D. Conklin, R. Ayala, N. A. O'Brien, C. Quadt, M. Akimov, D. J. Slamon and R. S. Finn, *Mol. Cancer Ther.*, 2013, **12**, 509.
127. W. M. Kuehl and P. L. Bergsagel, *J. Clin. Invest*, 2012, **122**, 3456.
128. D. Siegel, S. Jagannath, D. H. Vesole, I. Borello, A. Mazumder, C. Mitsiades, J. Goddard, J. Dunbar, E. Normant, J. Adams, D. Grayzel, K. C. Anderson and P. Richardson, *Leuk. Lymphoma*, 2011, **52**, 2308.
129. S. S. Cao and R. J. Kaufman, *Expert. Opin. Ther. Targets*, 2013, **17**, 437.
130. S. S. Cao and R. J. Kaufman, *Curr. Biol.*, 2012, **22**, R622.
131. A. H. Lee, N. N. Iwakoshi, K. C. Anderson and L. H. Glimcher, *Proc. Natl Acad. Sci. USA*, 2003, **100**, 9946.

132. J. Patterson, V. J. Palombella, C. Fritz and E. Normant, *Cancer Chemother. Pharmacol.*, 2008, **61**, 923.
133. M. A. Dimopoulos, C. S. Mitsiades, K. C. Anderson and P. G. Richardson, *Clin. Lymphoma Myeloma. Leuk.*, 2011, **11**, 17.
134. P. G. Richardson, A. A. Chanan-Khan, M. Alsina, M. Albitar, D. Berman, M. Messina, C. S. Mitsiades and K. C. Anderson, *Br. J. Haematol.*, 2010, **150**, 438.
135. T. Hideshima, P. Richardson, D. Chauhan, V. J. Palombella, P. J. Elliott, J. Adams and K. C. Anderson, *Cancer Res.*, 2001, **61**, 3071.
136. M. A. Dimopoulos, C. S. Mitsiades, K. C. Anderson and P. G. Richardson, *Clin. Lymphoma Myeloma. Leuk.*, 2011, **11**, 17.
137. P. G. Richardson, A. A. Chanan-Khan, S. Lonial, A. Y. Krishnan, M. P. Carroll, M. Alsina, M. Albitar, D. Berman, M. Messina and K. C. Anderson, *Br. J. Haematol.*, 2011, **153**, 729.
138. C. S. Mitsiades, N. S. Mitsiades, C. J. McMullan, V. Poulaki, A. L. Kung, F. E. Davies, G. Morgan, M. Akiyama, R. Shringarpure, N. C. Munshi, P. G. Richardson, T. Hideshima, D. Chauhan, X. Gu, C. Bailey, M. Joseph, T. A. Libermann, N. S. Rosen and K. C. Anderson, *Blood*, 2006, **107**, 1092.
139. R. Seggewiss-Bernhardt, R. Bargou, Y. T. Goh, A. K. Steward, A. Spenser, A. Alegre, J. Blade, O. Ottmann, A. Akimov, C. Fernandez-Ibarra, C. Kalmady and S. Padmanabhan-Iyer, *Haematologica*, 2012, **97**(suppl).
140. E. Ficker, A. T. Dennis, L. Wang and A. M. Brown, *Circ. Res.*, 2003, **92**, e87.
141. E. Ficker, A. Dennis, Y. Kuryshev, B. A. Wible and A. M. Brown, *Novartis Found. Symp.*, 2005, **266**, 57.
142. H. F. Mendes, R. Zaccarini and M. E. Cheetham, *Adv. Exp. Med. Biol.*, 2010, **664**, 317.
143. D. Zhou, F. Teofilovici, Y. Liu, J. Ye, W. Ying, L. S. Ogawa, T. Inoue, W. Lee, A. Adjiri-Awere, L. Kolodzieyski, N. Tatsuta, Y. Wada and A. J. Sonderfan, *J. Clin. Oncol.*, 2012, **30**(suppl.).
144. D. B. Solit, I. Osman, D. Polsky, K. S. Panageas, A. Daud, J. S. Goydos, J. Teitcher, J. D. Wolchok, F. J. Germino, S. E. Krown, D. Coit, N. Rosen and P. B. Chapman, *Clin. Cancer Res.*, 2008, **14**, 8302.
145. S. Luo, B. Zhang, X. P. Dong, Y. Tao, A. Ting, Z. Zhou, J. Meixiong, J. Luo, F. C. Chiu, W. C. Xiong and L. Mei, *Neuron*, 2008, **60**, 97.

CHAPTER 14

Heat-shock Protein 90 as an Antimalarial Target

ANKIT K. ROCHANI, MEETALI SINGH AND
UTPAL TATU*

Department of Biochemistry, Indian Institute of Science, Bangalore-12, India
*Email: tatu@biochem.iisc.ernet.in

14.1 History of Malaria

Malaria is Latin for "bad air". It is an ancient protozoan infection that has afflicted humans throughout the history of mankind resulting in high morbidity and mortality. However, the causative agent for malaria was not known until the late nineteenth century when the parasite responsible for causing malaria was discovered by French army surgeon Charles Louis Alphonse Laveran, who won the Nobel Prize for Physiology or Medicine in 1907. Sir Ronald Ross in the late 1890s discovered the role of mosquitos in disease transmission, for which he was awarded the Nobel Prize in Medicine in 1902. Sir Ronald Ross observed for the first time the parasite oocysts in the midgut of the mosquito in 1897. In 1948, Henry Shortt and Cyril Garnham identified the pre-erythrocytic phase of the life cycle of the malaria parasite. This set of events made a major contribution towards the understanding of the transmission of one of the most lethal protozoan infections.[1]

3.3 billion people worldwide are at risk of contracting malaria. WHO reports 219 million cases of malarial in 2010 and over 600,000 deaths.[2] Most infections and deaths, amounting to 91% of cases, happen in Africa. However, other regions including India, Latin America and some parts of the Middle East are also affected. The global economic growth due to international travel and interactions has led to a rampant increase in cases of malaria in non-endemic regions. Malaria poses a major risk to non-immune pregnant females, people with HIV/AIDS, young children and immigrants from endemic regions.[2] Over 86% of deaths globally due to malaria happen in children. The Centre for Disease Control (CDC) has recommended prophylactic treatment for malaria in order to prevent cross-over cases of malaria due to traveling to malarial endemic regions.[3]

Malaria is caused by a protozoan parasite belonging to *Plasmodium* genus. Five species of *Plasmodium* can infect humans, namely *P. falciparum*, *P. vivax*, *P. ovale*, *P. malariae* and *P. knowlesi*. The first two of these, *i.e.* Pf and Pv, account for the highest morbidity and mortality, whereas *P. ovale* and *P. malariae* generally cause a milder form of malaria. *P. knowlesi* generally infects monkeys but rarely can also cause infection in humans. Human malaria is transmitted by the female *Anopheles* mosquito.[4] Tropical areas with low socio-economic demography are most affected due to lack of hygiene and plenty of breeding grounds for mosquitos.

14.2 Malaria and Prevention

Malaria infection in humans is caused by sporozoite injection by feeding mosquitos. These sporozoites then, in turn, invade hepatocytes. In hepatocytes exoerythrocytic schizogony takes place, leading to the release of merozoites in the blood stream. These merozoites invade erythrocytes resulting in the ring stage of parasites. Active metabolism and proteolysis of hemoglobin is followed by development of trophozoites. Asexual division of trophozoites leads to formation of the schizont stage. Finally, merozoites are released by infected red blood cells (RBCs) and the cycle of erythrocyte invasion reinitiates. Most of the malaria symptoms are associated with the blood stage of the parasite. Fever episodes observed in malaria patients are due to synchronous lysis of infected erythrocytes.[4,5] If untreated, disease prognosis could lead to complicated malaria symptoms and in the case of *P. falciparum* advanced disease could result in manifestation of cerebral malaria.[6]

There are three strategies of combating malaria: a) elimination of vector, b) antimalarial drugs and c) vaccination against malaria. Elimination of vector is considered to be the most cost-effective way of combating malarial infection. On the other hand development of antimalarials has been a major challenge for the healthcare sector. Rapid emergence of drug resistance has made it necessary to develop new antimalarial drugs. An effective antimalarial vaccine is still futuristic and in the developmental pipeline.[4,7,8]

There are mainly two ways to control mosquitoes: a) preventing the contact of mosquitos to humans and b) elimination of mosquitoes using insecticides. Mosquito repellents are most commonly used for prevention of diseases. But this prevention is not sufficient for controlling malaria. On the other hand, use of insecticides had provided minor help in controlling the spread of disease and has also cleared infection in some regions such as North America, Russia and some parts of Europe. DDT (dichlorodiphenyltrichloroethane), the first effective insecticide, was invented by Dr. Paul Muller, for which he was awarded the Nobel Prize in Physiology or Medicine in 1948. There are reports of this molecule being a possible carcinogen.[9] DDT was one of the most used molecules in the antimalarial campaign by WHO (World Health Organization) before it was banned in the USA from 1972. It is commonly being used in India.

It seems that the one of the most environmental friendly strategies to fight malaria is to find new drug targets and their respective targeting molecules. The journey from the identification of molecules from cinchona bark to the *Artemisia annua* plant suggests that the answer for combating the disease in the human body comes from Nature. Current antimalarial drug therapy acts at merozoites, at primary and secondary schizonts and on the intra-erythrocyte developmental stages of merozoites or gametocytes.[4,10]

14.3 Current Antimalarial Drugs and Drug Targets

Malaria is an intracellular infection. The parasite resides in the RBCs of the host. An effective antimalarial drug has to pass through three cellular membranes (RBC, parasitophorus vacuole and parasite membrane) to have an antimalarial effect. Due to the complex biochemical properties of *Plasmodium* it has become difficult to make antimalarial drugs. The advent of "Omics" technologies like genomics, proteomics and metabolomics has relatively speeded up the methods of identification of novel drug targets. This in turn has provided momentum towards identification of active and potent antimalarial therapeutics.

The broad classification of available antimalarial scaffolds and their respective molecular drug targets is described in Table 14.1. The structures of the antimalarial drugs in clinical use are shown in Figure 14.1.

14.4 Heat-shock Protein 90 as Antimalarial Drug Target

Heat-shock protein 90 is a well-studied protein. It was first discovered as a protein up-regulated in response to heat shock. Decades of studies have now placed Hsp90 in the center of hub regulating key cellular processes like stress response, signal transduction, immunity, protein homeostasis, accumulation of drug resistance and many other processes.[11] Heat-shock protein is one of the most abundant proteins and belongs to the GHKL family of ATPases having three distinctive domains: N-terminal domain harboring ATP binding site joined to the middle domain by a charged linker region of variable length. The middle domain harbors the catalytic arginine residue and a C-terminal domain is responsible for dimer formation. All these domains have specific sites for binding of co-chaperones and clients.[12]

The chaperoning function of this protein depends on its ATPase activity. As indicated in Figure 14.2 the unfolded protein interacts with ADP bound Hsp90

Table 14.1 List of antimalarial scaffolds with their respective derivatives and their corresponding drug targets.

S. No.	Scaffolds	Drug Targets	Derivatives
1	Quinoline	Targets ferriprotoporphyrin IX	Quinine
			Quinidine
	8-Aminoquinoline		Chloroquine
			Hydroxychloroquine
			Amodiaquine
			Mefloquine
			Primaquine
2	Sulfonamides	PABA competitive inhibitor	Sulfadoxine
3	Guanidine	Dihydrofolate reductase inhibitor	Proguanil
4	Pyrmidine	Dihydrofolate reductase inhibitor	Pyrimethamine
5	Quinolones	Selective inhibitor of electron transport system of *Plasmodium*	Atovaquone
6	Polycyclic compounds	Binding to 30s ribosomal subunit and inhibits protein synthesis	Doxycycline
		Unknown	Halofantrine
		DNA intercalating agent, succinic acid dehydrogenase binding agent, mitochondrial electron transport chain inhibitor, cholinesterase inhibitor	Quinacrine
		Molecule reacts with iron of heme of infected erythrocyte, which forms reactive free radical species, which is postulated to be lethal for parasite	Artemisinin
			Dihydroartimisinin
			Artemether
			Artemotil
			Artesunate
		DOXP reductoisomerase inhibitor	Fosmidomycin

complex. ATP binding triggers conformational changes and subsequent ATP hydrolysis leads to release of mature or folded client protein. Hsp90 inhibitors like GA and its analogues competitively bind to the ATP binding site of the N-terminal domain of Hsp90 in turn compromising the folding of the client proteins. Unfolded or misfolded proteins are thus subjected to the proteasomal degradation pathway. This phenomenon ultimately leads to cell death.[11]

Plasmodium falciparum has a complex digenetic life cycle, where it has to transmit from its definitive host mosquito to its human host. During such a transition the parasite experiences a heat shock of 10 °C. During the asexual life cycle of *Plasmodium* in human erythrocytes, release of merozoites from infected RBCs is accompanied by high febrile episodes. The host immune system puts forward its best defense strategies to the incoming parasite, yet *Plasmodium* successfully manifests the disease.

Studies have shown that *Plasmodium* actually exploits these febrile episodes to its benefit. If parasites are subjected to a heat shock, then re-exposure to such a heat shock promotes stage transition from ring stage to trophozoite stage robustly during a second heat-shock episode.[13] In the same study, the authors have shown the involvement of a classical chaperone, heat-shock protein 90

Heat-shock Protein 90 as an Antimalarial Target

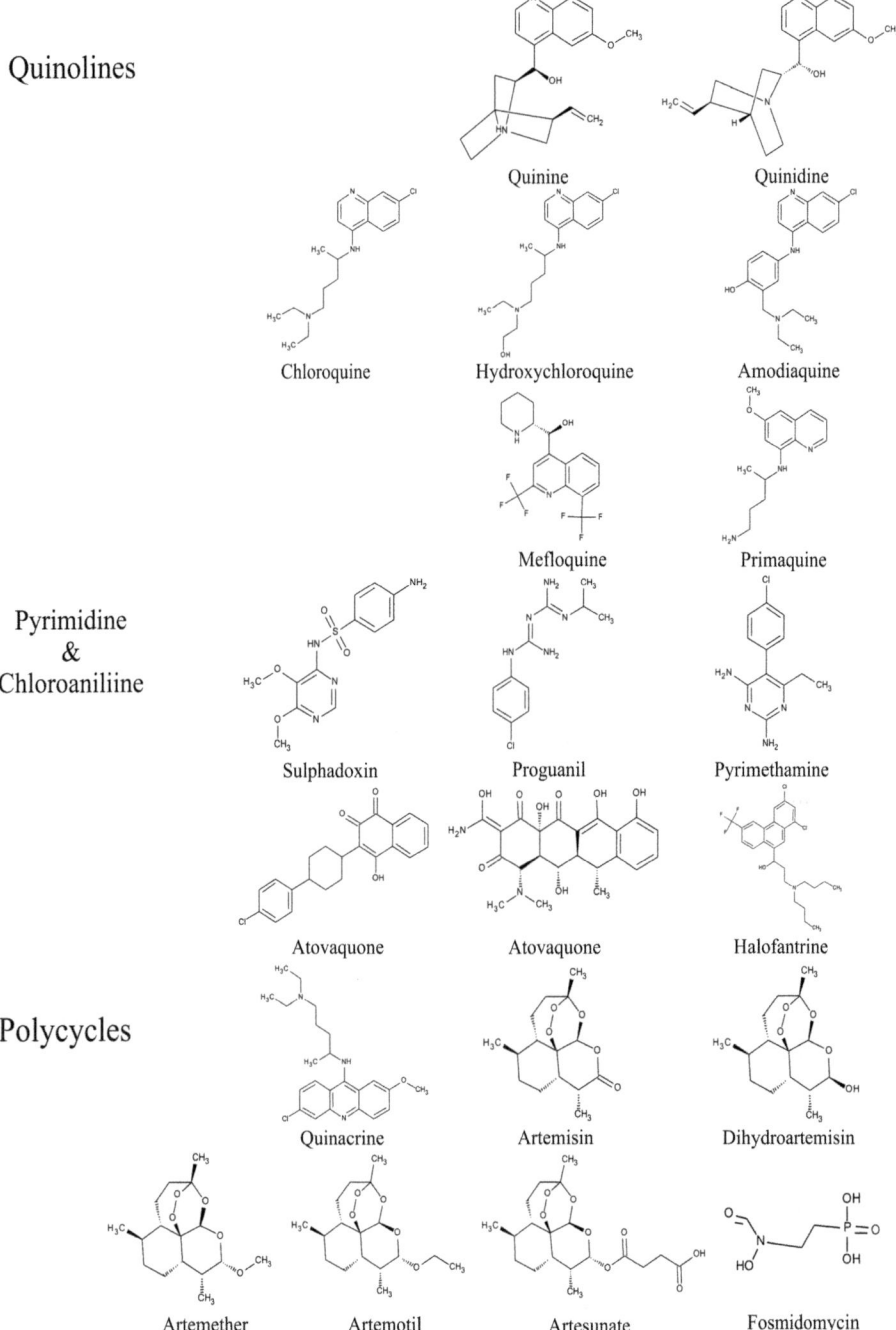

Figure 14.1 Structures of Anti-malarial Drugs.

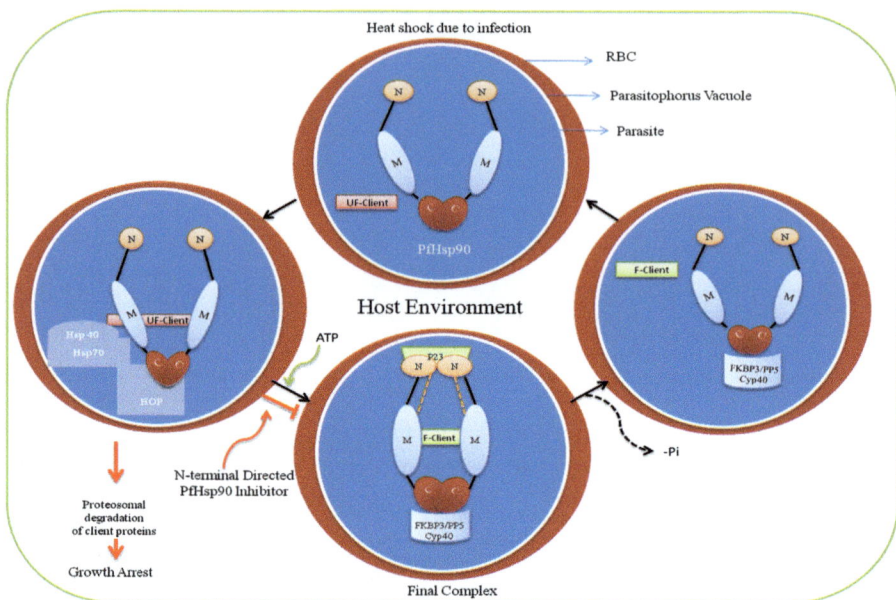

Figure 14.2 Mechanism of PfHsp90 Inhibitors.

(Hsp90), to be involved in a robust stage switch. Treatment of cells exposed to first heat shock with a specific Hsp90 inhibitor geldanamycin (GA) in turn interferes with parasite stage switch, thereby establishing the key role Hsp90 plays in survival during stress and exploiting these conditions for improved and faster manifestation of the infectious cycle by *Plasmodium*.

Hsp90 functions in association with its partner proteins known as co-chaperones, which modulate not only its ATPase activity but also prime Hsp90 to bind various different sets of client proteins. Pavithra *et al.* have shown inhibition of Hsp90 by GA in *Plasmodium falciparum* disrupts its complex with PfHsp70 and other co-chaperones, highlighting the importance of this multi-chaperone complex in a cyto-protective role.[13] Hsp90 function is essential for *Plasmodium* survival. Reduction in parasitemia was observed upon GA treatment with an LD_{50} of 0.2 µM. Studies have also shown that GA treatment abrogates stage transition from ring stage to trophozoites, leading to persistence of rings.[14]

At the preclinical level studies have shown the efficacy of GA derivative 17-AAG in mice models of malaria. Upon 17-AAG treatment (50 mg/kg b.w.) for *P. berghei* infection, mice survival was found to increase by up to two-fold in comparison to untreated mice.[15] Studies on another rodent malaria model of *P. yoelli* also support this observation.[16]

PfHsp90 shares ~70% identity with its human counterpart. One of the unique aspects of PfHsp90 is the presence of a long charged acidic linker region, which is 33 amino acids longer than the yeast linker region.[17,18] PfHsp90 nucleotide binding site shows significant conservation with respect to yeast or human Hsp90 but there are subtle differences in the binding pocket. Differences in the side change rotamer conformation of Met84 result in an altered ceiling shape in the case of PfHsp90. Other amino acid substitution in PfHsp90, like that of V173I and of S38A, results in a constricted and hydrophobic posterior end of the binding pocket.[17,18]

It is reported that PfHsp90 has a high affinity towards GA binding and is a hyperactive ATPase compared to other Hsp90s.[15] Also, yeast strain complemented with PfHsp90 is found to be more sensitive to Hsp90 inhibition-mediated growth arrest compared to yeast strain harboring wild-type Hsp90.[19]

These studies altogether provide proof of principle for the role of PfHsp90 as a potential antimalarial drug target.

Drug resistance is a major problem in disease treatment. Studies from various groups have shown the role of Hsp90 in the emergence of drug resistance. Hsp90 is expressed at very high levels in cells acting as a specialized reservoir for folding of metastable client proteins. Hsp90, thus, can potentiate folding of genetic variants leading to the emergence of newer phenotypes. Experimental evidence has shown that Hsp90 potentiates drug resistance in fungal species.[20–22] Therefore, the use of Hsp90 inhibitors can be exploited to overcome drug resistance and prospects of combination therapy in multi-drug resistant strains can be envisaged. The Hsp90 inhibitor GA is capable of arresting growth in a chloroquine-resistant strain of *Plasmodium*.[23] Synergistic action of GA and chloroquine has been observed in inhibiting *Plasmodium* growth.[23] Pallavi *et al.* have shown synergistic action of Trichostatin-A (histone deacetylase inhibitor) and GA in growth inhibition of *Plasmodium*, which is suggestive of the potential of combinatorial therapy with Hsp90 inhibitors and available drugs in the market.[15]

The approach of targeting PfHsp90 protein as a drug target has been one of the recent developments. The three major approaches adopted to find novel drug molecules targeting Plasmodium Hsp90 protein are a) repurposing strategy, b) high-throughput screening of molecules and c) rational approach to new drug discovery.

14.5 Hsp90 Targeted New Antimalarial Drug Discovery

The drug development process typically takes 10–15 years. Repurposing strategy is often used to speed up this process and has uncovered some of the blockbuster drugs. For example, the alternate use of Sidenafil citrate in erectile dysfunction became a drug called Viagra by Pfizer. Thalidomide was initially introduced as an anti-nausea and sedative drug in the 1950s, but it was withdrawn from the market in the 1960s during the Phase IV trial due to reports of birth defects in 10,000 children from nearly 46 countries. The same molecule was re-introduced for the treatment of erythema nodosum leprosum and was approved by the USFDA in 1998.[24] On similar lines, the use of Hsp90 inhibitors has recently been diversified from cancer to find a novel treatment solution for infections like malaria, surra, sleeping sickness, candidiasis, HIV, influenza, HBV and others.[25]

14.5.1 Repurposing for PfHsp90 Inhibitors

Using the repurposing strategy GA (geldanamycin) was identified as the first molecular lead targeting PfHsp90 protein in 2000.[14,25] Further in 2010 it was reported that 17-AAG (17-allyl amino 17-demethoxy geldanamycin), a semi-synthetic derivative of GA, showed high binding specificity to PfHsp90.[15] 17-AAG was the first human Hsp90 (hHsp90) inhibitor that entered Phase III clinical trials for treatment of cancer. By using the repurposing strategy it was observed that the binding affinity of 17-AAG towards PfHsp90 is greater than that of hHsp90.[15] This result was later confirmed by the treatment of malaria infection in mice model at 50 mg/kg b.w. with 40–50% survival. When this chapter was written there were 15 novel hHsp90 inhibitors in clinical trials for treatment of cancer from approximately 35 public and private ventures. Their use as antimalarial molecules by targeting PfHsp90 protein is still an unexplored area of research.

14.5.1.1 Ansamycins Antimalarial Antibiotic

GA and its analogue 17-AAG belong to a class of macrocyclic ansamycins antibiotics as shown in Figure 14.3a. Structurally they are benzoquinone ansamycin derivatives. They are naturally occurring polyketide compounds. Rifampicin was the first ansamycin antibiotic introduced as a drug for treatment of tuberculosis, leprosy and AIDS-associated mycobacterium infection.

GA is a fermentation product of a soil organism called *Streptomyces hygroscopicus* var *geldanus*. It was isolated in 1969 from fermentation broth.[26] 3-Amino 5-hydroxy benzoic acid (AHBA) is the building block of the antibiotic. It has been used extensively for carrying out structure activity relationships (SARs) and fragment-based drug discovery for binding toward hHsp90.[25,27,28] Clinically, GA has been reported to have hepatoxicity, which makes it unsuitable for human consumption even though it shows considerable therapeutic effect. Hence, 17-AAG was the first Hsp90 inhibitor to enter human trials due to relative safety. As indicated previously these molecules get positioned in the ATP binding pocket of PfHsp90 protein as shown in Figure 14.3b. The figure also shows comparatively the binding site for ATP and 17-AAG. Also, these molecules are reported to be negative toward mutagenicity studies. Also, from the clinical trial data of 17-AAG it is clear that the molecule has never been shown to have any life-threatening toxicity.[29] The most common drug related toxicities were nausea, vomiting, fatigue, pain and rise in liver transaminases like ALT & AST (alanine and aspartate transaminases) under high dosing regimen.[30–34] This provides an added advantage for Hsp90 targeted inhibitors over other classical anticancer drugs for repurposing them for the treatment of malaria.

14.5.1.2 Other Naturally Occurring Antimalarial Drugs

The naturally occurring plant *Azadirachta indica* (Indian neem tree) is known for its potent antimalarial activity. A tetranortriterpenoid called gedunin[35]

Geldanamycin 17-allyl amino 17-demethoxy geldanamycin

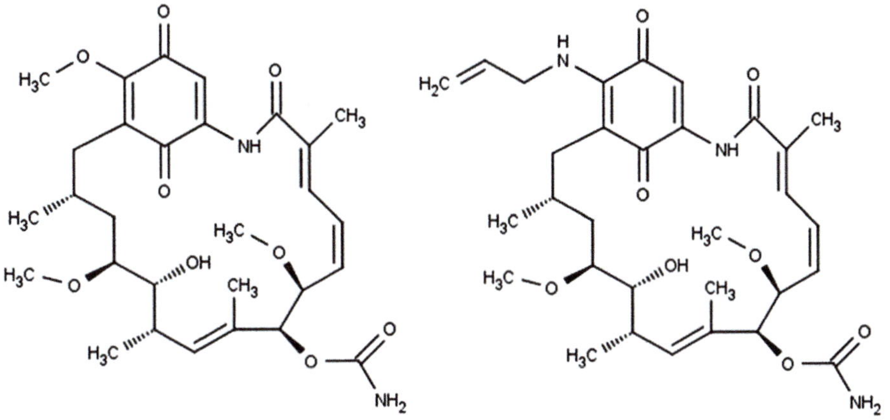

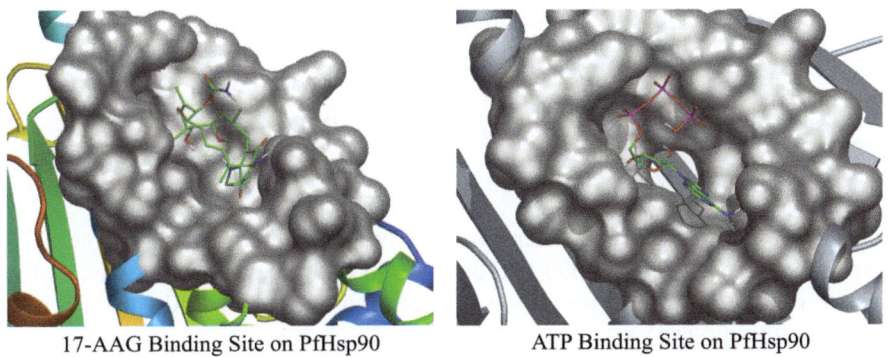

Figure 14.3a Geldanamycin and 17-allylamino 17-demethoxy geldanamycin.

17-AAG Binding Site on PfHsp90 ATP Binding Site on PfHsp90

Figure 14.3b Molecular docking of ansamycin scaffold to N-terminal domain of PfHsp90 protein.

isolated from *Azadirachta indica* is reported to have affinity towards hHsp90. The molecules have been reported to have antimalarial activity.[35] Hence it can be hypothesized that the molecule may exhibit affinity towards PfHsp90. This hypothesis remains to be validated by experimental evidence. Flavonoids were the first phytochemicals to be introduced as hHsp90 binding agents. However, none of these phytochemicals or their derivatives have been introduced for clinical applications due to their non-specificity. The idea of exploring these phytochemicals for their PfHsp90 inhibition remains a topic of discussion.

14.5.2 High-throughput Screening of PfHsp90 Inhibitors

High-throughput screening (HTS) of a molecular library is one of the most common and reliable strategies for carrying out new drug discovery for an

Table 14.2 List of chemical databases and libraries along with their URL (Uniform Resource Locator) addresses that provide updates on newly introduced information on PfHsp90 inhibitors.

S. No.	Source	URL	Type
1	PubChem	http://pubchem.ncbi.nlm.nih.gov/	Database
2	Chemspider	http://www.chemspider.com/	Database
3	Discovery Gate	https://www.discoverygate.com/dg3/DiscoveryGate/DiscoveryGate.jsp	Database
4	Zinc	http://zinc.docking.org/	Database
5	Vitasmlab	http://www.vitasmlab.com/downloads	Database and library
6	Sigma Lopac	http://www.sigmaaldrich.com/chemistry/drug-discovery/validation-libraries.html	Database and library
7	Prestwick Chemical	http://www.prestwickchemical.com/index.php?pa=3	Library
8	NIH	http://www.nihclinicalcollection.com/	Database and library
9	Chembridge	http://www.chembridge.com/screening_libraries/fragment_library/	Database and library
10	Asinex	http://www.asinex.com/download-zone.html	Database and library
11	Enamine	http://www.enamine.net/	Database and library
12	Maybridge	http://www.maybridge.com/portal/alias__Rainbow/lang__en/tabID__138/DesktopDefault.aspx	Database and library
13	Cerep	http://www.cerep.fr/cerep/Users/pages/ProductsServices/pharmacoetADME.asp	Database and library

identified novel drug target. There are nearly 5394 potential drug targets for *Plasmodium* species.[36] PfHsp90 (PDB ID: 3IED and 3k60) became one of the most commonly studied drug targets for carrying out HTS for identification of new drug-like molecules.

Benzoquinone ansamycin is the only well-characterized molecular scaffold that showed antimalarial activity and whose target is known to be PfHsp90 protein. HTS by some groups has shed light on some gray areas of molecular chemistry, which has provided around 1899 new molecular structures that have been found to have binding affinity towards PfHsp90 protein.[37] There are various databases available that are normally used for HTS of molecules targeting a variety of drug targets and these are listed in Table 14.2. All these databases and libraries can be used for carrying out necessary searches related to ligands binding PfHsp90 protein and other chemical properties.

14.5.3 Rational Approach to Hsp90 Targeted Drug Discovery

Understanding of chemistry of a molecular scaffold or drug-like molecule is the first step towards a rational approach to drug discovery. For a new drug target; it becomes difficult to design a derivative due to limited knowledge about the available molecular scaffolds. It is also a known fact that all molecules that bind to a drug target may not be a potential drug candidate. The Lipinski rule of five provides some theoretical understanding of molecular structure but its application is limited due to the dynamic nature of therapeutic effects.

GA derivatives bind to PfHsp90 at the ATP binding site of the target. There are many hHsp90 derivatives that are derived from purine and ansamycin scaffolds that compete with ATP for binding to Hsp90.[38] Few of these molecules show considerably good binding affinity towards the protein but whether these molecules prove to be better antimalarial drugs remains an open question.

14.6 Future of PfHsp90 Inhibitors as Antimalarials

Human Hsp90 inhibitors are now being examined under advanced clinical trials. The most studied hHsp90 inhibitors are 17-AAG, AUY922, IPI504 and STA9090.[25] The idea of using Hsp90 from malaria as a drug target has grown significantly since its inception in 2000.[39] The structure–activity relationship for scaffolds that can bind selectively to the N-terminal domain of PfHsp90 is in its nascence. From the knowledge at hand it is very clear that PfHsp90 proves to be a very valuable drug target towards the next generation of antimalarial therapy. This idea has also served as the first proof of principle that not only drugs but also the drug targets can be repurposed or researched for their alternative therapeutic implications. Hence, cancer is not the only condition in which Hsp90 can be used as drug target.[39]

Hsp90 has been proven as a new drug target for a variety of diseases from scientific findings. The summary of the last three decades suggests that we have

a) a next-generation antimalarial drug target and its potential inhibitor
b) around 15 highly potent hHsp90 inhibitors in advanced stages of clinical trials, which can be explored for their efficacy as antimalarial drugs
c) use of Hsp90 as a drug target from various infectious micro-organisms, which can be explored for finding novel treatment regimens.

17-AAG was the first reported antimalarial molecule and PfHsp90 inhibitor in 2010 under both *in vitro* and *in vivo* conditions. Formulation of this molecule has been a major issue due to its highly hydrophobic characteristic. 17-Dimethylaminoethylamino-17-demethoxygeldanamycin (17-DMAG) is a water-soluble molecule and shares the molecular ring structure with 17-AAG. When this chapter was written this molecule was under Phase I clinical trial examination for cancer treatment. Substitution of 17-allyl amine side-chain by 17-demethoxyaminoethylamino created a molecule that is water soluble and closer to having drug-like characteristics. But its use as an antimalarial candidate remains unexplored.

References

1. M. H. Azizi and M. Bahadori, *Arch. Iran. Med.*, 2013, **16**, 131–135.
2. http://www.who.int/features/factfiles/malaria/en/.

3. http://www.cdc.gov/malaria/malaria_worldwide/impact.html.
4. J. H. Block, in *Organic Medicinal and Pharmaceutical Chemistry*, J. M. Beale and J. H. Block, Lippincott Williams and Wilkins, USA, 2004, pp. 282–298.
5. T. N. Wells, P. L. Alonso and W. E. Gutteridge, *Nat. Rev. Drug Discov.*, 2009, **8**, 879–891.
6. N. Kheliouen, F. Viwami, F. Lalya, N. Tuikue-Ndam, E. C. Moukoko, C. Rogier, P. Deloron and A. Aubouy, *Malar. J.*, 2010, **9**, 220.
7. L. Schwartz, G. V. Brown, B. Genton and V. S. Moorthy, *Malar J.*, 2012, **11**, 11.
8. M. A. Thera and C. V. Plowe, *Annu. Rev. Med.*, 2012, **63**, 345–357.
9. http://www.atsdr.cdc.gov/toxprofiles/tp.asp?id=81&tid=20.
10. T. L. Lemke, in *Foye's Principles of Medicinal Chemistry*, ed. T. L. Lemke and D. A. Williams, Lippincott Williams & Wilkins, USA, 2008, pp. 1084–1111.
11. M. Taipale, D. F. Jarosz and S. Lindquist, *Nat. Rev. Mol. Cell Biol.*, 2010, **11**, 515–528.
12. L. H. Pearl and C. Prodromou, *Annu. Rev. Biochem.*, 2006, **75**, 271–294.
13. S. R. Pavithra, G. Banumathy, O. Joy, V. Singh and U. Tatu, *J. Biol. Chem.*, 2004, **279**, 46692–46699.
14. G. Banumathy, V. Singh, S. R. Pavithra and U. Tatu, *J. Biol. Chem.*, 2003, **278**, 18336–18345.
15. R. Pallavi, N. Roy, R. K. Nageshan, P. Talukdar, S. R. Pavithra, R. Reddy, S. Venketesh, R. Kumar, A. K. Gupta, R. K. Singh, S. C. Yadav and U. Tatu, *J. Biol. Chem.*, 2010, **285**, 37964–37975.
16. R. Mout, Z. D. Xu, A. K. Wolf, V. Jo Davisson and G. K. Jarori, *Malar. J.*, 2012, **11**, 54.
17. R. Kumar, S. R. Pavithra and U. Tatu, *J. Biosci.*, 2007, **32**, 531–536.
18. K. D. Corbett and J. M. Berger, *Proteins*, 2010, **78**, 2738–2744.
19. D. Wider, M. P. Peli-Gulli, P. A. Briand, U. Tatu and D. Picard, *Mol. Biochem. Parasitol.*, 2009, **164**, 147–152.
20. S. L. Rutherford and S. Lindquist, *Nature*, 1998, **396**, 336–342.
21. D. F. Jarosz and S. Lindquist, *Science*, 2010, **330**, 1820–1824.
22. L. E. Cowen, S. D. Singh, J. R. Kohler, C. Collins, A. K. Zaas, W. A. Schell, H. Aziz, E. Mylonakis, J. R. Perfect, L. Whitesell and S. Lindquist, *Proc. Natl Acad. Sci. USA*, 2009, **106**, 2818–2823.
23. R. Kumar, A. Musiyenko and S. Barik, *Mol. Biochem. Parasitol.*, 2005, **141**, 29–37.
24. http://www.nytimes.com/1998/07/17/us/thalidomide-approved-to-treat-leprosy-with-other-uses-seen.html.
25. A. K. Rochani, M. Singh and U. Tatu, *Curr. Pharm. Des.*, 2013, **19**, 377–386.
26. C. DeBoer, P. A. Meulman, R. J. Wnuk and D. H. Peterson, *J. Antibiot. (Tokyo)*, 1970, **23**, 442–447.
27. K. Lee, J. S. Ryu, Y. Jin, W. Kim, N. Kaur, S. J. Chung, Y. J. Jeon, J. T. Park, J. S. Bang, H. S. Lee, T. Y. Kim, J. J. Lee and Y. S. Hong, *Org. Biomol. Chem.*, 2008, **6**, 340–348.

28. S. Eichner, H. G. Floss, F. Sasse and A. Kirschning, *Chembiochem.*, 2009, **10**, 1801–1805.
29. K. Jhaveri, T. Taldone, S. Modi and G. Chiosis, *Biochim. Biophys. Acta*, 2012, 742–755.
30. R. Bagatell, L. Gore, M. J. Egorin, R. Ho, G. Heller, N. Boucher, E. G. Zuhowski, J. A. Whitlock, S. P. Hunger, A. Narendran, H. M. Katzenstein, R. J. Arceci, J. Boklan, C. E. Herzog, L. Whitesell, S. P. Ivy and T. M. Trippett, *Clin. Cancer Res.*, 2007, **13**, 1783–1788.
31. S. Modi, A. Stopeck, H. Linden, D. Solit, S. Chandarlapaty, N. Rosen, G. D'Andrea, M. Dickler, M. E. Moynahan, S. Sugarman, W. Ma, S. Patil, L. Norton, A. L. Hannah and C. Hudis, *Clin. Cancer Res.*, 2011, **17**, 5132–5139.
32. S. Pacey, M. Gore, D. Chao, U. Banerji, J. Larkin, S. Sarker, K. Owen, Y. Asad, F. Raynaud, M. Walton, I. Judson, P. Workman and T. Eisen, *Invest. New Drugs*, 2010, **30**, 341–349.
33. E. A. Ronnen, G. V. Kondagunta, N. Ishill, S. M. Sweeney, J. K. Deluca, L. Schwartz, J. Bacik and R. J. Motzer, *Invest. New Drugs*, 2006, **24**, 543–546.
34. E. I. Heath, M. Gaskins, H. C. Pitot, R. Pili, W. Tan, R. Marschke, G. Liu, D. Hillman, F. Sarkar, S. Sheng, C. Erlichman and P. Ivy, *Clin. Prostate Cancer*, 2005, **4**, 138–141.
35. G. E. Brandt, M. D. Schmidt, T. E. Prisinzano and B. S. Blagg, *J. Med. Chem.*, 2008, **51**, 6495–6502.
36. http://tdrtargets.org.
37. Y. Wang, J. Xiao, T. O. Suzek, J. Zhang, J. Wang, Z. Zhou, L. Han, K. Karapetyan, S. Dracheva, B. A. Shoemaker, E. Bolton, A. Gindulyte and S. H. Bryant, *Nucleic Acids Res.*, 2011, **40**, D400–412.
38. E. B. Garon, R. S. Finn, H. Hamidi, J. Dering, S. Pitts, N. Kamranpour, A. J. Desai, W. Hosmer, S. Ide, E. Avsar, M. Rugaard Jensen, C. Quadt, M. Liu, S. M. Dubinett and D. J. Slamon, *Mol. Cancer Ther.*, 2013, **12**(6), 890–900.
39. E. Dolgin and A. Motluk, *Nat. Med.*, 2011, **17**, 646–649.

CHAPTER 15

Molecular Chaperones as Potential Therapeutic Targets for Neurological Disorders

MARION DELENCLOS AND PAMELA J. McLEAN*

Department of Neuroscience, Mayo Clinic, 4500 San Pablo Road, Jacksonville, FL 32224, USA
*Email: Mclean.pamela@mayo.edu

15.1 Molecular Chaperones and Neurodegenerative Disorders

15.1.1 Protein Aggregation and Molecular Chaperones

Within cells, proteins are continually degraded into amino acids and replaced by newly synthesized proteins. However, in the cellular environment these newly synthesized proteins are at great risk of aberrant folding and aggregation. Protein misfolding can lead to the formation of toxic substrates including oligomers, protofibrils and fibrillars deposits. Therefore protein quality control and the maintenance of proteostasis are crucial steps for cellular health. To handle a build-up of abnormal proteins, cells employ different processes and machinery. Molecular chaperones and their regulators (co-chaperones) are a group of molecules that contribute to the prevention of aggregation and enhance the efficiency of *de novo* protein folding.[1] Molecular chaperones are composed of several distinct classes of sequence-conserved proteins, most of which are stress inducible like heat-shock proteins (Hsps). The two major chaperone systems in mammals are Hsp70 and Hsp90 but several other families exist and are named on the basis of their molecular mass: Hsp100, Hsp60, Hsp40 and the small Hsp (sHsp), which weigh less than 40 kDA. In addition to molecular chaperones, cells harbor two other mechanisms for the degradation of misfolded proteins: the ubiquitin–proteasome system (UPS) system, which is critical for reducing the level of misfolded proteins, and the lysosome-mediated autophagy pathway, which plays an important role in the clearance of these proteins. The chaperone system is intimately associated with both the UPS and the autophagy pathway; collectively they insure the maintenance of homeostasis in cells.

15.1.2 Molecular Chaperones in Neuronal Disorders

Deficiencies in proteostasis have been shown to facilitate the manifestation and/or progression of numerous diseases.[2-4] Indeed under certain conditions the production of misfolded proteins exceeds the degradative capacity of the cell and aggregates accumulate in the cells at dangerous levels.[1,5] Such accumulation characterizes several neurodegenerative disorders known as "protein-conformational disorders". The conversion of a protein from its functional conformation into a misfolded and toxic conformation constitutes the molecular basis of such diseases, including polyglutamine (polyQ) tract expansion diseases, Alzheimer's disease (AD), Parkinson's disease (PD), Amyotrophic Lateral Sclerosis (ALS) and prion diseases. Although the protein aggregates in these disorders are different and unrelated, the common feature of each disease is the presence of amyloid-like structures enriched in β-sheet. This characteristic has given rise to the hypothesis that protein aggregates may disturb the cellular homeostasis and perturb neuronal function. Even though the significance of protein aggregation in neuronal toxicity and cell death remains controversial, the pathways of protein repair and degradation appear to be involved in at least some aspect of the pathogenesis of the diseases.

Given that molecular chaperones play an important role in the folding of nascent proteins and prevent stress-induced misfolding, a key role in the development of neurodegenerative disorders is anticipated. Molecular chaperones as well as components of the UPS are typically found in neuronal inclusions (Table 15.1) and recent evidence indicates that chaperones can be potent suppressors of neurodegeneration and therefore promising therapeutic targets.[1]

15.1.2.1 Polyglutamine Diseases

The polyglutamine (polyQ) diseases are a group of nine hereditary neurodegenerative diseases, including Huntington's disease (HD), spinal bulbar muscular atrophy (SBMA), denatorubral-pallidoluysian atrophy and six forms of spinocerebellar ataxia (SCA1-3, 6, 17),[6] that are caused by abnormal expansions of CAG trinucleotide repeats codons (encoding the amino acid glutamine) in unrelated disease-causative proteins. With the exception of SBMA, these

Table 15.1 Neurodegenerative disorders associated with molecular chaperones. Aβ: Amyloid-β, SOD1: superoxide dismutase 1, Prp: prion disease protein.

Neurodegenerative disease	Misfolded protein	Pathology	Aggregate-associated chaperones	References
Alzheimer's disease (AD)	Aβ peptides, τ	Extracellular seniles plaques, neurofibrilatory tangles	Hsp27, Hsp70, Hsp90, GRP78, αB-crystallin	38–41
Parkinson's disease (PD)	α-synuclein	Intracellular Lewy bodies	Hsp70, Hsp40, αB-crystallin	70, 71
Amyotrophic lateral sclerosis (ALS)	SOD1	Intracellular inclusions bodies	Hsp70, Hsp25, αB-crystallin	89, 92
Polyglutamine diseases:				
Huntington's disease (HD)	Huntingtin (Htt)	Cytoplasmic and nuclear inclusions	Hsp40, Hsp70	12, 14, 20
Spinocerebellar ataxias (SCA1-3,7)	Ataxins			
Spinal and bulbar muscular atrophy (SBMA)	Androgen receptor			
Prion disease	Prp	Extra- and intracellular aggregates	Hsp104, Hsp70	106, 104

neurodegenerative disorders are inherited in an autosomal dominant manner. A characteristic feature of polyQ diseases is the formation of insoluble, granular and fibrous deposits in affected neurons leading to a progressive neuronal dysfunction and eventual neuronal loss. However, the diseases present with distinct pathologies since they affect different populations of neurons in different regions of the brain.[6,7] The symptoms of all nine disorders manifest in advanced age. They are progressive and worsen throughout the course of the disease. The diseases display distinct, yet overlapping, clinical and pathological findings. However, in each disease the phenotype and severity can vary greatly because of differences in repeat length. In general, the pathologic length is about 40 glutamines in the disease-proteins, with individuals expressing polyQ repeats with fewer residues not developing disease. Interestingly, the length of the CAG expansion is inversely correlated with age of disease onset; a longer polyQ expansion results in earlier onset of the disease and a more severe phenotype.

A number of investigators have suggested that unusually long polyQ tracts induce diseases because they interfere with the normal function of cellular proteins.[8,9] It remains controversial whether intranuclear inclusions (NI) contribute to the pathogenesis, are neuroprotective or are simply a by-product of the disease process;[10,11] however, several lines of evidence clearly link molecular chaperones and polyQ inclusions. In fact, polyQ proteins have been shown to interact with the Hsp70 and Hsp40 chaperone families.[12–14] Both Hsp70 and Hsp40 co-localize within NI in cellular and animal models of polyQ disease and also in patient tissue samples. The accumulation of chaperones in polyQ aggregates suggests that insufficient protein folding and degradation is implicated in the pathogenesis of polyQ disease.[15]

The work of Cummings and colleagues[12] in 1998 raised considerable interest in the role of chaperones in the polyQ neurotoxicity. In their study, the over-expression of HDJ-2/HSDJ, a member of the Hsp40 family, prevented aggregate formation of the SCA-1 protein in HeLa cells. Subsequent studies have examined the effect of over-expression of chaperones on NI formation and on the toxicity of pathogenic polyQ expansion. Experiments performed in yeast, fly or cellular models of polyQ diseases have shown the protective effects of chaperones on polyQ diseases. Over-expression of Hsp70 alone or with Hsp40 in HD, SBMA and SCA models have been reported to decrease the aggregate formation[16–18] and also to alter the biochemical properties of fibrillar aggregates such that detergent soluble amorphous structures are generated.[19] Furthermore it was found that over-expression of Hsp70 and Hsp40 chaperones acts synergistically to rescue neurodegeneration.[17,20] The increased level of expression of Hsp70 in a drosophila polyQ model was able to compensate for the deletions in the Hsp40 gene dHdj1, establishing a relationship between the two chaperones. In contrast, expression of a dominant negative mutant form for Hsp70 has been shown to increase polyQ toxicity.[21,22] Thus, expression of Hsp70 alone has a positive effect on the suppression of aggregates and a combination of Hsp70 and Hsp40 may be accumulative. In the R6/2 transgenic mouse model of HD and spinocerebellar ataxia, a progressive decline in Hsp70 and Hsp40 has been measured in brain tissue over time linking the chaperone's activity to disease progression.[23] However, the over-expression of Hsp70 in this mouse model had no effect on the neuropathological phenotypes.[24] On the other hand, in several other mouse models of polyQ diseases, such as SCA1 mice or an SMBA murine model, the chaperone Hsp70 could modulate neurodegeneration and, most importantly, could restore some behavioral phenotypes.[25,26]

Interestingly, it seems that only specific subclasses of molecular chaperones affect polyQ aggregation and toxicity. Using a *C. elegans* model, the Q35-YFP aggregation model, the entire genome of *C. elegans* was screened to find enhancers of polyQ aggregation by RNAi.[27] Only two members of the Hsp70 family and one member of the DnaJ domain protein (Hsp40 homologue) were identified as suppressors of aggregation among all the chaperone genes analyzed.[27–29] In addition to chaperones, other genes identified affected polyQ-induced aggregation, including those that have roles in RNA synthesis, splicing and processing, and in protein synthesis, transport and degradation.

Finally, all of the findings to date validate a role for molecular chaperones and their components in the development of polyQ diseases. Regulating their expression seems a promising target for a possible therapy for polyQ disorders.

15.1.2.2 Alzheimer's Disease

Alzheimer's disease (AD) is the most common neurodegenerative disorder and the most common form of dementia in the elderly. The likelihood of developing AD increases substantially after age 70 and may affect around 50% of people over the age of 85. Brain alterations arise in most cases sporadically (for unknown reasons) whereas only 2% of the cases represent familial forms of the disease. AD is characterized by a progressive decline of cognitive functions and the degeneration and loss of cholinergic neurons and synapses throughout the brain is a major pathological hallmark. Neuronal loss is most prominent in the basal forebrain, amygdala, hippocampus and cortical areas. Pathologic lesions of AD are characterized by the extracellular accumulation of senile plaques composed of numerous aggregated proteins with amyloid β-peptide (Aβ) as the main component. The other major pathological feature of AD is the presence of intracellular accumulations of hyper phosphorylated microtubule-associated protein Tau (τ), into both non-filamentous and filamentous inclusions known as neurofibrillary tangles (NFTs). The exact roles played by extracellular Aβ and intracellular τ are still elusive, but several lines of evidence from transgenic mouse models of AD demonstrate that Aβ initiates cellular dysfunction before the accumulation in senile plaques.[30–32] Aβ peptides are generated by the sequential action of three different groups of enzymes, α-, β- and γ-secretases on amyloid precursor protein (APP). Processing of APP by secretases occurs in several intracellular compartments including endoplasmic reticulum(ER)/golgi apparatus, endosomes and lipids rafts.[33–35] Several Aβ variants have been described that differ in their length, the most abundant being the Aβ peptide that is 40 residues long (Aβ 40), whereas a small proportion of the Aβ is 42 residues long (Aβ 42). Moreover, it has been shown that accumulation of Aβ *in vivo* can inhibit the proteasome, which leads to a build-up of abnormally phosphorylated τ protein.[36,37] Indeed, hyper phosphorylated τ found in pathogenic conditions has a propensity to aggregate and lose its ability to maintain the stability of the axonal microtubules, leading to disturbances in protein trafficking. Thus both Aβ and τ may be causally linked to AD. Aberrant processing of APP and modifications affecting τ may generate species that are susceptible to aggregation and are neurotoxic.

The involvement of chaperones in the pathogenesis of AD is well documented. When the clearance mechanisms become overwhelmed, insoluble fibrils and inclusions are formed that can include Hsps.[38] Histopathology studies and analyses of *post mortem* AD brains have shown elevated levels of Hsp27 and Hsp90 in specific regions such as hippocampus and cortical areas.[39–41] In cellular models it has been shown that a direct interaction between chaperones and Aβ might regulate the formation of toxic Aβ species. In fact, Yang and co-workers[42] have shown that the ER chaperone BIP/GRP78 (the ER isoform of Hsp70) was able to bind to APP and modulate its maturation and processing. Moreover, when GRP78 is over-expressed, APP maturation fails and the levels of Aβ40 and Aβ42 released into the medium decrease.[42–44] GRP78 might retain APP in the ER and/or protect APP from cleavage by secretase enzymes.[1,42]

αB-crystallin, a member of the sHsp family, was also shown to directly interact with Aβ in cell culture and worm models that express human Aβ intracellularly.[45–47] However, in this case the chaperones were not shown to increase the neurotoxicity of Aβ, possibly by preventing its aggregation into insoluble fibrils.[45,48] Another member of the sHsp family, Hsp27, appears to be associated with Aβ and NFTs. Hsp27 directly interacts with Aβ and inhibits fibril formation *in vitro*.[49] Hsp27 also interacts preferentially with hyperphosphorylated τ in brain samples and cell culture, and attenuates its toxicity by facilitating its degradation.[50] Interaction between τ, Hsp70 and Hsp90 has also been established. Dou and colleagues[51] reported an inverse relationship between aggregated τ and the levels of Hsp70/90 in τ transgenic mouse and AD brains. Increased levels of Hsp70 and Hsp90 promote τ solubility and τ binding to microtubules. Conversely, decreased levels of Hsp70 and Hsp90 resulted in the opposite effect. Lastly, a complex of Hsp70 and the co-chaperone CHIP (carboxy terminus of Hsp70-interacting protein) can regulate τ degradation. The co-operative activity of CHIP and Hsp70 decreases τ-mediated toxicity by facilitating the ubiquitylation of phosphorylated τ and leading to selective elimination of abnormal τ species.[52–54]

15.1.2.3 Parkinson's Disease

Parkinson's disease (PD) is the most common neurodegenerative disorder after AD. It affects 0.1% of the world's population and the average onset of the disease is between 50 and 60 years of age.[55] PD is characterized by a complex motor disorder known as parkinsonism, which is manifested by resting tremor, bradykinesia, rigidity and postural abnormalities. These manifestations are mainly due to the loss of dopaminergic neurons in the midbrain, which results in dysfunction of the nigrostriatal pathway. However, there is increasing evidence that PD is no longer a pure motor disorder but a more complex disorder that includes non-motor symptoms (NMS) such as sleep disorders, depression and gastrointestinal disturbances.[56,57] The pathological hallmark of PD is the presence of neuronal cytoplasmic inclusions called Lewy bodies (LBs). LB pathology is widespread in PD and it has been suggested that the pathology starts outside the central nervous system (CNS) in the lower brainstem nuclei and then progresses in an ascending path through the brainstem to the cortical areas.[58–60]

In PD, the majority of cases are sporadic but over the last 10 years several genes have been identified that are associated with familial forms of the disease. A key discovery in understanding PD was made when a mutation in the human α-synuclein gene (SNCA) was identified.[61] Now three point mutations (A30P, E46K, A53T) in α-synuclein (AS) are associated with rare familial form of PD[61,62] and duplication or triplication of the gene lead to familial form of the disease.[63–65] In animal models, over-expression of AS leads to its aggregation and toxicity.[66–68] Lastly, AS is abundant in LBs.[69] Interestingly, LBs are also positive for chaperone proteins like Hsp70, Hsp40 and Hsp27.[70,71] In the first *post-mortem* pathological studies that explored chaperones in PD, accumulation of αB-crystallin (sHsp family) or Hsp27 were found in neurons of PD patients but not in control cases.[72,73] The distribution of αB-crystallin positive neurons overlapped with the severity and regional spread of LB pathology but were not restricted to LB bearing neurons.[73,74] Subsequently, several studies have investigated the effects of molecular chaperones on AS aggregation and its toxicity. In cell culture models transfected with wild-type AS and co-transfected with synphilin-1 (AS interacting protein), the overexpression of Hsp70 or co-chaperone HDJ1 (Hsp40 family) decreased the number of cells that contained inclusion bodies by more than 50%.[71] In a similar study, Hsp70 over-expression reduced the amount of misfolded, aggregated AS, suggesting that the chaperone might enhance refolding and/or promote degradation of AS.[1,75] *In vivo*, crossbreeding AS transgenic mice with Hsp70 over-expressing mice[75] resulted in a reduction of the high molecular weight of AS. Thus, chaperones can modify AS aggregation *in vitro* as well as *in vivo*.[71,75,76] In a PD fly model, expression of wild-type or mutant AS resulted in inclusion body formation and neuronal loss. The co-expression of Hsp70 in this model was able to protect against AS-induced neurodegeneration,[70] although no effect was found on inclusion body formation. Two co-chaperones also seem to be intimately involved in PD pathology. CHIP (co-chaperone Hsp70 interacting protein) immunoreactivty is observed in LBs in *post-mortem* brain tissue and in a human neuroglioma cell line CHIP co-localizes with AS and Hsp70 in intracellular inclusions.[77] Furthermore, over-expression of CHIP inhibits AS inclusion formation and reduces AS protein levels. Finally, Hip (Hsp70 interacting protein), also known as ST13, was found under-expressed in PD patients.[78]

15.1.2.4 Amyotrophic Lateral Sclerosis

Amyotrophic lateral sclerosis (ALS), or Lou Gehrig's disease, is a fatal adult onset neurodegenerative disorder that impacts 1–2 people out of 100,000 per year.[79] The disease affects the upper and lower motoneurons in the brain and the spinal cord resulting in gradual muscle weakening and loss of motoneuron function, leading to paralysis and death.[80] The pathologic hallmarks of ALS are motoneuron atrophy, swelling of perykaria and proximal neurons and the presence of Bunina bodies, which are small eosinophilic intraneuronal inclusions in the remaining lower motor neurons. The vast majority of ALS cases are sporadic, with only 10% of cases being inherited familial cases (FALS). The exact mechanism of the disease is currently unknown and no effective treatments are available. However, the discovery of genetic mutations in FALS has provided invaluable insight into the pathogenesis of the disease. A major discovery was made 10 years ago when 20% of FALS were linked to a mutation in the Cn/Zn Superoxide Dismutase 1 (SOD1) gene.[81] Since then more than 100 different mutations in SOD1 have been made and 4 more genes have been reported to cause ALS.[79,82] With this breakthrough cellular models and transgenic mice over-expressing SOD1 have been developed to investigate the pathogenesis of this devastating disease. Functional deficits in axonal transport have been observed in SOD1 mice.[83,84] Also one prominent feature in the spinal cord of animals affected by ALS-like disease is related to alterations in protein folding and degradation pathways.[80] Indeed, as with many neurodegenerative disorders, intracellular protein aggregates are the hallmark of ALS. Proteinaceous aggregates of SOD1 are found in *post-mortem* brains of FALS[85,86] and have also been observed in the cytoplasm of cultured primary motoneurons of SOD1 mutant[87] and motoneurons of transgenic mice.[88] Evidence indicates that some toxic gain of function is responsible for motoneuron loss rather than decrease of activity of mutant SOD1.[89,90] Altered protein conformation of SOD1 may contribute to its toxic function by forming insoluble species or abnormal protein–protein interactions. Thus, alterations in protein chaperones that regulate protein folding and degradation pathways may play a role in the cascade of events leading to neurodegeneration.

In primary motoneurons expressing mutant SOD1, the co-expression of Hsp70 was able to prolong the viability of the cells and reduced the formation of aggregates compared to untransfected cells.[89] Moreover, in another cellular model of ALS, Takeuchi and colleagues[91] have shown a synergistic effect of Hsp70 and Hsp40 over-expression on neurite outgrowth and a reduction of intracytoplasmic aggregates. Additionally, in the mutant SOD1 transfected NIH3T3 cells line, an increase in survival correlated with an up-regulation in Hsp70, Hsp25 and αB-crystallin chaperone proteins.[92] Interestingly in this study the authors demonstrated that the Hsps co-immunoprecipitate with mutant SOD1. Since then Hsp105, Hsp40 and co-chaperone CHIP have all been shown to form a complex with mutated SOD1 but not with the wild-type SOD1.[80] A direct interaction between chaperones and mutant SOD1 could prevent aggregation and mediate neuroprotection. In addition, altered level of expression of Hsp25, Hsp27 and αB-crystallin is observed in transgenic mice model of ALS.[93] The level of chaperones is unchanged in pre-symptomatic SOD1 mice; however, in the late stage of the disease there is an absence of Hsp27 and αB-crystallin immunoreactivity in the motoneurons as well as a down-regulation of Hsp25.[94,95] The absence of a subgroup of chaperones during the later stages of the disease is likely to contribute to the motoneurons loss. Thus, depletion of chaperones could result in formation of SOD1 aggregates, increasing the vulnerability of the motoneurons to stress factors, axonal transport defects or others factors implicated in ALS pathogenesis.[89]

15.1.2.5 Prion Diseases

Prion diseases or transmissible spongiform encephalopathies (TSEs) are fatal neurodegenerative disorders and include Creutzfeldt–Jakob disease (CJD) in humans, or bovine spongiform encephalopathy (BSE) in cattle. The diseases can be sporadic, inherited or acquired by infection and are all associated with the misfolding of normal host-encoded protein, the cellular prion protein, PrP^C.[96–98] A prion is an infectious protein that is naturally transmissible and replicates into the cell. In prion diseases, PrP^C is subject to conformational modification from a soluble form to a pathogenic form that is aggregated and proteinase resistant (PrP^{SC}). PrP^{SC} becomes infectious and induces the same conformation changes as the host-encoded PrP^C. PrP^{SC} aggregates and accumulates in the nervous system, in particular in the neocortex, cerebellum and subcortical nuclei, leading to a rapid neurodegeneration.[99,100] The conformational change leading to infection and propagation of the disease suggests that molecular chaperones may be involved in the folding of PrP^C.[101,102] The discovery of prions in yeast, *Saccharomyces cerevisiae*, has been a breakthrough and has provided information on the basic mechanisms underlying prion assembly into amyloid-like fibrils and inheritance of the prion state. Moreover, many molecular chaperones are functionally conserved from yeast to humans, therefore studying how molecular chaperones modulate prion propagation in yeast yields substantial mechanistic insight in prion diseases.[103–105]

The role of chaperones in prion propagation was first demonstrated in the yeast prion [PSI$^+$]. Over-production or deletion of Hsp104, a yeast member of the Hsp100/ClpB family of AAA (ATPases associated with various cellular activities), was shown to cause the loss of [PSI$^+$].[106] However, the over-expression of Hsp104 could not prevent the formation of *de novo* [PSI$^+$].[104,107] In two other yeast prions, [PIN$^+$] and [URE3], the deletion or mutation of Hsp104 resulted in an inability of propagation of the prion, although over-expression of Hsp104 did not affect the stability of either [PIN$^+$] or [URE3].[107,108] Finally the interaction of Hsp104 with a second chaperone, GroEL protein, in a cellular model promoted the conversion of PrP^C to its protease-resistant state PrP^{SC}.[109] Hsp70 molecular chaperones also regulate prion propagation; in yeast they exist in a number of different variants with the Ssa family (Stress seventy subfamily A) representing the main class. Different effects and functions are found throughout the different classes. Mutations in Ssa1 destabilize [PSI$^+$] prion propagation while the Ssb family appears to cure the cells of [PSI$^+$] prion.[110–112] Interestingly, over-expression of Ssa1 has been shown to cure yeast of the prion [URE3] but not Ssa2p.[113,114] Hsp70 chaperone activity is tightly coordinated by the co-chaperone Hsp40. Hsp40 and Hsp70 chaperones co-operate with Hsp104 to refold aggregated proteins.[115] The Hsp40 class chaperone Ydj1p, along with the Hsp70 class Ssa1p, can renature prion proteins and overproduction of Ydj1p results in a complete loss of [URE3].[108] Subsequently, several studies have emphasized a fundamental role of Hsp40 in prion assembly and propagation.[105,116–118]

15.2 Modulation of Neurodegeneration by Molecular Chaperones

Molecular chaperones may contribute to the suppression of aggregates and the accumulation of damaged proteins. It has become more and more clear that there are several diseases in which chaperones have a causative role and where a cytoprotective effect of molecular chaperones is observed. In neurodegenerative diseases, chaperone over-expression has been shown to modify the phenotype of the disease in several models, possibly as a consequence of the ability of the chaperones to influence protein aggregation resulting in altered solubility of the mutant protein. Moderate over-expression of molecular chaperone results in an extended life span in *S. cerevisiae* or in fly;[119,120] yet, extreme over-expression of molecular chaperones has some deleterious effect on the growth of *C. elegans*[121] and undesirable effects such as alteration in cell cycle regulation are observed.[122] Still, therapeutic methods to influence chaperone responses and exploit the potential of these proteins to fold non-native protein or to initiate the degradation of abnormal proteins are needed.

15.2.1 Pharmacological Up-regulation of Molecular Chaperones

15.2.1.1 Hsp90 Inhibitors

Hsp90 is a molecular chaperone playing an important role in the pathogenesis of neurodegenerative disorders. Several studies suggest a role for Hsp90 in maintaining a functional stability of neuronal proteins over-expressed or mutated. Hsp90 is also well known to regulate the activity of transcription factor heat-shock factor-1 (Hsf1). In humans three heat-shock factors (Hsf) have been identified (Hsf1, Hsf2 and Hsf4) to play a role in the transcriptional control of Hsp expression with Hsf1 being the dominant factor controlling cellular responses to diverse stress. Under normal growth conditions, Hsf1 activity is repressed and exists in either the cytosol or nucleus in an inert monomeric state.[123] Heat shock and other stresses cause depression of Hsf1, initiating the events that lead to the appearance of the transcriptionally active form. Actually Hsp90 negatively regulates the activity of Hsf1 and the inhibition of Hsp90 activity leads to its recruitment.[124] The increase of activity of Hsf1 further leads to increased expression of Hsps such as Hsp70.

Hsf1 activation by Hsp90 inhibitors is well documented in neurodegenerative disease models. The antibiotic geldanamycin (GA) is a naturally occurring Hsp90 inhibitor. It acts by replacing ATP in the ATP binding pocket in the N-terminal domain of the chaperone.[125] GA has been found to induce an increase of Hsp70 in an AD cell model and therefore reduce the amount of insoluble τ proteins.[51] GA has also been well characterized in PD models. Positive effects on dopaminergic neuron cell death and prevention of AS aggregation were observed in PD cell culture.[126] Furthermore, the antibiotic fully protected against AS toxicity in drosophila model and in the 1-methyl-4-pheny-1,2,3,6-tetrahydropyridine (MPTP)-induced dopaminergic neurotoxicity mouse model of PD.[127,128] Lastly, inhibition of aggregate formation is also observed in HD and ALS models.[129,130] Unfortunately some properties of GA limit its clinical use. The compound does not cross the blood-brain barrier, has poor aqueous solubility and significant liver toxicity has been reported.[131] To counteract these effects several analogues have been designed. Among them, 17-AAG (17-allylamino-17-demethoxygeldanamycin) was effective against neurodegeneration in polyQ diseases such as HD or SBMA[132,133] and was able to up-regulate Hsp70 while reducing AS-induced toxicity in cellular model of PD.[134] Although 17-AAG has a higher affinity for Hsp90 and has less toxicity than GA, it has limited oral availability and causes hepatotoxicity in cancer trials.[135] Finally, radicicol, a natural antifungal, is also an Hsp90 inhibitor that has the same functional mechanism as GA but a 50-fold greater affinity for this chaperone.[131]

Other brain-permeable molecules with potent Hsp90 inhibitor activity are available and may potentially be used in therapy for neurodegenerative disorders. The SNX compounds derived from SNX-2112 represent a class of novel synthetic molecules chemically different from GA and its derivatives. SNX agents are orally available and display excellent blood-brain barrier permeability. They selectively inhibit Hsp90 even though their mechanism of action needs to be further elaborated. Moreover in a cellular model of PD these novel compounds could rescue AS-induced toxicity and decrease oligomerization in a dose-dependent manner at a lower dose than 17-AAG.[134] SNX-0723, was identified as a promising candidate because a preventive effect on AS oligomerization was observed and *in vivo* pharmacokinetic studies demonstrated excellent bioavailability and robust brain absorption.[74]

15.2.1.2 Modulation of the Heat-shock Transcription Factor 1

Until recently the activators of the transcription factor were compounds that promote Hsf1 activation through the inhibition of Hsp90. However Hsp90 is a major chaperone involved in many functions such as cell growth, signaling or proliferation. Thus, inhibition of this chaperone may have some deleterious effects on the whole organism. The expression of an activated form of the transcriptional factor has been shown to be protective in several models of protein misfolding. In a mouse model of HD disease, over-expression of Hsf1 suppresses polyglutamine aggregates and cell death.[136] Conversely, Hsf1$^{-/-}$ mice exhibited a shorter lifespan compared to control littermates, and in yeast cells expressing a mutated Hsf form, prion aggregate formation was observed.[137,138] Thus pharmacological chaperones activating human Hsf1 represent a potential therapy for neurodegenerative disorders. Novel compounds that activate Hsf1 without inhibiting Hsp90 have been screened. Arimoclomol is one candidate that can prolong the activation of Hsf1 and up-regulate the expression of Hsp70. Some interesting results were found in a transgenic ALS mouse where chronic arimoclomol treatment decreased the progression of the disease. It improved behavioral phenotypes and prevented neuronal loss resulting in a 22% increase in lifespan.[139] Another compound that is able to activate Hsf1 and up-regulate Hsp gene expression is celastrol. This natural product from southern China has shown some neuroprotective effects in *in vivo* models of AD, PD and ALS. Acute treatment with celastrol in Aβ transgenic mice reduces the β-amyloid pathology[140] and chronic treatment in SOD1 mice significantly increases the motor performance and delays the onset of ALS.[141] Lastly celastrol attenuated MPTP-induced cell death and partially restored striatal dopamine levels in the MPTP mouse model of PD.[142] Originally developed as an anti-ulcer drug, gerangylgeranylacetone (GGA) has shown some interesting effects on neurodegeneration as well. Indeed, GGA increased the level of Hsp70 and Hsp90 and inhibited the cell death in a SBMA cellular model and oral administration of the drug in SBMA transgenic mice suppressed the accumulation of the pathogenic protein.[143] It has been hypothesized that GGA promotes the expression of protein kinase C mediated phosphorylation of Hsf1 leading to increased expression of Hsp70.[144,145] Recently another molecule, Hsf1a, has been identified in a high-throughput screen.[146] Hsf1a is structurally different from the other characterized Hsf1 activators, yet the molecule is able to activate Hsf1 increasing the level of Hsp70 without binding or affecting Hsp90. Furthermore Hsf1a reduces cytotoxicity and suppresses aggregation in a fly model and a cellular model of polyQ disease.[146]

15.2.1.3 Hsp70 ATPase Modulators

As described previously in this chapter, chaperone and cytoprotective activities of Hsp70 represent a therapeutic target for neurodegenerative disorders. It is well known that the Hsp70 chaperone system can prevent and eventually reverse protein misfolding. Until recently pharmacological research has focused on the modulation of the level of Hsp70. However, Hsp70 is a chaperone that binds and releases polypeptides in an ATP-dependent cycle, therefore regulation of ATPase function could have a potential effect on amyloid formation. Due to low intrinsic ATP activity, modulators of Hsp70 have been challenging to identify, but in 2008 Chang and colleagues[147] observed several inhibitors and activators using a sensitive high-throughput screening assay. Two distinct chemical classes of inhibitors were observed, the benzothiazines and the flavones, and they were characterized in a cellular model of tauopathy. In HeLa cells over-expressing τ protein, treatment with compounds from the benzothiazines family led to a significant reduction of total τ level.[148] However, in a PD and ALS model the Hsp70 ATPase modulation did not affect the level of the disease-related proteins. Thus the effect of these drugs seems to be selective for τ regulation. The same authors also observed that the increased level of Hsp70 could improve ATPase inhibitor efficacy. A therapy where Hsp70 levels are induced in a first step and then followed by inhibition of its ATPase activity may be very effective. The ATPase domain of Hsp70 could be a potent target for the pharmacology of AD and tauopathies. In fact methylene blue, a benzothiazine compound, is entering Phase III trials for treatment of AD patients.[149]

15.2.2 Chemical Chaperones

Chemical chaperones are low-molecular-mass molecules with the ability to mimic chaperone function. They increase the stability of native proteins in a non-selective manner and rescue mutant proteins from aggregation. Most chemical chaperones are osmotically active and include the organic solvent dimethyl sulfoxide (DMSO), or cellular osmolytes glycerol, trimethylamine with trimethylamine N-oxide or trehalose and some amino acid derivatives, although other compounds such as 4-phenybutyric acid (PBA) or detergents are also members of the chemical chaperones group. The exact mechanism of action of these compounds is not fully understood, although it has been suggested that they could prevent non-productive interactions with other resident proteins and alter the activity of endogenous molecular chaperones to assist the transportation of mutant proteins to the appropriate intracellular or extracellular destination.[150,151] Numerous studies have reported their beneficial effects in different neurodegenerative disease models. Trehalose, a natural disaccharide, is accumulated under cellular stress[152] and has been shown to protect proteins from denaturation and aggregation.[153] In the R6/2 mouse model of HD, trehalose was able to inhibit polyQ aggregation[154] and oral administration led to improvement of motor dysfunction and extended lifespan of the animals. In support of this, positive effects on aggregation and its associated toxicity were reported in cellular models of AD,[155] prion[156] and SCA diseases.[157] Lastly, the deletion of a gene encoding the protein involved in trehalose synthesis enhances AS toxicity in yeast.[29] Due to the fact that PBA and tauroursodeodeoxycholic acid (TUDCA) can pass through the blood-brain barrier, these two chemical chaperones have also been investigated as potential protective agents in neurodegenerative disorders.[151,158] In a mouse model of PD, generated by transgenic over-expression of mutant AS, the administration of PBA resulted in improvement of the motor deterioration.[159] Furthermore, in a rotenone mouse model of PD, the authors observed a protective effect of PBA on nigral dopamine neurons.[160] In addition, TUDCA has also been shown to exert chaperone activity and could prevent Aβ peptide-induced neuronal death.[161]

Chemical chaperones are widely used in experimental systems;[162] however, the majority of these chaperones require high concentrations in order to induce a positive effect on the aggregates and to be able to appreciate the reduction of toxicity. Moreover, chemical chaperones are not selective molecules and generate a widespread effect on the whole organism. Thus, their use in human patients is limited, especially for chronic administration. At present, only the two chaperones PBA and TUDCA are approved by the US Food and Drug Administration (FDA) for use in humans.

15.2.3 Viral Mediated Strategies

Molecule-based therapies can sometimes be challenging and limiting due to poor stability *in vivo*, low bioavailability or lack of cellular uptake. Therefore, gene therapy may provide a new strategy to up-regulate the expression or the activity of molecular chaperones in targeted neurons vulnerable to neurodegeneration. New advances in recombinant viral technology have resulted in feasible gene therapy applications for delivery of large therapeutic molecules. The PD field has been pioneering in this respect and a Phase I study of viral delivery of aromatic amino acid decarboxylase (AACD) in PD patients provided proof of principle for the use of viral delivery systems in PD.[131,163] Recently, a randomized controlled double-blind clinical trial has been successful with the gene transfer of GAD (glutamic acid decarboxylase) in PD patients.[164] In pre-clinical studies, viral-mediated Hsp70 expression has provided some promising results. In the MPTP mouse model of PD, Hsp70 gene transfer to the striatal dopaminergic neurons inhibits MPTP-induced neurodegeneration[165] and in a rat model of PD, AAV (Adeno Associated Virus) vector-mediated over-expression of Hsp70 protected against dopaminergic cell death.[166] Recombinant viral vectors have also been used to investigate neuroprotection by expression of Hsp70 interacting proteins. Hsp104, a molecular chaperone in the AAA+ family of ATPase, can synergize with Hsp70 and rescue aggregated proteins.[167] Co-injection of Hsp104 and mutant AS in the substantia nigra of rats using a lentiviral vector reduced formation of AS inclusions and prevented cell death.[168] Finally, use of adeno-associated viral gene delivery systems has also been studied in *in vitro* models of SBMA. The expression of Hsj1 (Hsp40 family) *via* AAV was highly effective at reducing aggregation *in vitro*.[169] Furthermore, in the same study the authors used a polyQ disease rodent model where the polyglutamine-mediated inclusion formation was dramatically reduced when neurons were transduced with a lentivirus expressing Hsj1a. These experiments are promising and validate gene therapy as a new therapeutic approach for neurodegenerative disorders.

15.3 Concluding Remarks and Perspectives

To date abundant experimental evidence implicates the molecular chaperone system in the pathogenesis of neurodegenerative disorders. All the experimental results reviewed in this chapter document the crucial role played by chaperones and more specifically heat-shock proteins (Hsps) in amyloid formation. Converging data demonstrate that molecular chaperones have the capacity to rescue protein aggregation and can facilitate the clearance mechanisms. Molecular chaperones may convert proteins causing disease from toxic conformations to non-toxic species that can be tolerated by the neurons. It has become evident that protein aggregation leading to neurotoxicity is caused by an impairment of protein quality control and specifically by the decreased efficiency of Hsps to assist in protein folding. How molecular chaperones rescue this conformation is still uncertain and more experimentation is needed to clarify their precise role in neurodegenerative disorders. Furthermore, it is noteworthy that the neuroprotective chaperone effects observed *in vivo* or *in vitro* have no visible effect on neuronal inclusion formation.

As detailed in the first part of this chapter, the action of Hsp extends to various neurodegenerative disorders. The impact of molecular chaperones on protein misfolding diseases seems to be involved in more than one protein-causing disease. In this regard it will be interesting to know if molecular chaperones act with a common mechanism in these conformational disorders. Answering this question will be a great benefit for the design of effective therapies. Several therapeutic methods to influence chaperone responses and modulate neurodegeneration have been summarized in this chapter. Although not all the attempts were successful and are still far from being applicable to patients, they provide a rationale for the development of novel chaperone therapeutic strategies. Therapies should aim to inhibit or reverse the conformational changes in the proteins associated with diseases and up-regulating

Hsps in order to prevent aggregation and/or reverse misfolding has proven to be effective. In conclusion, by gaining insight into the mechanisms of molecular chaperones new effective strategies will certainly come to light and a new door could open into the clinical research of neurodegenerative disorders.

References

1. P. J. Muchowski and J. L. Wacker, *Nat. Rev. Neurosci.*, 2005, **6**, 11–22.
2. Y. A. Lam, C. M. Pickart, A. Alban, M. Landon, C. Jamieson, R. Ramage, R. J. Mayer and R. Layfield, *Proc. Natl Acad. Sci. USA*, 2000, **97**, 9902–9906.
3. N. F. Bence, R. M. Sampat and R. R. Kopito, *Science*, 2001, **292**, 1552–1555.
4. K. Lindsten, F. M. deVrij, L. G. Verhoef, D. F. Fischer, F. W. van Leeuwen, E. M. Hol, M. G. Masucci and N. P. Dantuma, *J. Cell Biol.*, 2002, **157**, 417–427.
5. A. B. Meriin and M. Y. Sherman, *Int. J. Hyperthermia*, 2005, **21**, 403–419.
6. H. T. Orr and H. Y. Zoghbi, *Annu. Rev. Neurosci.*, 2007, **30**, 575–621.
7. A. R. La Spada and J. P. Taylor, *Neuron*, 2003, **38**, 681–684.
8. S. W. Davies, M. Turmaine, B. A. Cozens, M. DiFiglia, A. H. Sharp, C. A. Ross, E. Scherzinger, E. E. Wanker, L. Mangiarini and G. P. Bates, *Cell*, 1997, **90**, 537–548.
9. C. A. Ross, R. L. Margolis, M. W. Becher, J. D. Wood, S. Engelender, J. K. Cooper and A. H. Sharp, *Prog. Brain Res.*, 1998, **117**, 397–419.
10. G. Bates, *Lancet*, 2003, **361**, 1642–1644.
11. A. Michalik and C. Van Broeckhoven, *Hum. Mol. Genet.*, 2003, **12**(2), R173–186.
12. C. J. Cummings, M. A. Mancini, B. Antalffy, D. B. DeFranco, H. T. Orr and H. Y. Zoghbi, *Nat. Genet.*, 1998, **19**, 148–154.
13. D. L. Stenoien, C. J. Cummings, H. P. Adams, M. G. Mancini, K. Patel, G. N. DeMartino, M. Marcelli, N. L. Weigel and M. A. Mancini, *Hum. Mol. Genet.*, 1999, **8**, 731–741.
14. N. R. Jana, M. Tanaka, G. Wang and N. Nukina, *Hum. Mol. Genet.*, 2000, **9**, 2009–2018.
15. C. Landles and G. P. Bates, *EMBO Rep.*, 2004, **5**, 958–963.
16. Y. Kobayashi and G. Sobue, *Brain Res. Bull.*, 2001, **56**, 165–168.
17. N. M. Bonini, *Proc. Natl Acad. Sci. USA*, 2002, **99**(4), 16407–16411.
18. S. Krobitsch and S. Lindquist, *Proc. Natl Acad. Sci. USA*, 2000, **97**, 1589–1594.
19. P. J. Muchowski, G. Schaffar, A. Sittler, E. E. Wanker, M. K. Hayer-Hartl and F. U. Hartl, *Proc. Natl Acad. Sci. USA*, 2000, **97**, 7841–7846.
20. Y. Chai, S. L. Koppenhafer, N. M. Bonini and H. L. Paulson, *J. Neurosci.*, 1999, **19**, 10338–10347.
21. J. M. Warrick, H. Y. Chan, G. L. Gray-Board, Y. Chai, H. L. Paulson and N. M. Bonini, *Nat. Genet.*, 1999, **23**, 425–428.

22. H. Sakahira, P. Breuer, M. K. Hayer-Hartl and F. U. Hartl, *Proc. Natl Acad. Sci. USA*, 2002, **99**(4), 16412–16418.
23. D. G. Hay, K. Sathasivam, S. Tobaben, B. Stahl, M. Marber, R. Mestril, A. Mahal, D. L. Smith, B. Woodman and G. P. Bates, *Hum. Mol. Genet.*, 2004, **13**, 1389–1405.
24. O. Hansson, J. Nylandsted, R. F. Castilho, M. Leist, M. Jaattela and P. Brundin, *Brain Res.*, 2003, **970**, 47–57.
25. C. J. Cummings, Y. Sun, P. Opal, B. Antalffy, R. Mestril, H. T. Orr, W. H. Dillmann and H. Y. Zoghbi, *Hum. Mol. Genet.*, 2001, **10**, 1511–1518.
26. H. Adachi, M. Katsuno, M. Minamiyama, C. Sang, G. Pagoulatos, C. Angelidis, M. Kusakabe, A. Yoshiki, Y. Kobayashi, M. Doyu and G. Sobue, *J. Neurosci.*, 2003, **23**, 2203–2211.
27. E. A. Nollen, S. M. Garcia, G. van Haaften, S. Kim, A. Chavez, R. I. Morimoto and R. H. Plasterk, *Proc. Natl Acad. Sci. USA*, 2004, **101**, 6403–6408.
28. P. Kazemi-Esfarjani and S. Benzer, *Science*, 2000, **287**, 1837–1840.
29. S. Willingham, T. F. Outeiro, M. J. DeVit, S. L. Lindquist and P. J. Muchowski, *Science*, 2003, **302**, 1769–1772.
30. F. M. LaFerla, K. N. Green and S. Oddo, *Nat. Rev. Neurosci.*, 2007, **8**, 499–509.
31. A. Y. Hsia, E. Masliah, L. McConlogue, G. Q. Yu, G. Tatsuno, K. Hu, D. Kholodenko, R. C. Malenka, R. A. Nicoll and L. Mucke, *Proc. Natl Acad. Sci. USA*, 1999, **96**, 3228–3233.
32. D. H. Chui, H. Tanahashi, K. Ozawa, S. Ikeda, F. Checler, O. Ueda, H. Suzuki, W. Araki, H. Inoue, K. Shirotani, K. Takahashi, F. Gallyas and T. Tabira, *Nat. Med.*, 1999, **5**, 560–564.
33. J. Hardy and D. J. Selkoe, *Science*, 2002, **297**, 353–356.
34. S. Kins, N. Lauther, A. Szodorai and K. Beyreuther, *Neurodegener. Dis.*, 2006, **3**, 218–226.
35. A. S. Chyung, B. D. Greenberg, D. G. Cook, R. W. Doms and V. M. Lee, *J. Cell Biol.*, 1997, **138**, 671–680.
36. B. P. Tseng, M. Kitazawa and F. M. LaFerla, *Curr. Alzheimer Res.*, 2004, **1**, 231–239.
37. S. Oddo, L. Billings, J. P. Kesslak, D. H. Cribbs and F. M. LaFerla, *Neuron*, 2004, **43**, 321–332.
38. J. Koren, 3rd, U. K. Jinwal, D. C. Lee, J. R. Jones, C. L. Shults, A. G. Johnson, L. J. Anderson and C. A. Dickey, *J. Cell. Mol. Med.*, 2009, **13**, 619–630.
39. B. C. Yoo, R. Seidl, N. Cairns and G. Lubec, *J. Neural Transm. Suppl.*, 1999, **57**, 315–322.
40. J. E. Hamos, B. Oblas, D. Pulaski-Salo, W. J. Welch, D. G. Bole and D. A. Drachman, *Neurology*, 1991, **41**, 345–350.
41. K. Renkawek, G. J. Bosman and M. Gaestel, *Neuroreport*, 1993, **5**, 14–16.
42. Y. Yang, R. S. Turner and J. R. Gaut, *J. Biol. Chem.*, 1998, **273**, 25552–25555.

43. T. Kudo, M. Okumura, K. Imaizumi, W. Araki, T. Morihara, H. Tanimukai, E. Kamagata, N. Tabuchi, R. Kimura, D. Kanayama, A. Fukumori, S. Tagami, M. Okochi, M. Kubo, H. Tanii, M. Tohyama, T. Tabira and M. Takeda, *Biochem. Biophys. Res. Commun.*, 2006, **344**, 525–530.
44. T. Hoshino, T. Nakaya, W. Araki, K. Suzuki, T. Suzuki and T. Mizushima, *Biochem. J.*, 2007, **402**, 581–589.
45. G. J. Stege, K. Renkawek, P. S. Overkamp, P. Verschuure, A. F. van Rijk, A. Reijnen-Aalbers, W. C. Boelens, G. J. Bosman and W. W. de Jong, *Biochem. Biophys. Res. Commun.*, 1999, **262**, 152–156.
46. J. J. Liang, *FEBS Lett.*, 2000, **484**, 98–101.
47. V. Fonte, V. Kapulkin, A. Taft, A. Fluet, D. Friedman and C. D. Link, *Proc. Natl Acad. Sci. USA*, 2002, **99**, 9439–9444.
48. S. Narayanan, B. Kamps, W. C. Boelens and B. Reif, *FEBS Lett.*, 2006, **580**, 5941–5946.
49. Y. C. Kudva, H. J. Hiddinga, P. C. Butler, C. S. Mueske and N. L. Eberhardt, *FEBS Lett.*, 1997, **416**, 117–121.
50. H. Shimura, Y. Miura-Shimura and K. S. Kosik, *J. Biol. Chem.*, 2004, **279**, 17957–17962.
51. F. Dou, W. J. Netzer, K. Tanemura, F. Li, F. U. Hartl, A. Takashima, G. K. Gouras, P. Greengard and H. Xu, *Proc. Natl Acad. Sci. USA*, 2003, **100**, 721–726.
52. H. Shimura, D. Schwartz, S. P. Gygi and K. S. Kosik, *J. Biol. Chem.*, 2004, **279**, 4869–4876.
53. L. Petrucelli, D. Dickson, K. Kehoe, J. Taylor, H. Snyder, A. Grover, M. De Lucia, E. McGowan, J. Lewis, G. Prihar, J. Kim, W. H. Dillmann, S. E. Browne, A. Hall, R. Voellmy, Y. Tsuboi, T. M. Dawson, B. Wolozin, J. Hardy and M. Hutton, *Hum. Mol. Genet.*, 2004, **13**, 703–714.
54. C. A. Dickey, M. Yue, W. L. Lin, D. W. Dickson, J. H. Dunmore, W. C. Lee, C. Zehr, G. West, S. Cao, A. M. Clark, G. A. Caldwell, K. A. Caldwell, C. Eckman, C. Patterson, M. Hutton and L. Petrucelli, *J. Neurosci.*, 2006, **26**, 6985–6996.
55. C. M. Tanner and S. M. Goldman, *Neurol. Clin.*, 1996, **14**, 317–335.
56. P. Calabresi, F. Galletti, E. Saggese, V. Ghiglieri and B. Picconi, *Parkinsonism Relat. Disord.*, 2007, **13**(3), S259–262.
57. K. R. Chaudhuri, D. G. Healy and A. H. Schapira, *Lancet Neurol.*, 2006, **5**, 235–245.
58. H. Braak and E. Braak, *J. Neurol.*, 2000, **247**(2), II3–10.
59. H. Braak, K. Del Tredici, H. Bratzke, J. Hamm-Clement, D. Sandmann-Keil and U. Rub, *J. Neurol.*, 2002, **249**(3), III/1–5.
60. H. Braak, K. Del Tredici, U. Rub, R. A. de Vos, E. N. Jansen Steur and E. Braak, *Neurobiol. Aging*, 2003, **24**, 197–211.
61. M. H. Polymeropoulos, C. Lavedan, E. Leroy, S. E. Ide, A. Dehejia, A. Dutra, B. Pike, H. Root, J. Rubenstein, R. Boyer, E. S. Stenroos, S. Chandrasekharappa, A. Athanassiadou, T. Papapetropoulos,

and R. L. Nussbaum, *Science*, 1997, **276**, 2045–2047.
62. R. Kruger, W. Kuhn, T. Muller, D. Woitalla, M. Graeber, S. Kosel, H. Przuntek, J. T. Epplen, L. Schols and O. Riess, *Nat. Genet.*, 1998, **18**, 106–108.
63. A. B. Singleton, M. Farrer, J. Johnson, A. Singleton, S. Hague, J. Kachergus, M. Hulihan, T. Peuralinna, A. Dutra, R. Nussbaum, S. Lincoln, A. Crawley, M. Hanson, D. Maraganore, C. Adler, M. R. Cookson, M. Muenter, M. Baptista, D. Miller, J. Blancato, J. Hardy and K. Gwinn-Hardy, *Science*, 2003, **302**, 841.
64. M. C. Chartier-Harlin, J. Kachergus, C. Roumier, V. Mouroux, X. Douay, S. Lincoln, C. Levecque, L. Larvor, J. Andrieux, M. Hulihan, N. Waucquier, L. Defebvre, P. Amouyel, M. Farrer and A. Destee, *Lancet*, 2004, **364**, 1167–1169.
65. P. Ibanez, A. M. Bonnet, B. Debarges, E. Lohmann, F. Tison, P. Pollak, Y. Agid, A. Durr and A. Brice, *Lancet*, 2004, **364**, 1169–1171.
66. D. Kirik, C. Rosenblad, C. Burger, C. Lundberg, T. E. Johansen, N. Muzyczka, R. J. Mandel and A. Bjorklund, *J. Neurosci.*, 2002, **22**, 2780–2791.
67. M. B. Feany and W. W. Bender, *Nature*, 2000, **404**, 394–398.
68. C. Lo Bianco, J. L. Ridet, B. L. Schneider, N. Deglon and P. Aebischer, *Proc. Natl Acad. Sci. USA*, 2002, **99**, 10813–10818.
69. M. G. Spillantini, M. L. Schmidt, V. M. Lee, J. Q. Trojanowski, R. Jakes and M. Goedert, *Nature*, 1997, **388**, 839–840.
70. P. K. Auluck, H. Y. Chan, J. Q. Trojanowski, V. M. Lee and N. M. Bonini, *Science*, 2002, **295**, 865–868.
71. P. J. McLean, H. Kawamata, S. Shariff, J. Hewett, N. Sharma, K. Ueda, X. O. Breakefield and B. T. Hyman, *J. Neurochem.*, 2002, **83**, 846–854.
72. A. Iwaki, T. Iwaki, J. E. Goldman, K. Ogomori, J. Tateishi and Y. Sakaki, *Neurosci. Lett.*, 1992, **140**, 89–92.
73. H. Braak, K. Del Tredici, D. Sandmann-Kiel, U. Rub and C. Schultz, *Acta Neuropathol.*, 2001, **102**, 449–454.
74. D. Ebrahimi-Fakhari, L. Wahlster and P. J. McLean, *J. Parkinsons Dis.*, 2011, **1**, 299–320.
75. J. Klucken, Y. Shin, E. Masliah, B. T. Hyman and P. J. McLean, *J. Biol. Chem.*, 2004, **279**, 25497–25502.
76. T. F. Outeiro, J. Klucken, K. E. Strathearn, F. Liu, P. Nguyen, J. C. Rochet, B. T. Hyman and P. J. McLean, *Biochem. Biophys. Res. Commun.*, 2006, **351**, 631–638.
77. Y. Shin, J. Klucken, C. Patterson, B. T. Hyman and P. J. McLean, *J. Biol. Chem.*, 2005, **280**, 23727–23734.
78. C. R. Scherzer, A. C. Eklund, L. J. Morse, Z. Liao, J. J. Locascio, D. Fefer, M. A. Schwarzschild, M. G. Schlossmacher, M. A. Hauser, J. M. Vance, L. R. Sudarsky, D. G. Standaert, J. H. Growdon, R. V. Jensen and S. R. Gullans, *Proc. Natl Acad. Sci. USA*, 2007, **104**, 955–960.

79. M. R. Jain, W. W. Ge, S. Elkabes and H. Li, *Proteomics Clin. Appl.*, 2008, **2**, 670–684.
80. J. P. Julien, *Cell*, 2001, **104**, 581–591.
81. D. R. Rosen, T. Siddique, D. Patterson, D. A. Figlewicz, P. Sapp, A. Hentati, D. Donaldson, J. Goto, J. P. O'Regan, H. X. Deng, Z. Rahmani, A. Krizus, D. McKenna-Yasek, A. Cayabyab, S. M. Gaston, R. Berger, R. E. Tanzi, J. J. Halperin, B. Herzfeld, R. Van den Bergh, W. Y. Hung, T. Bird, G. Deng, D. W. Mulder, C. Smyth, N. G. Laing, E. Soriano, M. A. Pericak-Vance, J. Haines, G. A. Rouleau, J. S. Gusella, H. R. Horvitz and R. H. J. Brown, *Nature*, 1993, **362**, 59–62.
82. P. Pasinelli and R. H. Brown, *Nat. Rev. Neurosci.*, 2006, **7**, 710–723.
83. T. L. Williamson and D. W. Cleveland, *Nat. Neurosci.*, 1999, **2**, 50–56.
84. L. A. Ligon, B. H. LaMonte, K. E. Wallace, N. Weber, R. G. Kalb and E. L. Holzbaur, *Neuroreport*, 2005, **16**, 533–536.
85. N. Shibata, A. Hirano, M. Kobayashi, T. Siddique, H. X. Deng, W. Y. Hung, T. Kato and K. Asayama, *J. Neuropathol. Exp. Neurol.*, 1996, **55**, 481–490.
86. S. Kato, M. Shimoda, Y. Watanabe, K. Nakashima, K. Takahashi and E. Ohama, *J. Neuropathol. Exp. Neurol.*, 1996, **55**, 1089–1101.
87. H. D. Durham, J. Roy, L. Dong and D. A. Figlewicz, *J. Neuropathol. Exp. Neurol.*, 1997, **56**, 523–530.
88. L. I. Bruijn, M. K. Houseweart, S. Kato, K. L. Anderson, S. D. Anderson, E. Ohama, A. G. Reaume, R. W. Scott and D. W. Cleveland, *Science*, 1998, **281**, 1851–1854.
89. W. Bruening, J. Roy, B. Giasson, D. A. Figlewicz, W. E. Mushynski and H. D. Durham, *J. Neurochem.*, 1999, **72**, 693–699.
90. A. G. Reaume, J. L. Elliott, E. K. Hoffman, N. W. Kowall, R. J. Ferrante, D. F. Siwek, H. M. Wilcox, D. G. Flood, M. F. Beal, R. H. Brown, Jr., R. W. Scott and W. D. Snider, *Nat. Genet.*, 1996, **13**, 43–47.
91. H. Takeuchi, Y. Kobayashi, T. Yoshihara, J. Niwa, M. Doyu, K. Ohtsuka and G. Sobue, *Brain Res.*, 2002, **949**, 11–22.
92. G. A. Shinder, M. C. Lacourse, S. Minotti and H. D. Durham, *J. Biol. Chem.*, 2001, **276**, 12791–12796.
93. A. Maatkamp, A. Vlug, E. Haasdijk, D. Troost, P. J. French and D. Jaarsma, *Eur. J. Neurosci.*, 2004, **20**, 14–28.
94. I. Pieri, C. Cifuentes-Diaz, J. P. Oudinet, B. Blondet, F. Rieger, S. Gonin, A. P. Arrigo and Y. Thomas, *J. Neurosci. Res.*, 2001, **65**, 247–253.
95. C. W. Strey, D. Spellman, A. Stieber, J. O. Gonatas, X. Wang, J. D. Lambris and N. K. Gonatas, *Am. J. Pathol.*, 2004, **165**, 1701–1718.
96. S. B. Prusiner, *Science*, 1982, **216**, 136–144.
97. B. Chesebro, *Neuron*, 1999, **24**, 503–506.
98. J. Collinge, *J. Neurol. Neurosurg. Psychiatry*, 2005, **76**, 906–919.
99. S. B. Prusiner, M. R. Scott, S. J. DeArmond and F. E. Cohen, *Cell*, 1998, **93**, 337–348.
100. J. Safar and S. B. Prusiner, *Prog. Brain Res.*, 1998, **117**, 421–434.

101. W. J. Welch and P. Gambetti, *Nature*, 1998, **392**, 23–24.
102. T. Jin, Y. Gu, G. Zanusso, M. Sy, A. Kumar, M. Cohen, P. Gambetti and N. Singh, *J. Biol. Chem.*, 2000, **275**, 38699–38704.
103. G. W. Jones and M. F. Tuite, *Bioessays*, 2005, **27**, 823–832.
104. E. G. Rikhvanov, N. V. Romanova and Y. O. Chernoff, *Prion*, 2007, **1**, 217–222.
105. D. W. Summers, P. M. Douglas and D. M. Cyr, *Prion*, 2009, **3**, 59–64.
106. Y. O. Chernoff, S. L. Lindquist, B. Ono, S. G. Inge-Vechtomov and S. W. Liebman, *Science*, 1995, **268**, 880–884.
107. P. Zhou, I. L. Derkatch and S. W. Liebman, *Mol. Microbiol.*, 2001, **39**, 37–46.
108. H. Moriyama, H. K. Edskes and R. B. Wickner, *Mol. Cell. Biol.*, 2000, **20**, 8916–8922.
109. S. K. DebBurman, G. J. Raymond, B. Caughey and S. Lindquist, *Proc. Natl Acad. Sci. USA*, 1997, **94**, 13938–13943.
110. A. Chacinska, B. Szczesniak, N. V. Kochneva-Pervukhova, V. V. Kushnirov, M. D. Ter-Avanesyan and M. Boguta, *Curr. Genet.*, 2001, **39**, 62–67.
111. Y. O. Chernoff, G. P. Newnam, J. Kumar, K. Allen and A. D. Zink, *Mol. Cell. Biol.*, 1999, **19**, 8103–8112.
112. K. D. Allen, R. D. Wegrzyn, T. A. Chernova, S. Muller, G. P. Newnam, P. A. Winslett, K. B. Wittich, K. D. Wilkinson and Y. O. Chernoff, *Genetics*, 2005, **169**, 1227–1242.
113. V. V. Kushnirov, D. S. Kryndushkin, M. Boguta, V. N. Smirnov and M. D. Ter-Avanesyan, *Curr. Biol.*, 2000, **10**, 1443–1446.
114. C. Schwimmer and D. C. Masison, *Mol. Cell. Biol.*, 2002, **22**, 3590–3598.
115. J. R. Glover and S. Lindquist, *Cell*, 1998, **94**, 73–82.
116. T. Higurashi, J. K. Hines, C. Sahi, R. Aron and E. A. Craig, *Proc. Natl Acad. Sci. USA*, 2008, **105**, 16596–16601.
117. R. Aron, T. Higurashi, C. Sahi and E. A. Craig, *EMBO J.*, 2007, **26**, 3794–3803.
118. D. W. Summers, P. M. Douglas, H. Y. Ren and D. M. Cyr, *J. Biol. Chem.*, 2009, **284**, 3628–3639.
119. M. Tatar, A. A. Khazaeli and J. W. Curtsinger, *Nature*, 1997, **390**, 30.
120. J. M. Smith, *Nature*, 1958, **181**, 496–497.
121. R. A. Krebs and M. E. Feder, *Cell Stress Chaperones*, 1997, **2**, 60–71.
122. D. D. Mosser and R. I. Morimoto, *Oncogene*, 2004, **23**, 2907–2918.
123. R. I. Morimoto, *Genes Dev.*, 1998, **12**, 3788–3796.
124. J. Zou, Y. Guo, T. Guettouche, D. F. Smith and R. Voellmy, *Cell*, 1998, **94**, 471–480.
125. L. Whitesell, E. G. Mimnaugh, B. De Costa, C. E. Myers and L. M. Neckers, *Proc. Natl Acad. Sci. USA*, 1994, **91**, 8324–8328.
126. P. J. McLean, J. Klucken, Y. Shin and B. T. Hyman, *Biochem. Biophys. Res. Commun.*, 2004, **321**, 665–669.
127. H. Y. Shen, J. C. He, Y. Wang, Q. Y. Huang and J. F. Chen, *J. Biol. Chem.*, 2005, **280**, 39962–39969.

128. P. K. Auluck and N. M. Bonini, *Nat. Med.*, 2002, **8**, 1185–1186.
129. A. Sittler, R. Lurz, G. Lueder, J. Priller, H. Lehrach, M. K. Hayer-Hartl, F. U. Hartl and E. E. Wanker, *Hum. Mol. Genet.*, 2001, **10**, 1307–1315.
130. Z. Batulan, D. M. Taylor, R. J. Aarons, S. Minotti, M. M. Doroudchi, J. Nalbantoglu and H. D. Durham, *Neurobiol. Dis.*, 2006, **24**, 213–225.
131. S. K. Kalia, L. V. Kalia and P. J. McLean, *CNS Neurol., Disord. Drug Targets*, 2010, **9**, 741–753.
132. M. Waza, H. Adachi, M. Katsuno, M. Minamiyama, C. Sang, F. Tanaka, A. Inukai, M. Doyu and G. Sobue, *Nat. Med.*, 2005, **11**, 1088–1095.
133. N. Fujikake, Y. Nagai, H. A. Popiel, Y. Okamoto, M. Yamaguchi and T. Toda, *J. Biol. Chem.*, 2008, **283**, 26188–26197.
134. P. Putcha, K. M. Danzer, L. R. Kranich, A. Scott, M. Silinski, S. Mabbett, C. D. Hicks, J. M. Veal, P. M. Steed, B. T. Hyman and P. J. McLean, *J. Pharmacol. Exp. Ther.*, 2010, **332**, 849–857.
135. R. L. Cysyk, R. J. Parker, J. J. Barchi, Jr., P. S. Steeg, N. R. Hartman and J. M. Strong, *Chem. Res. Toxicol.*, 2006, **19**, 376–381.
136. M. Fujimoto, E. Takaki, T. Hayashi, Y. Kitaura, Y. Tanaka, S. Inouye and A. Nakai, *J. Biol. Chem.*, 2005, **280**, 34908–34916.
137. K. W. Park, J. S. Hahn, Q. Fan, D. J. Thiele and L. Li, *Genetics*, 2006, **173**, 35–47.
138. A. D. Steele, G. Hutter, W. S. Jackson, F. L. Heppner, A. W. Borkowski, O. D. King, G. J. Raymond, A. Aguzzi and S. Lindquist, *Proc. Natl Acad. Sci. USA*, 2008, **105**, 13626–13631.
139. D. Kieran, B. Kalmar, J. R. Dick, J. Riddoch-Contreras, G. Burnstock and L. Greensmith, *Nat. Med.*, 2004, **10**, 402–405.
140. D. Paris, N. J. Ganey, V. Laporte, N. S. Patel, D. Beaulieu-Abdelahad, C. Bachmeier, A. March, G. Ait-Ghezala and M. J. Mullan, *J. Neuroinflammation*, 2010, **7**, 17.
141. M. Kiaei, K. Kipiani, S. Petri, J. Chen, N. Y. Calingasan and M. F. Beal, *Neurodegener. Dis.*, 2005, **2**, 246–254.
142. C. Cleren, N. Y. Calingasan, J. Chen and M. F. Beal, *J. Neurochem.*, 2005, **94**, 995–1004.
143. M. Katsuno, C. Sang, H. Adachi, M. Minamiyama, M. Waza, F. Tanaka, M. Doyu and G. Sobue, *Proc. Natl Acad. Sci. USA*, 2005, **102**, 16801–16806.
144. M. Fujiki, T. Hikawa, T. Abe, S. Uchida, M. Morishige, K. Sugita and H. Kobayashi, *J. Neurotrauma*, 2006, **23**, 1164–1178.
145. S. Uchida, M. Fujiki, Y. Nagai, T. Abe and H. Kobayashi, *Neurosci. Lett.*, 2006, **396**, 220–224.
146. D. W. Neef, M. L. Turski and D. J. Thiele, *PLoS Biol.*, 2010, **8**, e1000291.
147. L. Chang, E. B. Bertelsen, S. Wisen, E. M. Larsen, E. R. Zuiderweg and J. E. Gestwicki, *Anal. Biochem.*, 2008, **372**, 167–176.
148. U. K. Jinwal, J. Koren, J. C. O'Leary, J. R. Jones, J. F. Abisambra and C. A. Dickey, *Mol. Cell. Pharmacol.*, 2010, **2**, 43–46.
149. M. S. Rafii and P. S. Aisen, *BMC Med.*, 2009, **7**, 7.
150. D. H. Perlmutter, *Pediatr. Res.*, 2002, **52**, 832–836.

151. F. Engin and G. S. Hotamisligil, *Diabetes Obes. Metab.*, 2010, **12**(2), 108–115.
152. P. V. Attfield, *FEBS Lett.*, 1987, **225**, 259–263.
153. M. A. Singer and S. Lindquist, *Mol. Cell*, 1998, **1**, 639–648.
154. M. Tanaka, Y. Machida, S. Niu, T. Ikeda, N. R. Jana, H. Doi, M. Kurosawa, M. Nekooki and N. Nukina, *Nat. Med.*, 2004, **10**, 148–154.
155. R. Liu, H. Barkhordarian, S. Emadi, C. B. Park and M. R. Sierks, *Neurobiol. Dis.*, 2005, **20**, 74–81.
156. F. Beranger, C. Crozet, A. Goldsborough and S. Lehmann, *Biochem. Biophys. Res. Commun.*, 2008, **374**, 44–48.
157. T. Seki, N. Abe-Seki, T. Kikawada, H. Takahashi, K. Yamamoto, N. Adachi, S. Tanaka, I. Hide, N. Saito and N. Sakai, *J. Biol. Chem.*, 2010, **285**, 33252–33264.
158. S. Mimori, Y. Okuma, M. Kaneko, K. Kawada, T. Hosoi, K. Ozawa, Y. Nomura and H. Hamana, *Biol. Pharm. Bull.*, 2012, **35**, 84–90.
159. K. Ono, M. Ikemoto, T. Kawarabayashi, M. Ikeda, T. Nishinakagawa, M. Hosokawa, M. Shoji, M. Takahashi and M. Nakashima, *Parkinsonism Relat. Disord.*, 2009, **15**, 649–654.
160. M. Inden, Y. Kitamura, H. Takeuchi, T. Yanagida, K. Takata, Y. Kobayashi, T. Taniguchi, K. Yoshimoto, M. Kaneko, Y. Okuma, T. Taira, H. Ariga and S. Shimohama, *J. Neurochem.*, 2007, **101**, 1491–1504.
161. S. Sola, R. E. Castro, P. A. Laires, C. J. Steer and C. M. Rodrigues, *Mol. Med.*, 2003, **9**, 226–234.
162. D. F. Smith, L. Whitesell and E. Katsanis, *Pharmacol. Rev.*, 1998, **50**, 493–514.
163. C. W. Christine, P. A. Starr, P. S. Larson, J. L. Eberling, W. J. Jagust, R. A. Hawkins, H. F. VanBrocklin, J. F. Wright, K. S. Bankiewicz and M. J. Aminoff, *Neurology*, 2009, **73**, 1662–1669.
164. P. A. LeWitt, A. R. Rezai, M. A. Leehey, S. G. Ojemann, A. W. Flaherty, E. N. Eskandar, S. K. Kostyk, K. Thomas, A. Sarkar, M. S. Siddiqui, S. B. Tatter, J. M. Schwalb, K. L. Poston, J. M. Henderson, R. M. Kurlan, I. H. Richard, L. Van Meter, C. V. Sapan, M. J. During, M. G. Kaplitt and A. Feigin, *Lancet Neurol.*, 2011, **10**, 309–319.
165. Z. Dong, D. P. Wolfer, H. P. Lipp and H. Bueler, *Mol. Ther.*, 2005, **11**, 80–88.
166. A. E. Jung, H. L. Fitzsimons, R. J. Bland, M. J. During and D. Young, *Mol. Ther.*, 2008, **16**, 1048–1055.
167. D. D. Mosser, S. Ho and J. R. Glover, *Biochemistry*, 2004, **43**, 8107–8115.
168. C. Lo Bianco, J. Shorter, E. Regulier, H. Lashuel, T. Iwatsubo, S. Lindquist and P. Aebischer, *J. Clin. Invest.*, 2008, **118**, 3087–3097.
169. J. L. Howarth, S. Kelly, M. P. Keasey, C. P. Glover, Y. B. Lee, K. Mitrophanous, J. P. Chapple, J. M. Gallo, M. E. Cheetham and J. B. Uney, *Mol. Ther.*, 2007, **15**, 1100–1105.

Subject Index

A4 267
A107N 40, 42, 43
17-AAG *see* tanespimycin
acetylation 322–3
acute myelogenous leukemia 136
acyl benzamides 308, 311
AdaSGC 19, 21, 306
ADCK3 4
ADD70 20, 21
AEG3482 289
AHA1 17, 51–3, 130, 145
AHR 51
AHSP 4
AIP 51, 52
AKT 133, 262, 289, 292
alvespimycin 136, 324, 339–40
    adverse events 348
    chemical structure 348
    in malaria 389
    see also geldanamycin
Alzheimer's disease 1, 5, 91, 263, 291, 393, 394, 396–7
    neurofibrillary tangles 263, 396
amino-methyl functionalized diaryl pyrazoles
    binding assay 226, 229
    cell growth inhibition 226, 229
aminoacridine quinacrine (QC) 24
aminobenzamide derivatives 347, 352
    SNX5422 *see* SNX5422
2-aminopurines 161–2
amodiaquine 382, 383
amyloid β-peptide 396

amyloid precursor protein 396
amyotrophic lateral sclerosis 263, 393, 394, 398–9
ANKRD13 4
ansamycins 336
    in malaria 386, 387
    see also geldanamycin
anti-angiogenesis 183–4
apoptosis 3, 8, 9, 10, 11, 14, 17, 19–23, 66, 86, 104–8, 133, 136, 144, 184, 189, 191, 194, 242, 289, 310–11, 314, 345, 364, 365
apoptosis inducing factor (AIF) 21, 105
apoptozole 19, 307, 310–11
ARA9 50–1
arimoclomol 25
    neurological disorders 402
aromatic amino acid decarboxylase (AACD) 404
artemether 382, 383
artemisinin 382, 383
artemotil 382, 383
artesunate 382, 383
aryl hydrocarbon receptor (AhR) 290
AT13387 128, 140, 326, 344–5
    adverse events 350
    chemical structure 350
ATF6 100, 102, 103, 142
atovaquone 382, 383
ATPase activation 51–4
ATPase modulation
    Hsp70 403
    Hsp90 54–5, 242

Subject Index

autophagy 9, 20, 87, 91, 107–8, 129, 242, 313, 393
    chaperone-mediated 2, 26, 93
    endosomal microautophagy 2
    inhibition 94
auxilin 94–5
AUY-922 176–8, 326
Azadirachta indica 386–7

Bags 84, 86
Bag-1 18, 70, 85, 86, 87, 91, 92–3, 302
Bag-1M 70, 86
BAP 87, 100
BCR-ABL 133, 136, 142
Bergerat fold 162, 181
BEZ235 185
BIIB021 128, 140, 160–6, 169, 325, 330, 347
    adverse events 351
    chemical structure 351
    criteria 160
    design 160–2
    in vivo efficacy 165–6
    metabolism 164
    pharmacodynamics 164–5
    pharmacokinetics 163–4
    selectivity 162–3
    tolerability 176–8
BIIB028 140, 166–76
    criteria 166
    design 167–73
        potency 167–9
        selectivity 169–70
    efficacy 171
        in vivo 174–5, 176
    pharmacodynamics 174
    pharmacokinetics 170–1, 173–4
    tolerability 172–3, 176–8
bimoclomol 25
biomarkers 221–32
BiP 83, 100–2
bipolar disorder 5
bortezomib 329, 365
bovine spongiform encephalopathy 399
BRAF 37, 133, 147

breast cancer 132, 135, 137, 142, 143, 328
    HER2+ 357–9
    Hsp90 inhibitors 356–9
Bunina bodies 398

C9 13, 15, 17
cancer 126–57, 261–4
    driver oncoproteins 134–41
    drug targets 129
    future directions 142–5
    stress pathways 141–2
    treatment
        Hsp70 inhibitors 127
        Hsp90 C-terminus inhibitors 269–86
        Hsp90 inhibitors 127–8, 241–2, 261–4, 352–67
    see also specific cancer types
cardiotoxicity 190
caspases 9, 104, 106, 142, 183, 188, 192, 274, 280, 305, 313, 321, 345
cassette dosing studies 230–2
CCT0181059 216–17, 219, 234
    binding mode 222–3
Cdc37 16–17, 48–9, 50, 57, 130, 132, 145, 319
    phosphorylation/dephosphorylation 17
    siRNA 17
Cdk4 37, 49, 50, 57, 133, 135, 145
celastrol 15, 16–17
    HSF-1 pathway activation 25
chaperone-mediated autophagy 2, 93
chaperonins 3–4, 10, 90
    disease associations 5, 9
    mouse knockout data 7
Charcot-Marie-Tooth disease 5
chemical chaperones 403–4
CHIP 22, 50–1, 55–6, 57, 67, 132, 264
    protein degradation 92
chlorobiocin 141
    chemical structure 265
chloroquine 382, 383
CHOP 103, 142

chronic myelogenous leukemia 136, 329–30
cisplatin 16, 294–5
CK2 321–2
click chemistry 274–5
client proteins 11, 37, 48–51, 242, 256
    activation 18, 53, 56–7, 83, 92, 127, 130
    degradation 15, 16, 19, 21, 178, 274, 290–5
    inhibition of folding 164
    molecular profiling 138, 139
    oncogenic 132–4, 143, 183, 242, 255
    regulation of post-translational modification 323–4
    targeting by co-chaperones 55–6
    tumor drivers 325–8
CLIPS 89–90
clusterin 4
clusterin inhibitors 23
CNF2024 see BIIB021
co-chaperones
    Hsp70 79–89
        Hip 87–8, 94, 398
        Hop see Hop
        J proteins see J proteins
        NEFs 66, 84–7, 302
        targeting 312–14
    Hsp90
        ATPase activation 51–4
        ATPase modulation 54–5
        client protein delivery 48–51
        proteasome degradation 55–6
        regulation/modulation by 47–8
    see also individual co-chaperones
Conforma Therapeutics 158
coronary disease 5
coumarins 267
coumarin analogues 272
coumermycin A1 141, 286–7
    chemical structure 265

CP9 15
Cpr6 54
CRAF 133, 135, 145
Creutzfeldt-Jakob disease 399
crizotinib 185, 328, 356
Crohn's disease 5
CUDC-305 see Debio 0932
custirsen see OGX-011
cystic fibrosis transmembrane conductance regulator (CFTR) 92

D-peptides 21–2, 313–14
de novo protein folding 89–90
Debio 0932 128, 140, 327, 345–6
    adverse events 350
    chemical structure 350
dentorubral-pallidoluysian atrophy 393
15-deoxyspergualin (DSG) 308, 312
depression 5
diabetes mellitus 5
dihydroartimisinin 382, 383
dihydropyrimidines 22
dimethyl sulfoxide 403
"disconnect-rotate-reconnect" technique 160, 161
dissociation-enhanced lanthanide fluorescence immunoassay (DELFIA) 243
17-DMAG see alvespimycin
DNA gyrase 241, 264, 266
DNA gyrase inhibitors 269–70
DnaK 66, 74–5, 83
docetaxel 187, 192–3
doxycycline 382
driver oncoproteins 134–41
DS-2248 327

EBNA3 51
EC144 169
EGCG 16, 290–1, 305, 306
EGFR 147
ELISA 243
EML4-ALK 127, 133, 355–7
Emunin 24
ENCHANT trial 328

Subject Index

endocytosis 94–5
endonuclease-G 106
endoplasmic reticulum stress 102–4
endoplasmic reticulum-associated degradation (ERAD) 101, 102
endosomal microautophagy 2
EphA2 142
epidermal growth factor receptor (EGFR) 290, 353–5
epigallocatechin gallate see EGCG
ER see endoplasmic reticulum
ERα 51
ErbB2 see HER2
erlotinib 147
esanex see SNX5422
everolimus 329
exocytosis 95–6

favism 201
FLT3 136
fluorescence polarization 223, 304
fosmidomycin 382, 383

G proteins 51
G3129 217
G3130 217
GALAXY trial 192–3
gambogic acid 12, 14
ganetespib 128, 139–40, 180–97, 325, 326, 328, 344
    adverse events 350
    chemical description 180
    chemical structure 219, 350
    clinical applications
        breast cancer 358
        GIST 362
        NSCLC 184–6, 192–3, 354
    co-crystal structure 181–2
    efficacy
        clinical benefits 193–4
        hematological malignancies 186–7
        NSCLC 184–6, 192–3
    pharmacokinetics and toxicology 189–91
        lack of liver/cardiac toxicity 190
        lack of ocular toxicity 190–1
        tissue distribution in tumor-bearing mice 189–90
    pre-clinical pharmacology 182–9
        anti-angiogenic properties 183–4
        biological activity 188–9
        combination studies 187–8
        single agent activity 182–3
    screening process 180–1
gastric cancer 362–4
gastrointestinal stromal tumors (GIST) 138, 189, 330
    Hsp90 inhibitors 361–2
gedunin 386–7
geldanamycin 59–60, 127, 128, 134–5, 136–7, 214, 325, 336
    chemical structure 215, 244, 387
    clinical applications
        malaria 384, 385, 386
        neurological disorders 401
geldanamycin analogues 337–41
    alvespimycin 136, 324, 339–40
    retaspimycin 137, 327, 341
    tanespimycin see tanespimycin
genetic interference 17
geranylgeranylacetone 25, 402–3
GIST see gastrointestinal stromal tumors
glucocorticoid receptor 2, 56, 86, 88, 290, 322, 323
glucose-6-phosphate dehydrogenase deficiency 201
glycogen storage diseases 201
GroEL 3–4
GroES 3–4
GRP94 130, 142, 145, 198, 241
GRP170 100
GRPE 84

GTL16 cell line 252–4
gusperimus 308, 312

halofantrine 382, 383
Hch1 53–4
heat shock elements 68
heat shock proteins *see* Hsp
heat shock response 67–8
hematological malignancies 186–7
hepatic phosphorylase kinase
 deficiency 201
HER2 37, 127, 133, 135, 138, 145,
 147, 262, 292, 325
herbimycin 134
high-throughput screening 214, 242,
 304, 387–8
Hip 87–8, 94, 398
histone acetylases 322
histone chaperones 4, 322
histone deacetylases 322
holdases 65, 90
homogenous time-resolved
 fluorescence (HTRF) 243
Hop 17–18, 48, 52, 67, 88–9, 130, 264,
 302–3, 319
 siRNA 17
Hsc70 94–5
Hsf1 68, 130, 144
 pathway modulation 23–6,
  402–3
Hsf1A 25
Hsp1L 68, 69
Hsp10 4
Hsp27 144
Hsp27 inhibitors 22–3
 cancer treatment 127
Hsp40 *see* J proteins
Hsp40-Hsp70 interaction 21–2
Hsp60 3, 132
Hsp70 2, 3, 65–125, 130, 393
 allosteric mechanism 76–9
 background 66–8
 cellular functions 89–108
  apoptosis 104–8
  de novo protein folding
   89–90

endoplasmic reticulum
 stress 102–4
intracellular protein
 traffic 94–102
protein degradation 91–4
protein re-folding 90–1
senescence 108
classes 68–70
co-chaperones 79–89, 302–3
 Hip 87–8, 94, 398
 Hop *see* Hop
 J proteins *see* J proteins
 NEFs 66, 84–7, 302
disease associations 9
mitochondrial 98–100
mouse knockout data 6–7
properties 69
roles of 65–6
structure 67, 71–6
 disordered C-terminal
  domain 76
 interdomain linker 72–3
 NBD 67, 71–2, 303
 SBD 67, 73–4, 303
targeting 303–4
Hsp70 ATPase modulators 403
Hsp70 inhibitors 18–22, 302–17
 cancer treatment 127
 CHIP 22
 co-chaperone targeting 312–14
 future prospects 314
 Hsp40-Hsp70 interaction 21–2
 NBD targeting 19, 21, 304–11
 SBD targeting 21, 311–12
 see also specific inhibitors
Hsp72 135
Hsp75 *see* TRAP1
Hsp90 2–3, 10, 37–64, 219, 318–35,
 393
 biology 128–32, 318–19
 C-terminus 264–5
 chaperone cycle 131, 320
 charged linker 45–7
 client proteins 242
  delivery 48–51
  Hsp90 activation 56–7

Subject Index 419

oncogenic 133
protease degradation 55–6
co-chaperones 384
　ATPase activation 51–4
　ATPase modulation 54–5, 242
　client protein delivery 48–51
　proteasome degradation 55–6
　regulation/modulation by 47–8
disease associations 5, 8
　cancer 132–4
　malaria 379–91
domain structure 38–9
GHKL subfamily 241
long-range communication between domains 47
mouse knockout data 6
natural resistance mechanisms 59–60
nucleotide interactions 40–2
P. falciparum 384–5
phosphorylation 58–9
post-translational modifications 319–24
　acetylation 322–3
　client-dependent regulation 323–4
　nitrosylation/oxidation 322
　phosphorylation 320–2
protein folding 259–61
rate-limiting step 39–40
structural changes
　on ATP binding 42–3
　C-terminal domain 47
　co-chaperones 47–8
　in middle domain 44–5
targeting 324–30
　proteotoxic stress 328–9
　sensitive client tumor drivers 325–8
　tyrosine kinase inhibitors 329–30
　see also Hsp90 inhibitors

Hsp90 C-terminus inhibitors 15–16, 259–301
　AEG3482 289
　applications 295
　cisplatin 16, 294–5
　cytoprotective 266–9
　EGCG 290–1
　ITZ-1 289–90
　novobiocin-derived 266–87
　　amides 274–5
　　benzamide phenyl optimization 284–5
　　benzamide side-chain 270
　　cancer treatment 269–86
　　coumarin ring 270–4
　　dimeric inhibitors 285–6
　　DNA gyrase inhibitors 269–70
　　noviose sugar/sugar mimics 275–84
　novologues 287–8
　silybin 291–2
　silybin-novobiocin hybrids 292–4
Hsp90 inhibitors 10–18, 134–41, 242, 326–7
　adverse events 368
　aminobenzamide derivatives 347, 352
　assay development 242–3
　C-terminus see Hsp90 C-terminus inhibitors
　clinical applications
　　cancer treatment 127–8, 241–2, 261–4, 352–67
　　malaria 379–91
　　neurological disorders 261–4, 401–3
　displacement of structural waters 246–8
　"flexible loop" region 248–50
　geldanamycin analogues 337–41
　heat-shock response 261
　HSF-1 pathway activation 24–5
　hydrogen bond interactions 246

Hsp90 inhibitors (*continued*)
   lead optimization
      NVP-AUY922 218–19
      NVP-HSP990 252–6
      SNX5422 203–5
      testing cascade 252–4
   lead series selection 245
   library screening 243–5
   N-terminal ATP binding pocket 11, 14
   ocular toxicity 178
   optimization
      of binding affinity 250–2
      through Lys58 248
   orally bioavailable *see* NVP-HSP90; SNX5422
   protein interactions 246
   purine scaffold-based 244, 343–7
   resorcinol-based 218–19, 223, 244, 341–5
   structure analysis 245–6
   structure-activity relationships 247, 249, 250, 251
   tolerability 172–3, 176–8
   see also specific inhibitors
Hsp100 2
Hsp110 87
HspA1 70
HspA1A 68, 69, 70
HspA1B 68, 69, 70
HspA1L 68, 69, 70
HspA2 70
HspA5 68, 69
HspA6 68, 69, 70
HspA7 68, 69, 70
HspA8 68, 69, 70
HspA9 68, 69
HspA12A 68, 69
HspA12B 68, 69
HspA13 68, 69
HspA14 68, 69
HspBP1 86
Humicola fuscoatra 146
Huntington's disease 1, 91, 393, 394
hybrid Antp-TPR peptide 17

hydrogen bond interactions 246
hydroxychloroquine 382, 383
HYPK 5
hypoxia 106–7
hypoxia-inducible factors (HIFs) 106–7, 184

imatinib mesylate 330
interferon regulatory factor 1 (IRF-1) 146
IPI-504 *see* retaspimycin
IRE1 102–3
IRE1-dependent decay (RIDD) pathway 103
ITZ-1 289–90

J proteins 3, 79–84, 130, 302
   classes 82–3
   disease associations 9
   mouse knockout data 7
   types 80–1
JMAR cell line 280
JUH-011 cell line 280

K-Ras 329
Kaposi sarcoma 142
KNK437 24, 144
KRIBB3 22–3
KU-32 267, 268–9, 287
KU-135 141, 275, 279
KU-174 274, 287
KU-363 279, 281
KW-2478 128, 325

LA-12 13, 16
LANA 142
lapatinib 147
LATS1 146
LATS2 146
lead optimization of Hsp90 inhibitors
   NVP-AUY922 218–19
   NVP-HSP990 252–6
   SNX5422 203–5
   testing cascade 252–4
Leigh syndrome 201
Lewy bodies 263, 397

library screening 243–5
ligand efficiency 216
liver toxicity 190
LNCaP-LN3 277, 278, 280
Lou Gehrig's disease *see* amyotrophic lateral sclerosis
Lys58 248
lysosomal protein degradation 93–4

malachite green 303–4
malaria 379–91
   history 379
   Hsp90 inhibitors 381–5
      discovery 385–9
      future directions 389
      high-throughput screening 387–8
      rational approach 388–9
   Plasmodium spp. 380, 382–3
   prevalence 379–80
   prevention 380–1
   treatment 381, 382
      drug resistance 385
mammalian ribosome-associated complex (mRAC) 69
maple syrup urine disease 201
MCF-7 cell line 277, 278, 280, 281, 282, 283, 284, 285, 286
MDA-1986 cell line 280
mefloquine 382, 383
melanoma 136, 142
Ménière's disease 5
8-methyl-3-arylcoumarins 294
3-α-methylcrotonyl-CoA carboxylase 1 (MCC1) deficiency 201
mitochondria, pre-protein transport to 96–7
mitochondrial outer membrane permeabilization (MOMP) 104
MKT-007 19
MKT-077 305, 306, 309
MmDjC7 83
molecular chaperones
   central role 1–2
   disease associations 5–10
      cancer 132–4

      malaria 379–91
      neurological disorders 392–413
   major classes 3–4
   modulation 10–26
   protein aggregation 392–3
   see also specific molecular chaperones
molecular profiling 138, 139
molybdate 16
MPC-0767 346
MPC-3100 346
   adverse events 351
   chemical structure 351
MRC-5 cell line 280
mtHsp70 98–100
multiple myeloma 136, 329
   Hsp90 inhibitors 364–7

nascent polypeptide-associated complex (NAC) 89–90
NBD *see* nucleotide-binding domain
NEFs *see* nucleotide exchange factors
neurodegenerative diseases *see* neurological disorders
neurofibrillary tangles 263, 396
neurological disorders 261–4
   Alzheimer's disease 1, 91, 263, 291, 393, 394, 396–7
   amyotrophic lateral sclerosis 263, 393, 394, 398–9
   molecular chaperones in 392–413
   Parkinson's disease 1, 91, 263, 393, 394, 397–8
   polyglutamine diseases 393–6
   prion diseases 393, 394, 399–400
   treatment 400–5
      chemical chaperones 403–4
      Hsf1 modulation 402–3
      Hsp70 ATPase modulation 403
      Hsp90 inhibitors 401–2
      viral mediated 404–5

neuroprotective agents 267
nitrosylation/oxidation 322
non-aryl 4-substituted
  pyrazole-resorcinols 227
non-small-cell lung cancer
  (NSCLC) 132, 142, 329
    Hsp90 inhibitors 352–6
      ganetespib 184–6, 192–3
non-steroidal anti-inflammatory
  drugs see NSAIDs
noviose derivatives 275–84
    structure-activity relationships
      278, 279–81
novobiocin 141
    chemical structure 265
    derivatives 15
      Hsp90 C-terminus
        inhibitors 266–87
    DNA gyrase interactions 266
novobiocin-silybin hybrids 292–4
novologues 287–8
NSAIDs 26
NSC 630668-R/1 307, 310
NSCLC see non-small-cell lung
  cancer
nucleophosmin-anaplastic lymphoma
  kinase (NPM-ALK) 186
nucleotide exchange factors
  (NEFs) 66, 84–7, 302
    Bags see Bags
    BAP 87, 100
    GrpE 84
    Hsp110 87
    HspBP1 86
nucleotide-binding domain of
  Hsp70 67, 71–2, 303
NVP-AUY922 128, 137, 138–40, 146,
  213–40, 325, 328, 341–3
    adverse events 349
    chemical structure 214, 349
    clinical applications
      breast cancer 358–9
      gastric cancer 363–4
      multiple myeloma 366–7
      NSCLC 354–5, 356
    clinical development 236

compound optimization 234
history 213
hit identification 214–18
in vitro assays 219–21
in vitro biomarker discovery/
  development 221–32
in vivo characterization 232–6
lead optimization 218–19
oral bioavailability 232–3
pharmacokinetics 234
NVP-HSP990 241–58, 325, 343–4
    adverse events 349
    assay development 242–3
    binding affinity 250–2
    chemical structure 254, 349
    dose limiting toxicity 343
    efficacy 254–5
      in vivo 255–6
    lead optimization 252–6
    library screening 243–5
    pharmacodynamics 254–5
    pharmacokinetics 254–5
NZ28 24

ocular toxicity
    ganetespib 190–1
    Hsp90 inhibitors 178
OGX-011 23
OGX-042 23
OMIM (Online Mendelian
  Inheritance in Man) 199
oncogenic proteins 11, 106, 133, 262
    client proteins 132–4, 143, 183,
      242, 255
    see also individual proteins
organelles
    pre-protein transport to 96–8
    protein translocation into
      98–102
osteosarcoma 191–2

p23 18, 55, 130, 145
p53 9, 11, 21, 92, 99, 106, 108, 145,
  265, 269, 309, 313, 322, 323, 337
PA11 23
PA50 23

Subject Index

paclitaxel 187
Parkinson's disease 1, 5, 91, 263, 393, 394, 397–8
   Lewy bodies 263, 397
PC-3 cell line 207, 271, 272, 273, 276, 278, 280
PC3-MM2 cell line 274, 277, 281
PDE4A5 51
peptide aptamer A17 21
peroxisomes, pre-protein transport to 97–8
PF-04928473 *see* SNX5422
Pharmacological Audit Trail (PhAT) 135
4-phenylbutyric acid (PBA) 403, 404
2-phenylethynesulfonamide (PES) 21, 308, 313
pheochromocytoma 188
phosphatidylinositol-3 kinase-related protein kinases (PIKKs) 130
phosphorylation 320–2
pifithrin-mu 21
PKR-like ER kinase (PERK) 102, 103
placental growth factor (PGF) 184
Plasmodium spp. 380, 382–3
Plasmodium falciparum
   Hsp90 384–5
   Hsp90 inhibitors 384
polyglutamine diseases 393–5
   see also individual disorders
post-translational Hsp90 modifications 319–24
   acetylation 322–3
   client-dependent regulation 323–4
   nitrosylation/oxidation 322
   phosphorylation 320–2
PP5 130
pre-sequence translocase-associated motor (PAM) 98
primaquine 382, 383
prion diseases 393, 394, 399–400
PRMT5 146
proguanil 382, 383
prostate cancer 140, 189
   Hsp90 inhibitors 359–60

protein aggregation 392–3
protein degradation 91–4
   lysosomal 93–4
   26S proteasome 91–3
protein folding
   Hsp90-mediated 259–61
   inhibition 164
protein misfolding 90, 392
protein re-folding 90–1
protein-inhibitor interactions 246
proteome mining 199–203
proteostasis 1–2
proteotoxic stress 328–9
PU-H71 128, 140, 159, 325
PU3 158–9
   chemical structure 215
pulmonary edema 5
purine scaffold-based Hsp90 inhibitors 244, 343–7
   BIIB021 *see* BIIB021
   Debio 0932 128, 140, 327, 345–6
   MPC-0767 346
   MPC-3100 346
pyrimethamine 382, 383
pyruvate carboxylase deficiency 201

quantitative structure activity relationship (QSAR) 199
quinacrine 382, 383
quinidine 382, 383
quinine 382, 383

radicicol 59–60, 127, 128, 134–5, 137, 142, 214, 217–18, 336
   chemical structure 215, 217, 244
   fluorescence polarization 223
Raf 262, 292
Rar1 54, 55
reactive oxygen species 329
receptor interacting proteins (RIP)-1 290
resorcinol-based Hsp90 inhibitors 218–19, 341–5
   AT13387 128, 140, 326, 344–5
   chemical structure 244

resorcinol-based Hsp90 inhibitors (*continued*)
    fluorescence polarization 223
    ganetespib *see* ganetespib
    NVP-AUY922 *see* NVP-AUY922
    NVP-HSP990 *see* NVP-HSP990
    see also radicicol
RET 51
RET kinase 140
retaspimycin 128, 137, 327, 341
    adverse events 349, 360
    chemical structure 349
    clinical applications
        GIST 361–2
        multiple myeloma 366
        NSCLC 354, 356
        prostate cancer 359–60
riluzole 26
RNA-induced silencing complexes (RISC) 130
rous sarcoma virus 336

26S proteasome 91–3
sansalvamide A derivatives 14–15
Sba1 54–5, 56
SBD *see* substrate-binding domain
schizophrenia 5
secondary time-resolved fluorescence 243
senescence 108
Serenex *see* SNX5422
Sgt1 49–50
shepherdin 14
Si11 100
silybin 291–2
silybin-novobiocin hybrids 292–4
single nucleotide polymorphism (SNP) 199
SKBr3 cell line 277, 278, 280, 281, 282, 283, 284, 285, 286
small Hsps 4
    disease associations 5, 9
    mouse knockout data 7
small nucleolar ribonucleoprotein particle (snoRNP) 130

small-molecule proteostasis regulators 26
SNARES 95–6
SNX2112 203–5
    activity 207
SNX5422 128, 140, 176–8, 198–212, 325, 326, 347, 352
    adverse events 351
    chemical structure 351
    clinical studies 209–11
    development 205–8
    lead optimization 203–5
    pre-clinical studies 208–9
    proteome mining 199–203
    synthesis 208
sorafenib 147
spastic paraplegia 5
spinal bulbar muscular atrophy 393, 394
spinocerebellar ataxia 393, 394
spongiform encephalopathies 263
STA-1474 181, 191–2
STA-9090 *see* ganetespib
Streptomyces hydroscopicus 146, 336, 386
stress pathways 141–2
stromal cell-derived factor (SDF-1) 184
structure activity relationship (SAR) 202
structure-based drug design (SBDD) 213
substrate-binding domain
    Hsp70 67, 73–4, 303
    Hsp70 inhibitors 21
sulfadoxine 382, 383
sulfanyladenine 159
sulfogalactoglycerolipid (SGG) 309
sulfogalactosyl ceramide (SGC) 309
sulfoglycolipids 309
sunitinib 330
superoxide dismutase 263, 398–9
surface plasmon resonance (SPR) 221, 228
Survivin 51

Subject Index 425

T1D1 84
T790M 353–4
Tah1 54
tanespimycin 126, 135–6, 142, 143, 158, 262, 325, 328, 329
    activity 207
    adverse events 348, 365–6
    chemical structure 348, 387
    clinical applications 337–8
        breast cancer 357–8
        malaria 386, 387, 389
        multiple myeloma 364–6
        prostate cancer 359, 360
    dose limiting toxicity 337–8
    maximum tolerated dose 337
    structure-activity relationships 205, 206, 207
    see also geldanamycin
tauroursodeodeoxycholic acid (TUDCA) 404
tetratricopeptide repeat domain co-chaperones 302–3
therapeutic index 127
Tom20 96–7
TOMM20 50–1
TORIA 5
transmissible spongiform encephalopathies 393, 394, 399–400
TRAP1 130, 145, 198, 221, 241, 260, 319
triptolide 24
tubocapsenolide A 16
tumor necrosis factor (TNF) 104
tyrosine kinase inhibitors 329–30

ubiquitin-proteasome pathway 129, 130, 133, 134, 147, 393
UM-SCC-2 cell line 280

v-SRC 133
vascular endothelial growth factor (VEGF) 184
    gastric cancer 362–3
vemurafenib 147
VER-26617 215
VER-27042 215
VER-27129 215
VER-27193 215
VER-45864 220
VER-49009 137, 138, 218, 223, 234
    binding mode 224, 225
    chemical structure 227
VER-50589 137, 138
    chemical structure 227
VER-50598 218
VER-51001 220
VER-51047 234
VER-52296 see NVP-AUY922
VER-155008 19, 304, 307, 310
viral-mediated strategies 404–5

Wee1 145, 323–4
withaferin A 15–17

XAP2 50–1
XBP1 142
XL-888 326

ZAP70 136, 140
Zuotin 89